STATA BASE REFERENCE MANUAL
VOLUME 2
G–M
RELEASE 12

A Stata Press Publication
StataCorp LP
College Station, Texas

Title

> **gllamm** — Generalized linear and latent mixed models

Description

GLLAMM stands for generalized linear latent and mixed models, and `gllamm` is a Stata command for fitting such models written by Sophia Rabe-Hesketh (University of California–Berkeley) as part of joint work with Anders Skrondal (Norwegian Institute of Public Health) and Andrew Pickles (University of Manchester).

Remarks

Generalized linear latent and mixed models are a class of multilevel latent variable models, where a latent variable is a factor or a random effect (intercept or coefficient), or a disturbance (residual). The `gllamm` command for fitting such models is not an official command of Stata; it has been independently developed by highly regarded authors and is itself highly regarded. You can learn more about `gllamm` by visiting http://www.gllamm.org.

`gllamm` is available from the Statistical Software Components (SSC) archive. To install, type

    . ssc describe gllamm

    . ssc install gllamm

If you later wish to uninstall `gllamm`, type `ado uninstall gllamm`.

References

Miranda, A., and S. Rabe-Hesketh. 2006. Maximum likelihood estimation of endogenous switching and sample selection models for binary, ordinal, and count variables. *Stata Journal* 6: 285–308.

Rabe-Hesketh, S., and B. S. Everitt. 2007. *A Handbook of Statistical Analyses Using Stata*. 4th ed. Boca Raton, FL: Chapman & Hall/CRC.

Rabe-Hesketh, S., A. Pickles, and C. Taylor. 2000. sg129: Generalized linear latent and mixed models. *Stata Technical Bulletin* 53: 47–57. Reprinted in *Stata Technical Bulletin Reprints*, vol. 9, pp. 293–307. College Station, TX: Stata Press.

Rabe-Hesketh, S., and A. Skrondal. 2008. *Multilevel and Longitudinal Modeling Using Stata*. 2nd ed. College Station, TX: Stata Press.

Rabe-Hesketh, S., A. Skrondal, and A. Pickles. 2002. Reliable estimation of generalized linear mixed models using adaptive quadrature. *Stata Journal* 2: 1–21.

———. 2003. Maximum likelihood estimation of generalized linear models with covariate measurement error. *Stata Journal* 3: 386–411.

Skrondal, A., and S. Rabe-Hesketh. 2004. *Generalized Latent Variable Modeling: Multilevel, Longitudinal, and Structural Equation Models*. Boca Raton, FL: Chapman & Hall/CRC.

Zheng, X., and S. Rabe-Hesketh. 2007. Estimating parameters of dichotomous and ordinal item response models with gllamm. *Stata Journal* 7: 313–333.

The references above are restricted to works by the primary authors of `gllamm`. There are many other books and articles that use or discuss `gllamm`; see http://www.gllamm.org/pub.html for a list.

Also see

Title

> **glm** — Generalized linear models

Syntax

glm *depvar* [*indepvars*] [*if*] [*in*] [*weight*] [, *options*]

options	Description
Model	
<u>f</u>amily(*familyname*)	distribution of *depvar*; default is family(gaussian)
<u>l</u>ink(*linkname*)	link function; default is canonical link for family() specified
Model 2	
<u>noco</u>nstant	suppress constant term
<u>expo</u>sure(*varname*)	include ln(*varname*) in model with coefficient constrained to 1
<u>off</u>set(*varname*)	include *varname* in model with coefficient constrained to 1
<u>constraints</u>(*constraints*)	apply specified linear constraints
<u>col</u>linear	keep collinear variables
mu(*varname*)	use *varname* as the initial estimate for the mean of *depvar*
<u>init</u>(*varname*)	synonym for mu(*varname*)
SE/Robust	
vce(*vcetype*)	*vcetype* may be oim, <u>r</u>obust, <u>cl</u>uster *clustvar*, eim, opg, <u>boot</u>strap, <u>jack</u>knife, hac *kernel*, jackknife1, or <u>unb</u>iased
<u>vf</u>actor(*#*)	multiply variance matrix by scalar *#*
disp(*#*)	quasilikelihood multiplier
<u>sca</u>le(x2 | dev | *#*)	set the scale parameter
Reporting	
<u>l</u>evel(*#*)	set confidence level; default is level(95)
<u>ef</u>orm	report exponentiated coefficients
nocnsreport	do not display constraints
display_options	control column formats, row spacing, line width, and display of omitted variables and base and empty cells
Maximization	
ml	use maximum likelihood optimization; the default
irls	use iterated, reweighted least-squares optimization of the deviance
maximize_options	control the maximization process; seldom used
fisher(*#*)	use the Fisher scoring Hessian or expected information matrix (EIM)
search	search for good starting values
<u>nohe</u>ader	suppress header table from above coefficient table
<u>nota</u>ble	suppress coefficient table
<u>nodi</u>splay	suppress the output; iteration log is still displayed
<u>coefl</u>egend	display legend instead of statistics

familyname	Description
<u>ga</u>ussian	Gaussian (normal)
<u>i</u>gaussian	inverse Gaussian
<u>b</u>inomial $\left[\, varname_N \mid \#_N \,\right]$	Bernoulli/binomial
<u>p</u>oisson	Poisson
<u>nb</u>inomial $\left[\, \#_k \mid ml \,\right]$	negative binomial
<u>ga</u>mma	gamma

linkname	Description
<u>i</u>dentity	identity
<u>l</u>og	log
<u>l</u>ogit	logit
<u>p</u>robit	probit
<u>c</u>loglog	cloglog
<u>p</u>ower #	power
<u>o</u>power #	odds power
<u>n</u>binomial	negative binomial
<u>l</u>oglog	log-log
<u>l</u>ogc	log-complement

indepvars may contain factor variables; see [U] **11.4.3 Factor variables**.

depvar and *indepvars* may contain time-series operators; see [U] **11.4.4 Time-series varlists**.

bootstrap, by, fracpoly, jackknife, mfp, mi estimate, nestreg, rolling, statsby, stepwise, and svy are allowed; see [U] **11.1.10 Prefix commands**.

vce(bootstrap), vce(jackknife), and vce(jackknife1) are not allowed with the mi estimate prefix; see [MI] **mi estimate**.

Weights are not allowed with the bootstrap prefix; see [R] **bootstrap**.

aweights are not allowed with the jackknife prefix; see [R] **jackknife**.

vce(), vfactor(), disp(), scale(), irls, fisher(), noheader, notable, nodisplay, and weights are not allowed with the svy prefix; see [SVY] **svy**.

fweights, aweights, iweights, and pweights are allowed; see [U] **11.1.6 weight**.

noheader, notable, nodisplay, and coeflegend do not appear in the dialog box.

See [U] **20 Estimation and postestimation commands** for more capabilities of estimation commands.

Menu

Statistics > Generalized linear models > Generalized linear models (GLM)

Description

glm fits generalized linear models. It can fit models by using either IRLS (maximum quasilikelihood) or Newton–Raphson (maximum likelihood) optimization, which is the default.

See [I] **estimation commands** for a complete list of Stata's estimation commands, several of which fit models that can also be fit using glm.

Options

Model

family(*familyname*) specifies the distribution of *depvar*; family(gaussian) is the default.

link(*linkname*) specifies the link function; the default is the canonical link for the family() specified.

Model 2

noconstant, exposure(*varname*), offset(*varname*), constraints(*constraints*), collinear; see [R] **estimation options**. constraints(*constraints*) and collinear are not allowed with irls.

mu(*varname*) specifies *varname* as the initial estimate for the mean of *depvar*. This option can be useful with models that experience convergence difficulties, such as family(binomial) models with power or odds-power links. init(*varname*) is a synonym.

SE/Robust

vce(*vcetype*) specifies the type of standard error reported, which includes types that are derived from asymptotic theory, that are robust to some kinds of misspecification, that allow for intragroup correlation, and that use bootstrap or jackknife methods; see [R] *vce_option*.

In addition to the standard *vcetype*s, glm allows the following alternatives:

vce(eim) specifies that the EIM estimate of variance be used.

vce(jackknife1) specifies that the one-step jackknife estimate of variance be used.

vce(hac *kernel* $\lceil$#$\rceil$) specifies that a heteroskedasticity- and autocorrelation-consistent (HAC) variance estimate be used. HAC refers to the general form for combining weighted matrices to form the variance estimate. There are three kernels built into glm. *kernel* is a user-written program or one of

$$\underline{\text{nw}}\text{est} \mid \underline{\text{g}}\text{allant} \mid \underline{\text{an}}\text{derson}$$

   # specifies the number of lags. If # is not specified, $N - 2$ is assumed. If you wish to specify vce(hac ...), you must tsset your data before calling glm.

vce(unbiased) specifies that the unbiased sandwich estimate of variance be used.

vfactor(#) specifies a scalar by which to multiply the resulting variance matrix. This option allows you to match output with other packages, which may apply degrees of freedom or other small-sample corrections to estimates of variance.

disp(#) multiplies the variance of *depvar* by # and divides the deviance by #. The resulting distributions are members of the quasilikelihood family.

scale(x2 | dev | #) overrides the default scale parameter. This option is allowed only with Hessian (information matrix) variance estimates.

   By default, scale(1) is assumed for the discrete distributions (binomial, Poisson, and negative binomial), and scale(x2) is assumed for the continuous distributions (Gaussian, gamma, and inverse Gaussian).

   scale(x2) specifies that the scale parameter be set to the Pearson chi-squared (or generalized chi-squared) statistic divided by the residual degrees of freedom, which is recommended by McCullagh and Nelder (1989) as a good general choice for continuous distributions.

scale(dev) sets the scale parameter to the deviance divided by the residual degrees of freedom. This option provides an alternative to scale(x2) for continuous distributions and overdispersed or underdispersed discrete distributions.

scale(#) sets the scale parameter to #. For example, using scale(1) in family(gamma) models results in exponential-errors regression. Additional use of link(log) rather than the default link(power -1) for family(gamma) essentially reproduces Stata's streg, dist(exp) nohr command (see [ST] **streg**) if all the observations are uncensored.

> Reporting

level(#); see [R] **estimation options**.

eform displays the exponentiated coefficients and corresponding standard errors and confidence intervals. For family(binomial) link(logit) (that is, logistic regression), exponentiation results in odds ratios; for family(poisson) link(log) (that is, Poisson regression), exponentiated coefficients are rate ratios.

nocnsreport; see [R] **estimation options**.

display_options: noomitted, vsquish, noemptycells, baselevels, allbaselevels, cformat(*%fmt*), pformat(*%fmt*), sformat(*%fmt*), and nolstretch; see [R] **estimation options**.

> Maximization

ml requests that optimization be carried out using Stata's ml commands and is the default.

irls requests iterated, reweighted least-squares (IRLS) optimization of the deviance instead of Newton–Raphson optimization of the log likelihood. If the irls option is not specified, the optimization is carried out using Stata's ml commands, in which case all options of ml maximize are also available.

maximize_options: difficult, technique(*algorithm_spec*), iterate(#), [no]log, trace, gradient, showstep, hessian, showtolerance, tolerance(#), ltolerance(#), nrtolerance(#), nonrtolerance, and from(*init_specs*); see [R] **maximize**. These options are seldom used.

Setting the optimization type to technique(bhhh) resets the default *vcetype* to vce(opg).

fisher(#) specifies the number of Newton–Raphson steps that should use the Fisher scoring Hessian or EIM before switching to the observed information matrix (OIM). This option is useful only for Newton–Raphson optimization (and not when using irls).

search specifies that the command search for good starting values. This option is useful only for Newton–Raphson optimization (and not when using irls).

The following options are available with glm but are not shown in the dialog box:

noheader suppresses the header information from the output. The coefficient table is still displayed.

notable suppresses the table of coefficients from the output. The header information is still displayed.

nodisplay suppresses the output. The iteration log is still displayed.

coeflegend; see [R] **estimation options**.

Remarks

Remarks are presented under the following headings:

General use
Variance estimators
User-defined functions

General use

`glm` fits generalized linear models of y with covariates $\mathbf{x}$:

$$g\{E(y)\} = \mathbf{x}\boldsymbol{\beta}, \qquad y \sim F$$

$g(\ )$ is called the link function, and F is the distributional family. Substituting various definitions for $g(\ )$ and F results in a surprising array of models. For instance, if y is distributed as Gaussian (normal) and $g(\ )$ is the identity function, we have

$$E(y) = \mathbf{x}\boldsymbol{\beta}, \qquad y \sim \text{Normal}$$

or linear regression. If $g(\ )$ is the logit function and y is distributed as Bernoulli, we have

$$\text{logit}\{E(y)\} = \mathbf{x}\boldsymbol{\beta}, \qquad y \sim \text{Bernoulli}$$

or logistic regression. If $g(\ )$ is the natural log function and y is distributed as Poisson, we have

$$\ln\{E(y)\} = \mathbf{x}\boldsymbol{\beta}, \qquad y \sim \text{Poisson}$$

or Poisson regression, also known as the log-linear model. Other combinations are possible.

Although `glm` can be used to perform linear regression (and, in fact, does so by default), this regression should be viewed as an instructional feature; `regress` produces such estimates more quickly, and many postestimation commands are available to explore the adequacy of the fit; see [R] **regress** and [R] **regress postestimation**.

In any case, you specify the link function by using the `link()` option and specify the distributional family by using `family()`. The available link functions are

Link function	glm option
identity	link(identity)
log	link(log)
logit	link(logit)
probit	link(probit)
complementary log-log	link(cloglog)
odds power	link(opower #)
power	link(power #)
negative binomial	link(nbinomial)
log-log	link(loglog)
log-complement	link(logc)

Define $\mu = E(y)$ and $\eta = g(\mu)$, meaning that $g(\cdot)$ maps $E(y)$ to $\eta = \mathbf{x}\boldsymbol{\beta} + \text{offset}$.

Link functions are defined as follows:

identity is defined as $\eta = g(\mu) = \mu$.

log is defined as $\eta = \ln(\mu)$.

logit is defined as $\eta = \ln\{\mu/(1-\mu)\}$, the natural log of the odds.

probit is defined as $\eta = \Phi^{-1}(\mu)$, where $\Phi^{-1}(\,)$ is the inverse Gaussian cumulative.

cloglog is defined as $\eta = \ln\{-\ln(1-\mu)\}$.

opower is defined as $\eta = \left[\{\mu/(1-\mu)\}^n - 1\right]/n$, the power of the odds. The function is generalized so that link(opower 0) is equivalent to link(logit), the natural log of the odds.

power is defined as $\eta = \mu^n$. Specifying link(power 1) is equivalent to specifying link(identity). The power function is generalized so that $\mu^0 \equiv \ln(\mu)$. Thus link(power 0) is equivalent to link(log). Negative powers are, of course, allowed.

nbinomial is defined as $\eta = \ln\{\mu/(\mu+k)\}$, where $k = 1$ if family(nbinomial) is specified, $k = \#_k$ if family(nbinomial $\#_k$) is specified, and k is estimated via maximum likelihood if family(nbinomial ml) is specified.

loglog is defined as $\eta = -\ln\{-\ln(\mu)\}$.

logc is defined as $\eta = \ln(1-\mu)$.

The available distributional families are

Family	glm option
Gaussian (normal)	family(gaussian)
inverse Gaussian	family(igaussian)
Bernoulli/binomial	family(binomial)
Poisson	family(poisson)
negative binomial	family(nbinomial)
gamma	family(gamma)

family(normal) is a synonym for family(gaussian).

The binomial distribution can be specified as 1) family(binomial), 2) family(binomial $\#_N$), or 3) family(binomial $varname_N$). In case 2, $\#_N$ is the value of the binomial denominator N, the number of trials. Specifying family(binomial 1) is the same as specifying family(binomial). In case 3, $varname_N$ is the variable containing the binomial denominator, allowing the number of trials to vary across observations.

The negative binomial distribution can be specified as 1) family(nbinomial), 2) family(nbinomial $\#_k$), or 3) family(nbinomial ml). Omitting $\#_k$ is equivalent to specifying family(nbinomial 1). In case 3, the value of $\#_k$ is estimated via maximum likelihood. The value $\#_k$ enters the variance and deviance functions. Typical values range between 0.01 and 2; see the technical note below.

You do not have to specify both family() and link(); the default link() is the canonical link for the specified family() (except for nbinomial):

Family	Default link
family(gaussian)	link(identity)
family(igaussian)	link(power -2)
family(binomial)	link(logit)
family(poisson)	link(log)
family(nbinomial)	link(log)
family(gamma)	link(power -1)

If you specify both family() and link(), not all combinations make sense. You may choose from the following combinations:

	identity	log	logit	probit	cloglog	power	opower	nbinomial	loglog	logc
Gaussian	x	x				x				
inverse Gaussian	x	x				x				
binomial	x	x	x	x	x	x	x		x	x
Poisson	x	x				x				
negative binomial	x	x				x		x		
gamma	x	x				x				

❑ Technical note

Some family() and link() combinations result in models already fit by Stata. These are

family()	link()	Options	Equivalent Stata command
gaussian	identity	*nothing* | irls | irls vce(oim)	regress
gaussian	identity	t(*var*) vce(hac nwest #) vfactor($#_v$)	newey, t(*var*) lag(#) (see note 1)
binomial	cloglog	*nothing* | irls vce(oim)	cloglog (see note 2)
binomial	probit	*nothing* | irls vce(oim)	probit (see note 2)
binomial	logit	*nothing* | irls | irls vce(oim)	logit or logistic (see note 3)
poisson	log	*nothing* | irls | irls vce(oim)	poisson (see note 3)
nbinomial	log	*nothing* | irls vce(oim)	nbreg (see note 4)
gamma	log	scale(1)	streg, dist(exp) nohr (see note 5)

Notes:

1. The variance factor $\#_v$ should be set to $n/(n-k)$, where n is the number of observations and k the number of regressors. If the number of regressors is not specified, the estimated standard errors will, as a result, differ by this factor.

2. Because the link is not the canonical link for the binomial family, you must specify the vce(oim) option if using irls to get equivalent standard errors. If irls is used without vce(oim), the regression coefficients will be the same but the standard errors will be only asymptotically equivalent. If no options are specified (*nothing*), glm will optimize using Newton–Raphson, making it equivalent to the other Stata command.

   See [R] **cloglog** and [R] **probit** for more details about these commands.

3. Because the canonical link is being used, the standard errors will be equivalent whether the EIM or the OIM estimator of variance is used.

4. Family negative binomial, log-link models—also known as negative binomial regression models—are used for data with an overdispersed Poisson distribution. Although `glm` can be used to fit such models, using Stata's maximum likelihood `nbreg` command is probably better. In the GLM approach, you specify `family(nbinomial #_k)` and then search for a $\#_k$ that results in the deviance-based dispersion being 1. You can also specify `family(nbinomial ml)` to estimate $\#_k$ via maximum likelihood, which will report the same value returned from `nbreg`. However, `nbreg` also reports a confidence interval for it; see [R] **nbreg** and Rogers (1993). Of course, `glm` allows links other than `log`, and for those links, including the canonical `nbinomial` link, you will need to use `glm`.

5. `glm` can be used to estimate parameters from exponential regressions, but this method requires specifying `scale(1)`. However, censoring is not available. Censored exponential regression may be modeled using `glm` with `family(poisson)`. The log of the original response is entered into a Poisson model as an offset, whereas the new response is the censor variable. The result of such modeling is identical to the log relative hazard parameterization of `streg, dist(exp) nohr`. See [ST] **streg** for details about the `streg` command.

In general, where there is overlap between a capability of `glm` and that of some other Stata command, we recommend using the other Stata command. Our recommendation is not because of some inferiority of the GLM approach. Rather, those other commands, by being specialized, provide options and ancillary commands that are missing in the broader `glm` framework. Nevertheless, `glm` does produce the same answers where it should.

Special note. When equivalence is expected, for some datasets, you may still see very slight differences in the results, most often only in the later digits of the standard errors. When you compare `glm` output to an equivalent Stata command, these tiny discrepancies arise for many reasons:

a. `glm` uses a general methodology for starting values, whereas the equivalent Stata command may be more specialized in its treatment of starting values.

b. When using a canonical link, `glm, irls` should be equivalent to the maximum likelihood method of the equivalent Stata command, yet the convergence criterion is different (one is for deviance, the other for log likelihood). These discrepancies are easily resolved by adjusting one convergence criterion to correspond to the other.

c. When both `glm` and the equivalent Stata command use Newton–Raphson, small differences may still occur if the Stata command has a different default convergence criterion from that of `glm`. Adjusting the convergence criterion will resolve the difference. See [R] **ml** and [R] **maximize** for more details.

❏

▷ Example 1

In example 1 of [R] **logistic**, we fit a model based on data from a study of risk factors associated with low birthweight (Hosmer and Lemeshow 2000, 25). We can replicate the estimation by using `glm`:

```
. use http://www.stata-press.com/data/r12/lbw
(Hosmer & Lemeshow data)

. glm low age lwt i.race smoke ptl ht ui, family(binomial) link(logit)

Iteration 0:   log likelihood =  -101.0213
Iteration 1:   log likelihood = -100.72519
Iteration 2:   log likelihood =   -100.724
Iteration 3:   log likelihood =   -100.724
```

```
Generalized linear models                    No. of obs       =        189
Optimization      : ML                       Residual df      =        180
                                             Scale parameter  =          1
Deviance          =   201.4479911            (1/df) Deviance  =   1.119156
Pearson           =   182.0233425            (1/df) Pearson   =   1.011241

Variance function: V(u) = u*(1-u)            [Bernoulli]
Link function     : g(u) = ln(u/(1-u))       [Logit]

                                             AIC              =     1.1611
Log likelihood    = -100.7239956             BIC              =  -742.0665
```

low	Coef.	OIM Std. Err.	z	P>|z|	[95% Conf. Interval]	
age	-.0271003	.0364504	-0.74	0.457	-.0985418	.0443412
lwt	-.0151508	.0069259	-2.19	0.029	-.0287253	-.0015763
race						
2	1.262647	.5264101	2.40	0.016	.2309024	2.294392
3	.8620792	.4391532	1.96	0.050	.0013548	1.722804
smoke	.9233448	.4008266	2.30	0.021	.137739	1.708951
ptl	.5418366	.346249	1.56	0.118	-.136799	1.220472
ht	1.832518	.6916292	2.65	0.008	.4769494	3.188086
ui	.7585135	.4593768	1.65	0.099	-.1418484	1.658875
_cons	.4612239	1.20459	0.38	0.702	-1.899729	2.822176

glm, by default, presents coefficient estimates, whereas logistic presents the exponentiated coefficients—the odds ratios. glm's eform option reports exponentiated coefficients, and glm, like Stata's other estimation commands, replays results.

```
. glm, eform
```

```
Generalized linear models                        No. of obs      =       189
Optimization     : ML                            Residual df     =       180
                                                 Scale parameter =         1
Deviance         =  201.4479911                  (1/df) Deviance =  1.119156
Pearson          =  182.0233425                  (1/df) Pearson  =  1.011241
Variance function: V(u) = u*(1-u)                [Bernoulli]
Link function    : g(u) = ln(u/(1-u))            [Logit]
                                                 AIC             =    1.1611
Log likelihood   = -100.7239956                  BIC             = -742.0665
```

low	Odds Ratio	OIM Std. Err.	z	P>|z|	[95% Conf. Interval]	
age	.9732636	.0354759	-0.74	0.457	.9061578	1.045339
lwt	.9849634	.0068217	-2.19	0.029	.9716834	.9984249
race						
2	3.534767	1.860737	2.40	0.016	1.259736	9.918406
3	2.368079	1.039949	1.96	0.050	1.001356	5.600207
smoke	2.517698	1.00916	2.30	0.021	1.147676	5.523162
ptl	1.719161	.5952579	1.56	0.118	.8721455	3.388787
ht	6.249602	4.322408	2.65	0.008	1.611152	24.24199
ui	2.1351	.9808153	1.65	0.099	.8677528	5.2534
_cons	1.586014	1.910496	0.38	0.702	.1496092	16.8134

These results are the same as those reported in example 1 of [R] **logistic**.

Included in the output header are values for the Akaike (1973) information criterion (AIC) and the Bayesian information criterion (BIC) (Raftery 1995). Both are measures of model fit adjusted for the number of parameters that can be compared across models. In both cases, a smaller value generally indicates a better model fit. AIC is based on the log likelihood and thus is available only when Newton–Raphson optimization is used. BIC is based on the deviance and thus is always available. ◁

❑ Technical note

The values for AIC and BIC reported in the output after `glm` are different from those reported by `estat ic`:

```
. estat ic
```

Model	Obs	ll(null)	ll(model)	df	AIC	BIC
.	189	.	-100.724	9	219.448	248.6237

Note: N=Obs used in calculating BIC; see **[R] BIC note**

There are various definitions of these information criteria (IC) in the literature; `glm` and `estat ic` use different definitions. `glm` bases its computation of the BIC on deviance, whereas `estat ic` uses the likelihood. Both `glm` and `estat ic` use the likelihood to compute the AIC; however, the AIC from `estat ic` is equal to N, the number of observations, times the AIC from `glm`. Refer to *Methods and formulas* in this entry and [R] **estat** for the references and formulas used by `glm` and `estat`, respectively, to compute AIC and BIC. Inferences based on comparison of IC values reported by `glm` for different GLM models will be equivalent to those based on comparison of IC values reported by `estat ic` after `glm`.

❑

▷ Example 2

We use data from an early insecticide experiment, given in Pregibon (1980). The variables are ldose, the log dose of insecticide; n, the number of flour beetles subjected to each dose; and r, the number killed.

```
. use http://www.stata-press.com/data/r12/ldose
. list, sep(4)
```

	ldose	n	r
1.	1.6907	59	6
2.	1.7242	60	13
3.	1.7552	62	18
4.	1.7842	56	28
5.	1.8113	63	52
6.	1.8369	59	53
7.	1.861	62	61
8.	1.8839	60	60

The aim of the analysis is to estimate a dose–response relationship between p, the proportion killed, and X, the log dose.

As a first attempt, we will formulate the model as a linear logistic regression of p on ldose; that is, we will take the logit of p and represent the dose–response curve as a straight line in X:

$$\ln\{p/(1-p)\} = \beta_0 + \beta_1 X$$

Because the data are grouped, we cannot use Stata's logistic command to fit the model. Stata does, however, already have a command for performing logistic regression on data organized in this way, so we could type

```
. blogit r n ldose
  (output omitted)
```

Instead, we will fit the model by using glm:

```
. glm r ldose, family(binomial n) link(logit)
Iteration 0:   log likelihood = -18.824848
Iteration 1:   log likelihood = -18.715271
Iteration 2:   log likelihood = -18.715123
Iteration 3:   log likelihood = -18.715123
```

Generalized linear models			No. of obs	=	8
Optimization	: ML		Residual df	=	6
			Scale parameter	=	1
Deviance	=	11.23220702	(1/df) Deviance	=	1.872035
Pearson	=	10.0267936	(1/df) Pearson	=	1.671132

```
Variance function: V(u) = u*(1-u/n)              [Binomial]
Link function    : g(u) = ln(u/(n-u))            [Logit]
                                                 AIC           =   5.178781
Log likelihood    = -18.71512262                 BIC           =  -1.244442
```

		OIM				
r	Coef.	Std. Err.	z	P>|z|	[95% Conf. Interval]	
ldose	34.27034	2.912141	11.77	0.000	28.56265	39.97803
_cons	-60.71747	5.180713	-11.72	0.000	-70.87149	-50.56346

The only difference between `blogit` and `glm` here is how they obtain the answer. `blogit` expands the data to contain 481 observations (the sum of n) so that it can run Stata's standard, individual-level logistic command. `glm`, on the other hand, uses the information on the binomial denominator directly. We specified `family(binomial n)`, meaning that variable n contains the denominator. Parameter estimates and standard errors from the two approaches do not differ.

An alternative model, which gives asymmetric sigmoid curves for p, involves the complementary log-log, or cloglog, function:

$$\ln\{-\ln(1-p)\} = \beta_0 + \beta_1 X$$

We fit this model by using `glm`:

```
. glm r ldose, family(binomial n) link(cloglog)
Iteration 0:   log likelihood = -14.883594
Iteration 1:   log likelihood = -14.822264
Iteration 2:   log likelihood = -14.822228
Iteration 3:   log likelihood = -14.822228

Generalized linear models                      No. of obs        =          8
Optimization     : ML                          Residual df       =          6
                                               Scale parameter   =          1
Deviance         =  3.446418004                (1/df) Deviance   =   .574403
Pearson          =  3.294675153                (1/df) Pearson    =  .5491125

Variance function: V(u) = u*(1-u/n)            [Binomial]
Link function    : g(u) = ln(-ln(1-u/n))       [Complementary log-log]
                                               AIC               =   4.205557
Log likelihood    = -14.82222811               BIC               =  -9.030231
```

		OIM				
r	Coef.	Std. Err.	z	P>|z|	[95% Conf. Interval]	
ldose	22.04118	1.793089	12.29	0.000	18.52679	25.55557
_cons	-39.57232	3.229047	-12.26	0.000	-45.90114	-33.24351

The complementary log-log model is preferred; the deviance for the logistic model, 11.23, is much higher than the deviance for the cloglog model, 3.45. This change also is evident by comparing log likelihoods, or equivalently, AIC values.

This example also shows the advantage of the `glm` command—we can vary assumptions easily. Note the minor difference in what we typed to obtain the logistic and cloglog models:

```
. glm r ldose, family(binomial n) link(logit)
. glm r ldose, family(binomial n) link(cloglog)
```

If we were performing this work for ourselves, we would have typed the commands in a more abbreviated form:

```
. glm r ldose, f(b n) l(1)
. glm r ldose, f(b n) l(cl)
```

◁

❏ Technical note

Factor variables may be used with `glm`. Say that, in the example above, we had `ldose`, the log dose of insecticide; `n`, the number of flour beetles subjected to each dose; and `r`, the number killed—all as before—except that now we have results for three different kinds of beetles. Our hypothetical data include `beetle`, which contains the values 1, 2, and 3.

```
. use http://www.stata-press.com/data/r12/beetle
. list, sep(0)
```

	beetle	ldose	n	r
1.	1	1.6907	59	6
2.	1	1.7242	60	13
3.	1	1.7552	62	18
4.	1	1.7842	56	28
5.	1	1.8113	63	52
	(output omitted)			
23.	3	1.861	64	23
24.	3	1.8839	58	22

Let's assume that, at first, we wish merely to add a shift factor for the type of beetle. We could type

```
. glm r i.beetle ldose, f(bin n) l(cloglog)
Iteration 0:   log likelihood = -79.012269
Iteration 1:   log likelihood =  -76.94951
Iteration 2:   log likelihood = -76.945645
Iteration 3:   log likelihood = -76.945645
```

Generalized linear models				No. of obs	=	24
Optimization : ML				Residual df	=	20
				Scale parameter	=	1
Deviance = 73.76505595				(1/df) Deviance	=	3.688253
Pearson = 71.8901173				(1/df) Pearson	=	3.594506
Variance function: V(u) = u*(1-u/n)				[Binomial]		
Link function : g(u) = ln(-ln(1-u/n))				[Complementary log-log]		
				AIC	=	6.74547
Log likelihood = -76.94564525				BIC	=	10.20398

		OIM				
r	Coef.	Std. Err.	z	P>|z|	[95% Conf.	Interval]
beetle						
2	-.0910396	.1076132	-0.85	0.398	-.3019576	.1198783
3	-1.836058	.1307125	-14.05	0.000	-2.09225	-1.579867
ldose	19.41558	.9954265	19.50	0.000	17.46458	21.36658
_cons	-34.84602	1.79333	-19.43	0.000	-38.36089	-31.33116

We find strong evidence that the insecticide works differently on the third kind of beetle. We now check whether the curve is merely shifted or also differently sloped:

```
. glm r beetle##c.ldose, f(bin n) l(cloglog)
Iteration 0:    log likelihood = -67.270188
Iteration 1:    log likelihood = -65.149316
Iteration 2:    log likelihood = -65.147978
Iteration 3:    log likelihood = -65.147978

Generalized linear models                    No. of obs       =         24
Optimization      : ML                       Residual df      =         18
                                             Scale parameter =          1
Deviance          =  50.16972096             (1/df) Deviance =   2.787207
Pearson           =  49.28422567             (1/df) Pearson  =   2.738013
Variance function: V(u) = u*(1-u/n)          [Binomial]
Link function     : g(u) = ln(-ln(1-u/n))    [Complementary log-log]
                                             AIC              =   5.928998
Log likelihood    = -65.14797776             BIC              =  -7.035248
```

r	Coef.	OIM Std. Err.	z	P>|z|	[95% Conf. Interval]
beetle					
2	-.79933	4.470882	-0.18	0.858	-9.562098 7.963438
3	17.78741	4.586429	3.88	0.000	8.798172 26.77664
ldose	22.04118	1.793089	12.29	0.000	18.52679 25.55557
beetle#c.ldose					
2	.3838708	2.478477	0.15	0.877	-4.473855 5.241596
3	-10.726	2.526412	-4.25	0.000	-15.67768 -5.774321
_cons	-39.57232	3.229047	-12.26	0.000	-45.90114 -33.24351

We find that the (complementary log-log) dose–response curve for the third kind of beetle has roughly half the slope of that for the first kind.

See [U] **25 Working with categorical data and factor variables**; what is said there concerning linear regression is applicable to any GLM model.

❑

Variance estimators

glm offers many variance options and gives different types of standard errors when used in various combinations. We highlight some of them here, but for a full explanation, see Hardin and Hilbe (2007).

▷ Example 3

Continuing with our flour beetle data, we rerun the most recently displayed model, this time requesting estimation via IRLS.

```
. use http://www.stata-press.com/data/r12/beetle

. glm r beetle##c.ldose, f(bin n) l(cloglog) ltol(1e-15) irls
Iteration 1:    deviance =  54.41414
Iteration 2:    deviance =  50.19424
Iteration 3:    deviance =  50.16973
  (output omitted )
Iteration 17:   deviance =  50.16972
```

Generalized linear models		No. of obs	=	24
Optimization	: MQL Fisher scoring	Residual df	=	18
	(IRLS EIM)	Scale parameter	=	1
Deviance	= 50.16972096	(1/df) Deviance	=	2.787207
Pearson	= 49.28422567	(1/df) Pearson	=	2.738013
Variance function: V(u) = u*(1-u/n)		[Binomial]		
Link function	: g(u) = ln(-ln(1-u/n))	[Complementary log-log]		
		BIC	=	-7.035248

r	Coef.	EIM Std. Err.	z	P>|z|	[95% Conf. Interval]	
beetle						
2	-.79933	4.586649	-0.17	0.862	-9.788997	8.190337
3	17.78741	4.624834	3.85	0.000	8.7229	26.85192
ldose	22.04118	1.799356	12.25	0.000	18.5145	25.56785
beetle#c.ldose						
2	.3838708	2.544068	0.15	0.880	-4.602411	5.370152
3	-10.726	2.548176	-4.21	0.000	-15.72033	-5.731665
_cons	-39.57232	3.240274	-12.21	0.000	-45.92314	-33.2215

Note our use of the ltol() option, which, although unrelated to our discussion on variance estimation, was used so that the regression coefficients would match those of the previous Newton–Raphson (NR) fit.

Because IRLS uses the EIM for optimization, the variance estimate is also based on EIM. If we want optimization via IRLS but the variance estimate based on OIM, we specify glm, irls vce(oim):

```
. glm r beetle##c.ldose, f(b n) l(cl) ltol(1e-15) irls vce(oim) noheader nolog
```

r	Coef.	OIM Std. Err.	z	P>|z|	[95% Conf. Interval]	
beetle						
2	-.79933	4.470882	-0.18	0.858	-9.562098	7.963438
3	17.78741	4.586429	3.88	0.000	8.798172	26.77664
ldose	22.04118	1.793089	12.29	0.000	18.52679	25.55557
beetle#c.ldose						
2	.3838708	2.478477	0.15	0.877	-4.473855	5.241596
3	-10.726	2.526412	-4.25	0.000	-15.67768	-5.774321
_cons	-39.57232	3.229047	-12.26	0.000	-45.90114	-33.24351

This approach is identical to NR except for the convergence path. Because the cloglog link is not the canonical link for the binomial family, EIM and OIM produce different results. Both estimators, however, are asymptotically equivalent.

Going back to NR, we can also specify vce(robust) to get the Huber/White/sandwich estimator of variance:

```
. glm r beetle##c.ldose, f(b n) l(cl) vce(robust) noheader nolog
```

r	Coef.	Robust Std. Err.	z	P>|z|	[95% Conf. Interval]	
beetle						
2	-.79933	5.733049	-0.14	0.889	-12.0359	10.43724
3	17.78741	5.158477	3.45	0.001	7.676977	27.89784
ldose	22.04118	.8998551	24.49	0.000	20.27749	23.80486
beetle#c.ldose						
2	.3838708	3.174427	0.12	0.904	-5.837892	6.605633
3	-10.726	2.800606	-3.83	0.000	-16.21508	-5.236912
_cons	-39.57232	1.621306	-24.41	0.000	-42.75003	-36.39462

The sandwich estimator gets its name from the form of the calculation—it is the multiplication of three matrices, with the outer two matrices (the "bread") set to the OIM variance matrix. When irls is used along with vce(robust), the EIM variance matrix is instead used as the bread, and the resulting variance is labeled "Semirobust".

```
. glm r beetle##c.ldose, f(b n) l(cl) irls ltol(1e-15) vce(robust) noheader
> nolog
```

r	Coef.	Semirobust Std. Err.	z	P>|z|	[95% Conf. Interval]	
beetle						
2	-.79933	6.288963	-0.13	0.899	-13.12547	11.52681
3	17.78741	5.255307	3.38	0.001	7.487194	28.08762
ldose	22.04118	.9061566	24.32	0.000	20.26514	23.81721
beetle#c.ldose						
2	.3838708	3.489723	0.11	0.912	-6.455861	7.223603
3	-10.726	2.855897	-3.76	0.000	-16.32345	-5.128542
_cons	-39.57232	1.632544	-24.24	0.000	-42.77205	-36.3726

The outer product of the gradient (OPG) estimate of variance is one that avoids the calculation of second derivatives. It is equivalent to the "middle" part of the sandwich estimate of variance and can be specified by using glm, vce(opg), regardless of whether NR or IRLS optimization is used.

```
. glm r beetle##c.ldose, f(b n) l(cl) vce(opg) noheader nolog
```

r	Coef.	OPG Std. Err.	z	P>|z|	[95% Conf. Interval]	
beetle						
2	-.79933	6.664045	-0.12	0.905	-13.86062	12.26196
3	17.78741	6.838505	2.60	0.009	4.384183	31.19063
ldose	22.04118	3.572983	6.17	0.000	15.03826	29.0441
beetle#c.ldose						
2	.3838708	3.700192	0.10	0.917	-6.868372	7.636114
3	-10.726	3.796448	-2.83	0.005	-18.1669	-3.285097
_cons	-39.57232	6.433101	-6.15	0.000	-52.18097	-26.96368

The OPG estimate of variance is a component of the BHHH (Berndt et al. 1974) optimization technique. This method of optimization is also available with glm with the technique() option; however, the technique() option is not allowed with the irls option.

◁

Example 4

The Newey–West (1987) estimator of variance is a sandwich estimator with the "middle" of the sandwich modified to take into account possible autocorrelation between the observations. These estimators are a generalization of those given by the Stata command newey for linear regression. See [TS] **newey** for more details.

For example, consider the dataset given in [TS] **newey**, which has time-series measurements on usr and idle. We want to perform a linear regression with Newey–West standard errors.

```
. use http://www.stata-press.com/data/r12/idle2

. list usr idle time
```

	usr	idle	time
1.	0	100	1
2.	0	100	2
3.	0	97	3
4.	1	98	4
5.	2	94	5
	(output omitted)		
29.	1	98	29
30.	1	98	30

Examining *Methods and formulas* of [TS] **newey**, we see that the variance estimate is multiplied by a correction factor of $n/(n-k)$, where k is the number of regressors. glm, vce(hac ...) does not make this correction, so to get the same standard errors, we must use the vfactor() option within glm to make the correction manually.

```
. display 30/28
1.0714286

. tsset time
        time variable:  time, 1 to 30
                delta:  1 unit

. glm usr idle, vce(hac nwest 3) vfactor(1.0714286)

Iteration 0:   log likelihood = -71.743396

Generalized linear models                        No. of obs      =         30
Optimization     : ML                            Residual df     =         28
                                                 Scale parameter =   7.493297
Deviance         =  209.8123165                  (1/df) Deviance =   7.493297
Pearson          =  209.8123165                  (1/df) Pearson  =   7.493297

Variance function: V(u) = 1                      [Gaussian]
Link function    : g(u) = u                      [Identity]

HAC kernel (lags): Newey-West (3)
                                                 AIC             =   4.916226
Log likelihood   = -71.74339627                  BIC             =   114.5788
```

usr	Coef.	HAC Std. Err.	z	P>|z|	[95% Conf. Interval]	
idle	-.2281501	.0690928	-3.30	0.001	-.3635694	-.0927307
_cons	23.13483	6.327033	3.66	0.000	10.73407	35.53558

The glm command above reproduces the results given in [TS] **newey**. We may now generalize this output to models other than simple linear regression and to different kernel weights.

```
. glm usr idle, fam(gamma) link(log) vce(hac gallant 3)

Iteration 0:   log likelihood = -61.76593
Iteration 1:   log likelihood = -60.963233
Iteration 2:   log likelihood = -60.95097
Iteration 3:   log likelihood = -60.950965
```

```
Generalized linear models                    No. of obs      =         30
Optimization      : ML                       Residual df     =         28
                                             Scale parameter =   .431296
Deviance        =  9.908506707               (1/df) Deviance =  .3538752
Pearson         =  12.07628677               (1/df) Pearson  =   .431296

Variance function: V(u) = u^2                [Gamma]
Link function    : g(u) = ln(u)              [Log]

HAC kernel (lags): Gallant (3)
                                             AIC             =   4.196731
Log likelihood    = -60.95096484             BIC             = -85.32502
```

| | | HAC | | | | |
usr	Coef.	Std. Err.	z	P>|z|	[95% Conf.	Interval]
idle	-.0796609	.0184647	-4.31	0.000	-.115851	-.0434708
_cons	7.771011	1.510198	5.15	0.000	4.811078	10.73094

glm also offers variance estimators based on the bootstrap (resampling your data with replacement) and the jackknife (refitting the model with each observation left out in succession). Also included is the one-step jackknife estimate, which, instead of performing full reestimation when each observation is omitted, calculates a one-step NR estimate, with the full data regression coefficients as starting values.

```
. set seed 1
. glm usr idle, fam(gamma) link(log) vce(bootstrap, reps(100) nodots)

Generalized linear models                    No. of obs      =         30
Optimization      : ML                       Residual df     =         28
                                             Scale parameter =   .431296
Deviance        =  9.908506707               (1/df) Deviance =  .3538752
Pearson         =  12.07628677               (1/df) Pearson  =   .431296

Variance function: V(u) = u^2                [Gamma]
Link function    : g(u) = ln(u)              [Log]
                                             AIC             =   4.196731
Log likelihood    = -60.95096484             BIC             = -85.32502
```

| | Observed | Bootstrap | | | Normal-based | |
usr	Coef.	Std. Err.	z	P>|z|	[95% Conf.	Interval]
idle	-.0796609	.0216591	-3.68	0.000	-.1221119	-.0372099
_cons	7.771011	1.80278	4.31	0.000	4.237627	11.3044

◁

See Hardin and Hilbe (2007) for a full discussion of the variance options that go with glm and, in particular, of how the different variance estimators are modified when vce(cluster *clustvar*) is specified. Finally, not all variance options are supported with all types of weights. See help glm for a current table of the variance options that are supported with the different weights.

User-defined functions

`glm` may be called with a user-written link function, variance (family) function, Newey–West kernel-weight function, or any combination of the three.

Syntax of link functions

```
program progname
        version 12
        args todo eta mu return
        if 'todo' == -1 {
                /* Set global macros for output */
                global SGLM_lt " title for link function "
                global SGLM_lf " subtitle showing link definition "
                exit
        }
        if 'todo' == 0 {
                /* set  η=g(μ) */
                /* Intermediate calculations go here */
                generate double 'eta' = ...
                exit
        }
        if 'todo' == 1 {
                /* set  μ=g⁻¹(η) */
                /* Intermediate calculations go here */
                generate double 'mu' = ...
                exit
        }
        if 'todo' == 2 {
                /* set return = ∂μ/∂η */
                /* Intermediate calculations go here */
                generate double 'return' = ...
                exit
        }
        if 'todo' == 3 {
                /* set return = ∂²μ/∂η² */
                /* Intermediate calculations go here */
                generate double 'return' = ...
                exit
        }
        display as error "Unknown call to glm link function"
        exit 198
end
```

Syntax of variance functions

```
program progname
        version 12
        args todo eta mu return
        if 'todo' == -1 {
                /* Set global macros for output */
                /* Also check that depvar is in proper range */
                /* Note: For this call, eta contains indicator for whether each obs. is in est. sample */
                global SGLM_vt " title for variance function "
                global SGLM_vf " subtitle showing function definition "
                global SGLM_mu " program to call to enforce boundary conditions on μ "
                exit
        }
        if 'todo' == 0 {
                /* set η to initial value. */
                /* Intermediate calculations go here */
                generate double 'eta' = ...
                exit
        }
        if 'todo' == 1 {
                /* set return = V(μ) */
                /* Intermediate calculations go here */
                generate double 'return' = ...
                exit
        }
        if 'todo' == 2 {
                /* set return = ∂V(μ)/∂μ */
                /* Intermediate calculations go here */
                generate double 'return' = ...
                exit
        }
        if 'todo' == 3 {
                /* set return =  squared deviance (per observation) */
                /* Intermediate calculations go here */
                generate double 'return' = ...
                exit
        }
        if 'todo' == 4 {
                /* set return =  Anscombe residual */
                /* Intermediate calculations go here */
                generate double 'return' = ...
                exit
        }
        if 'todo' == 5 {
                /* set return =  log likelihood */
                /* Intermediate calculations go here */
                generate double 'return' = ...
                exit
        }
        if 'todo' == 6 {
                /* set return =  adjustment for deviance residuals */
                /* Intermediate calculations go here */
                generate double 'return' = ...
                exit
        }
        display as error "Unknown call to glm variance function"
        exit 198
end
```

Syntax of Newey–West kernel-weight functions

```
program progname, rclass
        version 12
        args G j
        /* G is the maximum lag */
        /* j is the current lag */
        /* Intermediate calculations go here */
        return scalar wt = computed weight
        return local setype "Newey-West"
        return local sewtype " name of kernel "
end
```

Global macros available for user-written programs

Global macro	Description
SGLM_V	program name of variance (family) evaluator
SGLM_L	program name of link evaluator
SGLM_y	dependent variable name
SGLM_m	binomial denominator
SGLM_a	negative binomial k
SGLM_p	power if power() or opower() is used, or an argument from a user-specified link function
SGLM_s1	indicator; set to one if scale is equal to one
SGLM_ph	value of scale parameter

▷ Example 5

Suppose that we wish to perform Poisson regression with a log-link function. Although this regression is already possible with standard glm, we will write our own version for illustrative purposes.

Because we want a log link, $\eta = g(\mu) = \ln(\mu)$, and for a Poisson family the variance function is $V(\mu) = \mu$.

The Poisson density is given by

$$f(y_i) = \frac{e^{-\exp(\mu_i)} e^{\mu_i y_i}}{y_i!}$$

resulting in a log likelihood of

$$L = \sum_{i=1}^{n} \{-e^{\mu_i} + \mu_i y_i - \ln(y_i!)\}$$

The squared deviance of the ith observation for the Poisson family is given by

$$d_i^2 = \begin{cases} 2\widehat{\mu}_i & \text{if } y_i = 0 \\ 2\{y_i \ln(y_i/\widehat{\mu}_i) - (y_i - \widehat{\mu}_i)\} & \text{otherwise} \end{cases}$$

We now have enough information to write our own Poisson-log `glm` module. We create the file `mylog.ado`, which contains

```
program mylog
        version 12
        args todo eta mu return
        if 'todo' == -1 {
                global SGLM_lt "My Log"              // Titles for output
                global SGLM_lf "ln(u)"
                exit
        }
        if 'todo' == 0  {
                gen double 'eta' = ln('mu')          //  η = ln(μ)
                exit
        }
        if 'todo' == 1  {
                gen double 'mu' = exp('eta')         //  μ = exp(η)
                exit
        }
        if 'todo' == 2 {
                gen double 'return' = 'mu'           //  ∂μ/∂η = exp(η) = μ
                exit
        }
        if 'todo' == 3 {
                gen double 'return' = 'mu'           //  ∂²μ/∂η² = exp(η) = μ
                exit
        }
        di as error "Unknown call to glm link function"
        exit 198
end
```

and we create the file `mypois.ado`, which contains

```
program mypois
        version 12
        args todo eta mu return
        if 'todo' == -1 {
                local y      "$SGLM_y"
                local touse "'eta'"                  // 'eta' marks estimation sample here
                capture assert 'y'>=0 if 'touse'     // check range of y
                if _rc {
                        di as error '"dependent variable 'y' has negative values"'
                        exit 499
                }
                global SGLM_vt "My Poisson"          // Titles for output
                global SGLM_vf "u"
                global SGLM_mu "glim_mu 0 ."         // see note 1
                exit
        }
        if 'todo' == 0 {                             // Initialization of η; see note 2
                gen double 'eta' = ln('mu')
                exit
        }
```

```
      if `todo' == 1 {
              gen double `return' = `mu'        //  V(μ) = μ
              exit
      }
      if `todo' == 2 {                          //  ∂ V(μ)/∂μ
              gen byte `return' = 1
              exit
      }
      if `todo' == 3 {                          // squared deviance, defined above
              local y "$SGLM_y"
              if "`y'" == "" {
                      local y "`e(depvar)'"
              }
              gen double `return' = cond(`y'==0, 2*`mu', /*
                      */ 2*(`y'*ln(`y'/`mu')-(`y'-`mu')))
              exit
      }
      if `todo' == 4 {                          // Anscombe residual; see note 3
              local y "$SGLM_y"
              if "`y'" == "" {
                      local y "`e(depvar)'"
              }
              gen double `return' = 1.5*(`y'^(2/3)-`mu'^(2/3)) / `mu'^(1/6)
              exit
      }
      if `todo' == 5 {                          // log likelihood; see note 4
              local y "$SGLM_y"
              if "`y'" == "" {
                      local y "`e(depvar)'"
              }
              gen double `return' = -`mu'+`y'*ln(`mu')-lngamma(`y'+1)
              exit
      }
      if `todo' == 6 {                          // adjustment to residual; see note 5
              gen double `return' = 1/(6*sqrt(`mu'))
              exit
      }
      di as error "Unknown call to glm variance function"
      error 198
end
```

Notes:

1. `glim_mu` is a Stata program that will, at each iteration, bring $\widehat{\mu}$ back into its plausible range, should it stray out of it. Here `glim_mu` is called with the arguments zero and missing, meaning that zero is the lower bound of $\widehat{\mu}$ and there exists no upper bound—such is the case for Poisson models.

2. Here the initial value of η is easy because we intend to fit this model with our user-defined log link. In general, however, the initialization may need to vary according to the link to obtain convergence. If so, the global macro SGLM_L is used to determine which link is being utilized.

3. The Anscombe formula is given here because we know it. If we were not interested in Anscombe residuals, we could merely set `return' to missing. Also, the local macro y is set either to SGLM_y if it is in current estimation or to e(depvar) if this function is being accessed by `predict`.

4. If we were not interested in ML estimation, we could omit this code entirely and just leave an `exit` statement in its place. Similarly, if we were not interested in deviance or IRLS optimization, we could set `return' in the deviance portion of the code (`todo'==3) to missing.

5. This code defines the term to be added to the predicted residuals if the `adjusted` option is specified. Again, if we were not interested, we could set 'return' to missing.

We can now test our Poisson-log module by running it on the airline data presented in [R] **poisson**.

```
. use http://www.stata-press.com/data/r12/airline
. list airline injuries n XYZowned
```

	airline	injuries	n	XYZowned
1.	1	11	0.0950	1
2.	2	7	0.1920	0
3.	3	7	0.0750	0
4.	4	19	0.2078	0
5.	5	9	0.1382	0
6.	6	4	0.0540	1
7.	7	3	0.1292	0
8.	8	1	0.0503	0
9.	9	3	0.0629	1

```
. gen lnN=ln(n)
. glm injuries XYZowned lnN, fam(mypois) link(mylog) scale(1)
Iteration 0:   log likelihood = -22.557572
Iteration 1:   log likelihood = -22.332861
Iteration 2:   log likelihood = -22.332276
Iteration 3:   log likelihood = -22.332276
```

```
Generalized linear models                      No. of obs      =         9
Optimization     : ML                          Residual df     =         6
                                               Scale parameter =         1
Deviance         =   12.70432823               (1/df) Deviance =  2.117388
Pearson          =   12.7695081                (1/df) Pearson  =  2.128251

Variance function: V(u) = u                    [My Poisson]
Link function    : g(u) = ln(u)                [My Log]
                                               AIC             =  5.629395
Log likelihood   = -22.33227605                BIC             = -.4790192
```

injuries	Coef.	OIM Std. Err.	z	P>|z|	[95% Conf. Interval]	
XYZowned	.6840668	.3895877	1.76	0.079	-.0795111	1.447645
lnN	1.424169	.3725155	3.82	0.000	.6940517	2.154286
_cons	4.863891	.7090501	6.86	0.000	3.474178	6.253603

(Standard errors scaled using dispersion equal to square root of 1.)

These are precisely the results given in [R] **poisson** and are those that would have been given had we run `glm, family(poisson) link(log)`. The only minor adjustment we needed to make was to specify the `scale(1)` option. If `scale()` is left unspecified, `glm` assumes `scale(1)` for discrete distributions and `scale(x2)` for continuous ones. By default, `glm` assumes that any user-defined family is continuous because it has no way of checking. Thus we needed to specify `scale(1)` because our model is discrete.

Because we were careful in defining the squared deviance, we could have fit this model with IRLS. Because log is the canonical link for the Poisson family, we would not only get the same regression coefficients but also the same standard errors.

◁

▷ Example 6

Suppose now that we wish to use our log link (`mylog.ado`) with `glm`'s binomial family. This task requires some modification because our current function is not equipped to deal with the binomial denominator, which we are allowed to specify. This denominator is accessible to our link function through the global macro `SGLM_m`. We now make the modifications and store them in `mylog2.ado`.

```
program mylog2                                          // <-- changed
        version 12
        args todo eta mu return
        if 'todo' == -1 {
                global SGLM_lt "My Log, Version 2"      // <-- changed
                if "$SGLM_m" == "1" {                   // <-- changed
                        global SGLM_lf "ln(u)"          // <-- changed
                }                                       // <-- changed
                else    global SGLM_lf "ln(u/$SGLM_m)"  // <-- changed
                exit
        }
        if 'todo' == 0  {
                gen double 'eta' = ln('mu'/$SGLM_m)     // <-- changed
                exit
        }
        if 'todo' == 1  {
                gen double 'mu' = $SGLM_m*exp('eta')    // <-- changed
                exit
        }
        if 'todo' == 2 {
                gen double 'return' = 'mu'
                exit
        }
        if 'todo' == 3 {
                gen double 'return' = 'mu'
                exit
        }
        di as error "Unknown call to glm link function"
        exit 198
end
```

We can now run our new log link with `glm`'s binomial family. Using the flour beetle data from earlier, we have

```
. use http://www.stata-press.com/data/r12/beetle, clear

. glm r ldose, fam(bin n) link(mylog2) irls
Iteration 1:   deviance =   2212.108
Iteration 2:   deviance =   452.9352
Iteration 3:   deviance =     429.95
Iteration 4:   deviance =   429.2745
Iteration 5:   deviance =   429.2192
Iteration 6:   deviance =   429.2082
Iteration 7:   deviance =   429.2061
Iteration 8:   deviance =   429.2057
Iteration 9:   deviance =   429.2056
Iteration 10:  deviance =   429.2056
Iteration 11:  deviance =   429.2056
Iteration 12:  deviance =   429.2056
```

```
Generalized linear models                 No. of obs      =         24
Optimization      : MQL Fisher scoring    Residual df     =         22
                    (IRLS EIM)            Scale parameter =          1
Deviance          =   429.205599          (1/df) Deviance =   19.50935
Pearson           =   413.088142          (1/df) Pearson  =   18.77673

Variance function: V(u) = u*(1-u/n)       [Binomial]
Link function    : g(u) = ln(u/n)         [My Log, Version 2]

                                          BIC             =   359.2884
```

r	Coef.	EIM Std. Err.	z	P>|z|	[95% Conf. Interval]	
ldose	8.478908	.4702808	18.03	0.000	7.557175	9.400642
_cons	-16.11006	.8723167	-18.47	0.000	-17.81977	-14.40035

◁

For a more detailed discussion on user-defined functions, and for an example of a user-defined Newey–West kernel weight, see Hardin and Hilbe (2007).

John Ashworth Nelder (1924–2010) was born in Somerset, England. He studied mathematics and statistics at Cambridge and worked as a statistician at the National Vegetable Research Station and then Rothamsted Experimental Station. In retirement, he was actively affiliated with Imperial College London. Nelder was especially well known for his contributions to the theory of linear models and to statistical computing. He was the principal architect of generalized and hierarchical generalized linear models and of the programs GenStat and GLIM.

Robert William Maclagan Wedderburn (1947–1975) was born in Edinburgh and studied mathematics and statistics at Cambridge. At Rothamsted Experimental Station, he developed the theory of generalized linear models with Nelder and originated the concept of quasilikelihood. He died of anaphylactic shock from an insect bite on a canal holiday.

Saved results

glm, ml saves the following in e():

Scalars
e(N)	number of observations
e(k)	number of parameters
e(k_eq)	number of equations in e(b)
e(k_eq_model)	number of equations in overall model test
e(k_dv)	number of dependent variables
e(df_m)	model degrees of freedom
e(df)	residual degrees of freedom
e(phi)	scale parameter
e(aic)	model AIC
e(bic)	model BIC
e(ll)	log likelihood, if NR
e(N_clust)	number of clusters
e(chi2)	χ^2
e(p)	significance
e(deviance)	deviance
e(deviance_s)	scaled deviance
e(deviance_p)	Pearson deviance
e(deviance_ps)	scaled Pearson deviance
e(dispers)	dispersion
e(dispers_s)	scaled dispersion
e(dispers_p)	Pearson dispersion
e(dispers_ps)	scaled Pearson dispersion
e(nbml)	1 if negative binomial parameter estimated via ML, 0 otherwise
e(vf)	factor set by vfactor(), 1 if not set
e(power)	power set by power(), opower()
e(rank)	rank of e(V)
e(ic)	number of iterations
e(rc)	return code
e(converged)	1 if converged, 0 otherwise

Macros
    e(cmd)   glm

e(cmd)	glm
e(cmdline)	command as typed
e(depvar)	name of dependent variable
e(varfunc)	program to calculate variance function
e(varfunct)	variance title
e(varfuncf)	variance function
e(link)	program to calculate link function
e(linkt)	link title
e(linkf)	link function
e(m)	number of binomial trials
e(wtype)	weight type
e(wexp)	weight expression
e(title)	title in estimation output
e(clustvar)	name of cluster variable
e(offset)	linear offset variable
e(chi2type)	Wald; type of model χ^2 test
e(cons)	set if noconstant specified
e(hac_kernel)	HAC kernel
e(hac_lag)	HAC lag
e(vce)	*vcetype* specified in vce()
e(vcetype)	title used to label Std. Err.
e(opt)	ml or irls
e(opt1)	optimization title, line 1
e(which)	max or min; whether optimizer is to perform maximization or minimization
e(ml_method)	type of ml method
e(user)	name of likelihood-evaluator program
e(technique)	maximization technique
e(properties)	b V
e(predict)	program used to implement predict
e(marginsok)	predictions allowed by margins
e(marginsnotok)	predictions disallowed by margins
e(asbalanced)	factor variables fvset as asbalanced
e(asobserved)	factor variables fvset as asobserved

Matrices

e(b)	coefficient vector
e(Cns)	constraints matrix
e(ilog)	iteration log (up to 20 iterations)
e(gradient)	gradient vector
e(V)	variance–covariance matrix of the estimators
e(V_modelbased)	model-based variance

Functions

e(sample)	marks estimation sample

`glm, irls` saves the following in `e()`:

Scalars
`e(N)`	number of observations
`e(k)`	number of parameters
`e(k_eq_model)`	number of equations in overall model test
`e(df_m)`	model degrees of freedom
`e(df)`	residual degrees of freedom
`e(phi)`	scale parameter
`e(disp)`	dispersion parameter
`e(bic)`	model BIC
`e(N_clust)`	number of clusters
`e(deviance)`	deviance
`e(deviance_s)`	scaled deviance
`e(deviance_p)`	Pearson deviance
`e(deviance_ps)`	scaled Pearson deviance
`e(dispers)`	dispersion
`e(dispers_s)`	scaled dispersion
`e(dispers_p)`	Pearson dispersion
`e(dispers_ps)`	scaled Pearson dispersion
`e(nbml)`	1 if negative binomial parameter estimated via ML, 0 otherwise
`e(vf)`	factor set by `vfactor()`, 1 if not set
`e(power)`	power set by `power()`, `opower()`
`e(rank)`	rank of `e(V)`
`e(rc)`	return code

Macros
`e(cmd)`	`glm`
`e(cmdline)`	command as typed
`e(depvar)`	name of dependent variable
`e(varfunc)`	program to calculate variance function
`e(varfunct)`	variance title
`e(varfuncf)`	variance function
`e(link)`	program to calculate link function
`e(linkt)`	link title
`e(linkf)`	link function
`e(m)`	number of binomial trials
`e(wtype)`	weight type
`e(wexp)`	weight expression
`e(clustvar)`	name of cluster variable
`e(offset)`	linear offset variable
`e(cons)`	set if `noconstant` specified
`e(hac_kernel)`	HAC kernel
`e(hac_lag)`	HAC lag
`e(vce)`	*vcetype* specified in `vce()`
`e(vcetype)`	title used to label Std. Err.
`e(opt)`	ml or irls
`e(opt1)`	optimization title, line 1
`e(opt2)`	optimization title, line 2
`e(properties)`	b V
`e(predict)`	program used to implement `predict`
`e(marginsok)`	predictions allowed by `margins`
`e(marginsnotok)`	predictions disallowed by `margins`
`e(asbalanced)`	factor variables `fvset` as asbalanced
`e(asobserved)`	factor variables `fvset` as asobserved

Matrices
`e(b)`	coefficient vector
`e(Cns)`	constraints matrix
`e(V)`	variance–covariance matrix of the estimators
`e(V_modelbased)`	model-based variance

Functions
`e(sample)`	marks estimation sample

Methods and formulas

glm is implemented as an ado-file.

The canonical reference on GLM is McCullagh and Nelder (1989). The term "generalized linear model" is from Nelder and Wedderburn (1972). Many people use the acronym GLIM for GLM models because of the classic GLM software tool GLIM, by Baker and Nelder (1985). See Dobson and Barnett (2008) for a concise introduction and overview. See Rabe-Hesketh and Everitt (2007) for more examples of GLM using Stata. Hoffmann (2004) focuses on applying generalized linear models, using real-world datasets, along with interpreting computer output, which for the most part is obtained using Stata.

This discussion highlights the details of parameter estimation and predicted statistics. For a more detailed treatment, and for information on variance estimation, see Hardin and Hilbe (2007). glm supports estimation with survey data. For details on VCEs with survey data, see[SVY] **variance estimation**.

glm obtains results by IRLS, as described in McCullagh and Nelder (1989), or by maximum likelihood using Newton–Raphson. The implementation here, however, allows user-specified weights, which we denote as v_j for the jth observation. Let M be the number of "observations" ignoring weights. Define

$$
w_j = \begin{cases} 1 & \text{if no weights are specified} \\ v_j & \text{if } \texttt{fweights} \text{ or } \texttt{iweights} \text{ are specified} \\ Mv_j/(\sum_k v_k) & \text{if } \texttt{aweights} \text{ or } \texttt{pweights} \text{ are specified} \end{cases}
$$

The number of observations is then $N = \sum_j w_j$ if fweights are specified and $N = M$ otherwise. Each IRLS step is performed by regress using w_j as the weights.

Let d_j^2 denote the squared deviance residual for the jth observation:

For the Gaussian family, $d_j^2 = (y_j - \widehat{\mu}_j)^2$.

For the Bernoulli family (binomial with denominator 1),

$$
d_j^2 = \begin{cases} -2\ln(1 - \widehat{\mu}_j) & \text{if } y_j = 0 \\ -2\ln(\widehat{\mu}_j) & \text{otherwise} \end{cases}
$$

For the binomial family with denominator m_j,

$$
d_j^2 = \begin{cases} 2y_j\ln(y_j/\widehat{\mu}_j) + 2(m_j - y_j)\ln\{(m_j - y_j)/(m_j - \widehat{\mu}_j)\} & \text{if } 0 < y_j < m_j \\ 2m_j\ln\{m_j/(m_j - \widehat{\mu}_j)\} & \text{if } y_j = 0 \\ 2y_j\ln(y_j/\widehat{\mu}_j) & \text{if } y_j = m_j \end{cases}
$$

For the Poisson family,

$$
d_j^2 = \begin{cases} 2\widehat{\mu}_j & \text{if } y_j = 0 \\ 2\{y_j\ln(y_j/\widehat{\mu}_j) - (y_j - \widehat{\mu}_j)\} & \text{otherwise} \end{cases}
$$

For the gamma family, $d_j^2 = -2\{\ln(y_j/\widehat{\mu}_j) - (y_j - \widehat{\mu}_j)/\widehat{\mu}_j\}$.

For the inverse Gaussian, $d_j^2 = (y_j - \widehat{\mu}_j)^2/(\widehat{\mu}_j^2 y_j)$.

For the negative binomial,

$$d_j^2 = \begin{cases} 2\ln(1 + k\widehat{\mu}_j)/k & \text{if } y_j = 0 \\ 2y_j\ln(y_j/\widehat{\mu}_j) - 2\{(1 + ky_j)/k\}\ln\{(1 + ky_j)/(1 + k\widehat{\mu}_j)\} & \text{otherwise} \end{cases}$$

Let $\phi = 1$ if the scale parameter is set to one; otherwise, define $\phi = \widehat{\phi}_0(n - k)/n$, where $\widehat{\phi}_0$ is the estimated scale parameter and k is the number of covariates in the model (including intercept).

Let $\ln L_j$ denote the log likelihood for the jth observation:

For the Gaussian family,

$$\ln L_j = -\frac{1}{2}\left[\left\{\frac{(y_j - \widehat{\mu}_j)^2}{\phi}\right\} + \ln(2\pi\phi)\right]$$

For the binomial family with denominator m_j (Bernoulli if all $m_j = 1$),

$$\ln L_j = \phi \times \begin{cases} \ln\{\Gamma(m_j + 1)\} - \ln\{\Gamma(y_j + 1)\} + \ln\{\Gamma(m_j - y_j + 1)\} & \text{if } 0 < y_j < m_j \\ \quad + (m_j - y_j)\ln(1 - \widehat{\mu}_j/m_j) + y_j\ln(\widehat{\mu}_j/m_j) & \\ m_j\ln(1 - \widehat{\mu}_j/m_j) & \text{if } y_j = 0 \\ m_j\ln(\widehat{\mu}_j/m_j) & \text{if } y_j = m_j \end{cases}$$

For the Poisson family,

$$\ln L_j = \phi\left[y_j\ln(\widehat{\mu}_j) - \widehat{\mu}_j - \ln\{\Gamma(y_j + 1)\}\right]$$

For the gamma family, $\ln L_j = -y_j/\widehat{\mu}_j + \ln(1/\widehat{\mu}_j)$.

For the inverse Gaussian,

$$\ln L_j = -\frac{1}{2}\left\{\frac{(y_j - \widehat{\mu}_j)^2}{y_j\widehat{\mu}_j^2} + 3\ln(y_j) + \ln(2\pi)\right\}$$

For the negative binomial (let $m = 1/k$),

$$\ln L_j = \phi\left[\ln\{\Gamma(m + y_j)\} - \ln\{\Gamma(y_j + 1)\} - \ln\{\Gamma(m)\}\right.$$
$$\left. - m\ln(1 + \widehat{\mu}_j/m) + y_j\ln\{\widehat{\mu}_j/(\widehat{\mu}_j + m)\}\right]$$

The overall deviance reported by `glm` is $D^2 = \sum_j w_j d_j^2$. The dispersion of the deviance is D^2 divided by the residual degrees of freedom.

The Akaike information criterion (AIC) and Bayesian information criterion (BIC) are given by

$$\text{AIC} = \frac{-2\ln L + 2k}{N}$$
$$\text{BIC} = D^2 - (N - k)\ln(N)$$

where $\ln L = \sum_j w_j \ln L_j$ is the overall log likelihood.

The Pearson deviance reported by glm is $\sum_j w_j r_j^2$. The corresponding Pearson dispersion is the Pearson deviance divided by the residual degrees of freedom. glm also calculates the scaled versions of all these quantities by dividing by the estimated scale parameter.

Acknowledgments

glm was written by James Hardin, University of South Carolina, and Joseph Hilbe, Arizona State University. The previous version of this routine was written by Patrick Royston, MRC Clinical Trials Unit, London. The original version of this routine was published in Royston (1994). Royston's work, in turn, was based on a prior implementation by Joseph Hilbe, first published in Hilbe (1993). Roger Newson wrote an early implementation (Newson 1999) of robust variance estimates for GLM. Parts of this entry are excerpts from Hardin and Hilbe (2007).

References

Akaike, H. 1973. Information theory and an extension of the maximum likelihood principle. In *Second International Symposium on Information Theory*, ed. B. N. Petrov and F. Csaki, 267–281. Budapest: Akailseoniai–Kiudo.

Anscombe, F. J. 1953. Contribution of discussion paper by H. Hotelling "New light on the correlation coefficient and its transforms". *Journal of the Royal Statistical Society, Series B* 15: 229–230.

Baker, R. J., and J. A. Nelder. 1985. *The Generalized Linear Interactive Modelling System, Release 3.77*. Oxford: Numerical Algorithms Group.

Basu, A. 2005. Extended generalized linear models: Simultaneous estimation of flexible link and variance functions. *Stata Journal* 5: 501–516.

Berndt, E. K., B. H. Hall, R. E. Hall, and J. A. Hausman. 1974. Estimation and inference in nonlinear structural models. *Annals of Economic and Social Measurement* 3/4: 653–665.

Cummings, P. 2009. Methods for estimating adjusted risk ratios. *Stata Journal* 9: 175–196.

Dobson, A. J., and A. G. Barnett. 2008. *An Introduction to Generalized Linear Models*. 3rd ed. Boca Raton, FL: Chapman & Hall/CRC.

Hardin, J. W., and J. M. Hilbe. 2007. *Generalized Linear Models and Extensions*. 2nd ed. College Station, TX: Stata Press.

Hilbe, J. M. 1993. sg16: Generalized linear models. *Stata Technical Bulletin* 11: 20–28. Reprinted in *Stata Technical Bulletin Reprints*, vol. 2, pp. 149–159. College Station, TX: Stata Press.

———. 2000. sg126: Two-parameter log-gamma and log-inverse Gaussian models. *Stata Technical Bulletin* 53: 31–32. Reprinted in *Stata Technical Bulletin Reprints*, vol. 9, pp. 273–275. College Station, TX: Stata Press.

———. 2009. *Logistic Regression Models*. Boca Raton, FL: Chapman & Hill/CRC.

Hoffmann, J. P. 2004. *Generalized Linear Models: An Applied Approach*. Boston: Pearson.

Hosmer, D. W., Jr., and S. Lemeshow. 2000. *Applied Logistic Regression*. 2nd ed. New York: Wiley.

McCullagh, P., and J. A. Nelder. 1989. *Generalized Linear Models*. 2nd ed. London: Chapman & Hall/CRC.

Nelder, J. A. 1975. Robert William MacLagan Wedderburn, 1947–1975. *Journal of the Royal Statistical Society, Series A* 138: 587.

Nelder, J. A., and R. W. M. Wedderburn. 1972. Generalized linear models. *Journal of the Royal Statistical Society, Series A* 135: 370–384.

Newey, W. K., and K. D. West. 1987. A simple, positive semi-definite, heteroskedasticity and autocorrelation consistent covariance matrix. *Econometrica* 55: 703–708.

Newson, R. 1999. sg114: rglm—Robust variance estimates for generalized linear models. *Stata Technical Bulletin* 50: 27–33. Reprinted in *Stata Technical Bulletin Reprints*, vol. 9, pp. 181–190. College Station, TX: Stata Press.

———. 2004. Generalized power calculations for generalized linear models and more. *Stata Journal* 4: 379–401.

Parner, E. T., and P. K. Andersen. 2010. Regression analysis of censored data using pseudo-observations. *Stata Journal* 10: 408–422.

Pregibon, D. 1980. Goodness of link tests for generalized linear models. *Applied Statistics* 29: 15–24.

Rabe-Hesketh, S., and B. S. Everitt. 2007. *A Handbook of Statistical Analyses Using Stata.* 4th ed. Boca Raton, FL: Chapman & Hall/CRC.

Rabe-Hesketh, S., A. Pickles, and C. Taylor. 2000. sg129: Generalized linear latent and mixed models. *Stata Technical Bulletin* 53: 47–57. Reprinted in *Stata Technical Bulletin Reprints*, vol. 9, pp. 293–307. College Station, TX: Stata Press.

Rabe-Hesketh, S., A. Skrondal, and A. Pickles. 2002. Reliable estimation of generalized linear mixed models using adaptive quadrature. *Stata Journal* 2: 1–21.

Raftery, A. 1995. Bayesian model selection in social research. In Vol. 25 of *Sociological Methodology*, ed. P. V. Marsden, 111–163. Oxford: Blackwell.

Rogers, W. H. 1993. sg16.4: Comparison of nbreg and glm for negative binomial. *Stata Technical Bulletin* 16: 7. Reprinted in *Stata Technical Bulletin Reprints*, vol. 3, pp. 82–84. College Station, TX: Stata Press.

Royston, P. 1994. sg22: Generalized linear models: Revision of glm. *Stata Technical Bulletin* 18: 6–11. Reprinted in *Stata Technical Bulletin Reprints*, vol. 3, pp. 112–121. College Station, TX: Stata Press.

Schonlau, M. 2005. Boosted regression (boosting): An introductory tutorial and a Stata plugin. *Stata Journal* 5: 330–354.

Senn, S. J. 2003. A conversation with John Nelder. *Statistical Science* 18: 118–131.

Williams, R. 2010. Fitting heterogeneous choice models with oglm. *Stata Journal* 10: 540–567.

Also see

[R] **glm postestimation** — Postestimation tools for glm

[R] **cloglog** — Complementary log-log regression

[R] **logistic** — Logistic regression, reporting odds ratios

[R] **nbreg** — Negative binomial regression

[R] **poisson** — Poisson regression

[R] **regress** — Linear regression

[MI] **estimation** — Estimation commands for use with mi estimate

[SVY] **svy estimation** — Estimation commands for survey data

[XT] **xtgee** — Fit population-averaged panel-data models by using GEE

[U] **20 Estimation and postestimation commands**

Title

> **glm postestimation** — Postestimation tools for glm

Description

The following postestimation commands are available after `glm`:

Command	Description
contrast	contrasts and ANOVA-style joint tests of estimates
estat	AIC, BIC, VCE, and estimation sample summary
estat (svy)	postestimation statistics for survey data
estimates	cataloging estimation results
lincom	point estimates, standard errors, testing, and inference for linear combinations of coefficients
linktest	link test for model specification
lrtest[1]	likelihood-ratio test
margins	marginal means, predictive margins, marginal effects, and average marginal effects
marginsplot	graph the results from margins (profile plots, interaction plots, etc.)
nlcom	point estimates, standard errors, testing, and inference for nonlinear combinations of coefficients
predict	predictions, residuals, influence statistics, and other diagnostic measures
predictnl	point estimates, standard errors, testing, and inference for generalized predictions
pwcompare	pairwise comparisons of estimates
suest	seemingly unrelated estimation
test	Wald tests of simple and composite linear hypotheses
testnl	Wald tests of nonlinear hypotheses

[1] `lrtest` is not appropriate with svy estimation results.

See the corresponding entries in the *Base Reference Manual* for details, but see [SVY] **estat** for details about `estat` (svy).

Syntax for predict

predict [*type*] *newvar* [*if*] [*in*] [, *statistic options*]

statistic	Description
Main	
mu	expected value of y; the default
xb	linear prediction $\eta = \mathbf{x}\widehat{\beta}$
eta	synonym of xb
stdp	standard error of the linear prediction
anscombe	Anscombe (1953) residuals
cooksd	Cook's distance
deviance	deviance residuals
hat	diagonals of the "hat" matrix
likelihood	a weighted average of standardized deviance and standardized Pearson residuals
pearson	Pearson residuals
response	differences between the observed and fitted outcomes
score	first derivative of the log likelihood with respect to $\mathbf{x}_j\beta$
working	working residuals

options	Description
Options	
nooffset	modify calculations to ignore offset variable
adjusted	adjust deviance residual to speed up convergence
standardized	multiply residual by the factor $(1 - h)^{-1/2}$
studentized	multiply residual by one over the square root of the estimated scale parameter
modified	modify denominator of residual to be a reasonable estimate of the variance of *depvar*

These statistics are available both in and out of sample; type predict ... if e(sample) ... if wanted only for the estimation sample.

mu, xb, stdp, and score are the only statistics allowed with svy estimation results.

Menu

Statistics > Postestimation > Predictions, residuals, etc.

Options for predict

⌐ **Main** ⌐

mu, the default, specifies that predict calculate the expected value of y, equal to $g^{-1}(\mathbf{x}\widehat{\beta})$ [$ng^{-1}(\mathbf{x}\widehat{\beta})$ for the binomial family].

xb calculates the linear prediction $\eta = \mathbf{x}\widehat{\beta}$.

eta is a synonym for xb.

stdp calculates the standard error of the linear prediction.

anscombe calculates the Anscombe (1953) residuals to produce residuals that closely follow a normal distribution.

cooksd calculates Cook's distance, which measures the aggregate change in the estimated coefficients when each observation is left out of the estimation.

deviance calculates the deviance residuals. Deviance residuals are recommended by McCullagh and Nelder (1989) and by others as having the best properties for examining the goodness of fit of a GLM. They are approximately normally distributed if the model is correct. They may be plotted against the fitted values or against a covariate to inspect the model's fit. Also see the pearson option below.

hat calculates the diagonals of the "hat" matrix as an analog to simple linear regression.

likelihood calculates a weighted average of standardized deviance and standardized Pearson residuals.

pearson calculates the Pearson residuals. Pearson residuals often have markedly skewed distributions for nonnormal family distributions. Also see the deviance option above.

response calculates the differences between the observed and fitted outcomes.

score calculates the equation-level score, $\partial \ln L / \partial (\mathbf{x}_j \boldsymbol{\beta})$.

working calculates the working residuals, which are response residuals weighted according to the derivative of the link function.

⌐ Options ⌐
---

nooffset is relevant only if you specified offset(*varname*) for glm. It modifies the calculations made by predict so that they ignore the offset variable; the linear prediction is treated as $\mathbf{x}_j \mathbf{b}$ rather than as $\mathbf{x}_j \mathbf{b} + \text{offset}_j$.

adjusted adjusts the deviance residual to speed up the convergence to the limiting normal distribution. The adjustment deals with adding to the deviance residual a higher-order term that depends on the variance function family. This option is allowed only when deviance is specified.

standardized requests that the residual be multiplied by the factor $(1 - h)^{-1/2}$, where h is the diagonal of the hat matrix. This operation is done to account for the correlation between *depvar* and its predicted value.

studentized requests that the residual be multiplied by one over the square root of the estimated scale parameter.

modified requests that the denominator of the residual be modified to be a reasonable estimate of the variance of *depvar*. The base residual is multiplied by the factor $(k/w)^{-1/2}$, where k is either one or the user-specified dispersion parameter and w is the specified weight (or one if left unspecified).

Remarks

Remarks are presented under the following headings:

> *Predictions*
> *Other postestimation commands*

Predictions

▷ Example 1

After glm estimation, predict may be used to obtain various predictions based on the model. In example 2 of [R] **glm**, we mentioned that the complementary log-log link seemed to fit the data better than the logit link. Now we go back and obtain the fitted values and deviance residuals:

```
. use http://www.stata-press.com/data/r12/ldose
. glm r ldose, f(binomial n) l(logit)
  (output omitted )
. predict mu_logit
(option mu assumed; predicted mean r)
. predict dr_logit, deviance
. quietly glm r ldose, f(binomial n) l(cloglog)
. predict mu_cl
(option mu assumed; predicted mean r)
. predict dr_cl, d
. format mu_logit dr_logit mu_cl dr_cl %9.5f
. list r mu_logit dr_logit mu_cl dr_cl, sep(4)
```

	r	mu_logit	dr_logit	mu_cl	dr_cl
1.	6	3.45746	1.28368	5.58945	0.18057
2.	13	9.84167	1.05969	11.28067	0.55773
3.	18	22.45139	-1.19611	20.95422	-0.80330
4.	28	33.89761	-1.59412	30.36942	-0.63439
5.	52	50.09584	0.60614	47.77644	1.28883
6.	53	53.29092	-0.12716	54.14273	-0.52366
7.	61	59.22216	1.25107	61.11331	-0.11878
8.	60	58.74297	1.59398	59.94723	0.32495

In six of the eight cases, $|dr\_logit| > |dr\_cl|$. The above represents only one of the many available options for predict. See Hardin and Hilbe (2007) for a more in-depth examination.

◁

Other postestimation commands

❏ Technical note

After glm estimation, you may perform any of the postestimation commands that you would perform after any other kind of estimation in Stata; see [U] **20 Estimation and postestimation commands**. Below we test the joint significance of all the interaction terms.

```
. use http://www.stata-press.com/data/r12/beetle, clear
. glm r beetle##c.ldose, f(bin n) l(cloglog)
  (output omitted )
. testparm i.beetle beetle#c.ldose
 ( 1)  [r]2.beetle = 0
 ( 2)  [r]3.beetle = 0
 ( 3)  [r]2.beetle#c.ldose = 0
 ( 4)  [r]3.beetle#c.ldose = 0
        chi2(  4) =   249.69
      Prob > chi2 =    0.0000
```

If you wanted to print the variance–covariance matrix of the estimators, you would type estat vce.

If you use the linktest postestimation command, you must also specify the family() and link() options; see [R] **linktest**.

❏

Methods and formulas

All postestimation commands listed above are implemented as ado-files.

We follow the terminology used in *Methods and formulas* of [R] **glm**.

The deviance residual calculated by `predict` following `glm` is $d_j = \text{sign}(y_j - \widehat{\mu}_j)\sqrt{d_j^2}$.

The Pearson residual calculated by `predict` following `glm` is

$$r_j = \frac{y_j - \widehat{\mu}_j}{\sqrt{V(\widehat{\mu}_j)}}$$

where $V(\widehat{\mu}_j)$ is the family-specific variance function.

$$V(\widehat{\mu}_j) = \begin{cases} \widehat{\mu}_j(1 - \widehat{\mu}_j/m_j) & \text{if binomial or Bernoulli } (m_j = 1) \\ \widehat{\mu}_j^2 & \text{if gamma} \\ 1 & \text{if Gaussian} \\ \widehat{\mu}_j^3 & \text{if inverse Gaussian} \\ \widehat{\mu}_j + k\widehat{\mu}_j^2 & \text{if negative binomial} \\ \widehat{\mu}_j & \text{if Poisson} \end{cases}$$

The response residuals are given by $r_i^R = y_i - \mu_i$. The working residuals are

$$r_i^W = (y_i - \widehat{\mu}_i)\left(\frac{\partial \eta}{\partial \mu}\right)_i$$

and the score residuals are

$$r_i^S = \frac{y_i - \widehat{\mu}_i}{V(\widehat{\mu}_i)}\left(\frac{\partial \eta}{\partial \mu}\right)_i^{-1}$$

Define $\widehat{W} = V(\widehat{\mu})$ and X to be the covariate matrix. h_i, then, is the ith diagonal of the hat matrix given by

$$\widehat{H} = \widehat{W}^{1/2}X(X^T\widehat{W}X)^{-1}X^T\widehat{W}^{1/2}$$

As a result, the likelihood residuals are given by

$$r_i^L = \text{sign}(y_i - \widehat{\mu}_i)\left\{h_i(r_i')^2 + (1 - h_i)(d_i')^2\right\}^{1/2}$$

where r_i' and d_i' are the standardized Pearson and standardized deviance residuals, respectively. By *standardized*, we mean that the residual is divided by $\{1 - h_i\}^{1/2}$.

Cook's distance is an overall measure of the change in the regression coefficients caused by omitting the ith observation from the analysis. Computationally, Cook's distance is obtained as

$$C_i = \frac{(r_i')^2 h_i}{k(1 - h_i)}$$

where k is the number of regressors, including the constant.

Anscombe residuals are given by

$$r_i^A = \frac{A(y_i) - A(\widehat{\mu}_i)}{A'(\widehat{\mu}_i)\{V(\widehat{\mu}_i)\}^{1/2}}$$

where

$$A(\cdot) = \int \frac{d\mu}{V^{1/3}(\mu)}$$

Deviance residuals may be adjusted (`predict, adjusted`) to make the following correction:

$$d_i^a = d_i + \frac{1}{6}\rho_3(\theta)$$

where $\rho_3(\theta)$ is a family-specific correction. See Hardin and Hilbe (2007) for the exact forms of $\rho_3(\theta)$ for each family.

References

Anscombe, F. J. 1953. Contribution of discussion paper by H. Hotelling "New light on the correlation coefficient and its transforms". *Journal of the Royal Statistical Society, Series B* 15: 229–230.

Hardin, J. W., and J. M. Hilbe. 2007. *Generalized Linear Models and Extensions*. 2nd ed. College Station, TX: Stata Press.

McCullagh, P., and J. A. Nelder. 1989. *Generalized Linear Models*. 2nd ed. London: Chapman & Hall/CRC.

Also see

Title

> **glogit** — Logit and probit regression for grouped data

Syntax

Logistic regression for grouped data

> blogit *pos_var pop_var* [*indepvars*] [*if*] [*in*] [, *blogit_options*]

Probit regression for grouped data

> bprobit *pos_var pop_var* [*indepvars*] [*if*] [*in*] [, *bprobit_options*]

Weighted least-squares logistic regression for grouped data

> glogit *pos_var pop_var* [*indepvars*] [*if*] [*in*] [, *glogit_options*]

Weighted least-squares probit regression for grouped data

> gprobit *pos_var pop_var* [*indepvars*] [*if*] [*in*] [, *gprobit_options*]

blogit_options	Description
Model	
noconstant	suppress constant term
asis	retain perfect predictor variables
offset(*varname*)	include *varname* in model with coefficient constrained to 1
constraints(*constraints*)	apply specified linear constraints
collinear	keep collinear variables
SE/Robust	
vce(*vcetype*)	*vcetype* may be oim, robust, cluster *clustvar*, bootstrap, or jackknife
Reporting	
level(#)	set confidence level; default is level(95)
or	report odds ratios
nocnsreport	do not display constraints
display_options	control column formats, row spacing, line width, and display of omitted variables and base and empty cells
Maximization	
maximize_options	control the maximization process; seldom used
nocoef	do not display coefficient table; seldom used
coeflegend	display legend instead of statistics

bprobit_options	Description
Model	
<u>nocon</u>stant	suppress constant term
asis	retain perfect predictor variables
<u>off</u>set(*varname*)	include *varname* in model with coefficient constrained to 1
<u>const</u>raints(*constraints*)	apply specified linear constraints
<u>coll</u>inear	keep collinear variables
SE/Robust	
vce(*vcetype*)	*vcetype* may be oim, <u>r</u>obust, <u>cl</u>uster *clustvar*, <u>boot</u>strap, or <u>jack</u>knife
Reporting	
<u>l</u>evel(*#*)	set confidence level; default is level(95)
<u>nocns</u>report	do not display constraints
display_options	control column formats, row spacing, line width, and display of omitted variables and base and empty cells
Maximization	
maximize_options	control the maximization process; seldom used
<u>nocoef</u>	do not display coefficient table; seldom used
<u>coefl</u>egend	display legend instead of statistics

glogit_options	Description
SE	
vce(*vcetype*)	*vcetype* may be ols, <u>boot</u>strap, or <u>jack</u>knife
Reporting	
<u>l</u>evel(*#*)	set confidence level; default is level(95)
or	report odds ratios
display_options	control column formats, row spacing, line width, and display of omitted variables and base and empty cells
<u>coefl</u>egend	display legend instead of statistics

gprobit_options	Description
SE	
vce(*vcetype*)	*vcetype* may be ols, <u>boot</u>strap, or <u>jack</u>knife
Reporting	
<u>l</u>evel(*#*)	set confidence level; default is level(95)
display_options	control column formats, row spacing, line width, and display of omitted variables and base and empty cells
<u>coefl</u>egend	display legend instead of statistics

indepvars may contain factor variables; see [U] **11.4.3 Factor variables**.

bootstrap, by, jackknife, rolling, and statsby are allowed; see [U] **11.1.10 Prefix commands**.

nocoef and coeflegend do not appear in the dialog box.

See [U] **20 Estimation and postestimation commands** for more capabilities of estimation commands.

Menu

blogit

Statistics > Binary outcomes > Grouped data > Logit regression for grouped data

bprobit

Statistics > Binary outcomes > Grouped data > Probit regression for grouped data

glogit

Statistics > Binary outcomes > Grouped data > Weighted least-squares logit regression

gprobit

Statistics > Binary outcomes > Grouped data > Weighted least-squares probit regression

Description

blogit and bprobit produce maximum-likelihood logit and probit estimates on grouped ("blocked") data; glogit and gprobit produce weighted least-squares estimates. In the syntax diagrams above, *pos_var* and *pop_var* refer to variables containing the total number of positive responses and the total population.

See [R] **logistic** for a list of related estimation commands.

Options for blogit and bprobit

    ⌐ Model ⌐

noconstant; see [R] **estimation options**.

asis forces retention of perfect predictor variables and their associated perfectly predicted observations and may produce instabilities in maximization; see [R] **probit**.

offset(*varname*), constraints(*constraints*), collinear; see [R] **estimation options**.

    ⌐ SE/Robust ⌐

vce(*vcetype*) specifies the type of standard error reported, which includes types that are derived from asymptotic theory, that are robust to some kinds of misspecification, that allow for intragroup correlation, and that use bootstrap or jackknife methods; see [R] *vce_option*.

    ⌐ Reporting ⌐

level(*#*); see [R] **estimation options**.

or (blogit only) reports the estimated coefficients transformed to odds ratios, that is, e^b rather than b. Standard errors and confidence intervals are similarly transformed. This option affects how results are displayed, not how they are estimated. or may be specified at estimation or when replaying previously estimated results.

nocnsreport; see [R] **estimation options**.

display_options: noomitted, vsquish, noemptycells, baselevels, allbaselevels, cformat(%*fmt*), pformat(%*fmt*), sformat(%*fmt*), and nolstretch; see [R] **estimation options**.

⌐ Maximization ⌐

maximize_options: difficult, technique(*algorithm_spec*), iterate(*#*), [no]log, trace, gradient, showstep, hessian, showtolerance, tolerance(*#*), ltolerance(*#*), nrtolerance(*#*), nonrtolerance, and from(*init_specs*); see [R] **maximize**. These options are seldom used.

The following options are available with blogit and bprobit but are not shown in the dialog box:

nocoef specifies that the coefficient table not be displayed. This option is sometimes used by program writers but is useless interactively.

coeflegend; see [R] **estimation options**.

Options for glogit and gprobit

⌐ SE ⌐

vce(*vcetype*) specifies the type of standard error reported, which includes types that are derived from asymptotic theory and that use bootstrap or jackknife methods; see [R] *vce_option*.

vce(ols), the default, uses the standard variance estimator for ordinary least-squares regression.

⌐ Reporting ⌐

level(*#*); see [R] **estimation options**.

or (glogit only) reports the estimated coefficients transformed to odds ratios, that is, e^b rather than b. Standard errors and confidence intervals are similarly transformed. This option affects how results are displayed, not how they are estimated. or may be specified at estimation or when replaying previously estimated results.

display_options: noomitted, vsquish, noemptycells, baselevels, allbaselevels, cformat(%*fmt*), pformat(%*fmt*), sformat(%*fmt*), and nolstretch; see [R] **estimation options**.

The following option is available with glogit and gprobit but is not shown in the dialog box:

coeflegend; see [R] **estimation options**.

Remarks

Remarks are presented under the following headings:

> *Maximum likelihood estimates*
> *Weighted least-squares estimates*

Maximum likelihood estimates

blogit produces the same results as logit and logistic, and bprobit produces the same results as probit, but the "blocked" commands accept data in a slightly different "shape". Consider the following two datasets:

```
. use http://www.stata-press.com/data/r12/xmpl1
. list, sepby(agecat)
```

	agecat	exposed	died	pop
1.	0	0	0	115
2.	0	0	1	5
3.	0	1	0	98
4.	0	1	1	8
5.	1	0	0	69
6.	1	0	1	16
7.	1	1	0	76
8.	1	1	1	22

```
. use http://www.stata-press.com/data/r12/xmpl2
. list
```

	agecat	exposed	deaths	pop
1.	0	0	5	120
2.	0	1	8	106
3.	1	0	16	85
4.	1	1	22	98

These two datasets contain the same information; observations 1 and 2 of xmpl1 correspond to observation 1 of xmpl2, observations 3 and 4 of xmpl1 correspond to observation 2 of xmpl2, and so on.

The first observation of xmpl1 says that for agecat==0 and exposed==0, 115 subjects did not die (died==0). The second observation says that for the same agecat and exposed groups, five subjects did die (died==1). In xmpl2, the first observation says that there were five deaths of a population of 120 in agecat==0 and exposed==0. These are two different ways of saying the same thing. Both datasets are transcriptions from the following table, reprinted in Rothman, Greenland, and Lash (2008, 260), for age-specific deaths from all causes for tolbutamide and placebo treatment groups (University Group Diabetes Program 1970):

	Age through 54		Age 55 and above	
	Tolbutamide	Placebo	Tolbutamide	Placebo
Dead	8	5	22	16
Surviving	98	115	76	79

The data in xmpl1 are said to be "fully relational", which is computer jargon meaning that each observation corresponds to one cell of the table. Stata typically prefers data in this format. The second form of storing these data in xmpl2 is said to be "folded", which is computer jargon for something less than fully relational.

blogit and bprobit deal with "folded" data and produce the same results that logit and probit would have if the data had been stored in the "fully relational" representation.

▷ Example 1

For the tolbutamide data, the fully relational representation is preferred. We could then use logistic, logit, and any of the epidemiological table commands; see [R] **logistic**, [R] **logit**, and [ST] **epitab**. Nevertheless, there are occasions when the folded representation seems more natural. With blogit and bprobit, we avoid the tedium of having to unfold the data:

```
. use http://www.stata-press.com/data/r12/xmpl2
. blogit deaths pop agecat exposed, or
```

Logistic regression for grouped data   Number of obs   =   409
                                                        LR chi2(2)   =   22.47
                                                        Prob > chi2   =   0.0000
Log likelihood = -142.6212   Pseudo R2   =   0.0730

_outcome	Odds Ratio	Std. Err.	z	P>|z|	[95% Conf. Interval]	
agecat	4.216299	1.431519	4.24	0.000	2.167361	8.202223
exposed	1.404674	.4374454	1.09	0.275	.7629451	2.586175
_cons	.0513818	.0170762	-8.93	0.000	.0267868	.0985593

If we had not specified the or option, results would have been presented as coefficients instead of as odds ratios. The estimated odds ratio of death for tolbutamide exposure is 1.40, although the 95% confidence interval includes 1. (By comparison, these data, in fully relational form and analyzed using the cs command [see [ST] **epitab**], produce a Mantel–Haenszel weighted odds ratio of 1.40 with a 95% confidence interval of 0.76 to 2.59.)

We can see the underlying coefficients by replaying the estimation results and not specifying the or option:

```
. blogit
```

Logistic regression for grouped data   Number of obs   =   409
                                                        LR chi2(2)   =   22.47
                                                        Prob > chi2   =   0.0000
Log likelihood = -142.6212   Pseudo R2   =   0.0730

_outcome	Coef.	Std. Err.	z	P>|z|	[95% Conf. Interval]	
agecat	1.438958	.3395203	4.24	0.000	.7735101	2.104405
exposed	.3398053	.3114213	1.09	0.275	-.2705692	.9501798
_cons	-2.968471	.33234	-8.93	0.000	-3.619846	-2.317097

◁

▷ Example 2

bprobit works like blogit, substituting the probit for the logit-likelihood function.

```
. bprobit deaths pop agecat exposed
```

Probit regression for grouped data

				Number of obs	=	409
				LR chi2(2)	=	22.58
				Prob > chi2	=	0.0000
Log likelihood = -142.56478				Pseudo R2	=	0.0734

_outcome	Coef.	Std. Err.	z	P>|z|	[95% Conf. Interval]	
agecat	.7542049	.1709692	4.41	0.000	.4191114	1.089298
exposed	.1906236	.1666059	1.14	0.253	-.1359179	.5171651
_cons	-1.673973	.1619594	-10.34	0.000	-1.991408	-1.356539

◁

Weighted least-squares estimates

▷ Example 3

We have state data for the United States on the number of marriages (marriage), the total population aged 18 years or more (pop18p), and the median age (medage). The dataset excludes Nevada, so it has 49 observations. We now wish to estimate a logit equation for the marriage rate. We will include age squared by specifying the term c.medage#c.medage:

```
. use http://www.stata-press.com/data/r12/census7
(1980 Census data by state)
. glogit marriage pop18p medage c.medage#c.medage
```

Weighted LS logistic regression for grouped data

Source	SS	df	MS		Number of obs	=	49
					F(2, 46)	=	12.89
Model	.71598314	2	.35799157		Prob > F	=	0.0000
Residual	1.27772858	46	.027776708		R-squared	=	0.3591
					Adj R-squared	=	0.3313
Total	1.99371172	48	.041535661		Root MSE	=	.16666

	Coef.	Std. Err.	t	P>|t|	[95% Conf. Interval]	
medage	-.6459349	.2828381	-2.28	0.027	-1.215258	-.0766114
c.medage# c.medage	.0095414	.0046608	2.05	0.046	.0001598	.0189231
_cons	6.503833	4.288977	1.52	0.136	-2.129431	15.1371

◁

▷ Example 4

We could just as easily have fit a grouped-probit model by typing gprobit rather than glogit:

```
. gprobit marriage pop18p medage c.medage#c.medage
```

Weighted LS probit regression for grouped data

Source	SS	df	MS		
Model	.108222962	2	.054111481		
Residual	.192322476	46	.004180923		
Total	.300545438	48	.006261363		

	Number of obs	=	49
	F(2, 46)	=	12.94
	Prob > F	=	0.0000
	R-squared	=	0.3601
	Adj R-squared	=	0.3323
	Root MSE	=	.06466

	Coef.	Std. Err.	t	P>|t|	[95% Conf. Interval]	
medage	-.2755007	.1121042	-2.46	0.018	-.5011548	-.0498466
c.medage# c.medage	.0041082	.0018422	2.23	0.031	.0004001	.0078163
_cons	2.357708	1.704446	1.38	0.173	-1.073164	5.788579

◁

Saved results

blogit and bprobit save the following in e():

Scalars

e(N)	number of observations
e(N_cds)	number of completely determined successes
e(N_cdf)	number of completely determined failures
e(k)	number of parameters
e(k_eq)	number of equations in e(b)
e(k_eq_model)	number of equations in overall model test
e(k_dv)	number of dependent variables
e(df_m)	model degrees of freedom
e(r2_p)	pseudo-R-squared
e(ll)	log likelihood
e(ll_0)	log likelihood, constant-only model
e(N_clust)	number of clusters
e(chi2)	χ^2
e(p)	significance of model test
e(rank)	rank of e(V)
e(ic)	number of iterations
e(rc)	return code
e(converged)	1 if converged, 0 otherwise

Macros

e(cmd)	blogit or bprobit
e(cmdline)	command as typed
e(depvar)	variable containing number of positive responses and variable containing population size
e(wtype)	weight type
e(wexp)	weight expression
e(title)	title in estimation output
e(clustvar)	name of cluster variable
e(offset)	linear offset variable
e(chi2type)	Wald or LR; type of model χ^2 test
e(vce)	*vcetype* specified in vce()
e(vcetype)	title used to label Std. Err.
e(opt)	type of optimization
e(which)	max or min; whether optimizer is to perform maximization or minimization
e(ml_method)	type of ml method
e(user)	name of likelihood-evaluator program
e(technique)	maximization technique
e(properties)	b V
e(predict)	program used to implement predict
e(marginsok)	predictions allowed by margins
e(asbalanced)	factor variables fvset as asbalanced
e(asobserved)	factor variables fvset as asobserved

Matrices

e(b)	coefficient vector
e(Cns)	constraints matrix
e(ilog)	iteration log (up to 20 iterations)
e(gradient)	gradient vector
e(mns)	vector of means of the independent variables
e(rules)	information about perfect predictors
e(V)	variance–covariance matrix of the estimators
e(V_modelbased)	model-based variance

Functions

e(sample)	marks estimation sample

glogit and gprobit save the following in e():

Scalars

e(N)	number of observations
e(mss)	model sum of squares
e(df_m)	model degrees of freedom
e(rss)	residual sum of squares
e(df_r)	residual degrees of freedom
e(r2)	R-squared
e(r2_a)	adjusted R-squared
e(F)	F statistic
e(rmse)	root mean squared error
e(rank)	rank of e(V)

Macros

e(cmd)	glogit or gprobit
e(cmdline)	command as typed
e(depvar)	variable containing number of positive responses and variable containing population size
e(model)	ols
e(title)	title in estimation output
e(vce)	*vcetype* specified in vce()
e(vcetype)	title used to label Std. Err.
e(properties)	b V
e(predict)	program used to implement predict
e(marginsok)	predictions allowed by margins
e(asbalanced)	factor variables fvset as asbalanced
e(asobserved)	factor variables fvset as asobserved

Matrices	
e(b)	coefficient vector
e(V)	variance–covariance matrix of the estimators
Functions	
e(sample)	marks estimation sample

Methods and formulas

blogit, bprobit, glogit, and gprobit are implemented as ado-files.

Methods and formulas are presented under the following headings:

> *Maximum likelihood estimates*
> *Weighted least-squares estimates*

Maximum likelihood estimates

The results reported by blogit and bprobit are obtained by maximizing a weighted logit- or probit-likelihood function. Let $F(\ )$ denote the normal- or logistic-likelihood function. The likelihood of observing each observation in the data is then

$$F(\beta x)^s \left\{1 - F(\beta x)\right\}^{t-s}$$

where s is the number of successes and t is the population. The term above is counted as contributing $s + (t - s) = t$ degrees of freedom. All of this follows directly from the definitions of logit and probit.

blogit and bprobit support the Huber/White/sandwich estimator of the variance and its clustered version using vce(robust) and vce(cluster *clustvar*), respectively. See [P] **_robust**, particularly *Maximum likelihood estimators* and *Methods and formulas*.

Weighted least-squares estimates

The logit function is defined as the log of the odds ratio. If there is one explanatory variable, the model can be written as

$$\log\left(\frac{p_j}{1 - p_j}\right) = \beta_0 + \beta_1 x_j + \epsilon_j \tag{1}$$

where p_j represents successes divided by population for the jth observation. (If there is more than one explanatory variable, we simply interpret β_1 as a row vector and x_j as a column vector.) The large-sample expectation of ϵ_j is zero, and its variance is

$$\sigma_j^2 = \frac{1}{n_j p_j (1 - p_j)}$$

where n_j represents the population for observation j. We can thus apply weighted least-squares to the observations, with weights proportional to $n_j p_j (1 - p_j)$.

As in any feasible generalized least-squares problem, estimation proceeds in two steps. First, we fit (1) by OLS and compute the predicted probabilities as

$$\widehat{p}_j = \frac{\exp(\widehat{\beta_0} + \widehat{\beta_1} x_j)}{1 + \exp(\widehat{\beta_0} + \widehat{\beta_1} x_j)}$$

In the second step, we fit (1) by using analytic weights equal to $n_j \widehat{p}_j (1 - \widehat{p}_j)$.

For `gprobit`, write $\Phi(\cdot)$ for the cumulative normal distribution, and define z_j implicitly by $\Phi(z_j) = p_j$, where p_j is the fraction of successes for observation j. The probit model for one explanatory variable can be written as

$$\Phi^{-1}(p_j) = \beta_0 + \beta_1 x_j + \epsilon_j$$

(If there is more than one explanatory variable, we simply interpret β_1 as a row vector and x_j as a column vector.)

The expectation of ϵ_j is zero, and its variance is given by

$$\sigma_j^2 = \frac{p_j(1-p_j)}{n_j \phi^2 \{\Phi^{-1}(p_j)\}}$$

where $\phi(\cdot)$ represents the normal density (Amemiya 1981, 1498). We can thus apply weighted least squares to the observations with weights proportional to $1/\sigma_j^2$. As for grouped logit, we use a two-step estimator to obtain the weighted least-squares estimates.

References

Amemiya, T. 1981. Qualitative response models: A survey. *Journal of Economic Literature* 19: 1483–1536.

Hosmer, D. W., Jr., and S. Lemeshow. 2000. *Applied Logistic Regression.* 2nd ed. New York: Wiley.

Judge, G. G., W. E. Griffiths, R. C. Hill, H. Lütkepohl, and T.-C. Lee. 1985. *The Theory and Practice of Econometrics.* 2nd ed. New York: Wiley.

Rothman, K. J., S. Greenland, and T. L. Lash. 2008. *Modern Epidemiology.* 3rd ed. Philadelphia: Lippincott Williams & Wilkins.

University Group Diabetes Program. 1970. A study of the effects of hypoglycemic agents on vascular complications in patients with adult-onset diabetes, II: Mortality results. *Diabetes* 19, supplement 2: 789–830.

Also see

[R] **glogit postestimation** — Postestimation tools for glogit, gprobit, blogit, and bprobit

[R] **logistic** — Logistic regression, reporting odds ratios

[R] **logit** — Logistic regression, reporting coefficients

[R] **probit** — Probit regression

[R] **scobit** — Skewed logistic regression

[U] **20 Estimation and postestimation commands**

Title

> **glogit postestimation** — Postestimation tools for glogit, gprobit, blogit, and bprobit

Description

The following postestimation commands are available after `glogit`, `gprobit`, `blogit`, and `bprobit`:

Command	Description
contrast	contrasts and ANOVA-style joint tests of estimates
*estat	AIC, BIC, VCE, and estimation sample summary
estimates	cataloging estimation results
lincom	point estimates, standard errors, testing, and inference for linear combinations of coefficients
*lrtest	likelihood-ratio test
margins	marginal means, predictive margins, marginal effects, and average marginal effects
marginsplot	graph the results from margins (profile plots, interaction plots, etc.)
nlcom	point estimates, standard errors, testing, and inference for nonlinear combinations of coefficients
predict	predictions, residuals, influence statistics, and other diagnostic measures
predictnl	point estimates, standard errors, testing, and inference for generalized predictions
pwcompare	pairwise comparisons of estimates
test	Wald tests of simple and composite linear hypotheses
testnl	Wald tests of nonlinear hypotheses

* estat ic and `lrtest` are not appropriate after `glogit` and `gprobit`.

See the corresponding entries in the *Base Reference Manual* for details.

Syntax for predict

> predict [*type*] *newvar* [*if*] [*in*] [, *statistic*]

statistic	Description
Main	
n	predicted count; the default
pr	probability of a positive outcome
xb	linear prediction
stdp	standard error of the linear prediction

These statistics are available both in and out of sample; type predict ... if e(sample) ... if wanted only for the estimation sample.

Menu

Statistics > Postestimation > Predictions, residuals, etc.

Options for predict

⌐ Main ⌐

n, the default, calculates the expected count, that is, the estimated probability times *pop_var*, which is the total population.

pr calculates the predicted probability of a positive outcome.

xb calculates the linear prediction.

stdp calculates the standard error of the linear prediction.

Methods and formulas

All postestimation commands listed above are implemented as ado-files.

Also see

[R] **glogit** — Logit and probit regression for grouped data

[U] **20 Estimation and postestimation commands**

Title

> **gmm** — Generalized method of moments estimation

Syntax

Interactive version

gmm ($[eqname_1:]<mexp_1>$) ($[eqname_2:]<mexp_2>$)...$[if]$ $[in]$ $[weight]$ $[, options]$

Moment-evaluator program version

gmm *moment_prog* $[if]$ $[in]$ $[weight]$, { <u>equat</u>ions(*namelist*) | <u>nequat</u>ions(*#*) }

{ <u>param</u>eters(*namelist*) | <u>nparam</u>eters(*#*) } $[options]$ $[program_options]$

where

mexp_j is the substitutable expression for the *j*th moment equation and

moment_prog is a moment-evaluator program.

options	Description
Model	
<u>deriv</u>ative($<dexp_{mn}>$)	specify derivative of *mexp_m* with respect to parameter *n*; can be specified more than once (interactive version only)
* <u>twostep</u>	use two-step GMM estimator; the default
* <u>onestep</u>	use one-step GMM estimator
* <u>igmm</u>	use iterative GMM estimator
Instruments	
<u>instruments</u>($[<eqlist>:]varlist[$, <u>noconst</u>ant$]$)	specify instruments; can be specified more than once
<u>xtinst</u>ruments($[<eqlist>:]varlist$, lags($\#_1/\#_2$))	specify panel-style instruments; can be specified more than once
Weight matrix	
<u>wmat</u>rix(*wmtype*$[$, <u>independent</u>$]$)	specify weight matrix; *wmtype* may be <u>robust</u>, <u>cl</u>uster *clustvar*, hac *kernel* $[lags]$, or <u>una</u>djusted
<u>center</u>	center moments in weight-matrix computation
<u>winitial</u>(*iwtype*$[$, <u>independent</u>$]$)	specify initial weight matrix; *iwtype* may be <u>identity</u>, <u>una</u>djusted, xt *xtspec*, or the name of a Stata matrix

Options

variables(*varlist*)	specify variables in model
nocommonesample	do not restrict estimation sample to be the same for all equations

SE/Robust

vce(*vcetype*[, independent])

vcetype may be robust, cluster *clustvar*, bootstrap,
     jackknife, hac *kernel lags*, or unadjusted

Reporting

level(#)	set confidence level; default is level(95)
title(*string*)	display *string* as title above the table of parameter estimates
title2(*string*)	display *string* as subtitle
display_options	control column formats and line width

Optimization

from(*initial_values*)	specify initial values for parameters
‡ igmmiterate(#)	specify maximum number of iterations for iterated GMM estimator
‡ igmmeps(#)	specify # for iterated GMM parameter convergence criterion; default is igmmeps(1e-6)
‡ igmmweps(#)	specify # for iterated GMM weight-matrix convergence criterion; default is igmmweps(1e-6)
optimization_options	control the optimization process; seldom used
coeflegend	display legend instead of statistics

* You can specify at most one of these options.

‡ These options may be specified only when igmm is specified.

program_options	Description
Model	
evaluator_options	additional options to be passed to the moment-evaluator program
hasderivatives	moment-evaluator program can calculate derivatives
* equations(*namelist*)	specify moment-equation names
* nequations(#)	specify number of moment equations
‡ parameters(*namelist*)	specify parameter names
‡ nparameters(#)	specify number of parameters

* You must specify equations(*namelist*) or nequations(#); you may specify both.

‡ You must specify parameters(*namelist*) or nparameters(#); you may specify both.

bootstrap, by, jackknife, rolling, statsby, and xi are allowed; see [U] **11.1.10 Prefix commands**.
Weights are not allowed with the bootstrap prefix; see [R] **bootstrap**.
aweights are not allowed with the jackknife prefix; see [R] **jackknife**.
aweights, fweights, and pweights are allowed; see [U] **11.1.6 weight**.
coeflegend does not appear in the dialog box.
See [U] **20 Estimation and postestimation commands** for more capabilities of estimation commands.

$<mexp_j>$ and $<dexp_{mn}>$ are extensions of valid Stata expressions that also contain parameters to be estimated. The parameters are enclosed in curly braces and must otherwise satisfy the naming requirements for variables; {beta} is an example of a parameter. Also allowed is a notation of the form {$<eqname>$:*varlist*} for linear combinations of multiple covariates and their parameters. For example, {xb: mpg price turn} defines a linear combination of the variables mpg, price, and turn. See *Substitutable expressions* under *Remarks* below.

Menu

Statistics > Endogenous covariates > Generalized method of moments estimation

Description

gmm performs generalized method of moments (GMM) estimation. With the interactive version of the command, you enter the moment equations directly into the dialog box or on the command line using substitutable expressions. The moment-evaluator program version gives you greater flexibility in exchange for increased complexity; with this version, you write a program in an ado-file that calculates the moments based on a vector of parameters passed to it.

gmm can fit both single- and multiple-equation models, and it allows moment conditions of the form $E\{\mathbf{z}_i u_i(\beta)\} = \mathbf{0}$, where $\mathbf{z}_i$ is a vector of instruments and $u_i(\beta)$ is often an additive regression error term, as well as more general moment conditions of the form $E\{\mathbf{h}_i(\mathbf{z}_i; \beta)\} = \mathbf{0}$. gmm works with cross-sectional, time-series, and longitudinal (panel) data.

Options

‾‾‾‾‾‾⌐ Model ⌐‾‾

derivative([*eqname* | #] /*name* = $<dexp_{mn}>$) specifies the derivative of moment equation *eqname* or # with respect to parameter *name*. If *eqname* or # is not specified, gmm assumes that the derivative applies to the first moment equation.

For a moment equation of the form $E\{\mathbf{z}_{mi} u_{mi}(\beta)\} = \mathbf{0}$, derivative($m/\beta_j$ = $<dexp_{mn}>$) is to contain a substitutable expression for $\partial u_{mi}/\partial \beta_j$.

For a moment equation of the form $E\{h_{mi}(\mathbf{z}_i; \beta)\} = \mathbf{0}$, derivative($m/\beta_j$ = $<dexp_{mn}>$) is to contain a substitutable expression for $\partial h_{mi}/\partial \beta_j$.

$<dexp_{mn}>$ uses the same substitutable expression syntax as is used to specify moment equations. If you declare a linear combination in a moment equation, you provide the derivative for the linear combination; gmm then applies the chain rule for you. See *Specifying derivatives* under *Remarks* below for examples.

If you do not specify the derivative() option, gmm calculates derivatives numerically. You must either specify no derivatives or specify all the derivatives that are not identically zero; you cannot specify some analytic derivatives and have gmm compute the rest numerically.

twostep, onestep, and igmm specify which estimator is to be used. You can specify at most one of these options. twostep is the default.

twostep requests the two-step GMM estimator. gmm obtains parameter estimates based on the initial weight matrix, computes a new weight matrix based on those estimates, and then reestimates the parameters based on that weight matrix.

onestep requests the one-step GMM estimator. The parameters are estimated based on an initial weight matrix, and no updating of the weight matrix is performed except when calculating the appropriate variance–covariance (VCE) matrix.

igmm requests the iterative GMM estimator. gmm obtains parameter estimates based on the initial weight matrix, computes a new weight matrix based on those estimates, reestimates the parameters based on that weight matrix, computes a new weight matrix, and so on, to convergence. Convergence is declared when the relative change in the parameter vector is less than igmmeps(), the relative change in the weight matrix is less than igmmweps(), or igmmiterate() iterations have been completed. Hall (2005, sec. 2.4 and 3.6) mentions that there may be gains to finite-sample efficiency from using the iterative estimator.

___

    Instruments

___

instruments([<eqlist>:]varlist[, noconstant]) specifies a list of instrumental variables to be used. If you specify a single moment equation, then you do not need to specify the equations to which the instruments apply; you can omit the *eqlist* and simply specify instruments(*varlist*). By default, a constant term is included in *varlist*; to omit the constant term, use the noconstant suboption: instruments(*varlist*, noconstant).

If you specify a model with multiple moment conditions of the form

$$
E \left\{ \begin{array}{c} \mathbf{z}_{1i} u_{1i}(\beta) \\ \cdots \\ \mathbf{z}_{qi} u_{qi}(\beta) \end{array} \right\} = \mathbf{0}
$$

then you can specify the equations to indicate the moment equations for which the list of variables is to be used as instruments if you do not want that list applied to all the moment equations. For example, you might type

    gmm (main:<*mexp₁*>) (<*mexp₂*>) (<*mexp₃*>), instruments(z1 z2) ///
        instruments(2: z3) instruments(main 3: z4)

Variables z1 and z2 will be used as instruments for all three equations, z3 will be used as an instrument for the second equation, and z4 will be used as an instrument for the first and third equations. Notice that we chose to supply a name for the first moment equation but not the second two.

xtinstruments([<eqlist>:]varlist, lags(#₁/#₂)) is for use with panel-data models in which the set of available instruments depends on the time period. As with instruments(), you can prefix the list of variables with equation names or numbers to target instruments to specific equations. Unlike with instruments(), a constant term is not included in *varlist*. You must xtset your data before using this option; see [XT] **xtset**.

If you specify

    gmm ..., xtinstruments(x, lags(1/.)) ...

then for panel i and period t, gmm uses as instruments $x_{i,t-1}, x_{i,t-2}, \ldots, x_{i1}$. More generally, specifying xtinstruments(x, lags(#₁, #₂)) uses as instruments $x_{i,t-\#_1}, \ldots, x_{i,t-\#_2}$; setting #₂ = . requests all available lags. #₁ and #₂ must be zero or positive integers.

gmm automatically excludes observations for which no valid instruments are available. It does, however, include observations for which only a subset of the lags is available. For example, if you request that lags one through three be used, then gmm will include the observations for the second and third time periods even though fewer than three lags are available as instruments.

wmatrix(*wmtype*[, independent]) specifies the type of weight matrix to be used in conjunction with the two-step and iterated GMM estimators.

Specifying wmatrix(robust) requests a weight matrix that is appropriate when the errors are independent but not necessarily identically distributed. wmatrix(robust) is the default.

Specifying wmatrix(cluster *clustvar*) requests a weight matrix that accounts for arbitrary correlation among observations within clusters identified by *clustvar*.

Specifying wmatrix(hac *kernel* #) requests a heteroskedasticity- and autocorrelation-consistent (HAC) weight matrix using the specified kernel (see below) with # lags. The bandwidth of a kernel is equal to the number of lags plus one.

Specifying wmatrix(hac *kernel* opt) requests an HAC weight matrix using the specified kernel, and the lag order is selected using Newey and West's (1994) optimal lag-selection algorithm.

Specifying wmatrix(hac *kernel*) requests an HAC weight matrix using the specified kernel and $N - 2$ lags, where N is the sample size.

There are three kernels available for HAC weight matrices, and you may request each one by using the name used by statisticians or the name perhaps more familiar to economists:

bartlett or nwest requests the Bartlett (Newey–West) kernel;

parzen or gallant requests the Parzen (Gallant) kernel; and

quadraticspectral or andrews requests the quadratic spectral (Andrews) kernel.

Specifying wmatrix(unadjusted) requests a weight matrix that is suitable when the errors are homoskedastic. In some applications, the GMM estimator so constructed is known as the (nonlinear) two-stage least-squares (2SLS) estimator.

Including the independent suboption creates a weight matrix that assumes moment equations are independent. This suboption is often used to replicate other models that can be motivated outside the GMM framework, such as the estimation of a system of equations by system-wide 2SLS. This suboption has no effect if only one moment equation is specified.

wmatrix() has no effect if onestep is also specified.

center requests that the sample moments be centered (demeaned) when computing GMM weight matrices. By default, centering is not done.

winitial(*wmtype*[, independent]) specifies the weight matrix to use to obtain the first-step parameter estimates.

Specifying winitial(unadjusted) requests a weighting matrix that assumes the moment equations are independent and identically distributed. This matrix is of the form $(\mathbf{Z}'\mathbf{Z})^{-1}$, where $\mathbf{Z}$ represents all the instruments specified in the instruments() option. To avoid a singular weight matrix, you should specify at least $q - 1$ moment equations of the form $E\{\mathbf{z}_{hi}u_{hi}(\beta)\} = \mathbf{0}$, where q is the number of moment equations, or you should specify the independent suboption.

Including the independent suboption creates a weight matrix that assumes moment equations are independent. Elements of the weight matrix corresponding to covariances between two moment equations are set equal to zero. This suboption has no effect if only one moment equation is specified.

winitial(unadjusted) is the default.

winitial(xt *xtspec*) is for use with dynamic panel-data models in which one of the moment equations is specified in first-differences form. *xtspec* is a string consisting of the letters "L" and "D", the length of which is equal to the number of moment equations in the model. You specify

"L" for a moment equation if that moment equation is written in levels, and you specify "D" for a moment equation if it is written in first-differences; *xtspec* is not case sensitive. When you specify this option, you can specify at most one moment equation in levels and one moment equation in first-differences. See the examples listed in *Dynamic panel-data models* under *Remarks* below.

winitial(identity) requests that the identity matrix be used.

winitial(*matname*) requests that Stata matrix *matname* be used. You cannot specify the independent suboption if you specify winitial(*matname*).

> Options |

variables(*varlist*) specifies the variables in the model. gmm ignores observations for which any of these variables has a missing value. If you do not specify variables(), then gmm assumes all the observations are valid and issues an error message with return code 480 if any moment equations evaluate to missing for any observations at the initial value of the parameter vector.

nocommonesample requests that gmm not restrict the estimation sample to be the same for all equations. By default, gmm will restrict the estimation sample to observations that are available for all equations in the model, mirroring the behavior of other multiple-equation estimators such as nlsur, sureg, or reg3. For certain models, however, different equations can have different numbers of observations. For these models, you should specify nocommonesample. See *Dynamic panel-data models* below for one application of this option. You cannot specify weights if you specify nocommonesample.

> SE/Robust |

vce(*vcetype* [, independent]) specifies the type of standard error reported, which includes types that are robust to some kinds of misspecification, that allow for intragroup correlation, and that use bootstrap or jackknife methods; see [R] *vce_option*.

vce(unadjusted) specifies that an unadjusted (nonrobust) VCE matrix be used; this, along with the twostep option, results in the "optimal two-step GMM" estimates often discussed in textbooks.

The default *vcetype* is based on the *wmtype* specified in the wmatrix() option. If wmatrix() is specified but vce() is not, then *vcetype* is set equal to *wmtype*. To override this behavior and obtain an unadjusted (nonrobust) VCE matrix, specify vce(unadjusted).

Specifying vce(bootstrap) or vce(jackknife) results in standard errors based on the bootstrap or jackknife, respectively. See [R] *vce_option*, [R] **bootstrap**, and [R] **jackknife** for more information on these VCEs.

The syntax for *vcetype*s other than bootstrap and jackknife are identical to those for wmatrix().

> Reporting |

level(*#*); see [R] **estimation options**.

title(*string*) specifies an optional title that will be displayed just above the table of parameter estimates.

title2(*string*) specifies an optional subtitle that will be displayed between the title specified in title() and the table of parameter estimates. If title2() is specified but title() is not, title2() has the same effect as title().

display_options: cformat(*%fmt*), pformat(*%fmt*), sformat(*%fmt*), and nolstretch; see [R] **estimation options**.

⌐ Optimization ⌐

from(*initial_values*) specifies the initial values to begin the estimation. You can specify a $1 \times k$ matrix, where k is the number of parameters in the model, or you can specify a parameter name, its initial value, another parameter name, its initial value, and so on. For example, to initialize alpha to 1.23 and delta to 4.57, you would type

  gmm ..., from(alpha 1.23 delta 4.57) ...

Initial values declared using this option override any that are declared within substitutable expressions. If you specify a parameter that does not appear in your model, gmm exits with error code 480. If you specify a matrix, the values must be in the same order in which the parameters are declared in your model. gmm ignores the row and column names of the matrix.

igmmiterate(*#*), igmmeps(*#*), and igmmweps(*#*) control the iterative process for the iterative GMM estimator. These options can be specified only if you also specify igmm.

  igmmiterate(*#*) specifies the maximum number of iterations to perform with the iterative GMM estimator. The default is the number set using set maxiter (set [R] **maximize**), which is 16,000 by default.

  igmmeps(*#*) specifies the convergence criterion used for successive parameter estimates when the iterative GMM estimator is used. The default is igmmeps(1e-6). Convergence is declared when the relative difference between successive parameter estimates is less than igmmeps() and the relative difference between successive estimates of the weight matrix is less than igmmweps().

  igmmweps(*#*) specifies the convergence criterion used for successive estimates of the weight matrix when the iterative GMM estimator is used. The default is igmmweps(1e-6). Convergence is declared when the relative difference between successive parameter estimates is less than igmmeps() and the relative difference between successive estimates of the weight matrix is less than igmmweps().

optimization_options: technique(), conv_maxiter(), conv_ptol(), conv_vtol(), conv_nrtol(), tracelevel(). technique() specifies the optimization technique to use; gn (the default), nr, dfp, and bfgs are allowed. conv_maxiter() specifies the maximum number of iterations; conv_ptol(), conv_vtol(), and conv_nrtol() specify the convergence criteria for the parameters, gradient, and scaled Hessian, respectively. tracelevel() allows you to obtain additional details during the iterative process. See [M-5] **optimize( )**.

The following options pertain only to the moment-evaluator program version of gmm.

⌐ Model ⌐

evaluator_options refer to any options allowed by your *moment_prog*.

hasderivatives indicates that you have written your moment-evaluator program to compute derivatives. If you do not specify this option, derivatives are computed numerically. If your moment-evaluator program does compute derivatives but you wish to use numerical derivatives instead (perhaps during debugging), do not specify this option.

equations(*namelist*) specifies the names of the moment equations in the model. If you specify both equations() and nequations(), the number of names in the former must match the number specified in the latter.

nequations(*#*) specifies the number of moment equations in the model. If you do not specify names with the equations() option, gmm numbers the moment equations 1, 2, 3, ... If you specify both equations() and nequations(), the number of names in the former must match the number specified in the latter.

parameters(*namelist*) specifies the names of the parameters in the model. The names of the parameters must adhere to the naming conventions of Stata's variables; see [U] **11.3 Naming conventions**. If you specify both parameters() and nparameters(), the number of names in the former must match the number specified in the latter.

nparameters(#) specifies the number of parameters in the model. If you do not specify names with the parameters() option, gmm names them b1, b2, ..., b#. If you specify both parameters() and nparameters(), the number of names in the former must match the number specified in the latter.

The following option is available with gmm but is not shown in the dialog box:

coeflegend; see [R] **estimation options**.

Remarks

Remarks are presented under the following headings:

> *Introduction*
> *Substitutable expressions*
> *The weight matrix and two-step estimation*
> *Obtaining standard errors*
> *Exponential (Poisson) regression models*
> *Specifying derivatives*
> *Exponential regression models with panel data*
> *Rational-expectations models*
> *System estimators*
> *Dynamic panel-data models*
> *Details of moment-evaluator programs*

Introduction

The generalized method of moments (GMM) estimator is a workhorse of modern econometrics and is discussed in all the leading textbooks, including Cameron and Trivedi (2005, 2010), Davidson and MacKinnon (1993, 2004), Greene (2012, 468–506), Ruud (2000), Hayashi (2000), Wooldridge (2010), Hamilton (1994), and Baum (2006). An excellent treatise on GMM with a focus on time-series applications is Hall (2005). The collection of papers by Mátyás (1999) provides both theoretical and applied aspects of GMM. Here we give a brief introduction to the methodology and emphasize how the various options of gmm are used.

The starting point for the generalized method of moments (GMM) estimator is the analogy principle, which says we can estimate a parameter by replacing a population moment condition with its sample analogue. For example, the mean of an independent and identically distributed (i.i.d.) population is defined as the value μ such that the first (central) population moment is zero; that is, μ solves $E(y - \mu) = 0$ where y is a random draw from the population. The analogy principle tells us that to obtain an estimate, $\widehat{\mu}$, of μ, we replace the population-expectations operator with its sample analogue (Manski 1988; Wooldridge 2010):

$$E(y - \mu) = 0 \quad \longrightarrow \quad \frac{1}{N} \sum_{i=1}^{N} (y_i - \widehat{\mu}) = 0 \quad \longrightarrow \quad \widehat{\mu} = \frac{1}{N} \sum_{i=1}^{N} y_i$$

where N denotes sample size and y_i represents the ith observation of y in our dataset. The estimator $\widehat{\mu}$ is known as the method of moments (MM) estimator, because we started with a population moment condition and then applied the analogy principle to obtain an estimator that depends on the observed data.

Ordinary least-squares (OLS) regression can also be viewed as an MM estimator. In the model

$$y = \mathbf{x}'\beta + u$$

we assume that u has mean zero conditional on $\mathbf{x}$: $E(u|\mathbf{x}) = 0$. This conditional expectation implies the unconditional expectation $E(\mathbf{x}u) = \mathbf{0}$ because, using the law of iterated expectations,

$$E(\mathbf{x}u) = E_{\mathbf{x}}\left\{E(\mathbf{x}u|\mathbf{x})\right\} = E_{\mathbf{x}}\left\{\mathbf{x}\,E(u|\mathbf{x})\right\} = \mathbf{0}$$

(Using the law of iterated expectations to derive unconditional expectations based on conditional expectations, perhaps motivated by subject theory, is extremely common in GMM estimation.) Continuing,

$$E(\mathbf{x}u) = E\left\{\mathbf{x}(y - \mathbf{x}'\beta)\right\} = \mathbf{0}$$

Applying the analogy principle,

$$E\left\{\mathbf{x}(y - \mathbf{x}'\beta)\right\} \; \longrightarrow \; \frac{1}{N}\sum_{i=1}^{N}\mathbf{x}_i(y_i - \mathbf{x}_i'\beta) = \mathbf{0}$$

so that

$$\widehat{\beta} = \left(\sum_i \mathbf{x}_i\mathbf{x}_i'\right)^{-1}\sum_i \mathbf{x}_i y_i$$

which is just the more familiar formula $\widehat{\beta} = (\mathbf{X}'\mathbf{X})^{-1}\mathbf{X}'\mathbf{y}$ written using summation notation.

In both the previous examples, the number of parameters we were estimating equaled the number of moment conditions. In the first example, we estimated one parameter, μ, and had one moment condition $E(y - \mu) = 0$. In the second example, the parameter vector β had k elements, as did the vector of regressors $\mathbf{x}$, yielding k moment conditions. Ignoring peculiar cases, a model of m equations in m unknowns has a unique solution, and because the moment equations in these examples were linear, we were able to solve for the parameters analytically. Had the moment conditions been nonlinear, we would have had to use numerical techniques to solve for the parameters, but that is not a significant limitation with modern computers.

What if we have more moment conditions than parameters? Say we have l moment conditions and k parameters. A model of $l > k$ equations in k unknowns does not have a unique solution. Any size-k subset of the moment conditions would yield a consistent parameter estimate, though the parameter estimate so obtained would in general be different based on which k moment conditions we used.

For concreteness, let's return to our regression model,

$$y = \mathbf{x}'\beta + u$$

but we no longer wish to assume that $E(\mathbf{x}u) = \mathbf{0}$; we suspect that the error term u affects one or more elements of $\mathbf{x}$. As a result, we can no longer use the OLS estimator. Suppose we have a vector $\mathbf{z}$ with the properties that $E(\mathbf{z}u) = \mathbf{0}$, that the rank of $E(\mathbf{z}'\mathbf{z})$ equals l, and that the rank of $E(\mathbf{z}'\mathbf{x}) = k$. The first assumption simply states that $\mathbf{z}$ is not correlated with the error term. The second assumption rules out perfect collinearity among the elements of $\mathbf{z}$. The third assumption, known as the *rank condition* in econometrics, ensures that $\mathbf{z}$ is sufficiently correlated with $\mathbf{x}$ and that the estimator is feasible. If some elements of $\mathbf{x}$ are not correlated with u, then they should also appear in $\mathbf{z}$.

If $l < k$, then the rank of $E(\mathbf{z}'\mathbf{x}) < k$, violating the rank condition.

If $l = k$, then we can use the simpler MM estimator we already discussed; we would obtain what is sometimes called the simple instrumental-variables estimator $\widehat{\beta} = (\sum_i \mathbf{z}_i \mathbf{x}_i')^{-1} \sum_i \mathbf{z}_i y_i$. The rank condition ensures that $\sum_i \mathbf{z}_i \mathbf{x}_i'$ is invertible, at least in the population.

If $l > k$, the GMM estimator chooses the value, $\widehat{\beta}$, that minimizes a quadratic function of the moment conditions. We could define

$$\widehat{\beta} \equiv \arg\min_\beta \left\{ \frac{1}{N} \sum_i \mathbf{z}_i u_i(\beta) \right\}' \left\{ \frac{1}{N} \sum_i \mathbf{z}_i u_i(\beta) \right\} \tag{1}$$

where for our linear regression example $u_i(\beta) = y_i - \mathbf{x}_i'\beta$. This estimator tries to make the moment conditions as close to zero as possible. This simple estimator, however, applies equal weight to each of the moment conditions; and as we shall see later, we can obtain more efficient estimators by choosing to weight some moment conditions more highly than others.

Consider the quadratic function

$$Q(\beta) = \left\{ \frac{1}{N} \sum_i \mathbf{z}_i u_i(\beta) \right\}' \mathbf{W} \left\{ \frac{1}{N} \sum_i \mathbf{z}_i u_i(\beta) \right\}$$

where $\mathbf{W}$ is a symmetric positive-definite matrix known as a weight matrix. Then we define the GMM estimator as

$$\widehat{\beta} \equiv \arg\min_\beta \ Q(\beta) \tag{2}$$

Continuing with our regression model example, if we choose

$$\mathbf{W} = \left(\frac{1}{N} \sum_i \mathbf{z}_i \mathbf{z}_i' \right)^{-1} \tag{3}$$

then we obtain

$$\widehat{\beta} = \left\{ \left(\frac{1}{N} \sum_i \mathbf{x}_i \mathbf{z}_i' \right) \left(\frac{1}{N} \sum_i \mathbf{z}_i \mathbf{z}_i' \right)^{-1} \left(\frac{1}{N} \sum_i \mathbf{z}_i \mathbf{x}_i' \right) \right\}^{-1} \times$$

$$\left(\frac{1}{N} \sum_i \mathbf{x}_i \mathbf{z}_i' \right) \left(\frac{1}{N} \sum_i \mathbf{z}_i \mathbf{z}_i' \right)^{-1} \left(\frac{1}{N} \sum_i \mathbf{z}_i y_i \right)$$

which is the well-known two-stage least-squares (2SLS) estimator. Our choice of weight matrix here was based on the assumption that u was homoskedastic. A feature of GMM estimation is that by selecting different weight matrices, we can obtain estimators that can tolerate heteroskedasticity, clustering, autocorrelation, and other features of u. See [R] **ivregress** for more information about the 2SLS and linear GMM estimators.

Returning to the case where the model is "just identified", meaning that $l = k$, if we apply the GMM estimator, we will obtain the same estimate, $\widehat{\beta}$, regardless of our choice of $\mathbf{W}$. Because $l = k$, if a unique solution exists, it will set all the sample moment conditions to zero jointly, so $\mathbf{W}$ has no impact on the value of β that minimizes the objective function.

We will highlight other features of the GMM estimator and the gmm command as we proceed through examples. First, though, we discuss how to specify moment equations by using substitutable expressions.

Substitutable expressions

To use the interactive version of gmm, you define the moment equations by using substitutable expressions. In most applications, your moment conditions are of the form $E\{\mathbf{z}_i u_i(\boldsymbol{\beta})\}$, where $u_i(\boldsymbol{\beta})$ is a residual term that depends on the parameter vector $\boldsymbol{\beta}$ as well as variables in your dataset, though we suppress expressing the variables for notational simplicity; we refer to $u_i(\boldsymbol{\beta})$ as the moment equation to differentiate it from the moment conditions $E\{z_i' u_i(\boldsymbol{\beta})\} = \mathbf{0}$.

Substitutable expressions in gmm work much like those used in nl and nlsur, though with one important difference. For the latter two commands, you type the name of the dependent variable, an equal sign, and then the regression function. For example, in nl, if you want to fit the function $y = f(\mathbf{x}; \boldsymbol{\beta}) + u$, you would type

    nl (y = <*expression for* $f(\mathbf{x};\ \beta)$>), ...

On the other hand, gmm requires you to write a substitutable expression for u; in this example, $u = y - f(\mathbf{x}; \boldsymbol{\beta})$, so you would type

    gmm (y - <*expression for* $f(\mathbf{x};\ \beta)$>), ...

The advantage of writing the substitutable expression directly in terms of u is that you are not restricted to fitting models with additive error terms as you are with nl and nlsur.

You specify substitutable expressions just like any other mathematical expression involving scalars and variables, such as those you would use with Stata's generate command, except that the parameters to be estimated are bound in braces. See [U] **13.2 Operators** and [U] **13.3 Functions** for more information on expressions. Parameter names must follow the same conventions as variable names. See [U] **11.3 Naming conventions**.

For example, say that the tth observation on a sample moment is

$$u_t = 1 - \beta\left\{(1 + r_{t+1})(c_{t+1}/c_t)^{-\gamma}\right\}$$

where t denotes time period, β and γ are the parameters to be estimated, and r and c are variables in your dataset. Then you would type

    gmm (1 - {beta}*((1 + F.r)*(F.c/c)^(-1*{gamma}))), ...

Because β and γ are parameters, we enclose them in braces. Also notice our use of the forward operator to refer to the values of r and c one period ahead; time-series operators are allowed in substitutable expressions as long as you have previously tsset (see [TS] **tsset**) your data. See [U] **13.9 Time-series operators** for more information on time-series operators.

To specify initial values for some parameters, you can include an equal sign and the initial value after a parameter:

    gmm (1 - {beta}*((1 + F.r)*(F.c/c)^(-1*{gamma=1}))), ...

would initialize γ to be one. If you do not specify an initial value for a parameter, it is initialized to zero.

Frequently, even nonlinear functions contain linear combinations of variables. As an example, suppose you have this moment equation:

$$u = \left\{y - \exp(\beta_1 x_1 + \beta_2 x_2 + \beta_3 x_3)\right\} / \exp(\beta_1 x_1 + \beta_2 x_2 + \beta_3 x_3)$$

Instead of typing

```
gmm ((y - exp({beta1}*x1 + {beta2}*x2 + {beta3}*x3)) /      ///
        exp({beta1}*x1 + {beta2}*x2 + {beta3}*x3)) ...
```

you can type

```
gmm ((y - exp({xb:x1 x2 x3})) / exp({xb:})) .....
```

The notation {xb:x1 x2 x3} tells gmm that you want a linear combination of the variables x1, x2, and x3. We named this linear combination xb, so gmm will name the three parameters corresponding to the three variables xb_x1, xb_x2, and xb_x3. You can name the linear combination anything you wish (subject to Stata's naming conventions for variable names); gmm then names the parameter corresponding to variable x lc_x, where lc is the name of your linear combination. You cannot use the same name for both an individual parameter and a linear combination. You can, however, refer to one parameter in a linear combination after it has been declared as you would any other parameter by using the notation {lc_x}. Linear combinations do not include a constant term.

Once we have declared the variables in the linear combination xb, we can subsequently refer to the linear combination in our substitutable expression by using the notation xb:. The colon is not optional; it tells gmm that you are referring to a previously declared linear combination, not an individual parameter. This shorthand notation is also handy when specifying derivatives, as we will show later.

In general, there are three rules to follow when defining substitutable expressions:

1. Parameters of the model are bound in braces: {b0}, {param}, etc.

2. Initial values for parameters are given by including an equal sign and the initial value inside the braces: {b0=1}, {param=3.571}, etc.

3. Linear combinations of variables can be included using the notation {*eqname*:*varlist*}: {xb: mpg price weight}, {score: w x z}, etc. Parameters of linear combinations are initialized to zero.

If you specify initial values by using the from() option, they override whatever initial values are given within the substitutable expression. Substitutable expressions are so named because, once values are assigned to the parameters, the resulting expressions can be handled by generate and replace.

Example 1: OLS regression

In *Introduction*, we stated that OLS is an MM estimator. Say that we want to fit the model

$$mpg = \beta_0 + \beta_1 weight + \beta_2 length + u$$

where u is an i.i.d. error term. We type

```
. use http://www.stata-press.com/data/r12/auto
(1978 Automobile Data)
. gmm (mpg - {b1}*weight - {b2}*length - {b0}), instruments(weight length)
Step 1
Iteration 0:   GMM criterion Q(b) =    475.4138
Iteration 1:   GMM criterion Q(b) =   2.696e-20
Iteration 2:   GMM criterion Q(b) =   3.329e-27
Step 2
Iteration 0:   GMM criterion Q(b) =   5.109e-28
Iteration 1:   GMM criterion Q(b) =   7.237e-32
```

```
GMM estimation

Number of parameters =    3
Number of moments    =    3
Initial weight matrix: Unadjusted                    Number of obs  =      74
GMM weight matrix:     Robust
```

	Coef.	Robust Std. Err.	z	P>|z|	[95% Conf.	Interval]
/b1	-.0038515	.0019472	-1.98	0.048	-.0076678	-.0000351
/b2	-.0795935	.0677528	-1.17	0.240	-.2123866	.0531996
/b0	47.88487	7.50599	6.38	0.000	33.1734	62.59634

Instruments for equation 1: weight length _cons

Recall that the moment condition for OLS regression is $E(\mathbf{x}u) = \mathbf{0}$, where $\mathbf{x}$, the list of instruments, is the same as the list of regressors in the model. In our command, we defined the residual term, u, inside parentheses by using a substitutable expression; because linear combinations declared in substitutable expressions do not include a constant term, we included our own (b0). Inside the `instruments()` option, we listed our instruments; by default, gmm includes a constant term among the instrument list.

Because the number of moments equals the number of parameters we are estimating, the model is said to be "just identified" or "exactly identified." Therefore, the choice of weight matrix has no impact on the solution to (2), and the criterion function $Q(\boldsymbol{\beta})$ achieves its minimum value at zero.

The OLS estimator is a one-step GMM estimator, but we did not bother to specify the `onestep` option because the model is just identified. Doing a second step of GMM estimation affects neither the point estimates nor the standard errors, so to keep the syntax as simple as possible, we did not include the `onestep` option. The first step of estimation resulted in $Q(\boldsymbol{\beta}) = 0$ as expected, and the second step of estimation did not change the minimized value of $Q(\boldsymbol{\beta})$. (3×10^{-27} and 7×10^{-32} are both zero for all practical purposes.)

When you do not specify either the `wmatrix()` or the `vce()` option, gmm reports heteroskedasticity-robust standard errors. The parameter estimates reported here match those that we would obtain from the command

```
. regress mpg weight length, vce(robust)
```

The standard errors reported by that `regress` command would be larger than those reported by gmm by a factor of sqrt($74/71$) because `regress` makes a small-sample adjustment to the estimated variance matrix while gmm does not. Likewise, had we specified the `vce(unadjusted)` option with our gmm command, then our standard errors would differ by a factor of sqrt($74/71$) from those reported by `regress` without the `vce(robust)` option.

Using the notation for linear combinations of parameters, we could have typed

```
. gmm (mpg - {xb: weight length} - {b0}), instruments(weight length)
```

and obtained identical results. Instead of having parameters b1 and b2, with this syntax we would have parameters xb_weight and xb_length.

◁

▷ Example 2: Instrumental-variables regression

In *Introduction*, we mentioned that 2SLS can be viewed as a GMM estimator. In example 1 of [R] **ivregress**, we fit by 2SLS a model of rental rates (`rent`) as a function of the value of owner-occupied housing (`hsngval`) and the percentage of the population living in urban areas (`pcturban`):

$$\texttt{rent} = \beta_0 + \beta_1 \texttt{hsngval} + \beta_2 \texttt{pcturban} + u$$

by 2SLS. We argued that random shocks that affect rental rates likely also affect housing values, so we treated `hsngval` as an endogenous variable. As additional instruments, we used family income, `faminc`, and three regional dummies (`reg2`–`reg4`).

To replicate the results of `ivregress 2sls` by using `gmm`, we type

```
. use http://www.stata-press.com/data/r12/hsng2
(1980 Census housing data)

. gmm (rent - {xb:hsngval pcturban} - {b0}),
> instruments(pcturban faminc reg2-reg4) vce(unadjusted) onestep

Step 1
Iteration 0:   GMM criterion Q(b) =   56115.03
Iteration 1:   GMM criterion Q(b) =   110.91583
Iteration 2:   GMM criterion Q(b) =   110.91583

GMM estimation

Number of parameters =    3
Number of moments    =    6
Initial weight matrix: Unadjusted                 Number of obs   =       50
```

	Coef.	Std. Err.	z	P>|z|	[95% Conf.	Interval]
/xb_hsngval	.0022398	.0003284	6.82	0.000	.0015961	.0028836
/xb_pcturban	.081516	.2987652	0.27	0.785	-.5040531	.667085
/b0	120.7065	15.22839	7.93	0.000	90.85942	150.5536

Instruments for equation 1: pcturban faminc reg2 reg3 reg4 _cons

We specified `vce(unadjusted)` so that we would obtain an unadjusted VCE matrix and our standard errors would match those reported in [R] **ivregress**.

Pay attention to how we specified the `instruments()` option. In *Introduction*, we mentioned that the moment conditions for the 2SLS estimator are $E(\mathbf{z}u) = \mathbf{0}$, and we mentioned that if some elements of $\mathbf{x}$ (the regressors) are not endogenous, then they should also appear in $\mathbf{z}$. In this model, we assume the regressor `pcturban` is exogenous, so we included it in the list of instrumental variables. Commands like `ivregress`, `ivprobit`, and `ivtobit` accept standard *varlists*, so they can deduce the exogenous regressors in the model. Because `gmm` accepts arbitrary functions in the form of substitutable expressions, it has no way of discerning the exogenous variables of the model on its own.

Also notice that we specified the `onestep` option. The 2SLS estimator is a one-step GMM estimator that is based on a weight matrix that assumes the error terms are i.i.d. Unlike the previous example, here we had more instruments than parameters, so the minimized value of $Q(\beta)$ is nonzero. We discuss the weight matrix and its relationship to two-step estimation next.

◁

he weight matrix and two-step estimation

Recall our definition of the GMM estimator given in (2). The estimator, $\widehat{\beta}$, depends on the choice of the weight matrix, $\mathbf{W}$. Under relatively mild assumptions, our estimator, $\widehat{\beta}$, is consistent regardless of the choice of $\mathbf{W}$, so how are we to decide what $\mathbf{W}$ to use? The most common solution is to use the two-step estimator, which we now describe.

A key result in Hansen's (1982) seminal paper is that if we denote by $\mathbf{S}$ the covariance matrix of the moment conditions, then the optimal (in a way we make precise later) GMM estimator is the one that uses a weight matrix equal to the inverse of the moment covariance matrix. That is, if we let $\mathbf{S} = \mathrm{Cov}(\mathbf{z}u)$, then we want to use $\mathbf{W} = \mathbf{S}^{-1}$. But how do we obtain $\mathbf{S}$ in the first place?

If we assume that the errors are i.i.d., then

$$\mathrm{Cov}(\mathbf{z}u) = E(u^2 \mathbf{z}\mathbf{z}') = \sigma^2 E(\mathbf{z}\mathbf{z}')$$

where σ^2 is the variance of u. Because σ^2 is a positive scalar, we can ignore it when solving (2). Thus we compute

$$\widehat{\mathbf{W}}_1 = \left(\frac{1}{N} \sum_i \mathbf{z}_i \mathbf{z}_i' \right)^{-1} \tag{4}$$

which does not depend on any unknown model parameters. (Notice that $\widehat{\mathbf{W}}_1$ is the same weight matrix used in 2SLS.) Given $\widehat{\mathbf{W}}_1$, we can solve (2) to obtain an initial estimate, say, $\widehat{\beta}_1$.

Our estimate, $\widehat{\beta}_1$, is consistent, so by Slutsky's theorem, the sample residuals $\widehat{u}$ computed at this value of β will also be consistent. Using virtually the same arguments used to justify the Huber/Eicker/White heteroskedasticity-robust VCE, if we assume that the residuals are independent though not identically distributed, we can estimate $\mathbf{S}$ as

$$\widehat{\mathbf{S}} = \frac{1}{N} \sum_i \widehat{u}_i^2 \mathbf{z}_i \mathbf{z}_i'$$

Then, in the second step, we re-solve (2), using $\widehat{\mathbf{W}}_2 = \widehat{\mathbf{S}}^{-1}$, yielding the two-step GMM estimate $\widehat{\beta}_2$. If the residuals exhibit clustering, you can specify `wmatrix(cluster varname)` so that `gmm` computes a weight matrix that does not assume the u_i's are independent within clusters identified by *varname*. You can specify `wmatrix(hac ...)` to obtain weight matrices that are suitable for when the u_i's exhibit autocorrelation as well as heteroskedasticity.

We could take the point estimates from the second round of estimation and use them to compute yet another weight matrix, $\widehat{\mathbf{W}}_3$, say, re-solve (2) yet again, and so on, stopping when the parameters or weight matrix do not change much from one iteration to the next. This procedure is known as the iterative GMM estimator and is obtained with the `igmm` option. Asymptotically, the two-step and iterative GMM estimators have the same distribution. However, Hall (2005, 90) suggests that the iterative estimator may have better finite-sample properties.

Instead of computing $\widehat{\mathbf{W}}_1$ as in (4), we could simply choose $\widehat{\mathbf{W}}_1 = \mathbf{I}$, the identity matrix. The initial estimate, $\widehat{\beta}_1$, would still be consistent. You can request this behavior by specifying `winitial(identity)` option. However, if you specify all your moment equations of the form $E(\mathbf{z}u) = \mathbf{0}$, we recommend using the default `winitial(unadjusted)` instead; the rescaling of the moment conditions implied by using a homoskedastic initial weight matrix makes the numerical routines used to solve (2) more stable.

If you fit a model with more than one of the moment equations of the form $E\{h(\mathbf{z};\beta)\} = \mathbf{0}$, then you must use `winitial(identity)` or `winitial(unadjusted, independent)`. With moment equations of that form, you do not specify a list of instruments, and `gmm` cannot evaluate (4)—the matrix expression in parentheses would necessarily be singular, so it cannot be inverted.

▷ Example 3: Two-step linear GMM estimator

From the previous discussion and the comments in *Introduction*, we see that the linear 2SLS estimator is a one-step GMM estimator where we use the weight matrix defined in (4) that assumes the errors are i.i.d. If we use the 2SLS estimate of β to obtain the sample residuals, compute a new weight matrix based on those residuals, and then do a second step of GMM estimation, we obtain the linear two-step GMM estimator as implemented by ivregress gmm.

In example 3 of [R] **ivregress**, we fit the model of rental rates as discussed in example 2 above. We now allow the residuals to be heteroskedastic, though we will maintain our assumption that they are independent. We type

```
. gmm (rent - {xb:hsngval pcturban} - {b0}), inst(pcturban faminc reg2-reg4)
Step 1
Iteration 0:    GMM criterion Q(b) =    56115.03
Iteration 1:    GMM criterion Q(b) =   110.91583
Iteration 2:    GMM criterion Q(b) =   110.91583

Step 2
Iteration 0:    GMM criterion Q(b) =    .2406087
Iteration 1:    GMM criterion Q(b) =   .13672801
Iteration 2:    GMM criterion Q(b) =   .13672801

GMM estimation

Number of parameters =    3
Number of moments    =    6
Initial weight matrix: Unadjusted                   Number of obs   =       50
GMM weight matrix:     Robust
```

	Coef.	Robust Std. Err.	z	P>|z|	[95% Conf. Interval]	
/xb_hsngval	.0014643	.0004473	3.27	0.001	.0005877	.002341
/xb_pcturban	.7615482	.2895105	2.63	0.009	.1941181	1.328978
/b0	112.1227	10.80234	10.38	0.000	90.95052	133.2949

Instruments for equation 1: pcturban faminc reg2 reg3 reg4 _cons

By default, gmm computes a heteroskedasticity-robust weight matrix before the second step of estimation, though we could have specified wmatrix(robust) if we wanted to be explicit. Because we did not specify the vce() option, gmm used a heteroskedasticity-robust one. Our results match those in example 3 of [R] **ivregress**. Moreover, the only difference between this example and the previous example of 2SLS is that here we did not use the onestep option.

◁

Obtaining standard errors

This section is a bit more theoretical and can be skipped on first reading. However, the information is sufficiently important that you should return to this section at some point.

So far in our discussion, we have focused on point estimation without much mention of how we obtain the standard errors of the estimates. We also mentioned that if we choose $\mathbf{W}$ to be the inverse of the covariance matrix of the moment conditions, then we obtain the "optimal" GMM estimator. We elaborate those points now.

Using mostly standard statistical arguments, we can show that for the GMM estimator defined in (2), the variance of $\widehat{\beta}$ is given by

$$\text{Var}(\widehat{\beta}) = \frac{1}{N} \left\{ \overline{\mathbf{G}}(\widehat{\beta})' \mathbf{W} \overline{\mathbf{G}}(\widehat{\beta}) \right\}^{-1} \overline{\mathbf{G}}(\widehat{\beta})' \mathbf{W} \mathbf{S} \mathbf{W} \overline{\mathbf{G}}(\widehat{\beta}) \left\{ \overline{\mathbf{G}}(\widehat{\beta})' \mathbf{W} \overline{\mathbf{G}}(\widehat{\beta}) \right\}^{-1} \tag{5}$$

where

$$\overline{\mathbf{G}}(\widehat{\beta}) = \frac{1}{N} \sum_i \mathbf{z}_i \left. \frac{\partial u_i}{\partial \beta} \right|_{\beta = \widehat{\beta}} \quad \text{or} \quad \overline{\mathbf{G}}(\widehat{\beta}) = \frac{1}{N} \sum_i \left. \frac{\partial \mathbf{h}_i}{\partial \beta} \right|_{\beta = \widehat{\beta}}$$

as the case may be and $\mathbf{S} = E(\mathbf{z} u u' \mathbf{z}')$.

Assuming the vce(unadjusted) option is not specified, gmm reports standard errors based on the robust variance matrix defined in (5). For the two-step estimator, $\mathbf{W}$ is the weight matrix requested using the wmatrix() option, and it is calculated based on the residuals obtained after the first estimation step. The second-step point estimates and residuals are obtained, and $\mathbf{S}$ is calculated based on the specification of the vce() option. For the iterated estimator, $\mathbf{W}$ is calculated based on the second-to-last round of estimation, while $\mathbf{S}$ is based on the residuals obtained after the last round of estimation. Computation of the covariance matrix for the one-step estimator is, perhaps surprisingly, more involved; we discuss the covariance matrix with the one-step estimator in the technical note at the end of this section.

If we choose the weight matrix to be the inverse of the covariance matrix of the moment conditions so that $\mathbf{W} = \mathbf{S}^{-1}$, then (5) simplifies substantially:

$$\text{Var}(\widehat{\beta}) = \frac{1}{N} \left\{ \overline{\mathbf{G}}(\widehat{\beta})' \mathbf{W} \overline{\mathbf{G}}(\widehat{\beta}) \right\}^{-1} \tag{6}$$

The GMM estimator constructed using this choice of weight matrix along with the covariance matrix in (6) is known as the "optimal" GMM estimator. One can show that if in fact $\mathbf{W} = \mathbf{S}^{-1}$, then the variance in (6) is smaller than the variance in (5) of any other GMM estimator based on the same moment conditions but with a different choice of weight matrix. Thus the optimal GMM estimator is also known as the efficient GMM estimator, because it has the smallest variance of any estimator based on the given moment conditions.

To obtain standard errors from gmm based on the optimal GMM estimator, you specify the vce(unadjusted) option. We call that VCE unadjusted because we do not recompute the residuals after estimation to obtain the matrix $\mathbf{S}$ required in (5) or allow for the fact that those residuals may not be i.i.d. Some statistical packages by default report standard errors based on (6) and offer standard errors based on (5) only as an option or not at all. While the optimal GMM estimator is theoretically appealing, Cameron and Trivedi (2005, 177) suggest that in finite samples it need not perform better than the GMM estimator that uses (5) to obtain standard errors.

❑ Technical note

Computing the covariance matrix of the parameters after using the one-step estimator is actually a bit more complex than after using the two-step or iterative estimators. We can illustrate most of the intricacies by using linear regression with moment conditions of the form $E\{\mathbf{x}(y - \mathbf{x}'\beta)\} = \mathbf{0}$.

If you specify winitial(unadjusted) and vce(unadjusted), then the initial weight matrix will be computed as

$$\widehat{\mathbf{W}}_1 = \left(\frac{1}{N} \sum_i \mathbf{x}_i \mathbf{x}_i' \right)^{-1} \tag{7}$$

Moreover, for linear regression, we can show that

$$\overline{\mathbf{G}}(\widehat{\beta}) = \frac{1}{N} \sum_i \mathbf{x}_i \mathbf{x}_i'$$

so that (6) becomes

$$\text{Var}(\widehat{\beta}) = \frac{1}{N}\left\{\left(\frac{1}{N}\sum_i \mathbf{x}_i\mathbf{x}_i'\right)\left(\frac{1}{N}\sum_i \mathbf{x}_i\mathbf{x}_i'\right)^{-1}\left(\frac{1}{N}\sum_i \mathbf{x}_i\mathbf{x}_i'\right)\right\}^{-1}$$

$$= \left(\sum_i \mathbf{x}_i\mathbf{x}_i'\right)^{-1}$$

$$= (\mathbf{X}'\mathbf{X})^{-1} \tag{8}$$

However, we know that the nonrobust covariance matrix for the OLS estimator is actually $\widehat{\sigma}^2(\mathbf{X}'\mathbf{X})^{-1}$. What is missing from (8) is the scalar $\widehat{\sigma}^2$, the estimated variance of the residuals. When you use the one-step estimator and specify `winitial(unadjusted)`, the weight matrix (7) does not include the $\widehat{\sigma}^2$ term because gmm does not have a consistent estimate of β from which it can then estimate σ^2. The point estimates are still correct, because multiplying the weight matrix by a scalar factor does not affect the solution to the minimization problem.

To circumvent this issue, if you specify `winitial(unadjusted)` and `vce(unadjusted)`, gmm uses the estimated $\widehat{\beta}$ (which is consistent) to obtain a new unadjusted weight matrix that does include the term $\widehat{\sigma}^2$ so that evaluating (6) will yield correct standard errors.

If you use the two-step or iterated GMM estimators, this extra effort is not needed to obtain standard errors because the first-step (and subsequent steps') estimate of β is consistent and can be used to estimate σ^2 or some other weight matrix based on the `wmatrix()` option. Straightforward algebra shows that this extra effort is also not needed if you request any type of adjusted (robust) covariance matrix with the one-step estimator.

A similar issue arises when you specify `winitial(identity)` and `vce(unadjusted)` with the one-step estimator. Again the solution is to compute an unadjusted weight matrix after obtaining $\widehat{\beta}$ so that (6) provides the correct standard errors.

We have illustrated the problem and solution using a single-equation linear model. However, the problem arises whenever you use the one-step estimator with an unadjusted VCE, regardless of the number of equations; and gmm handles all the details automatically. Computation of Hansen's J statistic presents an identical issue, and gmm takes care of that as well.

If you supply your own initial weight matrix by using `winitial(matname)`, then the standard errors (as well as the J statistic reported by `estat overid`) are based on that weight matrix. You should verify that the weight matrix you provide will yield appropriate statistics.

❑

Exponential (Poisson) regression models

Exponential regression models are frequently encountered in applied work. For example, they can be used as alternatives to linear regression models on log-transformed dependent variables, obviating the need for post-hoc transformations to obtain predicted values in the original metric of the dependent variable. When the dependent variable represents a discrete count variable, they are also known as Poisson regression models; see Cameron and Trivedi (1998).

For now, we consider models of the form

$$y = \exp(\mathbf{x}'\beta) + u \tag{9}$$

where u is a zero-mean additive error term so that $E(y) = \exp(\mathbf{x}'\beta)$. Because the error term is additive, if $\mathbf{x}$ represents strictly exogenous regressors, then we have the population moment condition

$$E[\mathbf{x}\{y - \exp(\mathbf{x}'\beta)\}] = \mathbf{0} \tag{10}$$

Moreover, because the number of parameters in the model is equal to the number of instruments, there is no point to using the two-step GMM estimator.

▷ Example 4: Exponential regression

Cameron and Trivedi (2010, 323) fit a model of the number of doctor visits based on whether the patient has private insurance, whether the patient has a chronic disease, gender, and income. Here we fit that model by using gmm. To allow for potential excess dispersion, we will obtain a robust VCE matrix, which is the default for gmm anyway. We type

```
. use http://www.stata-press.com/data/r12/docvisits

. gmm (docvis - exp({xb:private chronic female income}+{b0})),
> instruments(private chronic female income) onestep
Step 1
Iteration 0:   GMM criterion Q(b) =  16.853973
Iteration 1:   GMM criterion Q(b) =  2.2706472
Iteration 2:   GMM criterion Q(b) =  .19088097
Iteration 3:   GMM criterion Q(b) =  .00041101
Iteration 4:   GMM criterion Q(b) =  3.939e-09
Iteration 5:   GMM criterion Q(b) =  6.572e-19

GMM estimation

Number of parameters =   5
Number of moments    =   5
Initial weight matrix: Unadjusted              Number of obs   =     4412
```

	Coef.	Robust Std. Err.	z	P>|z|	[95% Conf. Interval]	
/xb_private	.7986654	.1089891	7.33	0.000	.5850507	1.01228
/xb_chronic	1.091865	.0559888	19.50	0.000	.9821291	1.201601
/xb_female	.4925481	.0585298	8.42	0.000	.3778317	.6072644
/xb_income	.003557	.0010824	3.29	0.001	.0014356	.0056784
/b0	-.2297263	.1108607	-2.07	0.038	-.4470093	-.0124434

```
Instruments for equation 1: private chronic female income _cons
```

Our point estimates agree with those reported by Cameron and Trivedi to at least six significant digits; the small discrepancies are attributable to different optimization techniques and convergence criteria being used by gmm and poisson. The standard errors differ by a factor of sqrt(4412/4411) because gmm uses N in the denominator of the formula for the robust covariance matrix, while the robust covariance matrix estimator used by poisson uses $N - 1$.

◁

❑ Technical note

That the GMM and maximum likelihood estimators of the exponential regression model coincide is not a general property of these two classes of estimators. The maximum likelihood estimator solves the score equations

$$\frac{1}{N} \sum_{i=1}^{N} \frac{\partial \ln \ell_i}{\partial \beta} = \mathbf{0}$$

where l_i is the likelihood for the ith observation. These score equations can be viewed as the sample analogues of the population moment conditions

$$E\left\{\frac{\partial \ln \ell_i}{\partial \beta}\right\} = \mathbf{0}$$

establishing that maximum likelihood estimators represent a subset of the class of GMM estimators.

For the Poisson model,

$$\ln \ell_i = -\exp(\mathbf{x}_i'\beta) + y_i \mathbf{x}_i'\beta - \ln y_i!$$

so the score equations are

$$\frac{1}{N}\sum_{i=1}^{N} \mathbf{x}_i \left\{y_i - \exp(\mathbf{x}_i'\beta)\right\} = \mathbf{0}$$

which are just the sample moment conditions implied by (10) that we used in the previous example. That is why our results using gmm match Cameron and Trivedi's results using poisson.

On the other hand, an intuitive set of moment conditions to consider for GMM estimation of a probit model is

$$E[\mathbf{x}\{y - \Phi(\mathbf{x}'\beta)\}] = \mathbf{0}$$

where $\Phi()$ is the standard normal cumulative distribution function. Differentiating the likelihood function for the maximum likelihood probit estimator, we can show that the corresponding score equations are

$$\frac{1}{N}\sum_{i=1}^{N} \left[\mathbf{x}_i \left\{y_i \frac{\phi(\mathbf{x}_i'\beta)}{\Phi(\mathbf{x}_i'\beta)} - (1-y_i)\frac{\phi(\mathbf{x}_i'\beta)}{1-\Phi(\mathbf{x}_i'\beta)}\right\}\right] = \mathbf{0}$$

where $\phi()$ is the standard normal density function. These two moment conditions are not equivalent, so the maximum likelihood and GMM probit estimators are distinct.

❑

Example 5: Comparison of GMM and maximum likelihood

Using the automobile dataset, here we fit a probit model of foreign on gear_ratio, length, and headroom using first the score equations and then the intuitive set of GMM equations. We type

```
. use http://www.stata-press.com/data/r12/auto
(1978 Automobile Data)
. global xb "{b1}*gear_ratio + {b2}*length + {b3}*headroom + {b0}"
. global phi "normalden($xb)"
. global Phi "normal($xb)"
. gmm (foreign*$phi/$Phi - (1-foreign)*$phi/(1-$Phi)),
> instruments(gear_ratio length headroom) onestep
  (output omitted )
. estimates store ml
. gmm (foreign - $Phi), instruments(gear_ratio length headroom) onestep
  (output omitted )
. estimates store gmm
```

```
. estimates table ml gmm, b se
```

Variable	ml	gmm
b1		
_cons	2.9586277	2.8489213
	.64042341	.63570246
b2		
_cons	-.02148933	-.02056033
	.01382043	.01396954
b3		
_cons	.01136927	.02240761
	.27278528	.2849891
b0		
_cons	-6.0222289	-5.8595615
	3.5594588	3.5188028

```
legend: b/se
```

The coefficients on gear_ratio and length are close for the two estimators. The GMM estimate of the coefficient on headroom is twice that of the maximum likelihood estimate, though the relatively large standard errors imply that this difference is not significant. You can verify that the coefficients in the column marked "ml" match those you would obtain using probit. We have not discussed the differences among standard errors based on the various GMM and maximum-likelihood covariance matrix estimators to avoid tedious algebra, though you can verify that the robust covariance matrix after one-step GMM estimation differs by only a finite-sample adjustment factor of $(N/N - 1)$ from the robust covariance matrix reported by probit. Both the maximum likelihood and GMM probit estimators require the normality assumption, and the maximum likelihood estimator is efficient if that normality assumption is correct; therefore, in this particular example, there is no reason to prefer the GMM estimator.

◁

We can modify (10) easily to allow for endogenous regressors. Suppose that x_j is endogenous in the sense that $E(u|x_j) \neq 0$. Then (10) is no longer a valid moment condition. However, suppose we have some variables other than $\mathbf{x}$ such that $E(u|\mathbf{z}) = 0$. We can instead use the moment conditions

$$E(\mathbf{z}u) = E[\mathbf{z}\{y - \exp(\mathbf{x}'\beta)\}] = \mathbf{0} \tag{11}$$

As usual, if some elements of $\mathbf{x}$ are exogenous, then they should appear in $\mathbf{z}$ as well.

▷ Example 6: Exponential regression with endogenous regressors

Returning to the model discussed in example 4, here we treat income as endogenous; unobservable factors that determine a person's income may also affect the number of times a person visits a doctor. We use a person's age and race as instruments. These are valid instruments if we believe that age and race influence a person's income but do not have a direct impact on the number of doctor visits. (Whether this belief is justified is another matter; we test that belief in [R] **gmm postestimation**.) Because we have more instruments (seven) than parameters (five), we have an overidentified model. Therefore, the choice of weight matrix does matter. We will utilize the default two-step GMM estimator. In the first step, we will use a weight matrix that assumes the errors are i.i.d. In the second step, we will use a weight matrix that assumes heteroskedasticity. When you specify twostep, these are the defaults for the first- and second-step weight matrices, so we do not have to use the winitial() or wmatrix() options. We will again obtain a robust VCE, which is also the default. We type

```
. use http://www.stata-press.com/data/r12/docvisits

. gmm (docvis - exp({xb:private chronic female income}+{b0})),
> instruments(private chronic female age black hispanic)
Step 1
Iteration 0:   GMM criterion Q(b) =   16.910173
Iteration 1:   GMM criterion Q(b) =   .82276104
Iteration 2:   GMM criterion Q(b) =   .21832032
Iteration 3:   GMM criterion Q(b) =   .12685935
Iteration 4:   GMM criterion Q(b) =   .12672369
Iteration 5:   GMM criterion Q(b) =   .12672365

Step 2
Iteration 0:   GMM criterion Q(b) =   .00234641
Iteration 1:   GMM criterion Q(b) =   .00215957
Iteration 2:   GMM criterion Q(b) =   .00215911
Iteration 3:   GMM criterion Q(b) =   .00215911

GMM estimation

Number of parameters =    5
Number of moments    =    7
Initial weight matrix: Unadjusted              Number of obs   =     4412
GMM weight matrix:     Robust
```

	Coef.	Robust Std. Err.	z	P>|z|	[95% Conf. Interval]	
/xb_private	.535335	.1599039	3.35	0.001	.2219291	.8487409
/xb_chronic	1.090126	.0617659	17.65	0.000	.9690668	1.211185
/xb_female	.6636579	.0959884	6.91	0.000	.4755241	.8517918
/xb_income	.0142855	.0027162	5.26	0.000	.0089618	.0196092
/b0	-.5983477	.138433	-4.32	0.000	-.8696713	-.327024

```
Instruments for equation 1: private chronic female age black hispanic _cons
```

Once we control for the endogeneity of income, we find that its coefficient has quadrupled in size. Additionally, access to private insurance has less of an impact on the number of doctor visits and gender has more of an impact.

◁

❑ Technical note

Although perhaps at first tempting, unlike the Poisson model, you cannot simply replace $\mathbf{x}$ in the moment conditions for the probit (or logit) model with a vector of instruments, $\mathbf{z}$, if you have endogenous regressors. See Wilde (2008).

❑

Mullahy (1997) considers a slightly more complicated version of the exponential regression model that incorporates nonadditive unobserved heterogeneity. His model can be written as

$$y_i = \exp(\mathbf{x}_i'\beta)\eta_i + \epsilon_i$$

where $\eta_i > 0$ is an unobserved heterogeneity term that may be correlated with $\mathbf{x}_i$. One result from his paper is that instead of using the additive moment condition (10), we can use the multiplicative moment condition

$$E\left\{\mathbf{z}\frac{y - \exp(\mathbf{x}'\beta)}{\exp(\mathbf{x}'\beta)}\right\} = E[\mathbf{z}\{y\exp(-\mathbf{x}'\beta) - 1\}] = \mathbf{0} \tag{12}$$

Windmeijer and Santos Silva (1997) discuss the use of additive versus multiplicative moment conditions with endogenous regressors and note that a set of instruments that satisfies the additive moment conditions will not also satisfy the multiplicative moment conditions. They remark that which to use is an empirical issue that can at least partially be settled by using the test of overidentifying restrictions that is implemented by `estat overid` after `gmm` to ascertain whether the instruments for a given model are valid. See [R] **gmm postestimation** for information on the test of overidentifying restrictions.

Specifying derivatives

By default, `gmm` calculates derivatives numerically, and the method used produces accurate results for the vast majority of applications. However, if you refit the same model repeatedly or else have the derivatives available, then `gmm` will run more quickly if you supply it with analytic derivatives.

When you use the interactive version of `gmm`, you specify derivatives using substitutable expressions in much the same way you specify the moment equations. There are three rules you must follow:

1. As with the substitutable expressions that define residual equations, you bind parameters of the model in braces: {b0}, {param}, etc.

2. You must specify a derivative for each parameter that appears in each moment equation. If a parameter does not appear in a moment equation, then you do not specify a derivative for that parameter in that moment equation.

3. If you declare a linear combination in an equation, then you specify a derivative with respect to that linear combination. `gmm` applies the chain rule to obtain the derivatives with respect to the individual parameters encompassed by that linear combination.

We illustrate with several examples.

▷ Example 7: Derivatives for a single-equation model

Consider a simple exponential regression model with one exogenous regressor and a constant term. We have

$$u = y - \exp(\beta_0 + \beta_1 x)$$

Now

$$\frac{\partial u}{\partial \beta_0} = -\exp(\beta_0 + \beta_1 x) \qquad \text{and} \qquad \frac{\partial u}{\partial \beta_1} = -x\exp(\beta_0 + \beta_1 x)$$

In Stata, we type

```
. gmm (docvis - exp({b0} + {b1}*income)), instruments(income)
> deriv(/b0 = -1*exp({b0} + {b1}*income))
> deriv(/b1 = -1*income*exp({b0}+{b1}*income)) onestep
Step 1
Iteration 0:   GMM criterion Q(b) =  9.1548611
Iteration 1:   GMM criterion Q(b) =  3.5146131
Iteration 2:   GMM criterion Q(b) =  .01344695
Iteration 3:   GMM criterion Q(b) =  3.690e-06
Iteration 4:   GMM criterion Q(b) =  4.606e-13
Iteration 5:   GMM criterion Q(b) =  1.502e-26
```

```
GMM estimation

Number of parameters =    2
Number of moments    =    2
Initial weight matrix: Unadjusted                     Number of obs  =    4412
```

	Coef.	Robust Std. Err.	z	P>|z|	[95% Conf. Interval]	
/b0	1.204888	.0462355	26.06	0.000	1.114268	1.295507
/b1	.0046702	.0009715	4.81	0.000	.0027662	.0065743

```
Instruments for equation 1: income _cons
```

Notice how we specified the derivative() option for each parameter. We simply specified a slash, the name of the parameter, an equal sign, then a substitutable expression that represents the derivative. Because our model has only one residual equation, we do not need to specify equation numbers in the derivative() options.

◁

When you specify a linear combination of variables, your derivative should be with respect to the entire linear combination. For example, say we have the residual equation

$$u = y - \exp(\mathbf{x}'\beta + \beta_0)$$

for which we would type

. gmm (y - exp({xb: x1 x2 x3} + {b0}) ...

Then in addition to the derivative $\partial u / \partial \beta_0$, we are to compute and specify

$$\frac{\partial u}{\partial (\mathbf{x}'\beta)} = -\exp(\mathbf{x}'\beta + \beta_0)$$

Using the chain rule, $\partial u / \partial \beta_j = \partial u / \partial (\mathbf{x}'\beta) \times \partial (\mathbf{x}'\beta) / \partial \beta_j = -x_j \exp(\mathbf{x}'\beta + \beta_0)$. Stata does this last calculation automatically. It knows the variables in the linear combination, so all it needs is the derivative of the residual function with respect to the linear combination. This allows you to change the variables in your linear combination without having to change the derivatives.

▷ Example 8: Derivatives with a linear combination

We refit the model described in the example illustrating exponential regression with endogenous regressors, now providing analytic derivatives. We type

```
. gmm (docvis - exp({xb:private chronic female income}+{b0})),
> instruments(private chronic female age black hispanic)
> derivative(/xb = -1*exp({xb:} + {b0}))
> derivative(/b0 = -1*exp({xb:} + {b0}))
Step 1
Iteration 0:   GMM criterion Q(b) =   16.910173
Iteration 1:   GMM criterion Q(b) =   .82270871
Iteration 2:   GMM criterion Q(b) =   .21831995
Iteration 3:   GMM criterion Q(b) =   .12685934
Iteration 4:   GMM criterion Q(b) =   .12672369
Iteration 5:   GMM criterion Q(b) =   .12672365
Step 2
Iteration 0:   GMM criterion Q(b) =   .00234641
Iteration 1:   GMM criterion Q(b) =   .00215957
Iteration 2:   GMM criterion Q(b) =   .00215911
Iteration 3:   GMM criterion Q(b) =   .00215911
```

```
GMM estimation
Number of parameters =    5
Number of moments    =    7
Initial weight matrix: Unadjusted                    Number of obs  =    4412
GMM weight matrix:     Robust
```

	Coef.	Robust Std. Err.	z	P>|z|	[95% Conf. Interval]	
/xb_private	.535335	.159904	3.35	0.001	.221929	.848741
/xb_chronic	1.090126	.0617659	17.65	0.000	.9690668	1.211185
/xb_female	.6636579	.0959885	6.91	0.000	.475524	.8517918
/xb_income	.0142855	.0027162	5.26	0.000	.0089618	.0196092
/b0	-.5983477	.138433	-4.32	0.000	-.8696714	-.327024

Instruments for equation 1: private chronic female age black hispanic _cons

In the first `derivative()` option, we specified the name of the linear combination, xb, instead of an individual parameter's name. We already declared the variables of our linear combination in the substitutable expression for the residual equation, so in our substitutable expressions for the derivatives, we can use the shorthand notation `{xb:}` to refer to it.

Our point estimates are identical to those we obtained earlier. The standard errors and confidence intervals differ by only trivial amounts.

◁

Exponential regression models with panel data

In addition to supporting cross-sectional and time-series data, `gmm` also works with panel-data models. Here we illustrate `gmm`'s panel-data capabilities by expanding our discussion of exponential regression models to allow for panel data. This also provides us the opportunity to demonstrate the moment-evaluator program version of `gmm`. Our discussion is based on Blundell, Griffith, and Windmeijer (2002). Also see Wooldridge (1999) for further discussion of nonlinear panel-data models.

First, we expand (9) for panel data. With individual heterogeneity term η_i, we have

$$E(y_{it}|\mathbf{x}_{it}, \eta_i) = \exp(\mathbf{x}'_{it}\beta + \eta_i) = \mu_{it}\nu_i$$

where $\mu_{it} = \exp(\mathbf{x}'_{it}\beta)$ and $\nu_i = \exp(\eta_i)$. Note that there is no constant term in this model because its effect cannot be disentangled from ν_i. With an additive idiosyncratic error term, we have the regression model

$$y_{it} = \mu_{it}\nu_i + \epsilon_{it}$$

We do not impose the assumption $E(\mathbf{x}_{it}\eta_i) = \mathbf{0}$, so η_i can be considered a fixed effect in the sense that it may be correlated with the regressors.

As discussed by Blundell, Griffith, and Windmeijer (2002), if $\mathbf{x}_{it}$ is strictly exogenous, meaning $E(\mathbf{x}_{it}\epsilon_{is}) = \mathbf{0}$ for all t and s, then we can estimate the parameters of the model by using the sample moment conditions

$$\sum_i \sum_t \mathbf{x}_{it} \left(y_{it} - \mu_{it}\frac{\overline{y}_i}{\overline{\mu}_i} \right) = \mathbf{0} \tag{13}$$

where $\overline{y}_i$ and $\overline{\mu}_i$ are the means of y_{it} and μ_{it} for panel i, respectively. Because $\overline{\mu}_i$ depends on the parameters of the model, it must be recomputed each time `gmm` needs to evaluate the residual equation. Therefore, we cannot use the substitutable expression version of `gmm`. Instead, we must use the moment-evaluator program version.

The moment-evaluator program version of gmm functions much like the function-evaluator program versions of nl and nlsur. The program you write is passed one or more variables to be filled in with the residuals evaluated at the parameter values specified in an option passed to your program. For the fixed-effects Poisson model with strictly exogenous regressors, our first crack at a function-evaluator program is

```
program gmm_poi
        version 12
        syntax varlist if, at(name)
        quietly {
                tempvar mu mubar ybar
                gen double 'mu' = exp(x1*'at'[1,1] + x2*'at'[1,2]        ///
                                        + x3*'at'[1,3]) 'if'
                egen double 'mubar' = mean('mu') 'if', by(id)
                egen double 'ybar' = mean(y) 'if', by(id)
                replace 'varlist' = y - 'mu'*'ybar'/'mubar' 'if'
        }
end
```

You can save your program in an ado-file named *name*.ado, where *name* is the name you use for your program; here we would save the program in the ado-file gmm_poi.ado. Alternatively, if you are working from within a do-file, you can simply define the program before calling gmm. The syntax statement declares that we are expecting to receive a *varlist*, containing the names of variables whose values we are to replace with the values of the residual equations, and an if expression that will mark the estimation sample; because our model has one residual equation, *varlist* will consist of one variable. at() is a required option to our program, and it will contain the name of a matrix containing the parameter values at which we are to evaluate the residual equation. All moment-evaluator programs must accept the *varlist*, if condition, and at() option.

The first part of our program computes μ_{it}. In the model we will fit shortly, we have three regressors, named x1, x2, and x3. The 'at' vector will have three elements, one for each of those variables. Notice that we included 'if' at the end of each statement that affects variables to restrict the computations to the relevant estimation sample. The two egen statements compute $\overline{\mu}_i$ and $\overline{y}_i$; in the example dataset we will use shortly, the panel variable is named id, and for simplicity we hardcoded that variable into our program as well. Finally, we compute the residual equation, which is the portion of (13) bound in parentheses.

Example 9: Panel poisson with strictly exogenous regressors

To fit our model, we type

```
. use http://www.stata-press.com/data/r12/poisson1

. gmm gmm_poi, nequations(1) parameters(b1 b2 b3)
> instruments(x1 x2 x3, noconstant) vce(cluster id) onestep
Step 1
Iteration 0:   GMM criterion Q(b) =   51.99142
Iteration 1:   GMM criterion Q(b) =  .04345191
Iteration 2:   GMM criterion Q(b) =  8.720e-06
Iteration 3:   GMM criterion Q(b) =  7.115e-13
Iteration 4:   GMM criterion Q(b) =  5.130e-27
```

```
GMM estimation

Number of parameters =   3
Number of moments     =   3
Initial weight matrix: Unadjusted                      Number of obs  =     409

                                        (Std. Err. adjusted for 45 clusters in id)
```

	Coef.	Robust Std. Err.	z	P>|z|	[95% Conf. Interval]	
/b1	1.94866	.1000265	19.48	0.000	1.752612	2.144709
/b2	-2.966119	.0923592	-32.12	0.000	-3.14714	-2.785099
/b3	1.008634	.1156561	8.72	0.000	.781952	1.235315

```
Instruments for equation 1: x1 x2 x3
```

All three of our regressors are strictly exogenous, so they can serve as their own regressors. There is no constant term in the model (it would be unidentified), so we exclude a constant term from our list of instruments. We have one residual equation as indicated by nequations(1), and we have three parameters, named b1, b2, and b3. The order in which you declare parameters in the parameters() option determines the order in which they appear in the 'at' vector in the moment-evaluator program. We specified vce(cluster id) to obtain standard errors that allow for correlation among observations within each panel.

◁

❑ Technical note

The program we just wrote is sufficient to fit the model to the poisson1 dataset, but if we want to fit that model to other datasets, we would need to change the variable names and perhaps account for having a different number of parameters as well. With a bit more programming, advanced users can write a more general program to work with arbitrary datasets. Any options not understood by gmm are passed along to the moment-evaluator program, and we can take advantage of that feature to pass the name of the dependent variable, list of regressors, and panel identifier variable. A better version of gmm_poi would be

```
program gmm_poi2
        version 12
        syntax varlist if, at(name) myrhs(varlist)  ///
               mylhs(varlist) myidvar(varlist)

        quietly {
                tempvar mu mubar ybar
                gen double `mu' = 0 `if'
                local j = 1
                foreach var of varlist `myrhs' {
                        replace `mu' = `mu' + `var'*`at'[1,`j'] `if'
                        local j = `j' + 1
                }
                replace `mu' = exp(`mu')
                egen double `mubar' = mean(`mu') `if', by(`myidvar')
                egen double `ybar' = mean(`mylhs') `if', by(`myidvar')
                replace `varlist' = `mylhs' - `mu'*`ybar'/`mubar' `if'
        }
end
```

Our program now accepts three more options. Our call to `gmm` is

```
. gmm gmm_poi2, mylhs(y) myrhs(x1 x2 x3) myidvar(id) nequations(1)
> parameters(b1 b2 b3) instruments(x1 x2 x3, noconstant) vce(cluster id) onestep
```

With `mylhs()`, `myrhs()`, and `myidvar()`, we are now able to fit our model to any panel dataset we wish without having to modify the `gmm_poi2` program. In the section *Details of moment-evaluator programs* below, we show how to incorporate weights and derivatives in moment-evaluator programs. ❏

When past values of the idiosyncratic error term affect the value of a regressor, we say that regressor is *predetermined*. When one or more regressors are predetermined, sample moment condition (10) is no longer valid. However, Chamberlain (1992) shows that a simple alternative is to consider moment conditions of the form

$$\sum_i \sum_{t=2}^{T} \mathbf{x}_{i,t-1} \left(y_{i,t-1} - \mu_{i,t-1} \frac{y_{it}}{\mu_{it}} \right) = \mathbf{0} \tag{14}$$

Also see Wooldridge (1997) and Windmeijer (2000) for other moment conditions that can be used with predetermined regressors.

⊳ Example 10: Panel Poisson with predetermined regressors

Here we refit the previous model, treating all the regressors as predetermined and using the moment conditions in (14). Our moment-evaluator program is

```
program gmm_poipre
        version 12
        syntax varlist if, at(name) myrhs(varlist) mylhs(varlist)
        quietly {
                tempvar mu mubar ybar
                gen double 'mu' = 0 'if'
                local j = 1
                foreach var of varlist 'myrhs' {
                        replace 'mu' = 'mu' + 'var'*'at'[1,'j'] 'if'
                        local j = 'j' + 1
                }
                replace 'mu' = exp('mu')
                replace 'varlist' = L.'mylhs' - L.'mu'*'mylhs'/'mu' 'if'
        }
end
```

As before, the first part of our program computes μ_{it}; the only difference is in how we compute the residual equation. We used lag-operator notation so that Stata properly handles gaps in our dataset. Equation (14) shows that we are to use the first lags of the regressors as instruments, so we type

```
. gmm gmm_poipre, mylhs(y) myrhs(x1 x2 x3) nequations(1) parameters(b1 b2 b3)
> instruments(L.(x1 x2 x3), noconstant) vce(cluster id) onestep
(obs = 364)

Step 1
Iteration 0:    GMM criterion Q(b) =  52.997808
Iteration 1:    GMM criterion Q(b) =  2.1678071
Iteration 2:    GMM criterion Q(b) =  .08716503
Iteration 3:    GMM criterion Q(b) =  .00007136
Iteration 4:    GMM criterion Q(b) =  4.699e-11
Iteration 5:    GMM criterion Q(b) =  1.932e-23

GMM estimation

Number of parameters =   3
Number of moments    =   3
Initial weight matrix: Unadjusted                    Number of obs  =     364
```

(Std. Err. adjusted for 45 clusters in id)

	Coef.	Robust Std. Err.	z	P>|z|	[95% Conf. Interval]
/b1	2.035125	.2662377	7.64	0.000	1.513308	2.556941
/b2	-2.929362	.2290397	-12.79	0.000	-3.378272	-2.480453
/b3	1.235219	.1673295	7.38	0.000	.9072596	1.563179

Instruments for equation 1: L.x1 L.x2 L.x3

Here, like earlier with strictly exogenous regressors, the number of instruments equals the number of parameters, so there is no gain to using the two-step or iterated estimators. However, if you do have more instruments than parameters, you will most likely want to use one of those other estimators instead.

◁

In the previous example, we used $x_{i,t-1}$ as instruments. A more efficient GMM estimator would also use $x_{i,t-2}, x_{i,t-3}, \ldots, x_{i,1}$ as instruments in period t as well. gmm's xtinstruments() option allows you to specify instrument lists that grow as t increases. Later we discuss the xtinstruments() option in detail in the context of linear dynamic panel-data models.

When a regressor is contemporaneously correlated with the idiosyncratic error term, we say that regressor is endogenous. Windmeijer (2000) shows that here we can use the moment condition

$$\sum_i \sum_{t=3}^{T} x_{i,t-2} \left(\frac{y_{it}}{\mu_{it}} - \frac{y_{i,t-1}}{\mu_{i,t-1}} \right)$$

Here we use the second lag of the endogenous regressor as an instrument. If a variable is strictly exogenous, it can of course serve as its own instrument.

▷ Example 11: Panel Poisson with endogenous regressors

Here we refit the model, treating x3 as endogenous and x1 and x2 as strictly exogenous. Our moment-evaluator program is

```
program gmm_poiend
        version 12
        syntax varlist if, at(name) myrhs(varlist) mylhs(varlist)
        quietly {
                tempvar mu mubar ybar
                gen double 'mu' = 0 'if'
                local j = 1
                foreach var of varlist 'myrhs' {
                        replace 'mu' = 'mu' + 'var'*'at'[1,'j'] 'if'
                        local j = 'j' + 1
                }
                replace 'mu' = exp('mu')
                replace 'varlist' = 'mylhs'/'mu' - L.'mylhs'/L.'mu' 'if'
        }
end
```

Now we call gmm using x1, x2, and L2.x3 as instruments:

```
. use http://www.stata-press.com/data/r12/poisson2
. gmm gmm_poiend, mylhs(y) myrhs(x1 x2 x3) nequations(1) parameters(b1 b2 b3)
> instruments(x1 x2 L2.x3, noconstant) vce(cluster id) onestep
Step 1
Iteration 0:   GMM criterion Q(b) =  47.376537
Iteration 1:   GMM criterion Q(b) =  .08115406
Iteration 2:   GMM criterion Q(b) =  .03477036
Iteration 3:   GMM criterion Q(b) =  .00041056
Iteration 4:   GMM criterion Q(b) =  1.189e-07
Iteration 5:   GMM criterion Q(b) =  1.298e-14
Iteration 6:   GMM criterion Q(b) =  1.574e-28

GMM estimation

Number of parameters =   3
Number of moments    =   3
Initial weight matrix: Unadjusted                Number of obs    =     3766
                                    (Std. Err. adjusted for 500 clusters in id)
```

	Coef.	Robust Std. Err.	z	P>|z|	[95% Conf.	Interval]
/b1	1.844082	.1515252	12.17	0.000	1.547098	2.141066
/b2	-2.904011	.108117	-26.86	0.000	-3.115916	-2.692105
/b3	3.277512	2.459066	1.33	0.183	-1.542169	8.097193

Instruments for equation 1: x1 x2 L2.x3

As with the predetermined case previously, instead of using just $x_{i,t-2}$ as an instrument, we could use all further lags of x_{it} as instruments as well.

◁

ational-expectations models

Macroeconomic models typically assume that agents' expectations about the future are formed rationally. By rational expectations, we mean that agents use all information available when forming their forecasts, so the forecast error is uncorrelated with the information available when the forecast was made. Say that at time t, people make a forecast, $\widehat{y}_{t+1}$, of variable y in the next period. If Ω_t denotes all available information at time t, then rational expectations implies that $E\left\{(\widehat{y}_{t+1} - y_{t+1})|\Omega_t\right\} = 0$. If Ω_t denotes observable variables such as interest rates or prices, then this conditional expectation can serve as the basis of a moment condition for GMM estimation.

▷ Example 12: Fitting a Euler equation

In a well-known article, Hansen and Singleton (1982) consider a model of portfolio decision making and discuss parameter estimation using GMM. We will consider a simple example with one asset in which the agent can invest. A consumer wants to maximize the present value of his lifetime utility derived from consuming the good. On the one hand, the consumer is impatient, so he would rather consume today than wait until tomorrow. On the other hand, if he consumes less today, he can invest more of his money, earning more interest that he can then use to consume more of the good tomorrow. Thus there is a tradeoff between having his cake today or sacrificing a bit today to have more cake tomorrow.

If we assume a specific form for the agent's utility function, known as the constant relative-risk aversion utility function, we can show that the Euler equation is

$$E\left[\mathbf{z}_t\left\{1 - \beta(1 + r_{t+1})(c_{t+1}/c_t)^{-\gamma}\right\}\right] = \mathbf{0}$$

where β and γ are the parameters to estimate, r_t is the return to the financial asset, and c_t is consumption in period t. β measures the agent's discount factor. If β is near one, the agent is patient and is more willing to forgo consumption this period. If β is close to zero, the agent is less patient and prefers to consume more now. The parameter γ characterizes the agent's utility function. If γ equals one, the utility function is linear. As γ tends toward zero, the utility function tends toward $u = \log(c)$.

We have data on 3-month Treasury bills (r_t) and consumption expenditures (c_t). As instruments, we will use lagged rates of return and past growth rates of consumption. We will use the two-step estimator and a weight matrix that allows for heteroskedasticity and autocorrelation up to four lags with the Bartlett kernel. In Stata, we type

```
. use http://www.stata-press.com/data/r12/cr
. generate cgrowth = c / L.c
(1 missing value generated)
. gmm (1 - {b=1}*(1+F.r)*(F.c/c)^(-1*{gamma=1})), inst(L.r L2.r cgrowth L.cgrowth)
> wmat(hac nw 4) twostep
warning: 1 missing value returned for equation 1 at initial values
Step 1
Iteration 0:   GMM criterion Q(b) =  .00226482
Iteration 1:   GMM criterion Q(b) =  .00054369
Iteration 2:   GMM criterion Q(b) =  .00053904
Iteration 3:   GMM criterion Q(b) =  .00053904
Step 2
Iteration 0:   GMM criterion Q(b) =  .0600729
Iteration 1:   GMM criterion Q(b) =  .0596369
Iteration 2:   GMM criterion Q(b) =  .0596369
GMM estimation
Number of parameters =   2
Number of moments    =   5
Initial weight matrix: Unadjusted              Number of obs   =      239
GMM weight matrix:    HAC Bartlett 4
```

	Coef.	HAC Std. Err.	z	P>|z|	[95% Conf. Interval]	
/b	.9204617	.0134646	68.36	0.000	.8940716	.9468518
/gamma	-4.222361	1.473895	-2.86	0.004	-7.111143	-1.333579

```
HAC standard errors based on Bartlett kernel with 4 lags
Instruments for equation 1: L.r L2.r cgrowth L.cgrowth _cons
```

The warning message at the top of the output appears because the forward operator in our substitutable expression says that residuals can be computed only for 239 observations; our dataset contains 240 observations. Our estimate of β is near one, in line with expectations and published results.

◁

System estimators

In many economic models, two or more variables are determined jointly through a system of simultaneous equations. Indeed, some of the earliest work in econometrics, including that of the Cowles Commission, was centered around estimation of the parameters of simultaneous equations. The 2SLS and IV estimators we have already discussed are used in some circumstances to estimate such parameters. Here we focus on the joint estimation of all the parameters of systems of equations, and we begin with the well-known three-stage least-squares (3SLS) estimator.

Recall that the 2SLS estimator is based on the moment conditions $E(\mathbf{z}u) = \mathbf{0}$. The 2SLS estimator can be used to estimate the parameters of one equation of a system of structural equations. Moreover, with the 2SLS estimator, we do not even need to specify the structural relationship among all the endogenous variables; we need to specify only the equation on which interest focuses and simply assume reduced-form relationships among the endogenous regressors of the equation of interest and the exogenous variables of the model. If we are willing to specify the complete system of structural equations, then assuming our model is correctly specified, by estimating all the equations jointly, we can obtain estimates that are more efficient than equation-by-equation 2SLS.

In [R] **reg3**, we fit a simple two-equation macroeconomic model:

$$\texttt{consump} = \beta_0 + \beta_1\texttt{wagepriv} + \beta_2\texttt{wagegovt} + \epsilon_1 \tag{15}$$

$$\texttt{wagepriv} = \beta_3 + \beta_4\texttt{consump} + \beta_5\texttt{govt} + \beta_6\texttt{capital1} + \epsilon_2 \tag{16}$$

where `consump` represents aggregate consumption; `wagepriv` and `wagegovt` are total wages paid by the private and government sectors, respectively; `govt` is government spending; and `capital1` is the previous period's capital stock. We are not willing to assume that ϵ_1 and ϵ_2 are independent, so we must treat both `consump` and `wagepriv` as endogenous. Suppose that a random shock makes ϵ_2 positive. Then by (16), `wagepriv` will be higher than it otherwise would. Moreover, ϵ_1 will either be higher or lower, depending on the correlation between it and ϵ_2. The shock to ϵ_2 has made both `wagepriv` and ϵ_1 move, implying that in (15) `wagepriv` is an endogenous regressor. A similar argument shows that `consump` is an endogenous regressor in the second equation. In our model, `wagegovt`, `govt`, and `capital1` are all exogenous variables.

Let $\mathbf{z}_1$ and $\mathbf{z}_2$ denote the instruments for the first and second equations, respectively; we will discuss what comprises them shortly. We have two sets of moment conditions:

$$E\left\{ \begin{array}{c} \mathbf{z}_1(\texttt{consump} - \beta_0 - \beta_1\texttt{wagepriv} - \beta_2\texttt{wagegovt}) \\ \mathbf{z}_2(\texttt{wagepriv} - \beta_3 - \beta_4\texttt{consump} - \beta_5\texttt{govt} - \beta_6\texttt{capital1}) \end{array} \right\} = \mathbf{0} \tag{17}$$

One of the defining characteristics of 3SLS is that the errors are homoskedastic conditional on the instrumental variables. Using this assumption, we have

$$E\left[\left\{ \begin{array}{c} \mathbf{z}_1\epsilon_1 \\ \mathbf{z}_2\epsilon_2 \end{array} \right\} \{ \mathbf{z}_1'\epsilon_1 \quad \mathbf{z}_2'\epsilon_2 \} \right] = \left\{ \begin{array}{cc} \sigma_{11}E(\mathbf{z}_1\mathbf{z}_1') & \sigma_{12}E(\mathbf{z}_1\mathbf{z}_2') \\ \sigma_{21}E(\mathbf{z}_2\mathbf{z}_1') & \sigma_{22}E(\mathbf{z}_2\mathbf{z}_2') \end{array} \right\} \tag{18}$$

where $\sigma_{ij} = \mathrm{cov}(\epsilon_i, \epsilon_j)$. Let $\boldsymbol{\Sigma}$ denote the 2×2 matrix with typical element σ_{ij}.

The second defining characteristic of the 3SLS estimator is that it uses all the exogenous variables as instruments for all equations; here $z_1 = z_2 = (\text{wagegovt}, \text{govt}, \text{capital1}, 1)$, where the 1 indicates a constant term. From our discussion on the weight matrix and two-step estimation, we want to use the sample analogue of the matrix inverse of the right-hand side of (18) as our weight matrix.

To implement the 3SLS estimator, we apparently need to know Σ or at least have a consistent estimator of it. The solution is to fit (15) and (16) by 2SLS, use the sample residuals $\widehat{\epsilon_1}$ and $\widehat{\epsilon_2}$ to estimate Σ, then estimate the parameters of (17) via GMM by using the weight matrix just discussed.

▷ Example 13: 3SLS estimation

3SLS is easier to do using gmm than it sounds. The 3SLS estimator is a two-step GMM estimator. In the first step, we do the equivalent of 2SLS on each equation, and then we compute a weight matrix based on (18). Finally, we perform a second step of GMM with this weight matrix.

In Stata, we type

```
. use http://www.stata-press.com/data/r12/klein, clear
. gmm (eq1:consump - {b0} - {xb: wagepriv wagegovt})
>      (eq2:wagepriv - {c0} - {xc: consump govt capital1}),
> instruments(eq1 eq2: wagegovt govt capital1) winitial(unadjusted, independent)
> wmatrix(unadjusted) twostep
Step 1
Iteration 0:    GMM criterion Q(b) =   4195.4487
Iteration 1:    GMM criterion Q(b) =   .22175631
Iteration 2:    GMM criterion Q(b) =   .22175631

Step 2
Iteration 0:    GMM criterion Q(b) =   .09716589
Iteration 1:    GMM criterion Q(b) =   .07028208
Iteration 2:    GMM criterion Q(b) =   .07028208

GMM estimation

Number of parameters =   7
Number of moments    =   8
Initial weight matrix: Unadjusted              Number of obs   =       22
GMM weight matrix:     Unadjusted
```

	Coef.	Std. Err.	z	P>|z|	[95% Conf. Interval]
/b0	19.3559	3.583772	5.40	0.000	12.33184 26.37996
/xb_wagepriv	.8012754	.1279329	6.26	0.000	.5505314 1.052019
/xb_wagegovt	1.029531	.3048424	3.38	0.001	.432051 1.627011
/c0	14.63026	10.26693	1.42	0.154	-5.492552 34.75306
/xc_consump	.4026076	.2567312	1.57	0.117	-.1005764 .9057916
/xc_govt	1.177792	.5421253	2.17	0.030	.1152461 2.240338
/xc_capital1	-.0281145	.0572111	-0.49	0.623	-.1402462 .0840173

```
Instruments for equation 1: wagegovt govt capital1 _cons
Instruments for equation 2: wagegovt govt capital1 _cons
```

The independent suboption of the winitial() option tells gmm to assume that the residuals are independent across equations; this suboption sets $\sigma_{21} = \sigma_{12} = 0$ in (18). Assuming both homoskedasticity and cross-equation independence is equivalent to fitting the two equations of our model independently by 2SLS. The wmatrix() option controls how the weight matrix is computed based on the first-step parameter estimates before the second step of estimation; here we request a weight matrix that assumes conditional homoskedasticity but that does not impose the cross-equation independence like the initial weight matrix we used. In this example, we also illustrated how to name equations and how equation names can be used in the `instruments()` option. Our results are identical to those in [R] **reg3**.

We could have specified our instruments with the syntax

```
instruments(wagegovt govt capital1)
```

because gmm uses the variables listed in the `instruments()` option for all equations unless you use the `equations()` suboption to restrict those variables to certain equations. However, we wanted to emphasize that the same instruments are being used for both equations; in a moment, we will discuss an estimator that does not use the same instruments in all equations.

◁

In the previous example, if we omit the `twostep` option, the resulting coefficients will be equivalent to equation-by-equation 2SLS, which Wooldridge (2010, 216) calls the "system 2SLS estimator". Eliminating the `twostep` option makes the `wmatrix()` option irrelevant, so that option can be eliminated as well.

So far, we have developed the traditional 3SLS estimator. Wooldridge (2010, chap. 8) discusses the "GMM 3SLS" estimator that extends the traditional 3SLS estimator by allowing for heteroskedasticity and different instruments for different equations.

Generalizing (18) to an arbitrary number of equations, we have

$$E\left(\mathbf{Z}'\epsilon\epsilon'\mathbf{Z}\right) = E\left(\mathbf{Z}'\mathbf{\Sigma}\mathbf{Z}\right) \tag{19}$$

where

$$\mathbf{Z} = \begin{bmatrix} \mathbf{z}_1 & \mathbf{0} & \cdots & \mathbf{0} \\ \mathbf{0} & \mathbf{z}_2 & \cdots & \mathbf{0} \\ \vdots & \vdots & \ddots & \vdots \\ \mathbf{0} & \mathbf{0} & \cdots & \mathbf{z}_m \end{bmatrix}$$

and $\mathbf{\Sigma}$ is now $m \times m$. Equation (19) is the multivariate analogue of a homoskedasticity assumption; for each equation, the error variance is constant for all observations, as is the covariance between any two equations' errors.

We can relax this homoskedasticity assumption by considering different weight matrices. For example, if we continue to assume that observations are independent but not necessarily identically distributed, then by specifying `wmatrix(robust)`, we would obtain a weight matrix that allows for heteroskedasticity:

$$\widehat{W} = \frac{1}{N}\sum_i \mathbf{Z}_i'\widehat{\epsilon}_i\widehat{\epsilon}_i'\mathbf{Z}_i$$

This is the weight matrix in Wooldridge's (2010, 218) Procedure 8.1, "GMM with Optimal Weighting Matrix". By default, gmm would report standard errors based on his covariance matrix (8.27); specifying `vce(unadjusted)` would provide the optimal GMM standard errors. If you have multiple observations for each individual or firm in your dataset, you could specify `wmatrix(cluster id)`, where *id* identifies individuals or firms. This would allow arbitrary within-individual correlation, though it does not account for an individual-specific fixed or random effect. In both cases, we would continue to use `winitial(unadjusted, independent)` so that the first-step estimates are the system 2SLS estimates.

Wooldridge (2010, sec. 9.6) discusses instances where it is necessary to use different instruments in different equations. The GMM 3SLS estimator with different instruments in different equations but with conditional homoskedasticity is what Hayashi (2000, 275) calls the "full-information instrumental variables efficient" (FIVE) estimator. Implementing the FIVE estimator is easy with gmm. For example, say we have a two-equation system, where kids, age, income, and education are all valid instruments for the first equation; but education is not a valid instrument for the second equation. Then our syntax would take the form

> gmm (<*mexp_1*>) (<*mexp_2*>), instruments(1:kids age income education)
>   instruments(2:kids age income)

The following syntax is equivalent:

> gmm (<*mexp_1*>) (<*mexp_2*>), instruments(kids age income)
>   instruments(1:education)

Because we did not specify a list of equations in the second example's first instruments() option, those variables are used as instruments in both equations. You can use whichever syntax you prefer. The first requires a bit more typing but is arguably more transparent.

If all the regressors in the model are exogenous, then the traditional 3SLS estimator is the seemingly unrelated regression (SUR) estimator. Here you would specify all the regressors as instruments.

Dynamic panel-data models

Commands in Stata that work with panel data expect the data to be in the "long" format, meaning that each row of the dataset consists of one subobservation that is a member of a logical observation (represented by the panel identifier variable). See [D] **reshape** for a discussion of the long versus "wide" data forms. gmm is no exception in this respect when used with panel data. From a theoretical perspective, however, it is sometimes easier to view GMM estimators for panel data as system estimators in which we have N observations on a system of T equations, where N and T are the number of observations and panels, respectively, rather than a single-equation estimator with NT observations. Usually, each of the T equations will in fact be the same, though we will want to specify different instruments for each of these equations.

In a dynamic panel-data model, lagged values of the dependent variable are included as regressors. Here we consider a simple model with one lag of the dependent variable y as a regressor and a vector of strictly exogenous regressors, $\mathbf{x}_{it}$:

$$y_{it} = \rho y_{i,t-1} + \mathbf{x}'_{it}\beta + u_i + \epsilon_{it} \tag{20}$$

u_i can be either a fixed- or a random-effect term, in the sense that we do not require $\mathbf{x}_{it}$ to be independent of it. Even with the assumption that ϵ_{it} is i.i.d., the presence of both $y_{i,t-1}$ and u_i in (20) renders both the standard fixed- and random-effects estimators to be inconsistent because of the well-known Nickell (1981) bias. OLS regression of y_{it} on $y_{i,t-1}$ and $\mathbf{x}_{it}$ also produces inconsistent estimates, because $y_{i,t-1}$ will be correlated with the error term.

❑ Technical note

Stata has the xtabond, xtdpd, and xtdpdsys commands (see [XT] **xtabond**, [XT] **xtdpd**, and [XT] **xtdpdsys**) to fit equations like (20), and for everyday use those commands are preferred because they offer features such as Windmeijer (2005) bias-corrected standard errors to account for the bias of traditional two-step GMM standard errors seen in dynamic panel-data models and, being linear estimators, only require you to specify variable names instead of complete equations. However, using gmm has several pedagogical advantages, including the ability to tie those model-specific commands into a more general framework, a clear illustration of how certain types of instrument matrices for panel-data models are formed, and demonstrations of several advanced features of gmm.

❑

First-differencing (20) removes the panel-specific u_i term:

$$y_{it} - y_{i,t-1} = \rho(y_{i,t-1} - y_{i,t-2}) + (\mathbf{x}_{it} - \mathbf{x}_{i,t-1})'\beta + (\epsilon_{it} - \epsilon_{i,t-1}) \tag{21}$$

However, now $(y_{i,t-1} - y_{i,t-2})$ is correlated with $(\epsilon_{it} - \epsilon_{i,t-1})$. Thus we need an instrument that is correlated with the former but not the latter. The lagged variables in (21) mean that equation is not estimable for $t < 3$, so consider when $t = 3$. We have

$$y_{i3} - y_{i2} = \rho(y_{i2} - y_{i1}) + (\mathbf{x}_{i3} - \mathbf{x}_{i2})'\beta + (\epsilon_{i3} - \epsilon_{i2}) \tag{22}$$

In the Arellano–Bond (1991) estimator, lagged levels of the dependent variable are used as instruments. With our assumption that the ϵ_{it} are i.i.d., (20) intimates that y_{i1} can serve as an instrumental variable when we fit (22).

Next consider (21) when $t = 4$. We have

$$y_{i4} - y_{i3} = \rho(y_{i3} - y_{i2}) + (\mathbf{x}_{i4} - \mathbf{x}_{i3})'\beta + (\epsilon_{i4} - \epsilon_{i3})$$

Now (20) shows that both y_{i1} and y_{i2} are uncorrelated with the error term $(\epsilon_{i4} - \epsilon_{i3})$, so we have two instruments available. For $t = 5$, you can show that y_{i1}, y_{i2}, and y_{i3} can serve as instruments. As may now be apparent, one of the key features of these dynamic panel-data models is that the available instruments depend on the time period, t, as was the case for some of the panel Poisson models we considered earlier. Because the $\mathbf{x}_{it}$ are strictly exogenous by assumption, they can serve as their own instruments.

The initial weight matrix that is appropriate for the GMM dynamic panel-data estimator is slightly more involved than the unadjusted matrix we have used in most of our previous examples that assumes the errors are i.i.d. First, rewrite (21) for panel i as

$$\mathbf{y}_i - \mathbf{y}_i^L = \rho\left(\mathbf{y}_i^L - \mathbf{y}_i^{LL}\right) + (\mathbf{X}_i - \mathbf{X}_i^L)\beta + (\epsilon_i - \epsilon_i^L)$$

where $\mathbf{y}_i = (y_{i3}, \ldots, y_{iT})$ and $\mathbf{y}_i^L = (y_{i2}, \ldots, y_{i,T-1})$, $\mathbf{y}_i^{LL} = (y_{i1}, \ldots, y_{i,T-2})$, and $\mathbf{X}_i$, $\mathbf{X}_i^L$, ϵ_i, and ϵ_i^L are defined analogously. Let $\mathbf{Z}$ denote the full matrix of instruments for panel i, including the variables specified in both the `instruments()` and `xtinstruments()` options; the exact structure is detailed in *Methods and formulas*.

By assumption, ϵ_{it} is i.i.d., so the first-difference $(\epsilon_{it} - \epsilon_{i,t-1})$ is necessarily autocorrelated with correlation -0.5. Therefore, we should not use a weight matrix that assumes the errors are independent. For dynamic panel-data models, we can show that the appropriate initial weight matrix is

$$\widehat{\mathbf{W}} = \left(\frac{1}{N} \sum_i \mathbf{Z}_i' \mathbf{H}_D \mathbf{Z}_i\right)^{-1}$$

where

$$\mathbf{H}_D = \begin{bmatrix} 1 & -0.5 & 0 & \ldots & 0 & 0 \\ -0.5 & 1 & -0.5 & \ldots & 0 & 0 \\ \vdots & \vdots & \vdots & \ddots & \vdots & \vdots \\ 0 & 0 & 0 & \ldots & 1 & -0.5 \\ 0 & 0 & 0 & \ldots & -0.5 & 1 \end{bmatrix}$$

We can obtain this initial weight matrix by specifying `winitial(xt D)`. The letter D indicates that the equation we are estimating is specified in first-differences.

▷ Example 14: Arellano–Bond estimator

Say we want to fit the model

$$n_{it} = \rho\, n_{i,t-1} + \beta_1 w_{it} + \beta_2 w_{i,t-1} + \beta_3 k_{it} + \beta_4 k_{i,t-1} + u_i + \epsilon_{it} \tag{23}$$

where we assume that w_{it} and k_{it} are strictly exogenous. First-differencing, our residual equation is

$$
\begin{aligned}
\epsilon_{it}^* = (\epsilon_{it} - \epsilon_{i,t-1}) =& n_{it} - n_{i,t-1} - \rho\,(n_{i,t-1} - n_{i,t-2}) - \beta_1(w_{it} - w_{i,t-1}) \\
& - \beta_2(w_{i,t-1} - w_{i,t-2}) - \beta_3(k_{it} - k_{i,t-1}) - \beta_4(k_{i,t-1} - k_{i,t-2}) \tag{24}
\end{aligned}
$$

In Stata, we type

```
. use http://www.stata-press.com/data/r12/abdata
. gmm (D.n - {rho}*LD.n - {xb:D.w LD.w D.k LD.k}), xtinstruments(n, lags(2/.))
> instruments(D.w LD.w D.k LD.k, noconstant) deriv(/rho = -1*LD.n)
> deriv(/xb = -1) winitial(xt D) onestep
Step 1
Iteration 0:   GMM criterion Q(b) =   .0011455
Iteration 1:   GMM criterion Q(b) =   .00009103
Iteration 2:   GMM criterion Q(b) =   .00009103

GMM estimation

Number of parameters =    5
Number of moments    =   32
Initial weight matrix: XT D                     Number of obs   =      751
```

	Coef.	Robust Std. Err.	z	P>|z|	[95% Conf. Interval]
/rho	.8041712	.1199819	6.70	0.000	.5690111	1.039331
/xb_D_w	-.5600476	.1619472	-3.46	0.001	-.8774583	-.242637
/xb_LD_w	.3946699	.1092229	3.61	0.000	.1805969	.6087429
/xb_D_k	.3520286	.0536546	6.56	0.000	.2468676	.4571897
/xb_LD_k	-.2160435	.0679689	-3.18	0.001	-.3492601	-.0828269

```
Instruments for equation 1:
      XT-style: L(2/.).n
      Standard: D.w LD.w D.k LD.k
```

Because w and k are strictly exogenous, we specified the variants of them that appear in (24) in the instruments() option; because there is no constant term in the model, we specified noconstant to omit the constant from the instrument list.

We specified xtinstruments(n, lags(2/.)) to tell gmm what instruments to use for the lagged dependent variable included as a regressor in (23). Based on our previous discussion, lags two and higher of n_{it} can serve as instruments. The lags(2/.) suboption tells gmm that the first available instrument for n_{it} is the lag-two value $n_{i,t-2}$. The "." tells gmm to use all further lags of n_{it} as instruments as well. The instrument matrices in dynamic panel-data models can become large if the dataset has many time periods per panel. In those cases, you could specify, for example, lags(2/4) to use just lags two through four instead of using all available lags.

Our results are identical to those we would obtain using xtabond with the syntax

```
xtabond n L(0/1).w L(0/1).k, lags(1) noconstant vce(robust)
```

Had we left off the vce(robust) option in our call to xtabond, we would have had to specify vce(unadjusted) in our call to gmm to obtain the same standard errors.

◁

Technical note

gmm automatically excludes observations for which there are no valid observations for the panel-style instruments. However, it keeps in the estimation sample those observations for which fewer than the maximum number of instruments you requested are available. For example, if you specify the lags(2/4) suboption, you have requested three instruments, but gmm will keep observations even if only one or two instruments are available.

❏

Example 15: Two-step Arellano–Bond estimator

Here we refit the model in the previous example, using the two-step GMM estimator.

```
. gmm (D.n - {rho}*LD.n - {xb:D.w LD.w D.k LD.k}),
> xtinstruments(n, lags(2/.)) instruments(D.w LD.w D.k LD.k, noconstant)
> deriv(/rho = -1*LD.n) deriv(/xb = -1) winitial(xt D) wmatrix(robust)
> vce(unadjusted)
Step 1
Iteration 0:   GMM criterion Q(b) =    .0011455
Iteration 1:   GMM criterion Q(b) =    .00009103
Iteration 2:   GMM criterion Q(b) =    .00009103

Step 2
Iteration 0:   GMM criterion Q(b) =    .44107941
Iteration 1:   GMM criterion Q(b) =    .4236729
Iteration 2:   GMM criterion Q(b) =    .4236729

GMM estimation

Number of parameters =   5
Number of moments    =  32
Initial weight matrix: XT D                     Number of obs   =      751
GMM weight matrix:      Robust
```

	Coef.	Std. Err.	z	P>|z|	[95% Conf. Interval]	
/rho	.8044783	.0534763	15.04	0.000	.6996667	.90929
/xb_D_w	-.5154978	.0335506	-15.36	0.000	-.5812557	-.4497399
/xb_LD_w	.4059309	.0637294	6.37	0.000	.2810235	.5308384
/xb_D_k	.3556204	.0390892	9.10	0.000	.2790071	.4322337
/xb_LD_k	-.2204521	.046439	-4.75	0.000	-.3114709	-.1294332

```
Instruments for equation 1:
      XT-style: L(2/.).n
      Standard: D.w LD.w D.k LD.k
```

Our results match those you would obtain with the command

```
xtabond n L(0/1).(w k), lags(1) noconstant twostep
```

◁

Technical note

Had we specified vce(robust) in our call to gmm, we would have obtained the traditional sandwich-based robust covariance matrix, but our standard errors would not match those we would obtain by specifying vce(robust) with the xtabond command. The xtabond, xtdpd, and xtdpdsys commands implement a bias-corrected robust VCE for the two-step GMM dynamic panel-data estimator. Traditional VCEs computed after the two-step dynamic panel-data estimator have been shown to exhibit often-severe bias; see Windmeijer (2005).

❏

Neither of the two dynamic panel-data examples we have fit so far include a constant term. When a constant term is included, the dynamic panel-data estimator is in fact a two-equation system estimator. For notational simplicity, consider a simple model containing just a constant term and one lag of the dependent variable:

$$y_{it} = \alpha + \rho y_{i,t-1} + u_i + \epsilon_{it}$$

First-differencing to remove the u_i term, we have

$$y_{it} - y_{i,t-1} = \rho(y_{i,t-1} - y_{i,t-2}) + (\epsilon_{it} - \epsilon_{i,t-1}) \tag{25}$$

This has also eliminated the constant term. If we assume $E(u_i) = 0$, which is reasonable if a constant term is included in the model, then we can recover α by including the moment condition

$$y_{it} = \alpha + \rho y_{i,t-1} + \epsilon'_{it} \tag{26}$$

where $\epsilon'_{it} = u_i + \epsilon_{it}$. The parameter ρ continues to be identified by (25), so the only instrument we use with (26) is a constant term. As before, the error term $(\epsilon_{i,t} - \epsilon_{i,t-1})$ is necessarily autocorrelated with correlation coefficient -0.5, though the error term ϵ'_{it} is white noise. Therefore, our initial weight matrix should be

$$\widehat{\mathbf{W}} = \left(\frac{1}{N} \sum_i \mathbf{Z}'_i \mathbf{H} \mathbf{Z}_i \right)^{-1}$$

where

$$\mathbf{H} = \begin{bmatrix} \mathbf{H}_D & \mathbf{0} \\ \mathbf{0} & \mathbf{I} \end{bmatrix}$$

and $\mathbf{I}$ is a conformable identity matrix.

One complication arises concerning the relevant estimation sample. Looking at (25), we apparently lose the first two observations from each panel because of the presence of $y_{i,t-2}$, but in (26) we need only to sacrifice one observation, for $y_{i,t-1}$. For most multiple-equation models, we need to use the same estimation sample for all equations. However, in dynamic panel-data models, we can use more observations to fit the equation in level form [(26) here] than the equation in first-differences [equation (25)]. To request this behavior, we specify the nocommonsample option to gmm. That option tells gmm to use as many observations as possible for each equation, ignoring the loss of observations due to lagging or differencing.

▷ Example 16: Arellano–Bond estimator with constant term

Here we fit the model

$$\mathbf{n}_{it} = \alpha + \rho \, \mathbf{n}_{i,t-1} + u_i + \epsilon_{it}$$

Without specifying derivatives, our command would be

```
. gmm (D.n - {rho}*LD.n) (n - {alpha} - {rho}*L.n),
> xtinstruments(1: n, lags(2/.)) instruments(1:, noconstant) onestep
> winitial(xt DL) vce(unadj) nocommonsample
```

We would specify winitial(xt DL) to obtain the required initial weight matrix. The notation DL indicates that our first moment equation is in first-differences and the second moment equation is in levels (not first-differenced). We exclude a constant in the instrument list for the first equation, because first-differencing removed the constant term. Because we do not specify the instruments() option for the second moment equation, a constant is used by default.

This example also provides us the opportunity to illustrate how to specify derivatives for multiple-equation GMM models. Within the `derivative()` option, instead of specifying just the parameter name, now you must specify the equation name or number, a slash, and the parameter name to which the derivative applies. In Stata, we type

```
. gmm (D.n - {rho}*LD.n) (n - {alpha} - {rho}*L.n),
> xtinstruments(1: n, lags(2/.)) instruments(1:, noconstant)
> derivative(1/rho = -1*LD.n) derivative(2/alpha = -1)
> derivative(2/rho = -1*L.n) winitial(xt DL) vce(unadj) nocommonsample onestep
Step 1
Iteration 0:   GMM criterion Q(b) =   .09894466
Iteration 1:   GMM criterion Q(b) =   .00023508
Iteration 2:   GMM criterion Q(b) =   .00023508

GMM estimation

Number of parameters =    2
Number of moments    =   29
Initial weight matrix: XT DL                    Number of obs  =   *
```

	Coef.	Std. Err.	z	P>|z|	[95% Conf. Interval]
/rho	1.023349	.0608293	16.82	0.000	.9041259	1.142572
/alpha	-.0690864	.0660343	-1.05	0.295	-.1985112	.0603384

```
* Number of observations for equation 1: 751
  Number of observations for equation 2: 891
```

```
Instruments for equation 1:
        XT-style: L(2/.).n
Instruments for equation 2:
        Standard: _cons
```

These results are identical to those we would obtain by typing

```
xtabond n, lags(1)
```

Because we specified `nocommonsample`, gmm did not report the number of observations used in the header of the output. In this dataset, there are in fact 1,031 observations on 140 panels. In the second equation, the presence of the lagged value of n reduces the sample size for that equation to $1031 - 140 = 891$. In the first equation, we lose the first two observations per panel due to lagging and differencing, leading to 751 usable observations. These tallies are listed after the coefficient table in the output.

◁

❏ Technical note

Specifying

```
xtinstruments(x1 x2 x3, lags(1/3))
```

differs from

```
instruments(L(1/3).(x1 x2 x3))
```

in how observations are excluded from the estimation sample. When you use the latter syntax, gmm must exclude the first three observations from each panel when computing the moment equation: you requested three lags of each regressor be used as instruments, so the first residual that could be interacted with those instruments is the one for $t = 4$. On the other hand, when you use `xtinstruments()`, you are telling gmm that you would like to use up to the first three lags of x1, x2, and x3 as instruments but that using just one lag is acceptable. Because most panel datasets have a relatively modest number of observations per panel, dynamic instrument lists are typically used so that the number of usable observations is maximized. Dynamic instrument lists also accommodate the fact that there are more valid instruments for later time periods than earlier time periods.

Specifying panel-style instruments using the `xtinstruments()` option also affects how the standard instruments specified in the `instruments()` option are treated. To illustrate, suppose we have a balanced panel dataset with $T = 5$ observations per panel and we specify

> . gmm ..., xtinstruments(w, lags(1/2)) instruments(x)

We will lose the first observation because we need at least one lag of `w` to serve as an instrument. Our instrument matrix for panel i will therefore be

$$
\mathbf{Z}_i = \begin{bmatrix}
w_{i1} & 0 & 0 & 0 \\
0 & w_{i1} & 0 & 0 \\
0 & w_{i2} & 0 & 0 \\
0 & 0 & w_{i2} & 0 \\
0 & 0 & w_{i3} & 0 \\
0 & 0 & 0 & w_{i3} \\
0 & 0 & 0 & w_{i4} \\
x_{i2} & x_{i3} & x_{i4} & x_{i5} \\
1 & 1 & 1 & 1
\end{bmatrix} \tag{27}
$$

The vector of ones in the final row represents the constant term implied by the `instruments()` option. Because we lost the first observation, the residual vector $\mathbf{u}_i$ will be 4×1. Thus our moment conditions for the ith panel can be written in matrix notation as

$$
E\{\mathbf{Z}_i\mathbf{u}_i(\beta)\} = E\left\{ \mathbf{Z}_i \begin{bmatrix} u_{i2}(\beta) \\ u_{i3}(\beta) \\ u_{i4}(\beta) \\ u_{i5}(\beta) \end{bmatrix} \right\} = \mathbf{0}
$$

The moment conditions corresponding to the final two rows of (27) say that

$$
E\left\{ \sum_{t=2}^{T=4} x_{it}u_{it}(\beta) \right\} = 0 \quad \text{and} \quad E\left\{ \sum_{t=2}^{T=4} u_{it}(\beta) \right\} = 0
$$

Because we specified panel-style instruments with the `xtinstruments()` option, gmm no longer uses moment conditions for strictly exogenous variables of the form $E\{x_{it}u_{it}(\beta)\} = 0$ for each t. Instead, the moment conditions now stipulate that the average (over t) of $x_{it}u_{it}(\beta)$ has expectation zero. This corresponds to the approach proposed by Arellano and Bond (1991, 280) and others.

When you request panel-style instruments with the `xtinstruments()` option, the number of instruments in the $\mathbf{Z}_i$ matrix increases quadratically in the number of periods. The dynamic panel-data estimators we have discussed in this section are designed for datasets that contain a large number of panels and a modest number of time periods. When the number of time periods is large, estimators that use standard (non–panel-style) instruments are more appropriate.

❏

We have focused on the Arellano–Bond dynamic panel-data estimator because of its relative simplicity. gmm can additionally fit any models that can be formulated using the xtdpd and xtdpdsys commands; see [XT] **xtdpd** and [XT] **xtdpdsys**. The key is to determine the appropriate instruments to use for the level and difference equations. You may find it useful to fit a version of your model with those commands to determine what instruments and XT-style instruments to use. We conclude this section with an example using the Arellano–Bover/Blundell–Bond estimator.

▷ Example 17: Arellano–Bover/Blundell–Bond estimator

We fit a small model that includes one lag of the dependent variable n as a regressor as well as the contemporaneous and first lag of w, which we assume are strictly exogenous. We could fit our model using xtdpdsys using the syntax

```
xtdpdsys n L(0/1).w, lags(1) twostep
```

Applying virtually all the syntax issues we have discussed so far, the equivalent gmm command is

```
. gmm (n - {rho}*L.n - {w}*w - {lagw}*L.w - {c})
>     (D.n - {rho}*LD.n - {w}*D.w - {lagw}*LD.w),
>     xtinst(1: D.n, lags(1/1)) xtinst(2: n, lags(2/.))
>     inst(2: D.w LD.w, noconstant)
>     deriv(1/rho = -1*L.n)
>     deriv(1/w = -1*w)
>     deriv(1/lagw = -1*L.w)
>     deriv(1/c = -1)
>     deriv(2/rho = -1*LD.n)
>     deriv(2/w = -1*D.w)
>     deriv(2/lagw = -1*LD.w)
>     winit(xt LD) wmatrix(robust) vce(unadjusted) nocommonesample
Step 1
Iteration 0:   GMM criterion Q(b) =  .10170339
Iteration 1:   GMM criterion Q(b) =  .00022772
Iteration 2:   GMM criterion Q(b) =  .00022772
Step 2
Iteration 0:   GMM criterion Q(b) =  .59965014
Iteration 1:   GMM criterion Q(b) =  .56578186
Iteration 2:   GMM criterion Q(b) =  .56578186
GMM estimation

Number of parameters =    4
Number of moments     =   39
Initial weight matrix: XT LD              Number of obs  =    *
GMM weight matrix:     Robust
```

	Coef.	Std. Err.	z	P>|z|	[95% Conf. Interval]	
/rho	1.122738	.0206512	54.37	0.000	1.082263	1.163214
/w	-.6719909	.0246148	-27.30	0.000	-.7202351	-.6237468
/lagw	.571274	.0403243	14.17	0.000	.4922398	.6503083
/c	.154309	.17241	0.90	0.371	-.1836084	.4922263

```
* Number of observations for equation 1: 891
  Number of observations for equation 2: 751

Instruments for equation 1:
        XT-style: LD.n
        Standard: _cons
Instruments for equation 2:
        XT-style: L(2/.).n
        Standard: D.w LD.w
```

◁

Details of moment-evaluator programs

In examples 9, 10, and 11, we used moment-evaluator programs to evaluate moment conditions that could not be specified using the interactive version of gmm. In the technical note after example 9, we showed how to make our program accept optional arguments so that we could pass the name of our dependent and independent variables and panel identifier variable. Here we discuss how to make moment-evaluator programs accept weights and provide derivatives.

The complete specification for a moment-evaluator program's syntax statement is

```
syntax varlist if [weight], at(name) options [derivatives(varlist)]
```

The macro 'varlist' contains the list of variables that we are to fill in with the values of our residual equations. The macro 'if' represents an if condition that restricts the estimation sample. The macro 'at' represents a vector containing the parameter values at which we are to evaluate our residual equations. options represent other options that you specify in your call to gmm and want to have passed to your moment-evaluator programs. In our previous examples, we included the mylhs(), myrhs(), and myidvar() options.

Two new elements of the syntax statement allow for weights and derivatives. weight specifies the types of weights your program allows. The interactive version of gmm allows for fweights, aweights, and pweights. However, unless you explicitly allow your moment evaluator program to accept weights, you cannot specify weights in your call to gmm with the moment-evaluator program version.

The derivatives() option is used to pass to your program a set of variables that you are to fill in with the derivatives of your residual equations with respect to the parameters. Say you specify k parameters in the nparameters() or parameters() option and q equations in the nequations() or equations() option. Then 'derivatives' will contain $k \times q$ variables. The first k variables are for the derivatives of the first residual equation with respect to the k parameters, the second k variables are for the derivatives of the second residual equation, and so on.

▷ Example 18: Panel Poisson with strictly exogenous regressors and derivatives

To focus on how to specify derivatives, we return to the simple moment-evaluator program we used in example 9, in which we had three regressors, and extend it to supply derivatives. The error equation corresponding to moment condition (13) is

$$u_{it}(\beta) = y_{it} - \mu_{it}\frac{\overline{y}_i}{\overline{\mu}_i}$$

where μ_{it}, $\overline{\mu}_i$, and $\overline{y}_i$ were defined previously. Now

$$\frac{\partial}{\partial \beta_j}u_{it}(\beta) = -\mu_{it}\frac{\overline{y}_i}{\overline{\mu}_i^2}\left(x_{it}^{(j)}\overline{\mu}_i - \frac{1}{T}\sum_{l=1}^{l=T}x_{il}^{(j)}\mu_{il}\right) \tag{28}$$

where $x_{it}^{(j)}$ represents the jth element of $\mathbf{x}_{it}$.

Our moment-evaluator program is

```
program gmm_poideriv
        version 12
        syntax varlist if, at(name) [derivatives(varlist)]
        quietly {
                // Calculate residuals as before
                tempvar mu mubar ybar
                gen double 'mu' = exp(x1*'at'[1,1] + x2*'at'[1,2]        ///
                                    + x3*'at'[1,3]) 'if'
                egen double 'mubar' = mean('mu') 'if', by(id)
                egen double 'ybar' = mean(y) 'if', by(id)
                replace 'varlist' = y - 'mu'*'ybar'/'mubar' 'if'
                // Did -gmm- request derivatives?
                if "'derivatives'" == "" {
                        exit                    // no, so we are done
                }
                // Calculate derivatives
                // We need the panel means of x1*mu, x2*mu, and x3*mu
                tempvar work x1mubar x2mubar x3mubar
                generate double 'work' = x1*'mu' 'if'
                egen double 'x1mubar' = mean('work') 'if', by(id)
                replace 'work' = x2*'mu' 'if'
                egen double 'x2mubar' = mean('work') 'if', by(id)
                replace 'work' = x3*'mu' 'if'
                egen double 'x3mubar' = mean('work') 'if', by(id)
                local d1: word 1 of 'derivatives'
                local d2: word 2 of 'derivatives'
                local d3: word 3 of 'derivatives'
                replace 'd1' = -1*'mu'*'ybar'/'mubar'^2*(x1*'mubar' - 'x1mubar')
                replace 'd2' = -1*'mu'*'ybar'/'mubar'^2*(x2*'mubar' - 'x2mubar')
                replace 'd3' = -1*'mu'*'ybar'/'mubar'^2*(x3*'mubar' - 'x3mubar')
        }
end
```

The derivatives() option is made optional in the syntax statement by placing it in square brackets. If gmm needs to evaluate your moment equations but does not need derivatives at that time, then the derivatives() option will be empty. In our program, we check to see if that is the case, and, if so, exit without calculating derivatives. As is often the case with [R] ml as well, the portion of our program devoted to derivatives is longer than the code to compute the objective function.

The first part of our derivative code computes the term

$$\frac{1}{T} \sum_{l=1}^{l=T} x_{il}^{(j)} \mu_{il} \tag{29}$$

for $x_{it}^{(j)} = $ x1, x2, and, x3. The 'derivatives' macro contains three variable names, corresponding to the three parameters of the 'at' matrix. We extract those names into local macros 'd1', 'd2', and 'd3', and then fill in the variables those macros represent with the derivatives shown in (28).

With our program written, we fit our model by typing

```
. use http://www.stata-press.com/data/r12/poisson1

. gmm gmm_poideriv, nequations(1) parameters(b1 b2 b3)
> instruments(x1 x2 x3, noconstant) vce(cluster id) onestep hasderivatives
Step 1
Iteration 0:   GMM criterion Q(b) =   51.99142
Iteration 1:   GMM criterion Q(b) =   .04345191
Iteration 2:   GMM criterion Q(b) =   8.720e-06
Iteration 3:   GMM criterion Q(b) =   7.115e-13
Iteration 4:   GMM criterion Q(b) =   5.130e-27

GMM estimation

Number of parameters =   3
Number of moments     =   3
Initial weight matrix: Unadjusted                Number of obs   =      409
                                        (Std. Err. adjusted for 45 clusters in id)
```

	Coef.	Robust Std. Err.	z	P>|z|	[95% Conf. Interval]	
/b1	1.94866	.1000265	19.48	0.000	1.752612	2.144709
/b2	-2.966119	.0923592	-32.12	0.000	-3.14714	-2.785099
/b3	1.008634	.1156561	8.72	0.000	.781952	1.235315

```
Instruments for equation 1: x1 x2 x3
```

Our results are identical to those in example 9. Another way to verify that our program calculates derivatives correctly would be to type

```
. gmm gmm_poideriv, nequations(1) parameters(b1 b2 b3)
>   instruments(x1 x2 x3, noconstant) vce(cluster id) onestep
```

Without the hasderivatives option, gmm will not request derivatives from your program, even if it contains code to compute them. If you have trouble obtaining convergence with the hasderivatives option but do not have trouble without that option, then you need to recheck your derivatives.

◁

In the technical note after example 9, we modified the gmm_poi program so that we could specify the name of our dependent variable and a list of regressors. Here is the analogous version that computes analytic derivatives:

```
program gmm_poideriv2
        version 12
        syntax varlist if, at(name) myrhs(varlist)                    ///
                            mylhs(varlist) myidvar(varlist)            ///
                            [ derivatives(varlist) ]

        quietly {
                tempvar mu mubar ybar

                gen double 'mu' = 0 'if'
                local j = 1
                foreach var of varlist 'myrhs' {
                        replace 'mu' = 'mu' + 'var'*'at'[1,'j'] 'if'
                        local j = 'j' + 1
                }
                replace 'mu' = exp('mu')
                egen double 'mubar' = mean('mu') 'if', by('myidvar')
                egen double 'ybar' = mean('mylhs') 'if', by('myidvar')
                replace 'varlist' = 'mylhs' - 'mu'*'ybar'/'mubar' 'if'

                if "'derivatives'" == "" {
                        exit
                }

                tempvar work xmubar
                local j = 1
                foreach var of varlist 'myrhs' {
                        generate double 'work' = 'var'*'mu' 'if'
                        egen double 'xmubar' = mean('work') 'if' 'wt''exp', ///
                                by('myidvar')
                        local deriv : word 'j' of 'derivatives'
                        replace 'deriv' = -1*'mu'*'ybar'/'mubar'^2*        ///
                                ('var'*'mubar' - 'xmubar')
                        local '++j'
                        drop 'work' 'xmubar'
                }
        }
end
```

To use this program, we type

```
gmm gmm_poideriv2, mylhs(y) myrhs(x1 x2 x3)
        nequations(1) parameters(b1 b2 b3)
        instruments(x1 x2 x3, noconstant)
        vce(cluster id) myidvar(id) onestep
        hasderivatives
```

We obtain results identical to those shown in example 18.

Depending on your model, allowing your moment-evaluator program to accept weights may be as easy as modifying the syntax command to allow them, or it may require significantly more work. If your program uses only commands like generate and replace, then just modifying the syntax command is all you need to do; gmm takes care of applying the weights to the observation-level residuals when computing the sample moments, derivatives, and weight matrices. On the other hand, if your moment-evaluator program computes residuals using statistics that depend on multiple observations, then you must apply the weights passed to your program when computing those statistics.

In our examples of panel Poisson with strictly exogenous regressors, we used the statistics $\bar{\mu}_i$ and $\bar{y}_i$ when computing the residuals. If we are to allow weights with our moment-evaluator program, then we must incorporate those weights when computing $\bar{\mu}_i$ and $\bar{y}_i$. Moreover, looking at the derivative in (28), the term highlighted in (29) is in fact a sample mean, so we must incorporate weights when computing it.

▷ Example 19: Panel Poisson with derivatives and weights

Here we modify the program in example 18. to accept frequency weights. One complication immediately arises: we had been using egen to compute $\overline{\mu}_i$ and $\overline{y}_i$. egen does not accept weights, so we must compute $\overline{\mu}_i$ and $\overline{y}_i$ ourselves, incorporating any weights the user may specify. Our program is

```
program gmm_poiderivfw
        version 12
        syntax varlist if [fweight/], at(name) [derivatives(varlist)]
        quietly {
                if "'exp'" == "" {          // no weights
                        local exp 1         // weight each observation equally
                }
                // Calculate residuals as before
                tempvar mu mubar ybar sumwt
                gen double 'mu' = exp(x1*'at'[1,1] + x2*'at'[1,2]      ///
                                        + x3*'at'[1,3]) 'if'
                bysort id: gen double 'sumwt' = sum('exp')
                by id: gen double 'mubar' = sum('mu'*'exp')
                by id: gen double 'ybar' = sum(y*'exp')
                by id: replace 'mubar' = 'mubar'[_N] / 'sumwt'[_N]
                by id: replace 'ybar' = 'ybar'[_N] / 'sumwt'[_N]

                replace 'varlist' = y - 'mu'*'ybar'/'mubar' 'if'

                // Did -gmm- request derivatives?
                if "'derivatives'" == "" {
                        exit         // no, so we are done
                }
                // Calculate derivatives
                // We need the panel means of x1*mu, x2*mu, and x3*mu
                tempvar work x1mubar x2mubar x3mubar
                generate double 'work' = x1*'mu' 'if'
                by id: generate double 'x1mubar' = sum('work'*'exp')
                by id: replace 'x1mubar' = 'x1mubar'[_N] / 'sumwt'[_N]

                replace 'work' = x2*'mu' 'if'
                by id: generate double 'x2mubar' = sum('work'*'exp')
                by id: replace 'x2mubar' = 'x2mubar'[_N] / 'sumwt'[_N]

                replace 'work' = x3*'mu' 'if'
                by id: generate double 'x3mubar' = sum('work'*'exp')
                by id: replace 'x3mubar' = 'x3mubar'[_N] / 'sumwt'[_N]

                local d1: word 1 of 'derivatives'
                local d2: word 2 of 'derivatives'
                local d3: word 3 of 'derivatives'
                replace 'd1' = -1*'mu'*'ybar'/'mubar'^2*(x1*'mubar' - 'x1mubar')
                replace 'd2' = -1*'mu'*'ybar'/'mubar'^2*(x2*'mubar' - 'x2mubar')
                replace 'd3' = -1*'mu'*'ybar'/'mubar'^2*(x3*'mubar' - 'x3mubar')
        }
end
```

Our syntax command now indicates that fweights are allowed. The first part of our code looks at the macro 'exp'. If it is empty, then the user did not specify weights in their call to gmm; and we set the macro equal to 1, so that we weight each observation equally. After we compute μ_{it}, we calculate $\overline{\mu}_i$ and $\overline{y}_i$, taking into account weights. To compute frequency-weighted means for each panel, we just multiply each observation by its respective weight, sum over all observations in the panel, then divide by the sum of the weights for the panel. (See [U] **20.22 Weighted estimation** for information on how to handle aweights and pweights.) We use the same procedure to compute the frequency-weighted variant of expression (29) in the derivative calculations.

To use our program, we type

```
. use http://www.stata-press.com/data/r12/poissonwts

. gmm gmm_poiderivfw [fw=fwt], nequations(1) parameters(b1 b2 b3)
> instruments(x1 x2 x3, noconstant) vce(cluster id) onestep hasderivatives
(sum of wgt is 819)
Step 1
Iteration 0:    GMM criterion Q(b) =    49.8292
Iteration 1:    GMM criterion Q(b) =  .11136736
Iteration 2:    GMM criterion Q(b) =  .00008519
Iteration 3:    GMM criterion Q(b) =  7.110e-11
Iteration 4:    GMM criterion Q(b) =  5.596e-23

GMM estimation

Number of parameters =   3
Number of moments    =   3
Initial weight matrix: Unadjusted                   Number of obs   =      819
                                       (Std. Err. adjusted for 45 clusters in id)
```

	Coef.	Robust Std. Err.	z	P>|z|	[95% Conf.	Interval]
/b1	1.967766	.111795	17.60	0.000	1.748652	2.186881
/b2	-3.060838	.0935561	-32.72	0.000	-3.244205	-2.877472
/b3	1.037594	.1184227	8.76	0.000	.80549	1.269698

```
Instruments for equation 1: x1 x2 x3
```

Testing whether our program works correctly with frequency weights is easy. A frequency-weighted dataset is just a compact form of a larger dataset in which identical observations are omitted and a frequency-weight variable is included to tell us how many times each observation in the smaller dataset appears in the larger dataset. Therefore, we can expand our smaller dataset by the frequency-weight variable and then refit our model without specifying frequency weights. If we obtain the same results, our program works correctly. When we type

```
. expand fw
(410 observations created)

. gmm gmm_poiderivfw, nequations(1) parameters(b1 b2 b3)
> instruments(x1 x2 x3, noconstant) vce(cluster id) onestep
```

we obtain the same results as before.

◁

Saved results

gmm saves the following in e():

Scalars
e(N)	number of observations
e(k)	number of parameters
e(k_eq)	number of equations in e(b)
e(k_eq_model)	number of equations in overall model test
e(k_aux)	number of auxiliary parameters
e(n_moments)	number of moments
e(n_eq)	number of equations in moment-evaluator program
e(Q)	criterion function
e(J)	Hansen J χ^2 statistic
e(J_df)	J statistic degrees of freedom
e(k_i)	number of parameters in equation i
e(has_xtinst)	1 if panel-style instruments specified, 0 otherwise
e(N_clust)	number of clusters
e(type)	1 if interactive version, 2 if moment-evaluator program version
e(rank)	rank of e(V)
e(ic)	number of iterations used by iterative GMM estimator
e(converged)	1 if converged, 0 otherwise

Macros
e(cmd)	gmm
e(cmdline)	command as typed
e(title)	title specified in title()
e(title_2)	title specified in title2()
e(clustvar)	name of cluster variable
e(inst_i)	equation i instruments
e(eqnames)	equation names
e(winit)	initial weight matrix used
e(winitname)	name of user-supplied initial weight matrix
e(estimator)	onestep, twostep, or igmm
e(rhs)	variables specified in variables()
e(params_i)	equation i parameters
e(wmatrix)	*wmtype* specified in wmatrix()
e(vce)	*vcetype* specified in vce()
e(vcetype)	title used to label Std. Err.
e(params)	parameter names
e(sexp_i)	substitutable expression for equation i
e(evalprog)	moment-evaluator program
e(evalopts)	options passed to moment-evaluator program
e(nocommonesample)	nocommonesample, if specified
e(technique)	optimization technique
e(properties)	b V
e(estat_cmd)	program used to implement estat
e(predict)	program used to implement predict
e(marginsnotok)	predictions disallowed by margins

Matrices
e(b)	coefficient vector
e(init)	initial values of the estimators
e(Wuser)	user-supplied initial weight matrix
e(W)	weight matrix used for final round of estimation
e(S)	moment covariance matrix used in robust VCE computations
e(N_byequation)	number of observations per equation, if nocommonesample specified
e(V)	variance–covariance matrix
e(V_modelbased)	model-based variance

Functions
e(sample)	marks estimation sample

Methods and formulas

gmm is implemented as an ado-file.

Let q denote the number of moment equations. For observation i, $i = 1, \ldots, N$, write the jth moment equation as $\mathbf{z}_{ij} u_{ij}(\beta_j)$ for $j = 1, \ldots, q$. $\mathbf{z}_{ij}$ is a $1 \times m_j$ vector, where m_j is the number of instruments specified for equation j. Let $m = m_1 + \cdots + m_q$.

Our notation can incorporate moment conditions of the form $h_{ij}(\mathbf{w}_{ij}; \beta_j)$ with instruments $\mathbf{w}_{ij}$ by defining $\mathbf{z}_{ij} = 1$ and $u_{ij}(\beta_j) = h_{ij}(\mathbf{w}_{ij}; \beta_j)$, so except when necessary we do not distinguish between the two types of moment conditions. We could instead use notation so that all our moment conditions are of the form $h_{ij}(\mathbf{w}_{ij}; \beta_j)$, or we could adopt notation that explicitly combines both forms of moment equations. However, because moment conditions of the form $\mathbf{z}'_{ij} u_{ij}(\beta_j)$ are arguably more common, we use that notation.

Let β denote a $k \times 1$ vector of parameters, consisting of all the unique parameters of $\beta_1, \ldots, \beta_q$. Then we can stack the moment conditions and write them more compactly as $\mathbf{Z}'_i \mathbf{u}_i(\beta)$, where

$$
\mathbf{Z}_i = \begin{bmatrix} \mathbf{z}_{i1} & \mathbf{0} & \cdots & \mathbf{0} \\ \mathbf{0} & \mathbf{z}_{i2} & \cdots & \mathbf{0} \\ \vdots & \vdots & \ddots & \vdots \\ \mathbf{0} & \mathbf{0} & \cdots & \mathbf{z}_{iq} \end{bmatrix} \quad \text{and} \quad \mathbf{u}_i(\beta) = \begin{bmatrix} u_{i1}(\beta_1) \\ u_{i2}(\beta_2) \\ \vdots \\ u_{iq}(\beta_j) \end{bmatrix}
$$

The GMM estimator $\widehat{\beta}$ is the value of β that minimizes

$$
Q(\beta) = \left\{ N^{-1} \sum_{i=1}^{N} \mathbf{Z}'_i \mathbf{u}_i(\beta) \right\}' \mathbf{W} \left\{ N^{-1} \sum_{i=1}^{N} \mathbf{Z}'_i \mathbf{u}_i(\beta) \right\} \tag{A1}
$$

for $q \times q$ weight matrix $\mathbf{W}$.

By default, gmm minimizes (A1) using the Gauss–Newton method. See Hayashi (2000, 498) for a derivation. This technique is typically faster than quasi-Newton methods and does not require second-order derivatives.

Methods and formulas are presented under the following headings:

> Initial weight matrix
> Weight matrix
> Variance–covariance matrix
> Hansen's J statistic
> Panel-style instruments

Initial weight matrix

If you specify winitial(identity), then we set $\mathbf{W} = \mathbf{I}_q$.

If you specify winitial(unadjusted), then we create matrix $\mathbf{\Lambda}$ with typical submatrix

$$
\mathbf{\Lambda}_{rs} = N^{-1} \sum_{i=1}^{N} \mathbf{z}'_{ir} \mathbf{z}_{is}
$$

for $r = 1, \ldots, q$ and $s = 1, \ldots, q$. If you include the independent suboption, then we set $\mathbf{\Lambda}_{rs} = \mathbf{0}$ for $r \neq s$. The weight matrix $\mathbf{W}$ equals $\mathbf{\Lambda}^{-1}$.

If you specify `winitial`(*matname*), then we set $\mathbf{W}$ equal to Stata matrix *matname*.

If you specify `winitial`(`xt` *xtspec*), then you must specify one or two items in *xtspec*, one for each equation. gmm allows you to specify at most two moment equations when you specify `winitial`(`xt` *xtspec*), one in first-differences and one in levels. We create the block-diagonal matrix $\mathbf{H}$ with typical block $\mathbf{H}_j$. If the jth element of *xtspec* is "L", then $\mathbf{H}_j$ is the identity matrix of suitable dimension. If the jth element of *xtspec* is "D", then

$$
\mathbf{H}_j = \begin{bmatrix} 1 & -0.5 & 0 & \dots & 0 & 0 \\ -0.5 & 1 & -0.5 & \dots & 0 & 0 \\ \vdots & \vdots & \vdots & \ddots & \vdots & \vdots \\ 0 & 0 & 0 & \dots & 1 & -0.5 \\ 0 & 0 & 0 & \dots & -0.5 & 1 \end{bmatrix}
$$

Then

$$
\boldsymbol{\Lambda}_H = N_g^{-1} \sum_{g=1}^{g=N_G} \mathbf{Z}_g' \mathbf{H} \mathbf{Z}_g
$$

where g indexes panels in the dataset, N_G is the number of panels, $\mathbf{Z}_g$ is the full instrument matrix for panel g, and $\mathbf{W} = \boldsymbol{\Lambda}_H^{-1}$. See *Panel-style instruments* below for a discussion of how $\mathbf{Z}_g$ is formed.

Weight matrix

Specification of the weight matrix applies only to the two-step and iterative estimators. When you use the `onestep` option, the `wmatrix()` option is ignored.

We first evaluate (A1) using the initial weight matrix described above and then compute $\mathbf{u}_i(\widehat{\beta})$. In all cases, $\mathbf{W} = \boldsymbol{\Lambda}^{-1}$. If you specify `wmatrix`(`unadjusted`), then we create $\boldsymbol{\Lambda}$ to have typical submatrix

$$
\boldsymbol{\Lambda}_{rs} = \sigma_{rs} N^{-1} \sum_{i=1}^{N} \mathbf{z}_{ir}' \mathbf{z}_{is}
$$

where

$$
\sigma_{rs} = N^{-1} \sum_{i=1}^{N} u_{ir}(\widehat{\beta}) u_{is}(\widehat{\beta})
$$

and r and s index moment equations. For all types of weight matrices, if the `independent` suboption is specified, then $\boldsymbol{\Lambda}_{rs} = \mathbf{0}$ for $r \neq s$, where $\boldsymbol{\Lambda}_{rs}$ measures the covariance between moment conditions for equations r and s.

If you specify `wmatrix`(`robust`), then

$$
\boldsymbol{\Lambda} = N^{-1} \sum_{i=1}^{N} \mathbf{Z}_i \mathbf{u}_i(\widehat{\beta}) \mathbf{u}_i'(\widehat{\beta}) \mathbf{Z}_i'
$$

If you specify `wmatrix`(`cluster` *clustvar*), then

$$
\boldsymbol{\Lambda} = N^{-1} \sum_{c=1}^{c=N_C} \mathbf{q}_c \mathbf{q}_c'
$$

where c indexes clusters, N_C is the number of clusters, and

$$\mathbf{q}_c = \sum_{i \in c_j} \mathbf{Z}_i \mathbf{u}_i(\widehat{\beta})$$

If you specify `wmatrix(hac` *kernel* $\left[\#\right]$`)`, then

$$\mathbf{\Lambda} = N^{-1} \sum_{i=1}^{N} \mathbf{Z}_i \mathbf{u}_i(\widehat{\beta}) \mathbf{u}_i(\widehat{\beta})' \mathbf{Z}_i' \quad +$$

$$N^{-1} \sum_{l=1}^{l=n-1} \sum_{i=l+1}^{N} K(l,m) \left\{ \mathbf{Z}_i \mathbf{u}_i(\widehat{\beta}) \mathbf{u}_{i-l}'(\widehat{\beta}) \mathbf{Z}_{i-l}' + \mathbf{Z}_{i-l} \mathbf{u}_{i-l}(\widehat{\beta}) \mathbf{u}_i'(\widehat{\beta}) \mathbf{Z}_i' \right\}$$

where $m = \#$ if $\#$ is specified and $m = N - 2$ otherwise. Define $z = l/(m + 1)$. If *kernel* is `bartlett` or `nwest`, then

$$K(l,m) = \begin{cases} 1 - z & 0 \leq z \leq 1 \\ 0 & \text{otherwise} \end{cases}$$

If *kernel* is `parzen` or `gallant`, then

$$K(l,m) = \begin{cases} 1 - 6z^2 + 6z^3 & 0 \leq z \leq 0.5 \\ 2(1 - z)^3 & 0.5 < z \leq 1 \\ 0 & \text{otherwise} \end{cases}$$

If *kernel* is `quadraticspectral` or `andrews`, then

$$K(l,m) = \begin{cases} 1 & z = 0 \\ 3\{\sin(\theta)/\theta - \cos(\theta)\}/\theta^2 & \text{otherwise} \end{cases}$$

where $\theta = 6\pi z/5$.

If `wmatrix(hac` *kernel* `opt)` is specified, then `gmm` uses Newey and West's (1994) automatic lag-selection algorithm, which proceeds as follows. Define $\mathbf{h}$ to be an $m \times 1$ vector of ones. Note that this definition of $\mathbf{h}$ is slightly different than the one used by `ivregress`. There, the element of $\mathbf{h}$ corresponding to the constant term equals zero, effectively ignoring the effect of the constant in determining the optimal lag length. Here we include the effect of the constant term. Now define

$$f_i = \{\mathbf{Z}_i' \mathbf{u}_i(\beta)\}' \mathbf{h}$$

$$\widehat{\sigma}_j = N^{-1} \sum_{i=j+1}^{N} f_i f_{i-j} \qquad j = 0, \ldots, m^*$$

$$\widehat{s}^{(q)} = 2 \sum_{j=1}^{j=m^*} \widehat{\sigma}_j j^q$$

$$\widehat{s}^{(0)} = \widehat{\sigma}_0 + 2 \sum_{j=1}^{j=m^*} \widehat{\sigma}_j$$

$$\widehat{\gamma} = c_\gamma \left\{ \left(\frac{\widehat{s}^{(q)}}{\widehat{s}^{(0)}} \right)^2 \right\}^{1/(2q+1)}$$

$$m = \widehat{\gamma} N^{1/(2q+1)}$$

where q, m^*, and c_γ depend on the kernel specified:

Kernel	q	m^*	c_γ
Bartlett/Newey–West	1	$\text{int}\left\{20(T/100)^{2/9}\right\}$	1.1447
Parzen/Gallant	2	$\text{int}\left\{20(T/100)^{4/25}\right\}$	2.6614
Quadratic spectral/Andres	2	$\text{int}\left\{20(T/100)^{2/25}\right\}$	1.3221

where $\text{int}(x)$ denotes the integer obtained by truncating x toward zero. For the Bartlett and Parzen kernels, the optimal lag is $\min\{\text{int}(m), m^*\}$. For the quadratic spectral kernel, the optimal lag is $\min\{m, m^*\}$.

Variance–covariance matrix

If you specify vce(unadjusted), then the VCE matrix is computed as

$$\text{Var}(\widehat{\beta}) = N^{-1}\left\{\overline{\mathbf{G}}(\widehat{\beta})'\mathbf{W}\overline{\mathbf{G}}(\widehat{\beta})\right\}^{-1} \tag{A2}$$

where

$$\overline{\mathbf{G}}(\widehat{\beta}) = N^{-1}\sum_{i=1}^{N}\mathbf{Z}_i'\left.\frac{\partial\mathbf{u}_i(\beta)}{\partial\beta'}\right|_{\beta=\widehat{\beta}}$$

For the two-step and iterated estimators, we use the weight matrix $\mathbf{W}$ that was used to compute the final-round estimate $\widehat{\beta}$.

For the one-step estimator, how the unadjusted VCE is computed depends on the type of initial weight matrix requested and the form of the moment equations. If you specify two or more moment equations of the form $h_{ij}(\mathbf{w}_{ij}; \beta_j)$, then gmm issues a warning message and computes a heteroskedasticity-robust VCE because here the matrix $\mathbf{Z}'\mathbf{Z}$ is necessarily singular; moreover, here you must use the identity matrix as the initial weight matrix. Otherwise, if you specify winitial(identity) or winitial(unadjusted), then gmm first computes an unadjusted weight matrix based on $\widehat{\beta}$ before evaluating (A2). If you specify winitial(*matname*), then (A2) is evaluated based on *matname*; the user is responsible for verifying that the VCE and other statistics so produced are appropriate.

All types of robust VCEs computed by gmm take the form

$$\text{Var}(\widehat{\beta}) = N^{-1}\left\{\overline{\mathbf{G}}(\widehat{\beta})'\mathbf{W}\overline{\mathbf{G}}(\widehat{\beta})\right\}^{-1}\overline{\mathbf{G}}(\widehat{\beta})'\mathbf{W}\mathbf{S}\mathbf{W}\overline{\mathbf{G}}(\widehat{\beta})\left\{\overline{\mathbf{G}}(\widehat{\beta})'\mathbf{W}\overline{\mathbf{G}}(\widehat{\beta})\right\}^{-1}$$

For the one-step estimator, $\mathbf{W}$ represents the initial weight matrix requested using the winitial() option, and $\mathbf{S}$ is computed based on the specification of the vce() option. The formulas for the $\mathbf{S}$ matrix are identical to the ones that define the $\mathbf{\Lambda}$ matrix in *Weight matrix* above, except that $\mathbf{S}$ is computed after the moment equations are reevaluated using the final estimate of $\widehat{\beta}$. For the two-step and iterated GMM estimators, computation of $\mathbf{W}$ is controlled by the wmatrix() option based on the penultimate estimate of $\widehat{\beta}$.

For details on computation of the VCE matrix with dynamic panel-data models, see *Panel-style instruments* below.

Hansen's J statistic

Hansen's (1982) J test of overidentifying restrictions is $J = N \times Q(\widehat{\beta})$. $J \sim \chi^2(m - k)$. If $m < k$, gmm issues an error message without estimating the parameters. If $m = k$, the model is just-identified and J is saved as missing ("."). For the two-step and iterated GMM estimators, the J statistic is based on the last-computed weight matrix as determined by the wmatrix() option. For the one-step estimator, gmm recomputes a weight matrix as described in the second paragraph of *Variance–covariance matrix* above. To obtain Hansen's J statistic, you use estat overid; see [R] **gmm postestimation**.

Panel-style instruments

Here we discuss several issues that arise only when you specify panel-style instruments by using the xtinstruments() option. When you specify the xtinstruments() option, we can no longer consider the instruments for one observation in isolation; instead, we must consider the instrument matrix for an entire panel at once. In the following discussion, we let T denote the number of time periods in a panel. To accommodate unbalanced datasets, conceptually we simply use zeros as instruments and residuals for time periods that are missing in a panel.

We consider the case where you specify both an equation in levels and an equation in differences, yielding two residual equations. Let $u_{pt}^L(\beta)$ denote the residual for the level equation for panel p in period t, and let $u_{pt}^D(\beta)$ denote the residual for the corresponding difference equation. Now define the $2T \times 1$ vector $\mathbf{u}_p(\beta)$ as

$$\mathbf{u}_p(\beta) = [u_{p1}^L(\beta), u_{p2}^L(\beta), \ldots, u_{pT}^L(\beta), u_{p2}^D(\beta), u_{p3}^D(\beta), \ldots, u_{pT}^D(\beta)]$$

The $T + 1$ element of $\mathbf{u}_p$ is $u_{p2}^D(\beta)$ since we lose the first observation of the difference equation because of differencing.

We write the moment conditions for the pth panel as $\mathbf{Z}_p \mathbf{u}_p(\beta)$. To see how $\mathbf{Z}_p$ is defined, let $\mathbf{w}_{pt}^L$ and $\mathbf{w}_{pt}^D$ denote the vectors of panel-style instruments for the level and difference equations, respectively, and let time be denoted by t; we discuss their dimensions momentarily. Also let $\mathbf{x}_{pt}^L$ and $\mathbf{x}_{pt}^D$ denote the vectors of instruments specified in instruments() for the level and difference equations at time t. Without loss of generality, for our discussion we assume that you specify the level equation first. Then $\mathbf{Z}_p$ has the form

$$
\mathbf{Z}_p =
\begin{bmatrix}
\mathbf{w}_1^L & \mathbf{0} & \cdots & \mathbf{0} & \mathbf{0} & \mathbf{0} & \cdots & \mathbf{0} \\
\mathbf{0} & \mathbf{w}_2^L & \cdots & \mathbf{0} & \mathbf{0} & \mathbf{0} & \cdots & \mathbf{0} \\
\vdots & \vdots & \ddots & \vdots & \vdots & \vdots & \ddots & \vdots \\
\mathbf{0} & \mathbf{0} & \cdots & \mathbf{w}_T^L & \mathbf{0} & \mathbf{0} & \cdots & \mathbf{0} \\
\mathbf{x}_1^L & \mathbf{x}_2^L & \cdots & \mathbf{x}_T^L & \mathbf{0} & \mathbf{0} & \cdots & \mathbf{0} \\
\mathbf{0} & \mathbf{0} & \cdots & \mathbf{0} & \mathbf{w}_1^D & \mathbf{0} & \cdots & \mathbf{0} \\
\mathbf{0} & \mathbf{0} & \cdots & \mathbf{0} & \mathbf{0} & \mathbf{w}_2^D & \cdots & \mathbf{0} \\
\vdots & \vdots & \ddots & \vdots & \vdots & \vdots & \ddots & \vdots \\
\mathbf{0} & \mathbf{0} & \cdots & \mathbf{0} & \mathbf{0} & \mathbf{0} & \cdots & \mathbf{w}_T^D \\
\mathbf{0} & \mathbf{0} & \cdots & \mathbf{0} & \mathbf{x}_1^D & \mathbf{x}_2^D & \cdots & \mathbf{x}_T^D
\end{bmatrix}
\tag{A3}
$$

To see how the $\mathbf{w}$ vectors are formed, suppose you specify

```
xtinstruments(eq(1): d, lags( a/ b))
```

Then $\mathbf{w}_t^L$ will be a $(b - a + 1) \times 1$ vector consisting of $d_{t-a}, \ldots, d_{t-b}$. If $(t - a) \leq 0$, then instead we set $\mathbf{w}_t^L = 0$. If $(t - a) > 0$ but $(t - b) \leq 0$, then we create $\mathbf{w}_t^L$ to consist of $d_{t-a}, \ldots, d_1$. With this definition, $(b - a + 1)$ defines the maximum number of lags of d used, but gmm will proceed with fewer lags if all $(b - a + 1)$ lags are not available. If you specify two panel-style instruments, d and e, say, then $\mathbf{w}_t^L$ will consist of $d_{t-a}, \ldots, d_{t-b}, e_{t-a}, \ldots, e_{t-b}$. $\mathbf{w}_t^D$ is handled analogously.

The $\mathbf{x}_t^L$ vectors are simply $j \times 1$ vectors, where j is the number of regular instruments specified with the `instruments()` option; these vectors include a "1" unless you specify the `noconstant` suboption.

Looking carefully at (A3), you will notice that for dynamic panel-data models, moment conditions corresponding to the instruments $\mathbf{x}_{pt}^L$ take the form

$$E\left[\sum_{t=1}^{t=T} \mathbf{x}_{pt}^L u_{pt}^L(\beta)\right] = \mathbf{0}$$

and likewise for $\mathbf{x}_{pt}^D$. Instead of having separate moment conditions for each time period, there is one moment condition equal to the average of individual periods' moments. See Arellano and Bond (1991, 280). To include separate moment conditions for each time period, instead of specifying, say,

        instruments(1: x)

you could instead first generate a variable called one equal to unity for all observations and specify

        xtinstruments(1: x one)

(Creating the variable one is necessary because a constant is not automatically included in variable lists specified in `xtinstruments()`.)

Unbalanced panels are essentially handled by including zeros rows and columns of $\mathbf{Z}_p$ and $\mathbf{u}_p(\beta)$ corresponding to missing time periods. However, the numbers of instruments and moment conditions reported by gmm do not reflect this trickery and instead reflect the numbers of instruments and moment conditions that are not manipulated in this way. Moreover, gmm includes code to work through these situations efficiently without actually having to fill in zeros.

When you specify `winitial(xt ...)`, the one-step unadjusted VCE is computed as

$$\text{Var}(\widehat{\beta}) = \widehat{\sigma}_1^2 \mathbf{\Lambda}_H$$

where $\mathbf{\Lambda}_H$ was defined previously,

$$\widehat{\sigma}_1^2 = (N - k)^{-1} \sum_{p=1}^{p=P} \mathbf{u}_p^D(\widehat{\beta})$$

and $\mathbf{u}_p^D(\widehat{\beta}) = [u_{p2}^D(\widehat{\beta}), \ldots, u_{pT}^D(\widehat{\beta})]$. Here we use $(N - k)^{-1}$ instead of N^{-1} to match xtdpd.

References

Arellano, M., and S. Bond. 1991. Some tests of specification for panel data: Monte Carlo evidence and an application to employment equations. *Review of Economic Studies* 58: 277–297.

Baum, C. F. 2006. *An Introduction to Modern Econometrics Using Stata.* College Station, TX: Stata Press.

Blundell, R., R. Griffith, and F. Windmeijer. 2002. Individual effects and dynamics in count data models. *Journal of Econometrics* 108: 113–131.

Cameron, A. C., and P. K. Trivedi. 1998. *Regression Analysis of Count Data*. Cambridge: Cambridge University Press.

——. 2005. *Microeconometrics: Methods and Applications*. New York: Cambridge University Press.

——. 2010. *Microeconometrics Using Stata*. Rev. ed. College Station, TX: Stata Press.

Chamberlain, G. 1992. Comment: Sequential moment restrictions in panel data. *Journal of Business and Economic Statistics* 10: 20–26.

Davidson, R., and J. G. MacKinnon. 1993. *Estimation and Inference in Econometrics*. New York: Oxford University Press.

——. 2004. *Econometric Theory and Methods*. New York: Oxford University Press.

Greene, W. H. 2012. *Econometric Analysis*. 7th ed. Upper Saddle River, NJ: Prentice Hall.

Hall, A. R. 2005. *Generalized Method of Moments*. Oxford: Oxford University Press.

Hamilton, J. D. 1994. *Time Series Analysis*. Princeton: Princeton University Press.

Hansen, L. P. 1982. Large sample properties of generalized method of moments estimators. *Econometrica* 50: 1029–1054.

Hansen, L. P., and K. J. Singleton. 1982. Generalized instrumental variables estimation of nonlinear rational expectations models. *Econometrica* 50: 1269–1286.

Hayashi, F. 2000. *Econometrics*. Princeton, NJ: Princeton University Press.

Manski, C. F. 1988. *Analog Estimation Methods in Econometrics*. New York: Chapman & Hall/CRC.

Mátyás, L. 1999. *Generalized Method of Moments Estimation*. Cambridge: Cambridge University Press.

Mullahy, J. 1997. Instrumental-variable estimation of count data models: Applications to models of cigarette smoking behavior. *Review of Economics and Statistics* 79: 586–593.

Newey, W. K., and K. D. West. 1994. Automatic lag selection in covariance matrix estimation. *Review of Economic Studies* 61: 631–653.

Nickell, S. J. 1981. Biases in dynamic models with fixed effects. *Econometrica* 49: 1417–1426.

Ruud, P. A. 2000. *An Introduction to Classical Econometric Theory*. New York: Oxford University Press.

Wilde, J. 2008. A note on GMM estimation of probit models with endogenous regressors. *Statistical Papers* 49: 471–484.

Windmeijer, F. 2000. Moment conditions for fixed effects count data models with endogenous regressors. *Economics Letters* 68: 21–24.

——. 2005. A finite sample correction for the variance of linear efficient two-step GMM estimators. *Journal of Econometrics* 126: 25–51.

Windmeijer, F., and J. M. C. Santos Silva. 1997. Endogeneity in count data models: An application to demand for health care. *Journal of Applied Econometrics* 12: 281–294.

Wooldridge, J. M. 1997. Multiplicative panel data models without the strict exogeneity assumption. *Econometric Theory* 13: 667–678.

——. 1999. Distribution-free estimation of some nonlinear panel data models. *Journal of Econometrics* 90: 77–97.

——. 2010. *Econometric Analysis of Cross Section and Panel Data*. 2nd ed. Cambridge, MA: MIT Press.

Also see

[R] **gmm postestimation** — Postestimation tools for gmm

[R] **ivregress** — Single-equation instrumental-variables regression

[R] **ivprobit** — Probit model with continuous endogenous regressors

[R] **ivtobit** — Tobit model with continuous endogenous regressors

[R] **nl** — Nonlinear least-squares estimation

[R] **nlsur** — Estimation of nonlinear systems of equations

[R] **regress** — Linear regression

[XT] **xtdpd** — Linear dynamic panel-data estimation

[U] **20 Estimation and postestimation commands**

Title

> **gmm postestimation** — Postestimation tools for gmm

Description

The following postestimation command is of special interest after gmm:

Command	Description
estat overid	perform test of overidentifying restrictions

For information about this command, see below.

The following standard postestimation commands are also available:

Command	Description
estat	VCE
estimates	cataloging estimation results
lincom	point estimates, standard errors, testing, and inference for linear combinations of coefficients
nlcom	point estimates, standard errors, testing, and inference for nonlinear combinations of coefficients
predict	residuals
predictnl	point estimates, standard errors, testing, and inference for generalized predictions
test	Wald tests of simple and composite linear hypotheses
testnl	Wald tests of nonlinear hypotheses

See the corresponding entries in the *Base Reference Manual* for details.

Special-interest postestimation command

estat overid reports Hansen's J statistic, which is used to determine the validity of the overidentifying restrictions in a GMM model. If the model is correctly specified in the sense that $E\{\mathbf{z}_i u_i(\beta)\} = \mathbf{0}$, then the sample analog to that condition should hold at the estimated value of β. Hansen's J statistic is valid only if the weight matrix is optimal, meaning that it equals the inverse of the covariance matrix of the moment conditions. Therefore, estat overid only reports Hansen's J statistic after two-step or iterated estimation, or if you specified winitial(*matname*) when calling gmm. In the latter case, it is your responsibility to determine the validity of the J statistic.

Syntax for predict

predict [*type*] *newvar* [*if*] [*in*] [, <u>equation</u>(#*eqno* | *eqname*)]

predict [*type*] { *stub** | *newvar*$_1$ ... *newvar*$_q$ } [*if*] [*in*]

Residuals are available both in and out of sample; type predict ... if e(sample) ... if wanted only for the estimation sample.

You specify one new variable and (optionally) equation(), or you specify *stub** or *q* new variables, where *q* is the number of moment equations.

Menu

Statistics > Postestimation > Predictions, residuals, etc.

Option for predict

⌐ Main ⌐

equation(*#eqno* | *eqname*) specifies the equation for which residuals are desired. Specifying equation(#1) indicates that the calculation is to be made for the first moment equation. Specifying equation(demand) would indicate that the calculation is to be made for the moment equation named demand, assuming there is an equation named demand in the model.

If you specify one new variable name and omit equation(), results are the same as if you had specified equation(#1).

For more information on using predict after multiple-equation estimation commands, see [R] **predict**.

Syntax for estat overid

    estat overid

Menu

Statistics > Postestimation > Reports and statistics

Remarks

As we noted in *Introduction* of [R] **gmm**, underlying generalized method of moments (GMM) estimators is a set of l moment conditions, $E\{\mathbf{z}_i u_i(\beta)\} = \mathbf{0}$. When l is greater than the number of parameters, k, any size-k subset of the moment conditions would yield a consistent parameter estimate. We remarked that the parameter estimates we would obtain would in general depend on which k moment conditions we used. However, if all our moment conditions are indeed valid, then the parameter estimates should not differ too much regardless of which k moment conditions we used to estimate the parameters. The test of overidentifying restrictions is a model specification test based on this observation. The test of overidentifying restrictions requires that the number of moment conditions be greater than the number of parameters in the model.

Recall that the GMM criterion function is

$$Q = \left\{ \frac{1}{N} \sum_i \mathbf{z}_i u_i(\beta) \right\}' \mathbf{W} \left\{ \frac{1}{N} \sum_i \mathbf{z}_i u_i(\beta) \right\}$$

The test of overidentifying restrictions is remarkably simple. If $\mathbf{W}$ is an optimal weight matrix, under the null hypothesis $H_0 : E\{\mathbf{z}_i u_i(\beta)\} = \mathbf{0}$, the test statistic $J = N \times Q \sim \chi^2(l - k)$. A large test statistic casts doubt on the null hypothesis.

For the test to be valid, $\mathbf{W}$ must be optimal, meaning that $\mathbf{W}$ must be the inverse of the covariance matrix of the moment conditions:

$$\mathbf{W}^{-1} = E\{\mathbf{z}_i u_i(\beta) u_i'(\beta) \mathbf{z}_i'\}$$

Therefore, `estat overid` works only after the two-step and iterated estimators, or if you supplied your own initial weight matrix by using the `winitial(matname)` option to `gmm` and used the one-step estimator.

Often the overidentifying restrictions test is interpreted as a test of the validity of the instruments **z**. However, other forms of model misspecification can sometimes lead to a significant test statistic. See Hall (2005, sec. 5.1) for a discussion of the overidentifying restrictions test and its behavior in correctly and misspecified models.

Example 1

In example 6 of [R] **gmm**, we fit an exponential regression model of the number of doctor visits based on the person's gender, income, possession of private health insurance, and presence of a chronic disease. We argued that the variable `income` may be endogenous; we used the person's age and race as additional instrumental variables. Here we refit the model and test the specification of the model. We type

```
. use http://www.stata-press.com/data/r12/docvisits
. gmm (docvis - exp({xb:private chronic female income} + {b0})),
> instruments(private chronic female age black hispanic)
 (output omitted )
. estat overid

  Test of overidentifying restriction:

  Hansen's J chi2(2) = 9.52598 (p = 0.0085)
```

The J statistic is significant even at the 1% significance level, so we conclude that our model is misspecified. One possibility is that age and race directly affect the number of doctor visits, so we are not justified in excluding them from the model.

A simple technique to explore whether any of the instruments is invalid is to examine the statistics

$$r_j = \mathbf{W}_{jj}^{1/2} \left\{ \frac{1}{N} \sum_{i=1}^{N} z_{ij} u_i(\widehat{\beta}) \right\}$$

for $j = 1, \ldots, k$, where $\mathbf{W}_{jj}$ denotes the jth diagonal element of $\mathbf{W}$, $u_i(\widehat{\beta})$ denotes the sample residuals, and k is the number of instruments. If all the instruments are valid, then the scaled sample moments should at least be on the same order of magnitude. If one (or more) instrument's r_j is large in absolute value relative to the others, then that could be an indication that instrument is not valid.

In Stata, we type

```
. predict double r if e(sample)    // obtain residual from the model
. matrix W = e(W)                  // retrieve weight matrix
. local i 1
. // loop over each instrument and compute r_j
. foreach var of varlist private chronic female age black hispanic {
  2.          generate double r'var' = r*'var'*sqrt(W['i', 'i'])
  3.          local '++i'
  4. }
```

```
. summarize r*
```

Variable	Obs	Mean	Std. Dev.	Min	Max
r	4412	.0344373	8.26176	-151.1847	113.059
rprivate	4412	.007988	3.824118	-72.66254	54.33852
rchronic	4412	.0026947	2.0707	-43.7311	32.703
rfemale	4412	.0028168	1.566397	-12.7388	24.43621
rage	4412	.0360978	4.752986	-89.74112	55.58143
rblack	4412	-.0379317	1.062027	-24.39747	27.34512
rhispanic	4412	-.017435	1.08567	-5.509386	31.53512

We notice that the r_j statistics for age, black, and hispanic are larger than those for the other instruments in our model, supporting our suspicion that age and race may have a direct impact on the number of doctor visits.

◁

Saved results

estat overid saves the following in r():

Scalars
r(J)	Hansen's J statistic
r(J_df)	J statistic degrees of freedom
r(J_p)	J statistic p-value

Methods and formulas

All postestimation commands listed above are implemented as ado-files.

Reference

Hall, A. R. 2005. *Generalized Method of Moments.* Oxford: Oxford University Press.

Also see

[R] **gmm** — Generalized method of moments estimation

[U] **20 Estimation and postestimation commands**

> **grmeanby** — Graph means and medians by categorical variables

yntax

grmeanby *varlist* [*if*] [*in*] [*weight*], <u>summarize</u>(*varname*) [*options*]

options	Description
Main	
* <u>summarize</u>(*varname*)	graph mean (or median) of *varname*
<u>med</u>ian	graph medians; default is to graph means
Plot	
cline_options	change the look of the lines
marker_options	change look of markers (color, size, etc.)
marker_label_options	add marker labels; change look or position
Y axis, X axis, Titles, Legend, Overall	
twoway_options	any options other than by() documented in [G-3] ***twoway_options***

* summarize(*varname*) is required.
aweights and fweights are allowed; see [U] **11.1.6 weight**.

enu

Statistics > Summaries, tables, and tests > Summary and descriptive statistics > Graph means/medians by groups

escription

grmeanby graphs the (optionally weighted) means or medians of *varname* according to the values of the variables in *varlist*. The variables in *varlist* may be string or numeric and, if numeric, may be labeled.

ptions

> **Main**

summarize(*varname*) is required; it specifies the name of the variable whose mean or median is to be graphed.

median specifies that the graph is to be of medians, not means.

> **Plot**

cline_options affect the rendition of the lines through the markers, including their color, pattern, and width; see [G-3] ***cline_options***.

marker_options affect the rendition of markers drawn at the plotted points, including their shape, size, color, and outline; see [G-3] ***marker_options***.

marker_label_options specify if and how the markers are to be labeled; see [G-3] ***marker_label_options***.

Y axis, X axis, Titles, Legend, Overall

twoway_options are any of the options documented in [G-3] ***twoway_options***, excluding by(). These include options for titling the graph (see [G-3] ***title_options***) and for saving the graph to disk (see [G-3] ***saving_option***).

Remarks

The idea of graphing means of categorical variables was shown in Chambers and Hastie (1992, 3). Because this was shown in the context of an S function for making such graphs, it doubtless has roots going back further than that. grmeanby is, in any case, another implementation of what we will assume is their idea.

▷ Example 1

Using a variation of our auto dataset, we graph the mean of mpg by foreign, rep77, rep78, and make:

```
. use http://www.stata-press.com/data/r12/auto1
(Automobile Models)
. grmeanby foreign rep77 rep78 make, sum(mpg)
```

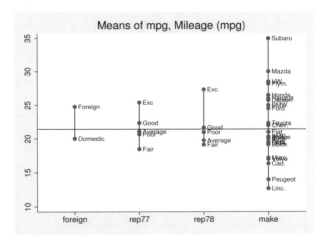

If we had wanted a graph of medians rather than means, we could have typed

```
. grmeanby foreign rep77 rep78 make, sum(mpg) median
```

◁

Methods and formulas

grmeanby is implemented as an ado-file.

References

Chambers, J. M., and T. J. Hastie, ed. 1992. *Statistical Models in S*. Pacific Grove, CA: Wadsworth and Brooks/Cole.

Gould, W. W. 1993. gr12: Graphs of means and medians by categorical variables. *Stata Technical Bulletin* 12: 13. Reprinted in *Stata Technical Bulletin Reprints*, vol. 2, pp. 44–45. College Station, TX: Stata Press.

Title

> **hausman** — Hausman specification test

Syntax

> hausman *name-consistent* [*name-efficient*] [, *options*]

options	Description
Main	
<u>c</u>onstant	include estimated intercepts in comparison; default is to exclude
<u>all</u>eqs	use all equations to perform test; default is first equation only
<u>sk</u>ipeqs(*eqlist*)	skip specified equations when performing test
equations(*matchlist*)	associate/compare the specified (by number) pairs of equations
force	force performance of test, even though assumptions are not met
df(*#*)	use *#* degrees of freedom
sigmamore	base both (co)variance matrices on disturbance variance estimate from efficient estimator
sigmaless	base both (co)variance matrices on disturbance variance estimate from consistent estimator
Advanced	
<u>tcon</u>sistent(*string*)	consistent estimator column header
<u>teff</u>icient(*string*)	efficient estimator column header

where *name-consistent* and *name-efficient* are names under which estimation results were saved via estimates store; see [R] **estimates store**.

A period (.) may be used to refer to the last estimation results, even if these were not already stored.

Not specifying *name-efficient* is equivalent to specifying the last estimation results as ".".

Menu

Statistics > Postestimation > Tests > Hausman specification test

Description

hausman performs Hausman's (1978) specification test.

Options

> ⌐ Main ⌐

constant specifies that the estimated intercept(s) be included in the model comparison; by default, they are excluded. The default behavior is appropriate for models in which the constant does not have a common interpretation across the two models.

706

alleqs specifies that all the equations in the models be used to perform the Hausman test; by default, only the first equation is used.

skipeqs(*eqlist*) specifies in *eqlist* the names of equations to be excluded from the test. Equation numbers are not allowed in this context, because the equation names, along with the variable names, are used to identify common coefficients.

equations(*matchlist*) specifies, by number, the pairs of equations that are to be compared.

The *matchlist* in equations() should follow the syntax

$$\#_c : \#_e \ \left[, \#_c : \#_e \left[, \ldots \right] \right]$$

where $\#_c$ ($\#_e$) is an equation number of the always-consistent (efficient under H_0) estimator. For instance, equations(1:1), equations(1:1, 2:2), or equations(1:2).

If equations() is not specified, then equations are matched on equation names.

equations() handles the situation in which one estimator uses equation names and the other does not. For instance, equations(1:2) means that equation 1 of the always-consistent estimator is to be tested against equation 2 of the efficient estimator. equations(1:1, 2:2) means that equation 1 is to be tested against equation 1 and that equation 2 is to be tested against equation 2. If equations() is specified, the alleqs and skipeqs options are ignored.

force specifies that the Hausman test be performed, even though the assumptions of the Hausman test seem not to be met, for example, because the estimators were pweighted or the data were clustered.

df(#) specifies the degrees of freedom for the Hausman test. The default is the matrix rank of the variance of the difference between the coefficients of the two estimators.

sigmamore and sigmaless specify that the two covariance matrices used in the test be based on a common estimate of disturbance variance (σ^2).

sigmamore specifies that the covariance matrices be based on the estimated disturbance variance from the efficient estimator. This option provides a proper estimate of the contrast variance for so-called tests of exogeneity and overidentification in instrumental-variables regression.

sigmaless specifies that the covariance matrices be based on the estimated disturbance variance from the consistent estimator.

These options can be specified only when both estimators save e(sigma) or e(rmse), or with the xtreg command. e(sigma_e) is saved after the xtreg command with the option fe or mle option. e(rmse) is saved after the xtreg command with the re option.

sigmamore or sigmaless are recommended when comparing fixed-effects and random-effects linear regression because they are much less likely to produce a non–positive-definite-differenced covariance matrix (although the tests are asymptotically equivalent whether or not one of the options is specified).

⌐ Advanced ⌐

tconsistent(*string*) and tefficient(*string*) are formatting options. They allow you to specify the headers of the columns of coefficients that default to the names of the models. These options will be of interest primarily to programmers.

Remarks

hausman is a general implementation of Hausman's (1978) specification test, which compares an estimator $\widehat{\theta}_1$ that is known to be consistent with an estimator $\widehat{\theta}_2$ that is efficient under the assumption being tested. The null hypothesis is that the estimator $\widehat{\theta}_2$ is indeed an efficient (and consistent) estimator of the true parameters. If this is the case, there should be no systematic difference between the two estimators. If there exists a systematic difference in the estimates, you have reason to doubt the assumptions on which the efficient estimator is based.

The assumption of efficiency is violated if the estimator is pweighted or the data are clustered, so hausman cannot be used. The test can be forced by specifying the force option with hausman. For an alternative to using hausman in these cases, see [R] **suest**.

To use hausman, you

```
.  (compute the always-consistent estimator)
.  estimates store name-consistent
.  (compute the estimator that is efficient under H_0)
.  hausman name-consistent  .
```

Alternatively, you can turn this around:

```
.  (compute the estimator that is efficient under H_0)
.  estimates store name-efficient
.  (fit the less-efficient model)
.  (compute the always-consistent estimator)
.  hausman  .  name-efficient
```

You can, of course, also compute and store both the always-consistent and efficient-under-H_0 estimators and perform the Hausman test with

```
.  hausman name-consistent name-efficient
```

▷ Example 1

We are studying the factors that affect the wages of young women in the United States between 1970 and 1988, and we have a panel-data sample of individual women over that time span.

```
. use http://www.stata-press.com/data/r12/nlswork4
(National Longitudinal Survey.  Young Women 14-26 years of age in 1968)

. describe

Contains data from http://www.stata-press.com/data/r12/nlswork4.dta
  obs:         28,534                     National Longitudinal Survey.
                                          Young Women 14-26 years of age
                                          in 1968
  vars:             6                     29 Jan 2011 16:35
  size:       370,942
```

variable name	storage type	display format	value label	variable label
idcode	int	%8.0g		NLS ID
year	byte	%8.0g		interview year
age	byte	%8.0g		age in current year
msp	byte	%8.0g		1 if married, spouse present
ttl_exp	float	%9.0g		total work experience
ln_wage	float	%9.0g		ln(wage/GNP deflator)

```
Sorted by:  idcode  year
```

We believe that a random-effects specification is appropriate for individual-level effects in our model. We fit a fixed-effects model that will capture all temporally constant individual-level effects.

```
. xtreg ln_wage age msp ttl_exp, fe

Fixed-effects (within) regression              Number of obs      =     28494
Group variable: idcode                         Number of groups   =      4710

R-sq:  within  = 0.1373                         Obs per group: min =         1
       between = 0.2571                                        avg =       6.0
       overall = 0.1800                                        max =        15

                                               F(3,23781)         =   1262.01
corr(u_i, Xb)  = 0.1476                         Prob > F           =    0.0000

------------------------------------------------------------------------------
     ln_wage |      Coef.   Std. Err.      t    P>|t|     [95% Conf. Interval]
-------------+----------------------------------------------------------------
         age |   -.005485    .000837     -6.55   0.000    -.0071256   -.0038443
         msp |   .0033427   .0054868      0.61   0.542    -.0074118    .0140971
     ttl_exp |   .0383604   .0012416     30.90   0.000     .0359268    .0407941
       _cons |   1.593953   .0177538     89.78   0.000     1.559154    1.628752
-------------+----------------------------------------------------------------
     sigma_u |  .37674223
     sigma_e |  .29751014
         rho |  .61591044   (fraction of variance due to u_i)
------------------------------------------------------------------------------
F test that all u_i=0:     F(4709, 23781) =      7.76        Prob > F = 0.0000
```

We assume that this model is consistent for the true parameters and save the results by using estimates store under a name, fixed:

```
. estimates store fixed
```

Now we fit a random-effects model as a fully efficient specification of the individual effects under the assumption that they are random and follow a normal distribution. We then compare these estimates with the previously saved results by using the hausman command.

```
. xtreg ln_wage age msp ttl_exp, re

Random-effects GLS regression                  Number of obs      =     28494
Group variable: idcode                         Number of groups   =      4710

R-sq:  within  = 0.1373                         Obs per group: min =         1
       between = 0.2552                                        avg =       6.0
       overall = 0.1797                                        max =        15

Random effects u_i ~ Gaussian                   Wald chi2(3)       =   5100.33
corr(u_i, X)       = 0 (assumed)                Prob > chi2        =    0.0000

------------------------------------------------------------------------------
     ln_wage |      Coef.   Std. Err.      z    P>|z|     [95% Conf. Interval]
-------------+----------------------------------------------------------------
         age |  -.0069749   .0006882    -10.13   0.000    -.0083238   -.0056259
         msp |   .0046594   .0051012      0.91   0.361    -.0053387    .0146575
     ttl_exp |   .0429635   .0010169     42.25   0.000     .0409704    .0449567
       _cons |   1.609916   .0159176    101.14   0.000     1.578718    1.641114
-------------+----------------------------------------------------------------
     sigma_u |  .32648519
     sigma_e |  .29751014
         rho |  .54633481   (fraction of variance due to u_i)
------------------------------------------------------------------------------
```

```
. hausman fixed ., sigmamore
                 —— Coefficients ——
               (b)          (B)           (b-B)      sqrt(diag(V_b-V_B))
              fixed          .          Difference          S.E.

       age   -.005485    -.0069749       .0014899         .0004803
       msp    .0033427    .0046594      -.0013167         .0020596
   ttl_exp    .0383604    .0429635      -.0046031         .0007181

                         b = consistent under Ho and Ha; obtained from xtreg
             B = inconsistent under Ha, efficient under Ho; obtained from xtreg
    Test:  Ho:  difference in coefficients not systematic

              chi2(3) = (b-B)'[(V_b-V_B)^(-1)](b-B)
                      =     260.40
            Prob>chi2 =      0.0000
```

Under the current specification, our initial hypothesis that the individual-level effects are adequately modeled by a random-effects model is resoundingly rejected. This result is based on the rest of our model specification, and random effects might be appropriate for some alternate model of wages.

◁

Jerry Allen Hausman was born in West Virginia in 1946. He studied economics at Brown and Oxford, has been at MIT since 1972, and has made many outstanding contributions to econometrics and applied microeconomics.

▷ Example 2

A stringent assumption of multinomial and conditional logit models is that outcome categories for the model have the property of independence of irrelevant alternatives (IIA). Stated simply, this assumption requires that the inclusion or exclusion of categories does not affect the relative risks associated with the regressors in the remaining categories.

One classic example of a situation in which this assumption would be violated involves the choice of transportation mode; see McFadden (1974). For simplicity, postulate a transportation model with the four possible outcomes: rides a train to work, takes a bus to work, drives the Ford to work, and drives the Chevrolet to work. Clearly, "drives the Ford" is a closer substitute to "drives the Chevrolet" than it is to "rides a train" (at least for most people). This means that excluding "drives the Ford" from the model could be expected to affect the relative risks of the remaining options and that the model would not obey the IIA assumption.

Using the data presented in [R] **mlogit**, we will use a simplified model to test for IIA. The choice of insurance type among indemnity, prepaid, and uninsured is modeled as a function of age and gender. The indemnity category is allowed to be the base category, and the model including all three outcomes is fit. The results are then stored under the name allcats.

```
. use http://www.stata-press.com/data/r12/sysdsn3
(Health insurance data)

. mlogit insure age male

Iteration 0:   log likelihood = -555.85446
Iteration 1:   log likelihood = -551.32973
Iteration 2:   log likelihood = -551.32802
Iteration 3:   log likelihood = -551.32802
```

Multinomial logistic regression

		Number of obs	=	615
		LR chi2(4)	=	9.05
		Prob > chi2	=	0.0598

Log likelihood = -551.32802   Pseudo R2   =   0.0081

insure	Coef.	Std. Err.	z	P>|z|	[95% Conf. Interval]	
Indemnity	(base outcome)					
Prepaid						
age	-.0100251	.0060181	-1.67	0.096	-.0218204	.0017702
male	.5095747	.1977893	2.58	0.010	.1219147	.8972346
_cons	.2633838	.2787575	0.94	0.345	-.2829708	.8097383
Uninsure						
age	-.0051925	.0113821	-0.46	0.648	-.0275011	.0171161
male	.4748547	.3618462	1.31	0.189	-.2343508	1.18406
_cons	-1.756843	.5309602	-3.31	0.001	-2.797506	-.7161803

```
. estimates store allcats
```

Under the IIA assumption, we would expect no systematic change in the coefficients if we excluded one of the outcomes from the model. (For an extensive discussion, see Hausman and McFadden [1984].) We reestimate the parameters, excluding the uninsured outcome, and perform a Hausman test against the fully efficient full model.

```
. mlogit insure age male if insure != "Uninsure":insure

Iteration 0:   log likelihood =  -394.8693
Iteration 1:   log likelihood =  -390.4871
Iteration 2:   log likelihood = -390.48643
Iteration 3:   log likelihood = -390.48643
```

Multinomial logistic regression

		Number of obs	=	570
		LR chi2(2)	=	8.77
		Prob > chi2	=	0.0125

Log likelihood = -390.48643   Pseudo R2   =   0.0111

insure	Coef.	Std. Err.	z	P>|z|	[95% Conf. Interval]	
Indemnity	(base outcome)					
Prepaid						
age	-.0101521	.0060049	-1.69	0.091	-.0219214	.0016173
male	.5144003	.1981735	2.60	0.009	.1259874	.9028133
_cons	.2678043	.2775563	0.96	0.335	-.276196	.8118046

```
. hausman . allcats, alleqs constant
                  ―― Coefficients ――
                 (b)           (B)          (b-B)      sqrt(diag(V_b-V_B))
                  .          allcats     Difference           S.E.
```

	(b) .	(B) allcats	(b-B) Difference	sqrt(diag(V_b-V_B)) S.E.
age	-.0101521	-.0100251	-.0001269	.
male	.5144003	.5095747	.0048256	.0123338
_cons	.2678043	.2633838	.0044205	.

```
                          b = consistent under Ho and Ha; obtained from mlogit
              B = inconsistent under Ha, efficient under Ho; obtained from mlogit
    Test:  Ho:  difference in coefficients not systematic
                  chi2(3) = (b-B)'[(V_b-V_B)^(-1)](b-B)
                        =         0.08
              Prob>chi2 =         0.9944
              (V_b-V_B is not positive definite)
```

The syntax of the if condition on the mlogit command simply identified the "Uninsured" category with the insure value label; see [U] **12.6.3 Value labels**. On examining the output from hausman, we see that there is no evidence that the IIA assumption has been violated.

Because the Hausman test is a standardized comparison of model coefficients, using it with mlogit requires that the base outcome be the same in both competing models. In particular, if the most-frequent category (the default base outcome) is being removed to test for IIA, you must use the baseoutcome() option in mlogit to manually set the base outcome to something else. Or you can use the equation() option of the hausman command to align the equations of the two models.

Having the missing values for the square root of the diagonal of the covariance matrix of the differences is not comforting, but it is also not surprising. This covariance matrix is guaranteed to be positive definite only asymptotically (it is a consequence of the assumption that one of the estimators is efficient), and assurances are not made about the diagonal elements. Negative values along the diagonal are possible, and the fourth column of the table is provided mainly for descriptive use.

We can also perform the Hausman IIA test against the remaining alternative in the model:

```
. mlogit insure age male if insure != "Prepaid":insure
Iteration 0:   log likelihood = -132.59913
Iteration 1:   log likelihood = -131.78009
Iteration 2:   log likelihood = -131.76808
Iteration 3:   log likelihood = -131.76807

Multinomial logistic regression              Number of obs  =       338
                                              LR chi2(2)     =      1.66
                                              Prob > chi2    =    0.4356
Log likelihood = -131.76807                   Pseudo R2      =    0.0063
```

insure	Coef.	Std. Err.	z	P>|z|	[95% Conf. Interval]	
Indemnity	(base outcome)					
Uninsure						
age	-.0041055	.0115807	-0.35	0.723	-.0268033	.0185923
male	.4591074	.3595663	1.28	0.202	-.2456296	1.163844
_cons	-1.801774	.5474476	-3.29	0.001	-2.874752	-.7287968

```
. hausman . allcats, alleqs constant
```

| | ─── Coefficients ─── | | | |
| | (b) | (B) | (b-B) | sqrt(diag(V_b-V_B)) |
	.	allcats	Difference	S.E.
age	-.0041055	-.0051925	.001087	.0021355
male	.4591074	.4748547	-.0157473	.
_cons	-1.801774	-1.756843	-.0449311	.1333421

```
                    b = consistent under Ho and Ha; obtained from mlogit
            B = inconsistent under Ha, efficient under Ho; obtained from mlogit
  Test:  Ho:  difference in coefficients not systematic

           chi2(3) = (b-B)'[(V_b-V_B)^(-1)](b-B)
                   =    -0.18    chi2<0 ==> model fitted on these
                                 data fails to meet the asymptotic
                                 assumptions of the Hausman test;
                                 see suest for a generalized test
```

Here the χ^2 statistic is actually negative. We might interpret this result as strong evidence that we cannot reject the null hypothesis. Such a result is not an unusual outcome for the Hausman test, particularly when the sample is relatively small—there are only 45 uninsured individuals in this dataset.

Are we surprised by the results of the Hausman test in this example? Not really. Judging from the z statistics on the original multinomial logit model, we were struggling to identify any structure in the data with the current specification. Even when we were willing to assume IIA and computed the efficient estimator under this assumption, few of the effects could be identified as statistically different from those on the base category. Trying to base a Hausman test on a contrast (difference) between two poor estimates is just asking too much of the existing data.

◁

In example 2, we encountered a case in which the Hausman was not well defined. Unfortunately, in our experience this happens fairly often. Stata provides an alternative to the Hausman test that overcomes this problem through an alternative estimator of the variance of the difference between the two estimators. This other estimator is guaranteed to be positive semidefinite. This alternative estimator also allows a widening of the scope of problems to which Hausman-type tests can be applied by relaxing the assumption that one of the estimators is efficient. For instance, you can perform Hausman-type tests to clustered observations and survey estimators. See [R] **suest** for details.

Saved results

hausman saves the following in r():

Scalars
    r(chi2) χ^2
    r(df)   degrees of freedom for the statistic
    r(p) p-value for the χ^2
    r(rank)   rank of (V_b-V_B)^(-1)

Methods and formulas

hausman is implemented as an ado-file.

The Hausman statistic is distributed as χ^2 and is computed as

$$H = (\beta_c - \beta_e)'(V_c - V_e)^{-1}(\beta_c - \beta_e)$$

where
β_c is the coefficient vector from the consistent estimator
β_e is the coefficient vector from the efficient estimator
V_c is the covariance matrix of the consistent estimator
V_e is the covariance matrix of the efficient estimator

When the difference in the variance matrices is not positive definite, a Moore–Penrose generalized inverse is used. As noted in Gourieroux and Monfort (1995, 125–128), the choice of generalized inverse is not important asymptotically.

The number of degrees of freedom for the statistic is the rank of the difference in the variance matrices. When the difference is positive definite, this is the number of common coefficients in the models being compared.

Acknowledgment

Portions of hausman are based on an early implementation by Jeroen Weesie, Utrecht University, The Netherlands.

References

Baltagi, B. H. 2008. *Econometrics*. 4th ed. Berlin: Springer.

Gourieroux, C., and A. Monfort. 1995. *Statistics and Econometric Models, Vol 2: Testing, Confidence Regions, Model Selection, and Asymptotic Theory*. Trans. Q. Vuong. Cambridge: Cambridge University Press.

Hausman, J. A. 1978. Specification tests in econometrics. *Econometrica* 46: 1251–1271.

Hausman, J. A., and D. L. McFadden. 1984. Specification tests for the multinomial logit model. *Econometrica* 52: 1219–1240.

McFadden, D. L. 1974. Measurement of urban travel demand. *Journal of Public Economics* 3: 303–328.

Also see

[R] **lrtest** — Likelihood-ratio test after estimation

[R] **suest** — Seemingly unrelated estimation

[R] **test** — Test linear hypotheses after estimation

[XT] **xtreg** — Fixed-, between-, and random-effects and population-averaged linear models

Title

| **heckman** — Heckman selection model |

Syntax

Basic syntax

> heckman *depvar* $\begin{bmatrix} indepvars \end{bmatrix}$, <u>sel</u>ect(*varlist$_s$*) $\begin{bmatrix} \underline{two}step \end{bmatrix}$

> or

> heckman *depvar* $\begin{bmatrix} indepvars \end{bmatrix}$, <u>sel</u>ect(*depvar$_s$* = *varlist$_s$*) $\begin{bmatrix} \underline{two}step \end{bmatrix}$

Full syntax for maximum likelihood estimates only

> heckman *depvar* $\begin{bmatrix} indepvars \end{bmatrix}$ $\begin{bmatrix} if \end{bmatrix}$ $\begin{bmatrix} in \end{bmatrix}$ $\begin{bmatrix} weight \end{bmatrix}$,
>
> > <u>sel</u>ect($\begin{bmatrix} depvar_s = \end{bmatrix}$ *varlist$_s$* $\begin{bmatrix} , \underline{off}set(varname) \underline{noc}onstant \end{bmatrix}$)
>
> > $\begin{bmatrix} heckman\_ml\_options \end{bmatrix}$

Full syntax for Heckman's two-step consistent estimates only

> heckman *depvar* $\begin{bmatrix} indepvars \end{bmatrix}$ $\begin{bmatrix} if \end{bmatrix}$ $\begin{bmatrix} in \end{bmatrix}$, <u>two</u>step
>
> > <u>sel</u>ect($\begin{bmatrix} depvar_s = \end{bmatrix}$ *varlist$_s$* $\begin{bmatrix} , \underline{noc}onstant \end{bmatrix}$) $\begin{bmatrix} heckman\_ts\_options \end{bmatrix}$

heckman_ml_options	Description
Model	
* <u>sel</u>ect()	specify selection equation: dependent and independent variables; whether to have constant term and offset variable
<u>noc</u>onstant	suppress constant term
<u>off</u>set(*varname*)	include *varname* in model with coefficient constrained to 1
<u>cons</u>traints(*constraints*)	apply specified linear constraints
<u>col</u>linear	keep collinear variables
SE/Robust	
vce(*vcetype*)	*vcetype* may be oim, <u>r</u>obust, <u>cl</u>uster *clustvar*, opg, <u>boot</u>strap, or jackknife
Reporting	
<u>l</u>evel(#)	set confidence level; default is level(95)
<u>first</u>	report first-step probit estimates
noskip	perform likelihood-ratio test
<u>nsh</u>azard(*newvar*)	generate nonselection hazard variable
<u>mi</u>lls(*newvar*)	synonym for nshazard()
nocnsreport	do not display constraints
display_options	control column formats, row spacing, line width, and display of omitted variables and base and empty cells

Maximization

maximize_options	control the maximization process; seldom used
<u>coefl</u>egend	display legend instead of statistics

*select( ) is required.

The full specification is <u>sel</u>ect($\left[depvar_s = \right]$ *varlist*$_s$ $\left[$, <u>off</u>set(*varname*) <u>noc</u>onstant $\right]$) .

heckman_ts_options	Description
Model	
*<u>sel</u>ect()	specify selection equation: dependent and independent variables; whether to have constant term
*<u>two</u>step	produce two-step consistent estimate
<u>noc</u>onstant	suppress constant term
<u>rhos</u>igma	truncate ρ to $\left[-1, 1 \right]$ with consistent σ
<u>rhot</u>runc	truncate ρ to $\left[-1, 1 \right]$
<u>rhol</u>imited	truncate ρ in limited cases
<u>rhof</u>orce	do not truncate ρ
SE	
vce(*vcetype*)	*vcetype* may be conventional, <u>boot</u>strap, or <u>jack</u>knife
Reporting	
<u>l</u>evel(#)	set confidence level; default is level(95)
<u>fir</u>st	report first-step probit estimates
<u>ns</u>hazard(*newvar*)	generate nonselection hazard variable
<u>mills</u>(*newvar*)	synonym for nshazard()
display_options	control column formats, row spacing, line width, and display of omitted variables and base and empty cells
<u>coefl</u>egend	display legend instead of statistics

*select( ) and twostep are required.

The full specification is <u>sel</u>ect($\left[depvar_s = \right]$ *varlist*$_s$ $\left[$, <u>noc</u>onstant $\right]$)

indepvars and *varlist*$_s$ may contain factor variables; see [U] **11.4.3 Factor variables**.
depvar, *indepvars*, *varlist*$_s$, and *depvar*$_s$ may contain time-series operators; see [U] **11.4.4 Time-series varlists**.
bootstrap, by, jackknife, rolling, statsby, and svy are allowed; see [U] **11.1.10 Prefix commands**.
Weights are not allowed with the bootstrap prefix; see [R] **bootstrap**.
aweights are not allowed with the jackknife prefix; see [R] **jackknife**.
twostep, vce(), first, noskip, and weights are not allowed with the svy prefix; see [SVY] **svy**.
pweights, aweights, fweights, and iweights are allowed with maximum likelihood estimation;
    see [U] **11.1.6 weight**. No weights are allowed if twostep is specified.
coeflegend does not appear in the dialog box.
See [U] **20 Estimation and postestimation commands** for more capabilities of estimation commands.

Menu

heckman for maximum likelihood estimates

Statistics > Sample-selection models > Heckman selection model (ML)

heckman for two-step consistent estimates

Statistics > Sample-selection models > Heckman selection model (two-step)

Description

heckman fits regression models with selection by using either Heckman's two-step consistent estimator or full maximum likelihood.

Options for Heckman selection model (ML)

⌐ Model ⌐

select(...) specifies the variables and options for the selection equation. It is an integral part of specifying a Heckman model and is required. The selection equation should contain at least one variable that is not in the outcome equation.

If *depvar*$_s$ is specified, it should be coded as 0 or 1, with 0 indicating an observation not selected and 1 indicating a selected observation. If *depvar*$_s$ is not specified, observations for which *depvar* is not missing are assumed selected, and those for which *depvar* is missing are assumed not selected.

noconstant, offset(*varname*), constraints(*constraints*), collinear; see [R] **estimation options**.

⌐ SE/Robust ⌐

vce(*vcetype*) specifies the type of standard error reported, which includes types that are derived from asymptotic theory, that are robust to some kinds of misspecification, that allow for intragroup correlation, and that use bootstrap or jackknife methods; see [R] **vce_option**.

⌐ Reporting ⌐

level(*#*); see [R] **estimation options**.

first specifies that the first-step probit estimates of the selection equation be displayed before estimation.

noskip specifies that a full maximum-likelihood model with only a constant for the regression equation be fit. This model is not displayed but is used as the base model to compute a likelihood-ratio test for the model test statistic displayed in the estimation header. By default, the overall model test statistic is an asymptotically equivalent Wald test that all the parameters in the regression equation are zero (except the constant). For many models, this option can substantially increase estimation time.

nshazard(*newvar*) and mills(*newvar*) are synonyms; either will create a new variable containing the nonselection hazard—what Heckman (1979) referred to as the inverse of the Mills' ratio—from the selection equation. The nonselection hazard is computed from the estimated parameters of the selection equation.

nocnsreport; see [R] **estimation options**.

display_options: <u>noomit</u>ted, vsquish, noemptycells, <u>base</u>levels, <u>allbase</u>levels,
    cformat(%*fmt*), pformat(%*fmt*), sformat(%*fmt*), and nolstretch; see [R] **estimation options**.

maximize_options: <u>dif</u>ficult, <u>tech</u>nique(*algorithm_spec*), <u>iter</u>ate(#), [<u>no</u>]<u>log</u>, <u>tra</u>ce,
    gradient, showstep, <u>hess</u>ian, <u>showtol</u>erance, <u>tol</u>erance(#), <u>ltol</u>erance(#),
    <u>nrtol</u>erance(#), nonrtolerance, and from(*init_specs*); see [R] **maximize**. These options are
    seldom used.

Setting the optimization type to technique(bhhh) resets the default *vcetype* to vce(opg).

The following option is available with heckman but is not shown in the dialog box:

coeflegend; see [R] **estimation options**.

Options for Heckman selection model (two-step)

select(...) specifies the variables and options for the selection equation. It is an integral part of
    specifying a Heckman model and is required. The selection equation should contain at least one
    variable that is not in the outcome equation.

If *depvar_s* is specified, it should be coded as 0 or 1, with 0 indicating an observation not selected
    and 1 indicating a selected observation. If *depvar_s* is not specified, observations for which *depvar*
    is not missing are assumed selected, and those for which *depvar* is missing are assumed not
    selected.

twostep specifies that Heckman's (1979) two-step efficient estimates of the parameters, standard
    errors, and covariance matrix be produced.

noconstant; see [R] **estimation options**.

rhosigma, rhotrunc, rholimited, and rhoforce are rarely used options to specify how the
    two-step estimator (option twostep) handles unusual cases in which the two-step estimate of ρ is
    outside the admissible range for a correlation, $[-1, 1]$. When abs$(\rho) > 1$, the two-step estimate of
    the coefficient variance–covariance matrix may not be positive definite and thus may be unusable
    for testing. The default is rhosigma.

rhosigma specifies that ρ be truncated, as with the rhotrunc option, and that the estimate of σ be
    made consistent with $\widehat{\rho}$, the truncated estimate of ρ. So, $\widehat{\sigma} = \beta_m \widehat{\rho}$; see *Methods and formulas* for
    the definition of β_m. Both the truncated ρ and the new estimate of $\widehat{\sigma}$ are used in all computations
    to estimate the two-step covariance matrix.

rhotrunc specifies that ρ be truncated to lie in the range $[-1, 1]$. If the two-step estimate is less
    than -1, ρ is set to -1; if the two-step estimate is greater than 1, ρ is set to 1. This truncated
    value of ρ is used in all computations to estimate the two-step covariance matrix.

rholimited specifies that ρ be truncated only in computing the diagonal matrix **D** as it enters
    $\mathbf{V}_{\text{twostep}}$ and $\mathbf{Q}$; see *Methods and formulas*. In all other computations, the untruncated estimate
    of ρ is used.

rhoforce specifies that the two-step estimate of ρ be retained, even if it is outside the admissible
    range for a correlation. This option may, in rare cases, lead to a non–positive-definite covariance
    matrix.

These options have no effect when estimation is by maximum likelihood, the default. They also have no effect when the two-step estimate of ρ is in the range $[-1, 1]$.

⌐ SE ⌐

vce(*vcetype*) specifies the type of standard error reported, which includes types that are derived from asymptotic theory and that use bootstrap or jackknife methods; see [R] *vce_option*.

vce(conventional), the default, uses the two-step variance estimator derived by Heckman.

⌐ Reporting ⌐

level(*#*); see [R] **estimation options**.

first specifies that the first-step probit estimates of the selection equation be displayed before estimation.

nshazard(*newvar*) and mills(*newvar*) are synonyms; either will create a new variable containing the nonselection hazard—what Heckman (1979) referred to as the inverse of the Mills' ratio—from the selection equation. The nonselection hazard is computed from the estimated parameters of the selection equation.

display_options: noomitted, vsquish, noemptycells, baselevels, allbaselevels, cformat(%*fmt*), pformat(%*fmt*), sformat(%*fmt*), and nolstretch; see [R] **estimation options**.

The following option is available with heckman but is not shown in the dialog box:

coeflegend; see [R] **estimation options**.

emarks

The Heckman selection model (Gronau 1974; Lewis 1974; Heckman 1976) assumes that there exists an underlying regression relationship,

$$y_j = \mathbf{x}_j\boldsymbol{\beta} + u_{1j} \qquad\qquad regression\ equation$$

The dependent variable, however, is not always observed. Rather, the dependent variable for observation j is observed if

$$\mathbf{z}_j\boldsymbol{\gamma} + u_{2j} > 0 \qquad\qquad selection\ equation$$

where

$$u_1 \sim N(0, \sigma)$$
$$u_2 \sim N(0, 1)$$
$$\mathrm{corr}(u_1, u_2) = \rho$$

When $\rho \neq 0$, standard regression techniques applied to the first equation yield biased results. heckman provides consistent, asymptotically efficient estimates for all the parameters in such models.

In one classic example, the first equation describes the wages of women. Women choose whether to work, and thus, from our point of view as researchers, whether we observe their wages in our data. If women made this decision randomly, we could ignore that not all wages are observed and use ordinary regression to fit a wage model. Such an assumption of random participation, however, is unlikely to be true; women who would have low wages may be unlikely to choose to work, and thus the sample of observed wages is biased upward. In the jargon of economics, women choose not to work when their personal reservation wage is greater than the wage offered by employers. Thus women who choose not to work might have even higher offer wages than those who do work—they may have high offer wages, but they have even higher reservation wages. We could tell a story that competency is related to wages, but competency is rewarded more at home than in the labor force.

In any case, in this problem—which is the paradigm for most such problems—a solution can be found if there are some variables that strongly affect the chances for observation (the reservation wage) but not the outcome under study (the offer wage). Such a variable might be the number of children in the home. (Theoretically, we do not need such identifying variables, but without them, we depend on functional form to identify the model. It would be difficult for anyone to take such results seriously because the functional-form assumptions have no firm basis in theory.)

▷ Example 1

In the syntax for `heckman`, *depvar* and *indepvars* are the dependent variable and regressors for the underlying regression model to be fit ($y = X\beta$), and *varlist$_s$* are the variables (Z) thought to determine whether *depvar* is observed or unobserved (selected or not selected). In our female wage example, the number of children at home would be included in the second list. By default, `heckman` assumes that missing values (see [U] **12.2.1 Missing values**) of *depvar* imply that the dependent variable is unobserved (not selected). With some datasets, it is more convenient to specify a binary variable (*depvar$_s$*) that identifies the observations for which the dependent is observed/selected (*depvar$_s \neq$ 0*) or not observed (*depvar$_s$ = 0*); `heckman` will accommodate either type of data.

We have a (fictional) dataset on 2,000 women, 1,343 of whom work:

```
. use http://www.stata-press.com/data/r12/womenwk
. summarize age educ married children wage
```

Variable	Obs	Mean	Std. Dev.	Min	Max
age	2000	36.208	8.28656	20	59
education	2000	13.084	3.045912	10	20
married	2000	.6705	.4701492	0	1
children	2000	1.6445	1.398963	0	5
wage	1343	23.69217	6.305374	5.88497	45.80979

We will assume that the hourly wage is a function of education and age, whereas the likelihood of working (the likelihood of the wage being observed) is a function of marital status, the number of children at home, and (implicitly) the wage (via the inclusion of age and education, which we think determine the wage):

```
. heckman wage educ age, select(married children educ age)
Iteration 0:   log likelihood = -5178.7009
Iteration 1:   log likelihood = -5178.3049
Iteration 2:   log likelihood = -5178.3045
```

Heckman selection model			Number of obs	=	2000
(regression model with sample selection)			Censored obs	=	657
			Uncensored obs	=	1343
			Wald chi2(2)	=	508.44
Log likelihood = -5178.304			Prob > chi2	=	0.0000

wage	Coef.	Std. Err.	z	P>|z|	[95% Conf. Interval]	
wage						
education	.9899537	.0532565	18.59	0.000	.8855729	1.094334
age	.2131294	.0206031	10.34	0.000	.1727481	.2535108
_cons	.4857752	1.077037	0.45	0.652	-1.625179	2.59673
select						
married	.4451721	.0673954	6.61	0.000	.3130794	.5772647
children	.4387068	.0277828	15.79	0.000	.3842534	.4931601
education	.0557318	.0107349	5.19	0.000	.0346917	.0767718
age	.0365098	.0041533	8.79	0.000	.0283694	.0446502
_cons	-2.491015	.1893402	-13.16	0.000	-2.862115	-2.119915
/athrho	.8742086	.1014225	8.62	0.000	.6754241	1.072993
/lnsigma	1.792559	.027598	64.95	0.000	1.738468	1.84665
rho	.7035061	.0512264			.5885365	.7905862
sigma	6.004797	.1657202			5.68862	6.338548
lambda	4.224412	.3992265			3.441942	5.006881

LR test of indep. eqns. (rho = 0):   chi2(1) =   61.20   Prob > chi2 = 0.0000

heckman assumes that wage is the dependent variable and that the first variable list (educ and age) are the determinants of wage. The variables specified in the select() option (married, children, educ, and age) are assumed to determine whether the dependent variable is observed (the selection equation). Thus we fit the model

$$\text{wage} = \beta_0 + \beta_1 \text{educ} + \beta_2 \text{age} + u_1$$

and we assumed that wage is observed if

$$\gamma_0 + \gamma_1 \text{married} + \gamma_2 \text{children} + \gamma_3 \text{educ} + \gamma_4 \text{age} + u_2 > 0$$

where u_1 and u_2 have correlation ρ.

The reported results for the wage equation are interpreted exactly as though we observed wage data for all women in the sample; the coefficients on age and education level represent the estimated marginal effects of the regressors in the underlying regression equation. The results for the two ancillary parameters require some explanation. heckman does not directly estimate ρ; to constrain ρ within its valid limits, and for numerical stability during optimization, it estimates the inverse hyperbolic tangent of ρ:

$$\text{atanh}\, \rho = \frac{1}{2} \ln\left(\frac{1+\rho}{1-\rho}\right)$$

This estimate is reported as /athrho. In the bottom panel of the output, heckman undoes this transformation for you: the estimated value of ρ is 0.7035061. The standard error for ρ is computed using the delta method, and its confidence intervals are the transformed intervals of /athrho.

Similarly, σ, the standard error of the residual in the wage equation, is not directly estimated; for numerical stability, heckman instead estimates $\ln\sigma$. The untransformed sigma is reported at the end of the output: 6.004797.

Finally, some researchers—especially economists—are used to the selectivity effect summarized not by ρ but by $\lambda = \rho\sigma$. heckman reports this, too, along with an estimate of the standard error and confidence interval.

◁

❑ Technical note

If each of the equations in the model had contained many regressors, the heckman command could have become long. An alternate way of specifying our wage model would be to use Stata's global macros. The following lines are an equivalent way of specifying our model:

```
. global wageeq "wage educ age"
. global seleq "married children educ age"
. heckman $wageeq, select($seleq)
  (output omitted )
```

❑

❑ Technical note

The reported model χ^2 test is a Wald test that all coefficients in the regression model (except the constant) are 0. heckman is an estimation command, so you can use test, testnl, or lrtest to perform tests against whatever nested alternate model you choose; see [R] **test**, [R] **testnl**, and [R] **lrtest**.

The estimation of ρ and σ in the forms atanh ρ and $\ln\sigma$ extends the range of these parameters to infinity in both directions, thus avoiding boundary problems during the maximization. Tests of ρ must be made in the transformed units. However, because atanh$(0) = 0$, the reported test for atanh $\rho = 0$ is equivalent to the test for $\rho = 0$.

The likelihood-ratio test reported at the bottom of the output is an equivalent test for $\rho = 0$ and is computationally the comparison of the joint likelihood of an independent probit model for the selection equation and a regression model on the observed wage data against the Heckman model likelihood. Because $\chi^2 = 61.20$, this clearly justifies the Heckman selection equation with these data.

❑

▷ Example 2

heckman supports the Huber/White/sandwich estimator of variance under the vce(robust) and vce(cluster *clustvar*) options or when pweights are used for population-weighted data; see [U] **20.20 Obtaining robust variance estimates**. We can obtain robust standard errors for our wage model by specifying clustering on county of residence (the county variable).

```
. heckman wage educ age, select(married children educ age) vce(cluster county)
Iteration 0:   log pseudolikelihood = -5178.7009
Iteration 1:   log pseudolikelihood = -5178.3049
Iteration 2:   log pseudolikelihood = -5178.3045
```

Heckman selection model		Number of obs	=	2000
(regression model with sample selection)		Censored obs	=	657
		Uncensored obs	=	1343
		Wald chi2(1)	=	.
Log pseudolikelihood = -5178.304		Prob > chi2	=	.

(Std. Err. adjusted for 10 clusters in county)

wage	Coef.	Robust Std. Err.	z	P>|z|	[95% Conf. Interval]	
wage						
education	.9899537	.0600061	16.50	0.000	.8723438	1.107564
age	.2131294	.020995	10.15	0.000	.17198	.2542789
_cons	.4857752	1.302103	0.37	0.709	-2.066299	3.03785
select						
married	.4451721	.0731472	6.09	0.000	.3018062	.5885379
children	.4387068	.0312386	14.04	0.000	.3774802	.4999333
education	.0557318	.0110039	5.06	0.000	.0341645	.0772991
age	.0365098	.004038	9.04	0.000	.0285954	.0444242
_cons	-2.491015	.1153305	-21.60	0.000	-2.717059	-2.264972
/athrho	.8742086	.1403337	6.23	0.000	.5991596	1.149258
/lnsigma	1.792559	.0258458	69.36	0.000	1.741902	1.843216
rho	.7035061	.0708796			.5364513	.817508
sigma	6.004797	.155199			5.708189	6.316818
lambda	4.224412	.5186709			3.207835	5.240988

Wald test of indep. eqns. (rho = 0): chi2(1) = 38.81   Prob > chi2 = 0.0000

The robust standard errors tend to be a bit larger, but we notice no systematic differences. This finding is not surprising because the data were not constructed to have any county-specific correlations or any other characteristics that would deviate from the assumptions of the Heckman model.

◁

> ## Example 3

Stata also produces Heckman's (1979) two-step efficient estimator of the model with the `twostep` option. Maximum likelihood estimation of the parameters can be time consuming with large datasets, and the two-step estimates may provide a good alternative in such cases. Continuing with the women's wage model, we can obtain the two-step estimates with Heckman's consistent covariance estimates by typing

```
. heckman wage educ age, select(married children educ age) twostep
```

Heckman selection model -- two-step estimates   Number of obs   =   2000
(regression model with sample selection)   Censored obs   =   657
Uncensored obs   =   1343

Wald chi2(2)   =   442.54
Prob > chi2   =   0.0000

wage	Coef.	Std. Err.	z	P>|z|	[95% Conf. Interval]	
wage						
education	.9825259	.0538821	18.23	0.000	.8769189	1.088133
age	.2118695	.0220511	9.61	0.000	.1686502	.2550888
_cons	.7340391	1.248331	0.59	0.557	-1.712645	3.180723
select						
married	.4308575	.074208	5.81	0.000	.2854125	.5763025
children	.4473249	.0287417	15.56	0.000	.3909922	.5036576
education	.0583645	.0109742	5.32	0.000	.0368555	.0798735
age	.0347211	.0042293	8.21	0.000	.0264318	.0430105
_cons	-2.467365	.1925635	-12.81	0.000	-2.844782	-2.089948
mills						
lambda	4.001615	.6065388	6.60	0.000	2.812821	5.19041
rho	0.67284					
sigma	5.9473529					

◁

❏ Technical note

The Heckman selection model depends strongly on the model being correct, much more so than ordinary regression. Running a separate probit or logit for sample inclusion followed by a regression, referred to in the literature as the two-part model (Manning, Duan, and Rogers 1987)—not to be confused with Heckman's two-step procedure—is an especially attractive alternative if the regression part of the model arose because of taking a logarithm of zero values. When the goal is to analyze an underlying regression model or to predict the value of the dependent variable that would be observed in the absence of selection, however, the Heckman model is more appropriate. When the goal is to predict an actual response, the two-part model is usually the better choice.

The Heckman selection model can be unstable when the model is not properly specified or if a specific dataset simply does not support the model's assumptions. For example, let's examine the solution to another simulated problem.

```
. use http://www.stata-press.com/data/r12/twopart

. heckman yt x1 x2 x3, select(z1 z2) nonrtol
Iteration 0:   log likelihood = -111.94996
Iteration 1:   log likelihood = -110.82258
Iteration 2:   log likelihood = -110.17707
Iteration 3:   log likelihood = -107.70663  (not concave)
Iteration 4:   log likelihood = -107.07729  (not concave)
  (output omitted)
Iteration 33:  log likelihood =  -104.0825  (not concave)
Iteration 34:  log likelihood =  -104.0825
```

```
Heckman selection model                    Number of obs      =        150
(regression model with sample selection)   Censored obs       =         87
                                           Uncensored obs     =         63

                                           Wald chi2(3)       =   8.64e+08
Log likelihood = -104.0825                 Prob > chi2        =     0.0000
```

yt	Coef.	Std. Err.	z	P>|z|	[95% Conf. Interval]	
yt						
x1	.8974192	.0002247	3994.69	0.000	.8969789	.8978595
x2	-2.525303	.0001472	-1.7e+04	0.000	-2.525591	-2.525014
x3	2.855786	.0004181	6829.86	0.000	2.854966	2.856605
_cons	.6975442	.0920515	7.58	0.000	.5171265	.8779619
select						
z1	-.6825988	.0900159	-7.58	0.000	-.8590267	-.5061709
z2	1.003605	.132347	7.58	0.000	.7442097	1.263
_cons	-.3604652	.1232778	-2.92	0.003	-.6020852	-.1188452
/athrho	16.19193	280.9822	0.06	0.954	-534.523	566.9069
/lnsigma	-.5396153	.1318714	-4.09	0.000	-.7980786	-.2811521
rho	1	9.73e-12			-1	1
sigma	.5829725	.0768774			.4501931	.7549135
lambda	.5829725	.0768774			.4322955	.7336494

```
LR test of indep. eqns. (rho = 0):   chi2(1) =    25.67   Prob > chi2 = 0.0000
```

The model has converged to a value of ρ that is 1.0—within machine-rounding tolerances. Given the form of the likelihood for the Heckman selection model, this implies a division by zero, and it is surprising that the model solution turns out as well as it does. Reparameterizing ρ has allowed the estimation to converge, but we clearly have problems with the estimates. Moreover, if this had occurred in a large dataset, waiting for convergence might take considerable time.

This dataset was not intentionally developed to cause problems. It is actually generated by a "Heckman process" and when generated starting from different random values can be easily estimated. The luck of the draw here merely led to data that, despite the source, did not support the assumptions of the Heckman model.

The two-step model is generally more stable when the data are problematic. It even tolerates estimates of ρ less than -1 and greater than 1. For these reasons, the two-step model may be preferred when exploring a large dataset. Still, if the maximum likelihood estimates cannot converge, or converge to a value of ρ that is at the boundary of acceptable values, there is scant support for fitting a Heckman selection model on the data. Heckman (1979) discusses the implications of ρ being exactly 1 or 0, together with the implications of other possible covariance relationships among the model's determinants.

❏

James Joseph Heckman was born in Chicago in 1944 and studied mathematics at Colorado College and economics at Princeton. He has taught economics at Columbia and (since 1973) at the University of Chicago. He has worked on developing a scientific basis for economic policy evaluation, with emphasis on models of individuals or disaggregated groups and the problems and possibilities created by heterogeneity, diversity, and unobserved counterfactual states. In 2000, he shared the Nobel Prize in Economics with Daniel L. McFadden.

Saved results

heckman (maximum likelihood) saves the following in e():

Scalars

e(N)	number of observations
e(N_cens)	number of censored observations
e(k)	number of parameters
e(k_eq)	number of equations in e(b)
e(k_eq_model)	number of equations in overall model test
e(k_aux)	number of auxiliary parameters
e(k_dv)	number of dependent variables
e(df_m)	model degrees of freedom
e(ll)	log likelihood
e(ll_0)	log likelihood, constant-only model
e(N_clust)	number of clusters
e(lambda)	λ
e(selambda)	standard error of λ
e(sigma)	sigma
e(chi2)	χ^2
e(chi2_c)	χ^2 for comparison test
e(p_c)	p-value for comparison test
e(p)	significance of comparison test
e(rho)	ρ
e(rank)	rank of e(V)
e(rank0)	rank of e(V) for constant-only model
e(ic)	number of iterations
e(rc)	return code
e(converged)	1 if converged, 0 otherwise

Macros
    e(cmd)   heckman
    e(cmdline)   command as typed
    e(depvar)   names of dependent variables
    e(wtype)   weight type
    e(wexp)   weight expression
    e(title)   title in estimation output
    e(title2)   secondary title in estimation output
    e(clustvar)   name of cluster variable
    e(offset1)   offset for regression equation
    e(offset2)   offset for selection equation
    e(mills)   variable containing nonselection hazard (inverse of Mills')
    e(chi2type)   Wald or LR; type of model χ^2 test
    e(chi2_ct)   Wald or LR; type of model χ^2 test corresponding to e(chi2_c)
    e(vce)   *vcetype* specified in vce()
    e(vcetype)   title used to label Std. Err.
    e(opt)   type of optimization
    e(which)   max or min; whether optimizer is to perform maximization or minimization
    e(method)   ml
    e(ml_method)   type of ml method
    e(user)   name of likelihood-evaluator program
    e(technique)   maximization technique
    e(properties)   b V
    e(predict)   program used to implement predict
    e(marginsok)   predictions allowed by margins
    e(asbalanced)   factor variables fvset as asbalanced
    e(asobserved)   factor variables fvset as asobserved
Matrices
    e(b)   coefficient vector
    e(Cns)   constraints matrix
    e(ilog)   iteration log (up to 20 iterations)
    e(gradient)   gradient vector
    e(V)   variance–covariance matrix of the estimators
    e(V_modelbased)   model-based variance
Functions
    e(sample)   marks estimation sample

`heckman` (two-step) saves the following in `e()`:

Scalars

`e(N)`	number of observations
`e(N_cens)`	number of censored observations
`e(df_m)`	model degrees of freedom
`e(lambda)`	λ
`e(selambda)`	standard error of λ
`e(sigma)`	sigma
`e(chi2)`	χ^2
`e(p)`	significance of comparison test
`e(rho)`	ρ
`e(rank)`	rank of `e(V)`

Macros

`e(cmd)`	heckman
`e(cmdline)`	command as typed
`e(depvar)`	names of dependent variables
`e(title)`	title in estimation output
`e(title2)`	secondary title in estimation output
`e(mills)`	variable containing nonselection hazard (inverse of Mills')
`e(chi2type)`	Wald or LR; type of model χ^2 test
`e(vce)`	*vcetype* specified in `vce()`
`e(vcetype)`	title used to label Std. Err.
`e(rhometh)`	rhosigma, rhotrunc, rholimited, or rhoforce
`e(method)`	twostep
`e(properties)`	b V
`e(predict)`	program used to implement `predict`
`e(marginsok)`	predictions allowed by margins
`e(marginsnotok)`	predictions disallowed by margins
`e(asbalanced)`	factor variables fvset as asbalanced
`e(asobserved)`	factor variables fvset as asobserved

Matrices

`e(b)`	coefficient vector
`e(Cns)`	constraints matrix
`e(V)`	variance–covariance matrix of the estimators

Functions

`e(sample)`	marks estimation sample

Methods and formulas

`heckman` is implemented as an ado-file. Cameron and Trivedi (2010, 556–562) and Greene (2012, 873–880) provide good introductions to the Heckman selection model. Adkins and Hill (2008, 395–400) describe the two-step estimator with an application using Stata. Jones (2007, 35–40) illustrates Heckman estimation with an application to health economics.

Regression estimates using the nonselection hazard (Heckman 1979) provide starting values for maximum likelihood estimation.

The regression equation is

$$y_j = \mathbf{x}_j\boldsymbol{\beta} + u_{1j}$$

The selection equation is

$$\mathbf{z}_j\boldsymbol{\gamma} + u_{2j} > 0$$

where

$$u_1 \sim N(0, \sigma)$$

$$u_2 \sim N(0, 1)$$

$$\mathrm{corr}(u_1, u_2) = \rho$$

The log likelihood for observation j, $\ln L_j = l_j$, is

$$
l_j = \begin{cases} w_j \ln\Phi\left\{\dfrac{\mathbf{z}_j\boldsymbol{\gamma} + (y_j - \mathbf{x}_j\boldsymbol{\beta})\rho/\sigma}{\sqrt{1-\rho^2}}\right\} - \dfrac{w_j}{2}\left(\dfrac{y_j - \mathbf{x}_j\boldsymbol{\beta}}{\sigma}\right)^2 - w_j \ln(\sqrt{2\pi}\sigma) & y_j \ observed \\[2em] w_j \ln\Phi(-\mathbf{z}_j\boldsymbol{\gamma}) & y_j \ not\ observed \end{cases}
$$

where $\Phi(\cdot)$ is the standard cumulative normal and w_j is an optional weight for observation j.

In the maximum likelihood estimation, σ and ρ are not directly estimated. Directly estimated are $\ln\sigma$ and $\mathrm{atanh}\,\rho$:

$$\mathrm{atanh}\,\rho = \frac{1}{2}\ln\left(\frac{1+\rho}{1-\rho}\right)$$

The standard error of $\lambda = \rho\sigma$ is approximated through the propagation of error (delta) method; that is,

$$\mathrm{Var}(\lambda) \approx \mathbf{D}\,\mathrm{Var}\left\{(\mathrm{atanh}\,\rho\ \ \ln\sigma)\right\}\mathbf{D}'$$

where $\mathbf{D}$ is the Jacobian of λ with respect to $\mathrm{atanh}\,\rho$ and $\ln\sigma$.

With maximum likelihood estimation, this command supports the Huber/White/sandwich estimator of the variance and its clustered version using vce(robust) and vce(cluster *clustvar*), respectively. See [P] _robust, particularly *Maximum likelihood estimators* and *Methods and formulas*.

The maximum likelihood version of heckman also supports estimation with survey data. For details on VCEs with survey data, see [SVY] **variance estimation**.

The two-step estimates are computed using Heckman's (1979) procedure.

Probit estimates of the selection equation

$$\Pr(y_j\ observed \mid \mathbf{z}_j) = \Phi(\mathbf{z}_j\boldsymbol{\gamma})$$

are obtained. From these estimates, the nonselection hazard—what Heckman (1979) referred to as the inverse of the Mills' ratio, m_j—for each observation j is computed as

$$m_j = \frac{\phi(\mathbf{z}_j\widehat{\boldsymbol{\gamma}})}{\Phi(\mathbf{z}_j\widehat{\boldsymbol{\gamma}})}$$

where ϕ is the normal density. We also define

$$\delta_j = m_j(m_j + \widehat{\boldsymbol{\gamma}}\mathbf{z}_j)$$

Following Heckman, the two-step parameter estimates of β are obtained by augmenting the regression equation with the nonselection hazard $\mathbf{m}$. Thus the regressors become $[\,\mathbf{X}\ \mathbf{m}\,]$, and we obtain the additional parameter estimate β_m on the variable containing the nonselection hazard.

A consistent estimate of the regression disturbance variance is obtained using the residuals from the augmented regression and the parameter estimate on the nonselection hazard,

$$\widehat{\sigma}^2 = \frac{\mathbf{e}'\mathbf{e} + \beta_m^2 \sum_{j=1}^{N} \delta_j}{N}$$

The two-step estimate of ρ is then

$$\widehat{\rho} = \frac{\beta_m}{\widehat{\sigma}}$$

Heckman derived consistent estimates of the coefficient covariance matrix on the basis of the augmented regression.

Let $\mathbf{W} = [\,\mathbf{X}\ \mathbf{m}\,]$ and $\mathbf{R}$ be a square, diagonal matrix of dimension N, with $(1 - \widehat{\rho}^2 \delta_j)$ as the diagonal elements. The conventional VCE is

$$\mathbf{V}_{\text{twostep}} = \widehat{\sigma}^2 (\mathbf{W}'\mathbf{W})^{-1}(\mathbf{W}'\mathbf{R}\mathbf{W} + \mathbf{Q})(\mathbf{W}'\mathbf{W})^{-1}$$

where

$$\mathbf{Q} = \widehat{\rho}^2 (\mathbf{W}'\mathbf{D}\mathbf{Z})\mathbf{V_p}(\mathbf{Z}'\mathbf{D}\mathbf{W})$$

where $\mathbf{D}$ is the square, diagonal matrix of dimension N with δ_j as the diagonal elements; $\mathbf{Z}$ is the data matrix of selection equation covariates; and $\mathbf{V}_p$ is the variance–covariance estimate from the probit estimation of the selection equation.

References

Adkins, L. C., and R. C. Hill. 2008. *Using Stata for Principles of Econometrics.* 3rd ed. Hoboken, NJ: Wiley.

Baum, C. F. 2006. *An Introduction to Modern Econometrics Using Stata.* College Station, TX: Stata Press.

Cameron, A. C., and P. K. Trivedi. 2010. *Microeconometrics Using Stata.* Rev. ed. College Station, TX: Stata Press.

Chiburis, R., and M. Lokshin. 2007. Maximum likelihood and two-step estimation of an ordered-probit selection model. *Stata Journal* 7: 167–182.

Greene, W. H. 2012. *Econometric Analysis.* 7th ed. Upper Saddle River, NJ: Prentice Hall.

Gronau, R. 1974. Wage comparisons: A selectivity bias. *Journal of Political Economy* 82: 1119–1143.

Heckman, J. 1976. The common structure of statistical models of truncation, sample selection and limited dependent variables and a simple estimator for such models. *Annals of Economic and Social Measurement* 5: 475–492.

———. 1979. Sample selection bias as a specification error. *Econometrica* 47: 153–161.

Jones, A. 2007. *Applied Econometrics for Health Economists: A Practical Guide.* 2nd ed. Abingdon, UK: Radcliffe.

Lewis, H. G. 1974. Comments on selectivity biases in wage comparisons. *Journal of Political Economy* 82: 1145–1155.

Manning, W. G., N. Duan, and W. H. Rogers. 1987. Monte Carlo evidence on the choice between sample selection and two-part models. *Journal of Econometrics* 35: 59–82.

Also see

[R] **heckman postestimation** — Postestimation tools for heckman

[R] **heckprob** — Probit model with sample selection

[R] **regress** — Linear regression

[R] **tobit** — Tobit regression

[R] **treatreg** — Treatment-effects model

[SVY] **svy estimation** — Estimation commands for survey data

[U] **20 Estimation and postestimation commands**

Title

> **heckman postestimation** — Postestimation tools for heckman

Description

The following postestimation commands are available after `heckman`:

Command	Description
contrast	contrasts and ANOVA-style joint tests of estimates
estat[1]	AIC, BIC, VCE, and estimation sample summary
estat (svy)	postestimation statistics for survey data
estimates	cataloging estimation results
lincom	point estimates, standard errors, testing, and inference for linear combinations of coefficients
lrtest[2]	likelihood-ratio test; not available with two-step estimator
margins	marginal means, predictive margins, marginal effects, and average marginal effects
marginsplot	graph the results from margins (profile plots, interaction plots, etc.)
nlcom	point estimates, standard errors, testing, and inference for nonlinear combinations of coefficients
predict	predictions, residuals, influence statistics, and other diagnostic measures
predictnl	point estimates, standard errors, testing, and inference for generalized predictions
pwcompare	pairwise comparisons of estimates
suest[1]	seemingly unrelated estimation
test	Wald tests of simple and composite linear hypotheses
testnl	Wald tests of nonlinear hypotheses

[1] `estat ic` and `suest` are not appropriate after `heckman, twostep`.

[2] `lrtest` is not appropriate with svy estimation results.

See the corresponding entries in the *Base Reference Manual* for details, but see [SVY] **estat** for details about `estat` (svy).

Syntax for predict

After ML or twostep

>   predict $\big[$ *type* $\big]$ *newvar* $\big[$ *if* $\big]$ $\big[$ *in* $\big]$ $\big[$, *statistic* <u>nooff</u>set $\big]$

After ML

>   predict $\big[$ *type* $\big]$ $\big\{$ *stub** | *newvar*_{reg} *newvar*_{sel} *newvar*_{athrho} *newvar*_{lnsigma} $\big\}$
>
> $\big[$ *if* $\big]$ $\big[$ *in* $\big]$, <u>sc</u>ores

statistic	Description
Main	
xb	linear prediction; the default
stdp	standard error of the prediction
stdf	standard error of the forecast
<u>xb</u>sel	linear prediction for selection equation
stdpsel	standard error of the linear prediction for selection equation
<u>pr</u>(*a*,*b*)	$\Pr(y_j \mid a < y_j < b)$
e(*a*,*b*)	$E(y_j \mid a < y_j < b)$
<u>ys</u>tar(*a*,*b*)	$E(y_j^*)$, $y_j^* = \max\{a, \min(y_j, b)\}$
<u>yc</u>ond	$E(y_j | y_j \text{ observed})$
<u>ye</u>xpected	$E(y_j^*)$, y_j taken to be 0 where unobserved
<u>ns</u>hazard or <u>mills</u>	nonselection hazard (also called the inverse of Mills' ratio)
psel	$\Pr(y_j \text{ observed})$

These statistics are available both in and out of sample; type predict ... if e(sample) ... if wanted only for the estimation sample.

stdf is not allowed with svy estimation results.

where *a* and *b* may be numbers or variables; *a* missing (*a* ≥ .) means $-\infty$, and *b* missing (*b* ≥ .) means $+\infty$; see [U] **12.2.1 Missing values**.

Menu

Statistics > Postestimation > Predictions, residuals, etc.

Options for predict

┌─ Main └─

xb, the default, calculates the linear prediction $\mathbf{x}_j \mathbf{b}$.

stdp calculates the standard error of the prediction, which can be thought of as the standard error of the predicted expected value or mean for the observation's covariate pattern. The standard error of the prediction is also referred to as the standard error of the fitted value.

stdf calculates the standard error of the forecast, which is the standard error of the point prediction for 1 observation. It is commonly referred to as the standard error of the future or forecast value. By construction, the standard errors produced by stdf are always larger than those produced by stdp; see *Methods and formulas* in [R] **regress postestimation**.

xbsel calculates the linear prediction for the selection equation.

stdpsel calculates the standard error of the linear prediction for the selection equation.

pr(a,b) calculates $\Pr(a < \mathbf{x}_j\mathbf{b} + u_1 < b)$, the probability that $y_j|\mathbf{x}_j$ would be observed in the interval (a, b).

a and b may be specified as numbers or variable names; *lb* and *ub* are variable names; pr(20,30) calculates $\Pr(20 < \mathbf{x}_j\mathbf{b} + u_1 < 30)$; pr(*lb*,*ub*) calculates $\Pr(lb < \mathbf{x}_j\mathbf{b} + u_1 < ub)$; and pr(20,*ub*) calculates $\Pr(20 < \mathbf{x}_j\mathbf{b} + u_1 < ub)$.

a missing ($a \geq .$) means $-\infty$; pr(.,30) calculates $\Pr(-\infty < \mathbf{x}_j\mathbf{b} + u_j < 30)$; pr(*lb*,30) calculates $\Pr(-\infty < \mathbf{x}_j\mathbf{b} + u_j < 30)$ in observations for which $lb \geq .$ and calculates $\Pr(lb < \mathbf{x}_j\mathbf{b} + u_j < 30)$ elsewhere.

b missing ($b \geq .$) means $+\infty$; pr(20,.) calculates $\Pr(+\infty > \mathbf{x}_j\mathbf{b} + u_j > 20)$; pr(20,*ub*) calculates $\Pr(+\infty > \mathbf{x}_j\mathbf{b} + u_j > 20)$ in observations for which $ub \geq .$ and calculates $\Pr(20 < \mathbf{x}_j\mathbf{b} + u_j < ub)$ elsewhere.

e(a,b) calculates $E(\mathbf{x}_j\mathbf{b} + u_1 \mid a < \mathbf{x}_j\mathbf{b} + u_1 < b)$, the expected value of $y_j|\mathbf{x}_j$ conditional on $y_j|\mathbf{x}_j$ being in the interval (a, b), meaning that $y_j|\mathbf{x}_j$ is truncated. a and b are specified as they are for pr().

ystar(a,b) calculates $E(y_j^*)$, where $y_j^* = a$ if $\mathbf{x}_j\mathbf{b} + u_j \leq a$, $y_j^* = b$ if $\mathbf{x}_j\mathbf{b} + u_j \geq b$, and $y_j^* = \mathbf{x}_j\mathbf{b} + u_j$ otherwise, meaning that y_j^* is censored. a and b are specified as they are for pr().

ycond calculates the expected value of the dependent variable conditional on the dependent variable being observed, that is, selected; $E(y_j \mid y_j \text{ observed})$.

yexpected calculates the expected value of the dependent variable (y_j^*), where that value is taken to be 0 when it is expected to be unobserved; $y_j^* = \Pr(y_j \text{ observed})E(y_j \mid y_j \text{ observed})$.

The assumption of 0 is valid for many cases where nonselection implies nonparticipation (for example, unobserved wage levels, insurance claims from those who are uninsured) but may be inappropriate for some problems (for example, unobserved disease incidence).

nshazard and mills are synonyms; both calculate the nonselection hazard—what Heckman (1979) referred to as the inverse of the Mills' ratio—from the selection equation.

psel calculates the probability of selection (or being observed): $\Pr(y_j \text{ observed}) = \Pr(\mathbf{z}_j\boldsymbol{\gamma} + u_{2j} > 0)$.

nooffset is relevant when you specify offset(*varname*) for heckman. It modifies the calculations made by predict so that they ignore the offset variable; the linear prediction is treated as $\mathbf{x}_j\mathbf{b}$ rather than as $\mathbf{x}_j\mathbf{b} + \text{offset}_j$.

scores, not available with twostep, calculates equation-level score variables.

The first new variable will contain $\partial \ln L / \partial(\mathbf{x}_j\boldsymbol{\beta})$.

The second new variable will contain $\partial \ln L / \partial(\mathbf{z}_j\boldsymbol{\gamma})$.

The third new variable will contain $\partial \ln L / \partial(\text{atanh } \rho)$.

The fourth new variable will contain $\partial \ln L / \partial(\ln \sigma)$.

Remarks

> ## Example 1

The default statistic produced by predict after heckman is the expected value of the dependent variable from the underlying distribution of the regression model. In the wage model of [R] **heckman**, this is the expected wage rate among all women, regardless of whether they were observed to participate in the labor force:

```
. use http://www.stata-press.com/data/r12/womenwk
. heckman wage educ age, select(married children educ age) vce(cluster county)
(output omitted )
. predict heckwage
(option xb assumed; fitted values)
```

It is instructive to compare these predicted wage values from the Heckman model with an ordinary regression model—a model without the selection adjustment:

```
. regress wage educ age
```

Source	SS	df	MS		Number of obs =	1343
					F(2, 1340) =	227.49
Model	13524.0337	2	6762.01687		Prob > F =	0.0000
Residual	39830.8609	1340	29.7245231		R-squared =	0.2535
					Adj R-squared =	0.2524
Total	53354.8946	1342	39.7577456		Root MSE =	5.452

wage	Coef.	Std. Err.	t	P>|t|	[95% Conf. Interval]	
education	.8965829	.0498061	18.00	0.000	.7988765	.9942893
age	.1465739	.0187135	7.83	0.000	.109863	.1832848
_cons	6.084875	.8896182	6.84	0.000	4.339679	7.830071

```
. predict regwage
(option xb assumed; fitted values)
. summarize heckwage regwage
```

Variable	Obs	Mean	Std. Dev.	Min	Max
heckwage	2000	21.15532	3.83965	14.6479	32.85949
regwage	2000	23.12291	3.241911	17.98218	32.66439

Since this dataset was concocted, we know the true coefficients of the wage regression equation to be 1, 0.2, and 1, respectively. We can compute the true mean wage for our sample.

```
. generate truewage = 1 + .2*age + 1*educ
. summarize truewage
```

Variable	Obs	Mean	Std. Dev.	Min	Max
truewage	2000	21.3256	3.797904	15	32.8

Whereas the mean of the predictions from heckman is within 18 cents of the true mean wage, ordinary regression yields predictions that are on average about $1.80 per hour too high because of the selection effect. The regression predictions also show somewhat less variation than the true wages.

The coefficients from heckman are so close to the true values that they are not worth testing. Conversely, the regression equation is significantly off but seems to give the right sense. Would we be led far astray if we relied on the OLS coefficients? The effect of age is off by more than 5 cents per year of age, and the coefficient on education level is off by about 10%. We can test the OLS coefficient on education level against the true value by using test.

```
. test educ = 1
( 1)  education = 1
       F(  1,  1340) =      4.31
             Prob > F =    0.0380
```

Not only is the OLS coefficient on education substantially lower than the true parameter, but the difference from the true parameter is also statistically significant beyond the 5% level. We can perform a similar test for the OLS age coefficient:

```
. test age = .2
( 1)  age = .2
       F(  1,  1340) =      8.15
             Prob > F =    0.0044
```

We find even stronger evidence that the OLS regression results are biased away from the true parameters.

◁

▷ Example 2

Several other interesting aspects of the Heckman model can be explored with `predict`. Continuing with our wage model, we can obtain the expected wages for women conditional on participating in the labor force with the `ycond` option. Let's get these predictions and compare them with actual wages for women participating in the labor force.

```
. use http://www.stata-press.com/data/r12/womenwk, clear
. heckman wage educ age, select(married children educ age)
  (output omitted )
. predict hcndwage, ycond
. summarize wage hcndwage if wage != .
```

Variable	Obs	Mean	Std. Dev.	Min	Max
wage	1343	23.69217	6.305374	5.88497	45.80979
hcndwage	1343	23.68239	3.335087	16.18337	33.7567

We see that the average predictions from `heckman` are close to the observed levels but do not have the same mean. These conditional wage predictions are available for all observations in the dataset but can be directly compared only with observed wages, where individuals are participating in the labor force.

What if we were interested in making predictions about mean wages for all women? Here the expected wage is 0 for those who are not expected to participate in the labor force, with expected participation determined by the selection equation. These values can be obtained with the `yexpected` option of `predict`. For comparison, a variable can be generated where the wage is set to 0 for nonparticipants.

```
. predict hexpwage, yexpected
. generate wage0 = wage
(657 missing values generated)
. replace wage0 = 0 if wage == .
(657 real changes made)
```

```
. summarize hexpwage wage0
    Variable |        Obs        Mean    Std. Dev.        Min         Max
-------------+--------------------------------------------------------
    hexpwage |       2000    15.92511    5.979336    2.492469    32.45858
       wage0 |       2000    15.90929    12.27081           0    45.80979
```

Again we note that the predictions from `heckman` are close to the observed mean hourly wage rate for all women. Why aren't the predictions using `ycond` and `yexpected` equal to their observed sample equivalents? For the Heckman model, unlike linear regression, the sample moments implied by the optimal solution to the model likelihood do not require that these predictions match observed data. Properly accounting for the additional variation from the selection equation requires that the model use more information than just the sample moments of the observed wages.

◁

Methods and formulas

All postestimation commands listed above are implemented as ado-files.

Reference

Heckman, J. 1979. Sample selection bias as a specification error. *Econometrica* 47: 153–161.

Also see

[R] **heckman** — Heckman selection model

[U] **20 Estimation and postestimation commands**

Title

> **heckprob** — Probit model with sample selection

Syntax

> heckprob *depvar indepvars* $\begin{bmatrix} if \end{bmatrix}$ $\begin{bmatrix} in \end{bmatrix}$ $\begin{bmatrix} weight \end{bmatrix}$,
>
> <u>sel</u>ect($\begin{bmatrix} depvar_s \ = \end{bmatrix}$ *varlist$_s$* $\begin{bmatrix} , & \underline{\text{off}}\text{set}(varname) & \underline{\text{nocon}}\text{stant} \end{bmatrix}$) $\begin{bmatrix} options \end{bmatrix}$

options	Description
Model	
* <u>sel</u>ect( )	specify selection equation: dependent and independent variables; whether to have constant term and offset variable
<u>nocon</u>stant	suppress constant term
<u>off</u>set(*varname*)	include *varname* in model with coefficient constrained to 1
<u>const</u>raints(*constraints*)	apply specified linear constraints
<u>col</u>linear	keep collinear variables
SE/Robust	
vce(*vcetype*)	*vcetype* may be oim, <u>r</u>obust, <u>cl</u>uster *clustvar*, opg, <u>boot</u>strap, or <u>jack</u>knife
Reporting	
<u>level</u>(#)	set confidence level; default is level(95)
<u>first</u>	report first-step probit estimates
noskip	perform likelihood-ratio test
nocnsreport	do not display constraints
display_options	control column formats, row spacing, line width, and display of omitted variables and base and empty cells
Maximization	
maximize_options	control the maximization process; seldom used
<u>coef</u>legend	display legend instead of statistics

* select( ) is required.

  The full specification is <u>sel</u>ect($\begin{bmatrix} depvar_s \ = \end{bmatrix}$ *varlist$_s$* $\begin{bmatrix} , & \underline{\text{off}}\text{set}(varname) & \underline{\text{nocon}}\text{stant} \end{bmatrix}$).

indepvars and *varlist$_s$* may contain factor variables; see [U] **11.4.3 Factor variables**.

depvar, *indepvars*, *depvar$_s$*, and *varlist$_s$* may contain time-series operators; see [U] **11.4.4 Time-series varlists**.

bootstrap, by, jackknife, rolling, statsby, and svy are allowed; see [U] **11.1.10 Prefix commands**.

Weights are not allowed with the bootstrap prefix; see [R] **bootstrap**.

vce(), first, noskip, and weights are not allowed with the svy prefix; see [SVY] **svy**.

pweights, fweights, and iweights are allowed; see [U] **11.1.6 weight**.

coeflegend does not appear in the dialog box.

See [U] **20 Estimation and postestimation commands** for more capabilities of estimation commands.

Menu

Statistics > Sample-selection models > Probit model with selection

Description

heckprob fits maximum-likelihood probit models with sample selection.

Options

────────┐ Model └──

select(...) specifies the variables and options for the selection equation. It is an integral part of specifying a selection model and is required. The selection equation should contain at least one variable that is not in the outcome equation.

If *depvar_s* is specified, it should be coded as 0 or 1, 0 indicating an observation not selected and 1 indicating a selected observation. If *depvar_s* is not specified, observations for which *depvar* is not missing are assumed selected, and those for which *depvar* is missing are assumed not selected.

noconstant, offset(*varname*), constraints(*constraints*), collinear; see [R] **estimation options**.

────────┐ SE/Robust └───

vce(*vcetype*) specifies the type of standard error reported, which includes types that are derived from asymptotic theory, that are robust to some kinds of misspecification, that allow for intragroup correlation, and that use bootstrap or jackknife methods; see [R] *vce_option*.

────────┐ Reporting └───

level(*#*); see [R] **estimation options**.

first specifies that the first-step probit estimates of the selection equation be displayed before estimation.

noskip specifies that a full maximum-likelihood model with only a constant for the regression equation be fit. This model is not displayed but is used as the base model to compute a likelihood-ratio test for the model test statistic displayed in the estimation header. By default, the overall model test statistic is an asymptotically equivalent Wald test that all the parameters in the regression equation are zero (except the constant). For many models, this option can substantially increase estimation time.

nocnsreport; see [R] **estimation options**.

display_options: <u>noomitted</u>, vsquish, <u>noempty</u>cells, <u>base</u>levels, <u>allbase</u>levels, cformat(*%fmt*), pformat(*%fmt*), sformat(*%fmt*), and nolstretch; see [R] **estimation options**.

────────┐ Maximization └───

maximize_options: <u>dif</u>ficult, <u>tech</u>nique(*algorithm_spec*), <u>iter</u>ate(*#*), [<u>no</u>]<u>log</u>, <u>tra</u>ce, gradient, showstep, <u>hess</u>ian, showtolerance, <u>tol</u>erance(*#*), <u>ltol</u>erance(*#*), <u>nrtol</u>erance(*#*), <u>nonrtol</u>erance, and from(*init_specs*); see [R] **maximize**. These options are seldom used.

Setting the optimization type to technique(bhhh) resets the default *vcetype* to vce(opg).

The following option is available with `heckprob` but is not shown in the dialog box:

`coeflegend`; see [R] **estimation options**.

Remarks

The probit model with sample selection (Van de Ven and Van Pragg 1981) assumes that there exists an underlying relationship

$$y_j^* = \mathbf{x}_j\boldsymbol{\beta} + u_{1j} \qquad\qquad latent\ equation$$

such that we observe only the binary outcome

$$y_j^{\text{probit}} = (y_j^* > 0) \qquad\qquad probit\ equation$$

The dependent variable, however, is not always observed. Rather, the dependent variable for observation j is observed if

$$y_j^{\text{select}} = (\mathbf{z}_j\boldsymbol{\gamma} + u_{2j} > 0) \qquad\qquad selection\ equation$$

where

$$u_1 \sim N(0,1)$$
$$u_2 \sim N(0,1)$$
$$\text{corr}(u_1, u_2) = \rho$$

When $\rho \neq 0$, standard probit techniques applied to the first equation yield biased results. `heckprob` provides consistent, asymptotically efficient estimates for all the parameters in such models.

For the model to be well identified, the selection equation should have at least one variable that is not in the probit equation. Otherwise, the model is identified only by functional form, and the coefficients have no structural interpretation.

▷ Example 1

We use the data from Pindyck and Rubinfeld (1998). In this dataset, the variables are whether children attend private school (`private`), number of years the family has been at the present residence (`years`), log of property tax (`logptax`), log of income (`loginc`), and whether one voted for an increase in property taxes (`vote`).

In this example, we alter the meaning of the data. Here we assume that we observe whether children attend private school only if the family votes for increasing the property taxes. This assumption is not true in the dataset, and we make it only to illustrate the use of this command.

We observe whether children attend private school only if the head of household voted for an increase in property taxes. We assume that the vote is affected by the number of years in residence, the current property taxes paid, and the household income. We wish to model whether children are sent to private school on the basis of the number of years spent in the current residence and the current property taxes paid.

```
. use http://www.stata-press.com/data/r12/school

. heckprob private years logptax, select(vote=years loginc logptax)

Fitting probit model:

Iteration 0:   log likelihood = -17.122381
Iteration 1:   log likelihood = -16.243974
  (output omitted)
Iteration 5:   log likelihood = -15.883655

Fitting selection model:

Iteration 0:   log likelihood = -63.036914
Iteration 1:   log likelihood = -58.534843
Iteration 2:   log likelihood = -58.497292
Iteration 3:   log likelihood = -58.497288

Comparison:    log likelihood = -74.380943

Fitting starting values:

Iteration 0:   log likelihood = -40.895684
Iteration 1:   log likelihood = -16.654497
  (output omitted)
Iteration 6:   log likelihood = -15.753765

Fitting full model:

Iteration 0:   log likelihood = -75.010619  (not concave)
Iteration 1:   log likelihood = -74.287786
Iteration 2:   log likelihood = -74.250137
Iteration 3:   log likelihood = -74.245088
Iteration 4:   log likelihood = -74.244973
Iteration 5:   log likelihood = -74.244973
```

Probit model with sample selection

Number of obs	=	95
Censored obs	=	36
Uncensored obs	=	59
Wald chi2(2)	=	1.04
Prob > chi2	=	0.5935

Log likelihood = -74.24497

	Coef.	Std. Err.	z	P>|z|	[95% Conf. Interval]	
private						
years	-.1142597	.1461717	-0.78	0.434	-.400751	.1722317
logptax	.3516098	1.016485	0.35	0.729	-1.640665	2.343884
_cons	-2.780665	6.905838	-0.40	0.687	-16.31586	10.75453
vote						
years	-.0167511	.0147735	-1.13	0.257	-.0457067	.0122045
loginc	.9923024	.4430009	2.24	0.025	.1240366	1.860568
logptax	-1.278783	.5717545	-2.24	0.025	-2.399401	-.1581647
_cons	-.545821	4.070418	-0.13	0.893	-8.523694	7.432052
/athrho	-.8663156	1.450028	-0.60	0.550	-3.708318	1.975687
rho	-.6994973	.7405343			-.9987984	.962269

LR test of indep. eqns. (rho = 0):   chi2(1) =   0.27   Prob > chi2 = 0.6020

The output shows several iteration logs. The first iteration log corresponds to running the probit model for those observations in the sample where we have observed the outcome. The second iteration log corresponds to running the selection probit model, which models whether we observe our outcome of interest. If $\rho = 0$, the sum of the log likelihoods from these two models will equal the log likelihood of the probit model with sample selection; this sum is printed in the iteration log as the comparison log likelihood. The third iteration log shows starting values for the iterations.

The final iteration log is for fitting the full probit model with sample selection. A likelihood-ratio test of the log likelihood for this model and the comparison log likelihood is presented at the end of the output. If we had specified the vce(robust) option, this test would be presented as a Wald test instead of as a likelihood-ratio test.

◁

▷ Example 2

In example 1, we could have obtained robust standard errors by specifying the vce(robust) option. We do this here and also eliminate the iteration logs by using the nolog option:

```
. heckprob private years logptax, sel(vote=years loginc logptax) vce(robust) nolog
```

Probit model with sample selection				Number of obs	=	95
				Censored obs	=	36
				Uncensored obs	=	59
				Wald chi2(2)	=	2.55
Log pseudolikelihood = -74.24497				Prob > chi2	=	0.2798

	Coef.	Robust Std. Err.	z	P>|z|	[95% Conf. Interval]	
private						
years	-.1142597	.1113977	-1.03	0.305	-.3325951	.1040758
logptax	.3516098	.7358265	0.48	0.633	-1.090584	1.793803
_cons	-2.780665	4.786678	-0.58	0.561	-12.16238	6.601051
vote						
years	-.0167511	.0173344	-0.97	0.334	-.0507259	.0172237
loginc	.9923024	.4228044	2.35	0.019	.1636209	1.820984
logptax	-1.278783	.5095156	-2.51	0.012	-2.277415	-.2801508
_cons	-.545821	4.543892	-0.12	0.904	-9.451686	8.360044
/athrho	-.8663156	1.630643	-0.53	0.595	-4.062318	2.329687
rho	-.6994973	.8327753			-.9994079	.981233

Wald test of indep. eqns. (rho = 0): chi2(1) =   0.28   Prob > chi2 = 0.5952

Regardless of whether we specify the vce(robust) option, the outcome is not significantly different from the outcome obtained by fitting the probit and selection models separately. This result is not surprising because the selection mechanism estimated was invented for the example rather than borne from any economic theory.

◁

Saved results

heckprob saves the following in e():

Scalars

e(N)	number of observations
e(N_cens)	number of censored observations
e(k)	number of parameters
e(k_eq)	number of equations in e(b)
e(k_eq_model)	number of equations in overall model test
e(k_aux)	number of auxiliary parameters
e(k_dv)	number of dependent variables
e(df_m)	model degrees of freedom
e(ll)	log likelihood
e(ll_0)	log likelihood, constant-only model
e(ll_c)	log likelihood, comparison model
e(N_clust)	number of clusters
e(chi2)	χ^2
e(chi2_c)	χ^2 for comparison test
e(p_c)	p-value for comparison test
e(p)	significance of comparison test
e(rho)	ρ
e(rank)	rank of e(V)
e(rank0)	rank of e(V) for constant-only model
e(ic)	number of iterations
e(rc)	return code
e(converged)	1 if converged, 0 otherwise

Macros

e(cmd)	heckprob
e(cmdline)	command as typed
e(depvar)	names of dependent variables
e(wtype)	weight type
e(wexp)	weight expression
e(title)	title in estimation output
e(clustvar)	name of cluster variable
e(offset1)	offset for regression equation
e(offset2)	offset for selection equation
e(chi2type)	Wald or LR; type of model χ^2 test
e(chi2_ct)	type of comparison χ^2 test
e(vce)	*vcetype* specified in vce()
e(vcetype)	title used to label Std. Err.
e(opt)	type of optimization
e(which)	max or min; whether optimizer is to perform maximization or minimization
e(ml_method)	type of ml method
e(user)	name of likelihood-evaluator program
e(technique)	maximization technique
e(properties)	b V
e(predict)	program used to implement predict
e(asbalanced)	factor variables fvset as asbalanced
e(asobserved)	factor variables fvset as asobserved

Matrices

e(b)	coefficient vector
e(Cns)	constraints matrix
e(ilog)	iteration log (up to 20 iterations)
e(gradient)	gradient vector
e(V)	variance–covariance matrix of the estimators
e(V_modelbased)	model-based variance

Functions

e(sample)	marks estimation sample

Methods and formulas

`heckprob` is implemented as an ado-file. Van de Ven and Van Pragg (1981) provide an introduction and an explanation of this model.

The probit equation is

$$y_j = (\mathbf{x}_j\boldsymbol{\beta} + u_{1j} > 0)$$

The selection equation is

$$\mathbf{z}_j\boldsymbol{\gamma} + u_{2j} > 0$$

where

$$u_1 \sim N(0,1)$$
$$u_2 \sim N(0,1)$$
$$\mathrm{corr}(u_1, u_2) = \rho$$

The log likelihood is

$$
\begin{aligned}
\ln L = &\sum_{\substack{j \in S \\ y_j \neq 0}} w_j \ln \left\{ \Phi_2 \left(x_j\beta + \mathrm{offset}_j^\beta, z_j\gamma + \mathrm{offset}_j^\gamma, \rho \right) \right\} \\
&+ \sum_{\substack{j \in S \\ y_j = 0}} w_j \ln \left\{ \Phi_2 \left(-x_j\beta + \mathrm{offset}_j^\beta, z_j\gamma + \mathrm{offset}_j^\gamma, -\rho \right) \right\} \\
&+ \sum_{j \notin S} w_j \ln \left\{ 1 - \Phi \left(z_j\gamma + \mathrm{offset}_j^\gamma \right) \right\}
\end{aligned}
$$

where S is the set of observations for which y_j is observed, $\Phi_2(\cdot)$ is the cumulative bivariate normal distribution function (with mean $[0\ \ 0]'$), $\Phi(\cdot)$ is the standard cumulative normal, and w_j is an optional weight for observation j.

In the maximum likelihood estimation, ρ is not directly estimated. Directly estimated is atanh ρ:

$$\mathrm{atanh}\,\rho = \frac{1}{2}\ln\left(\frac{1+\rho}{1-\rho}\right)$$

From the form of the likelihood, it is clear that if $\rho = 0$, the log likelihood for the probit model with sample selection is equal to the sum of the probit model for the outcome y and the selection model. We can perform a likelihood-ratio test by comparing the likelihood of the full model with the sum of the log likelihoods for the probit and selection models.

This command supports the Huber/White/sandwich estimator of the variance and its clustered version using `vce(robust)` and `vce(cluster clustvar)`, respectively. See [P] **_robust**, particularly *Maximum likelihood estimators* and *Methods and formulas*.

`heckprob` also supports estimation with survey data. For details on VCEs with survey data, see [SVY] **variance estimation**.

References

Baum, C. F. 2006. *An Introduction to Modern Econometrics Using Stata.* College Station, TX: Stata Press.

Chiburis, R., and M. Lokshin. 2007. Maximum likelihood and two-step estimation of an ordered-probit selection model. *Stata Journal* 7: 167–182.

De Luca, G. 2008. SNP and SML estimation of univariate and bivariate binary-choice models. *Stata Journal* 8: 190–220.

Greene, W. H. 2012. *Econometric Analysis.* 7th ed. Upper Saddle River, NJ: Prentice Hall.

Heckman, J. 1979. Sample selection bias as a specification error. *Econometrica* 47: 153–161.

Miranda, A., and S. Rabe-Hesketh. 2006. Maximum likelihood estimation of endogenous switching and sample selection models for binary, ordinal, and count variables. *Stata Journal* 6: 285–308.

Muro, J., C. Suárez, and M. del Mar Zamora. 2010. Computing Murphy–Topel-corrected variances in a heckprobit model with endogeneity. *Stata Journal* 10: 252–258.

Pindyck, R. S., and D. L. Rubinfeld. 1998. *Econometric Models and Economic Forecasts.* 4th ed. New York: McGraw–Hill.

Van de Ven, W. P. M. M., and B. M. S. Van Pragg. 1981. The demand for deductibles in private health insurance: A probit model with sample selection. *Journal of Econometrics* 17: 229–252.

Also see

[R] **heckprob postestimation** — Postestimation tools for heckprob

[R] **heckman** — Heckman selection model

[R] **probit** — Probit regression

[R] **treatreg** — Treatment-effects model

[SVY] **svy estimation** — Estimation commands for survey data

[U] **20 Estimation and postestimation commands**

Title

Description

The following postestimation commands are available after `heckprob`:

Command	Description
contrast	contrasts and ANOVA-style joint tests of estimates
estat	AIC, BIC, VCE, and estimation sample summary
estat (svy)	postestimation statistics for survey data
estimates	cataloging estimation results
lincom	point estimates, standard errors, testing, and inference for linear combinations of coefficients
lrtest[1]	likelihood-ratio test
margins	marginal means, predictive margins, marginal effects, and average marginal effects
marginsplot	graph the results from margins (profile plots, interaction plots, etc.)
nlcom	point estimates, standard errors, testing, and inference for nonlinear combinations of coefficients
predict	predictions, residuals, influence statistics, and other diagnostic measures
predictnl	point estimates, standard errors, testing, and inference for generalized predictions
pwcompare	pairwise comparisons of estimates
suest	seemingly unrelated estimation
test	Wald tests of simple and composite linear hypotheses
testnl	Wald tests of nonlinear hypotheses

[1] `lrtest` is not appropriate with svy estimation results.

See the corresponding entries in the *Base Reference Manual* for details, but see [SVY] **estat** for details about `estat` (svy).

Syntax for predict

> predict [*type*] *newvar* [*if*] [*in*] [, *statistic* <u>nooff</u>set]

> predict [*type*] { *stub** | *newvar*_{reg} *newvar*_{sel} *newvar*_{athrho} } [*if*] [*in*] , <u>sc</u>ores

statistic	Description
Main	
<u>pmargin</u>	$\Phi(\mathbf{x}_j \mathbf{b})$, success probability; the default
p11	$\Phi_2(\mathbf{x}_j \mathbf{b}, \mathbf{z}_j \mathbf{g}, \rho)$, predicted probability $\Pr(y_j^{\text{probit}} = 1, y_j^{\text{select}} = 1)$
p10	$\Phi_2(\mathbf{x}_j \mathbf{b}, -\mathbf{z}_j \mathbf{g}, -\rho)$, predicted probability $\Pr(y_j^{\text{probit}} = 1, y_j^{\text{select}} = 0)$
p01	$\Phi_2(-\mathbf{x}_j \mathbf{b}, \mathbf{z}_j \mathbf{g}, -\rho)$, predicted probability $\Pr(y_j^{\text{probit}} = 0, y_j^{\text{select}} = 1)$
p00	$\Phi_2(-\mathbf{x}_j \mathbf{b}, -\mathbf{z}_j \mathbf{g}, \rho)$, predicted probability $\Pr(y_j^{\text{probit}} = 0, y_j^{\text{select}} = 0)$
<u>psel</u>	$\Phi(\mathbf{z}_j \mathbf{g})$, selection probability
<u>pcond</u>	$\Phi_2(\mathbf{x}_j \mathbf{b}, \mathbf{z}_j \mathbf{g}, \rho)/\Phi(\mathbf{z}_j \mathbf{g})$, probability of success conditional on selection
xb	linear prediction
stdp	standard error of the linear prediction
<u>xbsel</u>	linear prediction for selection equation
stdpsel	standard error of the linear prediction for selection equation

where $\Phi(\cdot)$ is the standard normal distribution function and $\Phi_2(\cdot)$ is the bivariate normal distribution function.

These statistics are available both in and out of sample; type predict ... if e(sample) ... if wanted only for the estimation sample.

Menu

Statistics > Postestimation > Predictions, residuals, etc.

Options for predict

    └ Main └

pmargin, the default, calculates the univariate (marginal) predicted probability of success $\Pr(y_j^{\text{probit}} = 1)$.

p11 calculates the bivariate predicted probability $\Pr(y_j^{\text{probit}} = 1, y_j^{\text{select}} = 1)$.

p10 calculates the bivariate predicted probability $\Pr(y_j^{\text{probit}} = 1, y_j^{\text{select}} = 0)$.

p01 calculates the bivariate predicted probability $\Pr(y_j^{\text{probit}} = 0, y_j^{\text{select}} = 1)$.

p00 calculates the bivariate predicted probability $\Pr(y_j^{\text{probit}} = 0, y_j^{\text{select}} = 0)$.

psel calculates the univariate (marginal) predicted probability of selection $\Pr(y_j^{\text{select}} = 1)$.

pcond calculates the conditional (on selection) predicted probability of success $\Pr(y_j^{\text{probit}} = 1, y_j^{\text{select}} = 1)/\Pr(y_j^{\text{select}} = 1)$.

xb calculates the probit linear prediction $\mathbf{x}_j \mathbf{b}$.

stdp calculates the standard error of the prediction, which can be thought of as the standard error of the predicted expected value or mean for the observation's covariate pattern. The standard error of the prediction is also referred to as the standard error of the fitted value.

xbsel calculates the linear prediction for the selection equation.

stdpsel calculates the standard error of the linear prediction for the selection equation.

nooffset is relevant only if you specified offset(*varname*) for heckprob. It modifies the calculations made by predict so that they ignore the offset variable; the linear prediction is treated as $\mathbf{x}_j\mathbf{b}$ rather than as $\mathbf{x}_j\mathbf{b} + \text{offset}_j$.

scores calculates equation-level score variables.

The first new variable will contain $\partial \ln L/\partial(\mathbf{x}_j\boldsymbol{\beta})$.

The second new variable will contain $\partial \ln L/\partial(\mathbf{z}_j\boldsymbol{\gamma})$.

The third new variable will contain $\partial \ln L/\partial(\text{atanh}\,\rho)$.

Remarks

▷ Example 1

It is instructive to compare the marginal predicted probabilities with the predicted probabilities that we would obtain by ignoring the selection mechanism. To compare the two approaches, we will synthesize data so that we know the "true" predicted probabilities.

First, we need to generate correlated error terms, which we can do using a standard Cholesky decomposition approach. For our example, we will clear any data from memory and then generate errors that have a correlation of 0.5 by using the following commands. We set the seed so that interested readers can type in these same commands and obtain the same results.

```
. set seed 12309
. set obs 5000
obs was 0, now 5000
. gen c1 = rnormal()
. gen c2 = rnormal()
. matrix P = (1,.5\.5,1)
. matrix A = cholesky(P)
. local fac1 = A[2,1]
. local fac2 = A[2,2]
. gen u1 = c1
. gen u2 = 'fac1'*c1 + 'fac2'*c2
```

We can check that the errors have the correct correlation by using the correlate command. We will also normalize the errors so that they have a standard deviation of one, so we can generate a bivariate probit model with known coefficients. We do that with the following commands:

```
. correlate u1 u2
(obs=5000)
                 |      u1       u2
      -----------+------------------
            u1 |  1.0000
            u2 |  0.5020   1.0000

. summarize u1
  (output omitted )
. replace u1 = u1/r(sd)
(5000 real changes made)
. summarize u2
  (output omitted )
. replace u2 = u2/r(sd)
(5000 real changes made)
. drop c1 c2
. gen x1 = runiform()-.5
. gen x2 = runiform()+1/3
. gen y1s = 0.5 + 4*x1 + u1
. gen y2s = 3 - 3*x2 + .5*x1 + u2
. gen y1 = (y1s>0)
. gen y2 = (y2s>0)
```

We have now created two dependent variables, y1 and y2, which are defined by our specified coefficients. We also included error terms for each equation, and the error terms are correlated. We run heckprob to verify that the data have been correctly generated according to the model

$$y_1 = .5 + 4x_1 + u_1$$
$$y_2 = 3 + .5x_1 - 3x_2 + u_2$$

where we assume that y_1 is observed only if $y_2 = 1$.

```
. heckprob y1 x1, sel(y2 = x1 x2) nolog
Probit model with sample selection          Number of obs    =      5000
                                            Censored obs     =      1762
                                            Uncensored obs   =      3238

                                            Wald chi2(1)     =    953.71
Log likelihood =    -3679.5                 Prob > chi2      =    0.0000
```

	Coef.	Std. Err.	z	P>|z|	[95% Conf. Interval]
y1						
x1	3.784705	.1225532	30.88	0.000	3.544505	4.024905
_cons	.4630922	.0453952	10.20	0.000	.3741192	.5520653
y2						
x1	.3693052	.0721694	5.12	0.000	.2278558	.5107547
x2	-3.05069	.0832424	-36.65	0.000	-3.213842	-2.887538
_cons	3.037696	.0777733	39.06	0.000	2.885263	3.190128
/athrho	.5186232	.083546	6.21	0.000	.354876	.6823705
rho	.4766367	.0645658			.3406927	.5930583

```
LR test of indep. eqns. (rho = 0):    chi2(1) =    40.43   Prob > chi2 = 0.0000
```

Now that we have verified that we have generated data according to a known model, we can obtain and then compare predicted probabilities from the probit model with sample selection and a (usual) probit model.

```
. predict pmarg
(option pmargin assumed; Pr(y1=1))
. probit y1 x1 if y2==1
(output omitted)
. predict phat
(option pr assumed; Pr(y1))
```

Using the (marginal) predicted probabilities from the probit model with sample selection (`pmarg`) and the predicted probabilities from the (usual) probit model (`phat`), we can also generate the "true" predicted probabilities from the synthesized `y1s` variable and then compare the predicted probabilities:

```
. gen ptrue = normal(y1s)
. summarize pmarg ptrue phat
```

Variable	Obs	Mean	Std. Dev.	Min	Max
pmarg	5000	.6071226	.3147861	.0766334	.9907113
ptrue	5000	.5974195	.348396	5.53e-06	.9999999
phat	5000	.6568175	.3025085	.1059824	.9954919

Here we see that ignoring the selection mechanism (comparing the `phat` variable with the true `ptrue` variable) results in predicted probabilities that are much higher than the true values. Looking at the marginal predicted probabilities from the model with sample selection, however, results in more accurate predictions.

◁

Methods and formulas

All postestimation commands listed above are implemented as ado-files.

Also see

[R] **heckprob** — Probit model with sample selection

[U] **20 Estimation and postestimation commands**

help — Display online help

Display help information in Viewer

> <u>he</u>lp [*command_or_topic_name*] [, <u>non</u>ew name(*viewername*) <u>marker</u>(*markername*)]

Display help information in Results window

> <u>ch</u>elp [*command_or_topic_name*]

Help > Stata Command...

The help command displays help information about the specified command or topic.

Stata for Mac, Stata for Unix(GUI), and Stata for Windows:

help launches a new Viewer to display help for the specified command or topic. If help is not followed by a command or a topic name, Stata launches the Viewer and displays help contents, the table of contents for the online help.

Help may be accessed either by selecting **Help > Stata Command...** and filling in the desired command name or by typing help followed by a command or topic name.

chelp will display help in the Results window.

Stata for Unix(console):

Typing help followed by a command name or a topic name will display help on the console.

If help is not followed by a command or a topic name, a description of how to use the help system is displayed.

Stata for Unix(both GUI and console):

man is a synonym for chelp.

nonew specifies that a new Viewer window not be opened for the help topic if a Viewer window is already open. The default is for a new Viewer window to be opened each time help is typed so that multiple help files may be viewed at once. nonew causes the help file to be displayed in the topmost open Viewer.

name(*viewername*) specifies that help be displayed in a Viewer window named *viewername*. If the named window already exists, its contents will be replaced. If the named window does not exist, it will be created.

`marker(`*markername*`)` specifies that the help file be opened to the position of *markername* within the help file.

Remarks

To obtain help for any Stata command, type `help` *command* or select **Help > Stata Command...** and fill in *command*.

`help` is best explained by examples.

To obtain help for ...	type
`regress`	`help regress`
postestimation tools for `regress`	`help regress postestimation` or
	`help regress post`
graph option `xlabel()`	`help graph xlabel()`
Stata function `strpos()`	`help strpos()`
Mata function `optimize()`	`help mata optimize()`

Tips:

- `help` displays a subject table of contents for the online help.

- `help guide` displays a table of contents for basic Stata concepts.

- `help estimation commands` displays an alphabetical listing of all Stata estimation commands.

- `help functions` displays help on Stata functions by category.

- `help mata functions` displays a subject table of contents for Mata's functions.

- `help ts glossary` displays the glossary for the time-series manual, and similarly for the other Stata specialty manuals.

See [U] **4 Stata's help and search facilities** for a complete description of how to use `help`.

❑ Technical note

When you type `help` *topic*, Stata first looks along the adopath for *topic*`.sthlp`; see [U] **17.5 Where does Stata look for ado-files?**. ❑

Also see

[R] **hsearch** — Search help files

[R] **search** — Search Stata documentation

[R] **net search** — Search the Internet for installable packages

[GSM] **4 Getting help**

[GSW] **4 Getting help**

[GSU] **4 Getting help**

[U] **4 Stata's help and search facilities**

Title

hetprob — Heteroskedastic probit model

Syntax

hetprob *depvar* [*indepvars*] [*if*] [*in*] [*weight*] ,

     het(*varlist*[, <u>off</u>set(*varname*)]) [*options*]

options	Description
Model	
* het(*varlist*[...])	independent variables to model the variance and possible offset variable
<u>nocon</u>stant	suppress constant term
<u>off</u>set(*varname*)	include *varname* in model with coefficient constrained to 1
asis	retain perfect predictor variables
<u>constraints</u>(*constraints*)	apply specified linear constraints
<u>coll</u>inear	keep collinear variables
SE/Robust	
vce(*vcetype*)	*vcetype* may be oim, <u>r</u>obust, <u>cl</u>uster *clustvar*, opg, <u>boot</u>strap, or <u>jack</u>knife
Reporting	
<u>level</u>(#)	set confidence level; default is level(95)
noskip	perform likelihood-ratio test
<u>nolr</u>test	perform Wald test on variance
<u>nocns</u>report	do not display constraints
display_options	control column formats, row spacing, line width, and display of omitted variables and base and empty cells
Maximization	
maximize_options	control the maximization process; seldom used
<u>coefl</u>egend	display legend instead of statistics

*het() is required. The full specification is het(*varlist* [, <u>off</u>set(*varname*)]).

indepvars and *varlist* may contain factor variables; see [U] **11.4.3 Factor variables**.

depvar, *indepvars*, and *varlist* may contain time-series operators; see [U] **11.4.4 Time-series varlists**.

bootstrap, by, jackknife, rolling, statsby, and svy are allowed; see [U] **11.1.10 Prefix commands**.

Weights are not allowed with the bootstrap prefix; see [R] **bootstrap**.

vce(), noskip, and weights are not allowed with the svy prefix; see [SVY] **svy**.

fweights, iweights, and pweights are allowed; see [U] **11.1.6 weight**.

coeflegend does not appear in the dialog box.

See [U] **20 Estimation and postestimation commands** for more capabilities of estimation commands.

Menu

Statistics > Binary outcomes > Heteroskedastic probit regression

Description

hetprob fits a maximum-likelihood heteroskedastic probit model.

See [R] **logistic** for a list of related estimation commands.

Options

    ⌐Model⌐

het(*varlist* [, offset(*varname*)]) specifies the independent variables and the offset variable, if there is one, in the variance function. het() is required.

noconstant, offset(*varname*); see [R] **estimation options**.

asis forces the retention of perfect predictor variables and their associated perfectly predicted observations and may produce instabilities in maximization; see [R] **probit**.

constraints(*constraints*), collinear; see [R] **estimation options**.

    ⌐SE/Robust⌐

vce(*vcetype*) specifies the type of standard error reported, which includes types that are derived from asymptotic theory, that are robust to some kinds of misspecification, that allow for intragroup correlation, and that use bootstrap or jackknife methods; see [R] *vce_option*.

    ⌐Reporting⌐

level(*#*); see [R] **estimation options**.

noskip requests fitting of the constant-only model and calculation of the corresponding likelihood-ratio χ^2 statistic for testing significance of the full model. By default, a Wald χ^2 statistic is computed for testing the significance of the full model.

nolrtest specifies that a Wald test of whether lnsigma2 = 0 be performed instead of the LR test.

nocnsreport; see [R] **estimation options**.

display_options: noomitted, vsquish, noemptycells, baselevels, allbaselevels, cformat(%*fmt*), pformat(%*fmt*), sformat(%*fmt*), and nolstretch; see [R] **estimation options**.

    ⌐Maximization⌐

maximize_options: difficult, technique(*algorithm_spec*), iterate(*#*), [no]log, trace, gradient, showstep, hessian, showtolerance, tolerance(*#*), ltolerance(*#*), nrtolerance(*#*), nonrtolerance, and from(*init_specs*); see [R] **maximize**. These options are seldom used.

Setting the optimization type to technique(bhhh) resets the default *vcetype* to vce(opg).

The following option is available with hetprob but is not shown in the dialog box:

coeflegend; see [R] **estimation options**.

Remarks

Remarks are presented under the following headings:

Introduction
Robust standard errors

Introduction

hetprob fits a maximum-likelihood heteroskedastic probit model, which is a generalization of the probit model. Let $y_j, j = 1, \ldots, N$, be a binary outcome variable taking on the value 0 (failure) or 1 (success). In the probit model, the probability that y_j takes on the value 1 is modeled as a nonlinear function of a linear combination of the k independent variables $\mathbf{x}_j = (x_{1j}, x_{2j}, \ldots, x_{kj})$,

$$\Pr(y_j = 1) = \Phi(\mathbf{x}_j \mathbf{b})$$

in which $\Phi()$ is the cumulative distribution function (CDF) of a standard normal random variable, that is, a normally distributed (Gaussian) random variable with mean 0 and variance 1. The linear combination of the independent variables, $\mathbf{x}_j \mathbf{b}$, is commonly called the *index function*, or *index*. Heteroskedastic probit generalizes the probit model by generalizing $\Phi()$ to a normal CDF with a variance that is no longer fixed at 1 but can vary as a function of the independent variables. hetprob models the variance as a multiplicative function of these m variables $\mathbf{z}_j = (z_{1j}, z_{2j}, \ldots, z_{mj})$, following Harvey (1976):

$$\sigma_j^2 = \{\exp(\mathbf{z}_j \boldsymbol{\gamma})\}^2$$

Thus the probability of success as a function of all the independent variables is

$$\Pr(y_j = 1) = \Phi\left\{\mathbf{x}_j \mathbf{b} / \exp(\mathbf{z}_j \boldsymbol{\gamma})\right\}$$

From this expression, it is clear that, unlike the index $\mathbf{x}_j \mathbf{b}$, no constant term can be present in $\mathbf{z}_j \boldsymbol{\gamma}$ if the model is to be identifiable.

Suppose that the binary outcomes y_j are generated by thresholding an unobserved random variable, w, which is normally distributed with mean $\mathbf{x}_j \mathbf{b}$ and variance 1 such that

$$y_j = \begin{cases} 1 & \text{if } w_j > 0 \\ 0 & \text{if } w_j \leq 0 \end{cases}$$

This process gives the probit model:

$$\Pr(y_j = 1) = \Pr(w_j > 0) = \Phi(\mathbf{x}_j \mathbf{b})$$

Now suppose that the unobserved w_j are heteroskedastic with variance

$$\sigma_j^2 = \{\exp(\mathbf{z}_j \boldsymbol{\gamma})\}^2$$

Relaxing the homoskedastic assumption of the probit model in this manner yields our multiplicative heteroskedastic probit model:

$$\Pr(y_j = 1) = \Phi\left\{\mathbf{x}_j \mathbf{b} / \exp(\mathbf{z}_j \boldsymbol{\gamma})\right\}$$

▷ Example 1

For this example, we generate simulated data for a simple heteroskedastic probit model and then estimate the coefficients with `hetprob`:

```
. set obs 1000
obs was 0, now 1000
. set seed 1234567
. gen x = 1-2*runiform()
. gen xhet = runiform()
. gen sigma = exp(1.5*xhet)
. gen p = normal((0.3+2*x)/sigma)
. gen y = cond(runiform()<=p,1,0)
. hetprob y x, het(xhet)
Fitting probit model:
Iteration 0:   log likelihood = -688.53208
Iteration 1:   log likelihood = -591.59895
Iteration 2:   log likelihood = -591.50674
Iteration 3:   log likelihood = -591.50674
Fitting full model:
Iteration 0:   log likelihood = -591.50674
Iteration 1:   log likelihood = -572.12219
Iteration 2:   log likelihood =  -570.7742
Iteration 3:   log likelihood = -569.48921
Iteration 4:   log likelihood = -569.47828
Iteration 5:   log likelihood = -569.47827
```

Heteroskedastic probit model				Number of obs	=	1000
				Zero outcomes	=	452
				Nonzero outcomes	=	548
				Wald chi2(1)	=	78.66
Log likelihood = -569.4783				Prob > chi2	=	0.0000

y	Coef.	Std. Err.	z	P>|z|	[95% Conf. Interval]	
y						
x	2.228031	.2512073	8.87	0.000	1.735673	2.720388
_cons	.2493822	.0862833	2.89	0.004	.08027	.4184943
lnsigma2						
xhet	1.602537	.2640131	6.07	0.000	1.085081	2.119993

```
Likelihood-ratio test of lnsigma2=0:  chi2(1) =    44.06   Prob > chi2 = 0.0000
```

Above we created two variables, x and xhet, and then simulated the model

$$\Pr(y = 1) = F\Big\{(\beta_0 + \beta_1 x)/\exp(\gamma_1 \text{xhet})\Big\}$$

for $\beta_0 = 0.3$, $\beta_1 = 2$, and $\gamma_1 = 1.5$. According to `hetprob`'s output, all coefficients are significant, and, as we would expect, the Wald test of the full model versus the constant-only model—for example, the index consisting of $\beta_0 + \beta_1 x$ versus that of just β_0—is significant with $\chi^2(1) = 79$. Likewise, the likelihood-ratio test of heteroskedasticity, which tests the full model with heteroskedasticity against the full model without, is significant with $\chi^2(1) = 44$. See [R] **maximize** for more explanation of the output. For this simple model, `hetprob` took five iterations to converge. As stated elsewhere (Greene 2012, 714), this is a difficult model to fit, and it is not uncommon for it to require many iterations or for the optimizer to print out warnings and informative messages during the optimization. Slow convergence is especially common for models in which one or more of the independent variables appear in both the index and variance functions.   ◁

❏ Technical note

Stata interprets a value of 0 as a negative outcome (failure) and treats all other values (except missing) as positive outcomes (successes). Thus if your dependent variable takes on the values 0 and 1, then 0 is interpreted as failure and 1 as success. If your dependent variable takes on the values 0, 1, and 2, then 0 is still interpreted as failure, but both 1 and 2 are treated as successes.

❏

Robust standard errors

If you specify the vce(robust) option, hetprob reports robust standard errors as described in [U] **20.20 Obtaining robust variance estimates**. To illustrate the effect of this option, we will reestimate our coefficients by using the same model and data in our example, this time adding vce(robust) to our hetprob command.

▷ Example 2

```
. hetprob y x, het(xhet) vce(robust) nolog
Heteroskedastic probit model                Number of obs   =      1000
                                            Zero outcomes   =       452
                                            Nonzero outcomes =      548

                                            Wald chi2(1)    =     65.23
Log pseudolikelihood = -569.4783            Prob > chi2     =    0.0000
```

	y	Coef.	Robust Std. Err.	z	P>|z|	[95% Conf.	Interval]
y							
	x	2.22803	.2758597	8.08	0.000	1.687355	2.768705
	_cons	.2493821	.0843367	2.96	0.003	.0840853	.4146789
lnsigma2							
	xhet	1.602537	.2671326	6.00	0.000	1.078967	2.126107

```
Wald test of lnsigma2=0:              chi2(1) =    35.99   Prob > chi2 = 0.0000
```

The vce(robust) standard errors for two of the three parameters are larger than the previously reported conventional standard errors. This is to be expected, even though (by construction) we have perfect model specification because this option trades off efficient estimation of the coefficient variance–covariance matrix for robustness against misspecification.

◁

Specifying the vce(cluster *clustvar*) option relaxes the usual assumption of independence between observations to the weaker assumption of independence just between clusters; that is, hetprob, vce(cluster *clustvar*) is robust with respect to within-cluster correlation. This option is less efficient than the xtgee population-averaged models because hetprob inefficiently sums within cluster for the standard-error calculation rather than attempting to exploit what might be assumed about the within-cluster correlation.

Saved results

hetprob saves the following in e():

Scalars
e(N)	number of observations
e(N_f)	number of zero outcomes
e(N_s)	number of nonzero outcomes
e(k)	number of parameters
e(k_eq)	number of equations in e(b)
e(k_eq_model)	number of equations in overall model test
e(k_dv)	number of dependent variables
e(df_m)	model degrees of freedom
e(ll)	log likelihood
e(ll_0)	log likelihood, constant-only model
e(ll_c)	log likelihood, comparison model
e(N_clust)	number of clusters
e(chi2)	χ^2
e(chi2_c)	χ^2 for heteroskedasticity LR test
e(p_c)	p-value for heteroskedasticity LR test
e(df_m_c)	degrees of freedom for heteroskedasticity LR test
e(p)	significance
e(rank)	rank of e(V)
e(rank0)	rank of e(V) for constant-only model
e(ic)	number of iterations
e(rc)	return code
e(converged)	1 if converged, 0 otherwise

Macros
e(cmd)	hetprob
e(cmdline)	command as typed
e(depvar)	name of dependent variable
e(wtype)	weight type
e(wexp)	weight expression
e(title)	title in estimation output
e(clustvar)	name of cluster variable
e(offset1)	offset for probit equation
e(offset2)	offset for variance equation
e(chi2type)	Wald or LR; type of model χ^2 test
e(chi2_ct)	Wald or LR; type of model χ^2 test corresponding to e(chi2_c)
e(vce)	*vcetype* specified in vce()
e(vcetype)	title used to label Std. Err.
e(opt)	type of optimization
e(which)	max or min; whether optimizer is to perform maximization or minimization
e(method)	requested estimation method
e(ml_method)	type of ml method
e(user)	name of likelihood-evaluator
e(technique)	maximization technique
e(properties)	b V
e(predict)	program used to implement predict
e(asbalanced)	factor variables fvset as asbalanced
e(asobserved)	factor variables fvset as asobserved

Matrices
e(b)	coefficient vector
e(Cns)	constraints matrix
e(ilog)	iteration log (up to 20 iterations)
e(gradient)	gradient vector
e(V)	variance–covariance matrix of the estimators
e(V_modelbased)	model-based variance

Functions
e(sample)	marks estimation sample

Methods and formulas

`hetprob` is implemented as an ado-file.

The heteroskedastic probit model is a generalization of the probit model because it allows the scale of the inverse link function to vary from observation to observation as a function of the independent variables.

The log-likelihood function for the heteroskedastic probit model is

$$\ln L = \sum_{j \in S} w_j \ln\Phi\{\mathbf{x}_j\boldsymbol{\beta}/\exp(\mathbf{z}\boldsymbol{\gamma})\} + \sum_{j \notin S} w_j \ln\left[1 - \Phi\{\mathbf{x}_j\boldsymbol{\beta}/\exp(\mathbf{z}\boldsymbol{\gamma})\}\right]$$

where S is the set of all observations j such that $y_j \neq 0$ and w_j denotes the optional weights. $\ln L$ is maximized as described in [R] **maximize**.

This command supports the Huber/White/sandwich estimator of the variance and its clustered version using `vce(robust)` and `vce(cluster` *clustvar*`)`, respectively. See [P] **_robust**, particularly *Maximum likelihood estimators* and *Methods and formulas*.

`hetprob` also supports estimation with survey data. For details on VCEs with survey data, see [SVY] **variance estimation**.

References

Greene, W. H. 2012. *Econometric Analysis*. 7th ed. Upper Saddle River, NJ: Prentice Hall.

Harvey, A. C. 1976. Estimating regression models with multiplicative heteroscedasticity. *Econometrica* 44: 461–465.

Also see

[R] **hetprob postestimation** — Postestimation tools for hetprob

[R] **logistic** — Logistic regression, reporting odds ratios

[R] **probit** — Probit regression

[SVY] **svy estimation** — Estimation commands for survey data

[XT] **xtprobit** — Random-effects and population-averaged probit models

[U] **20 Estimation and postestimation commands**

Title

hetprob postestimation — Postestimation tools for hetprob

Description

The following postestimation commands are available after `hetprob`:

Command	Description
contrast	contrasts and ANOVA-style joint tests of estimates
estat	AIC, BIC, VCE, and estimation sample summary
estat (svy)	postestimation statistics for survey data
estimates	cataloging estimation results
lincom	point estimates, standard errors, testing, and inference for linear combinations of coefficients
linktest	link test for model specification
lrtest[1]	likelihood-ratio test
margins	marginal means, predictive margins, marginal effects, and average marginal effects
marginsplot	graph the results from margins (profile plots, interaction plots, etc.)
nlcom	point estimates, standard errors, testing, and inference for nonlinear combinations of coefficients
predict	predictions, residuals, influence statistics, and other diagnostic measures
predictnl	point estimates, standard errors, testing, and inference for generalized predictions
pwcompare	pairwise comparisons of estimates
suest	seemingly unrelated estimation
test	Wald tests of simple and composite linear hypotheses
testnl	Wald tests of nonlinear hypotheses

[1] `lrtest` is not appropriate with svy estimation results.

See the corresponding entries in the *Base Reference Manual* for details, but see [SVY] **estat** for details about `estat` (svy).

Syntax for predict

predict [*type*] *newvar* [*if*] [*in*] [, *statistic* <u>nooff</u>set]

predict [*type*] { *stub** | *newvar*$_{reg}$ *newvar*$_{lnsigma2}$ } [*if*] [*in*] , <u>sc</u>ores

statistic	Description
Main	
<u>pr</u>	probability of a positive outcome; the default
xb	linear prediction
<u>sigma</u>	standard deviation of the error term

These statistics are available both in and out of sample; type `predict ... if e(sample) ...` if wanted only for the estimation sample.

Menu

Statistics > Postestimation > Predictions, residuals, etc.

Options for predict

⌐‾‾‾⌐ Main ⌐‾‾‾

pr, the default, calculates the probability of a positive outcome.

xb calculates the linear prediction.

sigma calculates the standard deviation of the error term.

nooffset is relevant only if you specified offset(*varname*) for hetprob. It modifies the calculations made by predict so that they ignore the offset variable; the linear prediction is treated as $\mathbf{x}_j\mathbf{b}$ rather than as $\mathbf{x}_j\mathbf{b} + \text{offset}_j$.

scores calculates equation-level score variables.

The first new variable will contain $\partial \ln L / \partial(\mathbf{x}_j\boldsymbol{\beta})$.

The second new variable will contain $\partial \ln L / \partial(\mathbf{z}_j\boldsymbol{\gamma})$.

Remarks

Once you have fit a model, you can use the predict command to obtain the predicted probabilities for both the estimation sample and other samples; see [U] **20 Estimation and postestimation commands** and [R] **predict**. predict without arguments calculates the predicted probability of a positive outcome. With the xb option, predict calculates the index function combination, $\mathbf{x}_j\mathbf{b}$, where $\mathbf{x}_j$ are the independent variables in the jth observation and $\mathbf{b}$ is the estimated parameter vector. With the sigma option, predict calculates the predicted standard deviation, $\sigma_j = \exp(\mathbf{z}_j\boldsymbol{\gamma})$.

▷ Example 1

We use predict to compute the predicted probabilities and standard deviations based on the model in example 2 in [R] **hetprob** to compare these with the actual values:

```
. predict phat
(option pr assumed; Pr(y))
. gen diff_p = phat - p
. summarize diff_p
```

Variable	Obs	Mean	Std. Dev.	Min	Max
diff_p	1000	-.0107081	.0131869	-.0466331	.010482

```
. predict sigmahat, sigma
. gen diff_s = sigmahat - sigma
. summarize diff_s
```

Variable	Obs	Mean	Std. Dev.	Min	Max
diff_s	1000	.1558881	.1363698	.0000417	.4819107

◁

Methods and formulas

All postestimation commands listed above are implemented as ado-files.

Also see

[R] **hetprob** — Heteroskedastic probit model

[U] **20 Estimation and postestimation commands**

Title

> **histogram** — Histograms for continuous and categorical variables

Syntax

<u>histo</u>gram *varname* [*if*] [*in*] [*weight*] [, [*continuous_opts* | *discrete_opts*] *options*]

continuous_opts	Description
Main	
bin(*#*)	set number of bins to *#*
<u>w</u>idth(*#*)	set width of bins to *#*
start(*#*)	set lower limit of first bin to *#*

discrete_opts	Description
Main	
<u>discrete</u>	specify that data are discrete
<u>w</u>idth(*#*)	set width of bins to *#*
start(*#*)	set theoretical minimum value to *#*

options	Description
Main	
<u>den</u>sity	draw as density; the default
<u>frac</u>tion	draw as fractions
<u>freq</u>uency	draw as frequencies
<u>per</u>cent	draw as percentages
bar_options	rendition of bars
<u>addl</u>abels	add height labels to bars
addlabopts(*marker_label_options*)	affect rendition of labels
Density plots	
<u>normal</u>	add a normal density to the graph
normopts(*line_options*)	affect rendition of normal density
<u>kden</u>sity	add a kernel density estimate to the graph
kdenopts(*kdensity_options*)	affect rendition of kernel density
Add plots	
addplot(*plot*)	add other plots to the histogram
Y axis, X axis, Titles, Legend, Overall, By	
twoway_options	any options documented in [G-3] ***twoway_options***

fweights are allowed; see [U] **11.1.6 weight**.

763

Menu

Graphics > Histogram

Description

histogram draws histograms of *varname*, which is assumed to be the name of a continuous variable unless the discrete option is specified.

Options for use in the continuous case

    ⌐ Main ⌐

bin(*#*) and width(*#*) are alternatives. They specify how the data are to be aggregated into bins: bin() by specifying the number of bins (from which the width can be derived) and width() by specifying the bin width (from which the number of bins can be derived).

If neither option is specified, results are the same as if bin(*k*) had been specified, where

$$k = \min\left\{\operatorname{sqrt}(N), 10\ln(N)/\ln(10)\right\}$$

and where N is the (weighted) number of observations.

start(*#*) specifies the theoretical minimum of *varname*. The default is start(*m*), where *m* is the observed minimum value of *varname*.

Specify start() when you are concerned about sparse data, for instance, if you know that *varname* can have a value of 0, but you are concerned that 0 may not be observed.

start(*#*), if specified, must be less than or equal to *m*, or else an error will be issued.

Options for use in the discrete case

    ⌐ Main ⌐

discrete specifies that *varname* is discrete and that you want each unique value of *varname* to have its own bin (bar of histogram).

width(*#*) is rarely specified in the discrete case; it specifies the width of the bins. The default is width(*d*), where *d* is the observed minimum difference between the unique values of *varname*.

Specify width() if you are concerned that your data are sparse. For example, in theory *varname* could take on the values, say, 1, 2, 3, ..., 9, but because of the sparseness, perhaps only the values 2, 4, 7, and 8 are observed. Here the default width calculation would produce width(2), and you would want to specify width(1).

start(*#*) is also rarely specified in the discrete case; it specifies the theoretical minimum value of *varname*. The default is start(*m*), where *m* is the observed minimum value.

As with width(), specify start(*#*) if you are concerned that your data are sparse. In the previous example, you might also want to specify start(1). start() does nothing more than add white space to the left side of the graph.

The value of *#* in start() must be less than or equal to *m*, or an error will be issued.

Options for use in both the continuous and discrete cases

⌐ Main ⌐

density, fraction, frequency, and percent specify whether you want the histogram scaled to density units, fractional units, frequencies, or percentages. density is the default.

density scales the height of the bars so that the sum of their areas equals 1.

fraction scales the height of the bars so that the sum of their heights equals 1.

frequency scales the height of the bars so that each bar's height is equal to the number of observations in the category. Thus the sum of the heights is equal to the total number of observations.

percent scales the height of the bars so that the sum of their heights equals 100.

bar_options are any of the options allowed by graph twoway bar; see [G-2] **graph twoway bar**.

One of the most useful *bar_options* is barwidth(#), which specifies the width of the bars in *varname* units. By default, histogram draws the bars so that adjacent bars just touch. If you want gaps between the bars, do not specify histogram's width() option—which would change how the histogram is calculated—but specify the *bar_option* barwidth() or the histogram option gap, both of which affect only how the bar is rendered.

The *bar_option* horizontal cannot be used with the addlabels option.

addlabels specifies that the top of each bar be labeled with the density, fraction, or frequency, as determined by the density, fraction, and frequency options.

addlabopts(*marker_label_options*) specifies how to render the labels atop the bars. See [G-3] **marker_label_options**. Do not specify the *marker_label_option* mlabel(*varname*), which specifies the variable to be used; this is specified for you by histogram.

addlabopts() will accept more options than those documented in [G-3] **marker_label_options**. All options allowed by twoway scatter are also allowed by addlabopts(); see [G-2] **graph twoway scatter**. One particularly useful option is yvarformat(); see [G-3] **advanced_options**.

⌐ Density plots ⌐

normal specifies that the histogram be overlaid with an appropriately scaled normal density. The normal will have the same mean and standard deviation as the data.

normopts(*line_options*) specifies details about the rendition of the normal curve, such as the color and style of line used. See [G-2] **graph twoway line**.

kdensity specifies that the histogram be overlaid with an appropriately scaled kernel density estimate of the density. By default, the estimate will be produced using the Epanechnikov kernel with an "optimal" half-width. This default corresponds to the default of kdensity; see [R] **kdensity**. How the estimate is produced can be controlled using the kdenopts() option described below.

kdenopts(*kdensity_options*) specifies details about how the kernel density estimate is to be produced along with details about the rendition of the resulting curve, such as the color and style of line used. The kernel density estimate is described in [G-2] **graph twoway kdensity**. As an example, if you wanted to produce kernel density estimates by using the Gaussian kernel with optimal half-width, you would specify kdenopts(gauss) and if you also wanted a half-width of 5, you would specify kdenopts(gauss width(5)).

⌐ Add plots ⌐

addplot(*plot*) allows adding more graph twoway plots to the graph; see [G-3] *addplot_option*.

⌐ Y axis, X axis, Titles, Legend, Overall, By ⌐

twoway_options are any of the options documented in [G-3] *twoway_options*. This includes, most importantly, options for titling the graph (see [G-3] *title_options*), options for saving the graph to disk (see [G-3] *saving_option*), and the by() option, which will allow you to simultaneously graph histograms for different subsets of the data (see [G-3] *by_option*).

Remarks

Remarks are presented under the following headings:

> *Histograms of continuous variables*
> *Overlaying normal and kernel density estimates*
> *Histograms of discrete variables*
> *Use with by()*

For an example of editing a histogram with the Graph Editor, see Pollock (2011, 29–31).

Histograms of continuous variables

histogram assumes that the variable is continuous, so you need type only histogram followed by the variable name:

```
. use http://www.stata-press.com/data/r12/sp500
(S&P 500)

. histogram volume
(bin=15, start=4103, width=1280.3533)
```

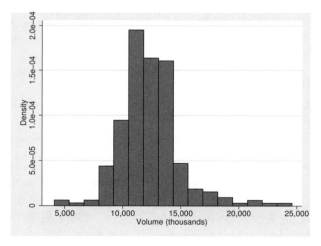

The small values reported for density on the *y* axis are correct; if you added up the area of the bars, you would get 1. Nevertheless, many people are used to seeing histograms scaled so that the bar heights sum to 1,

```
. histogram volume, fraction
(bin=15, start=4103, width=1280.3533)
```

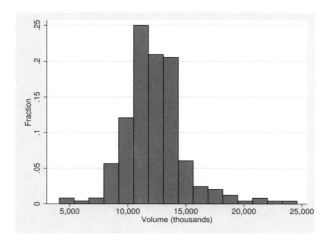

and others are used to seeing histograms so that the bar height reflects the number of observations,

```
. histogram volume, frequency
(bin=15, start=4103, width=1280.3533)
```

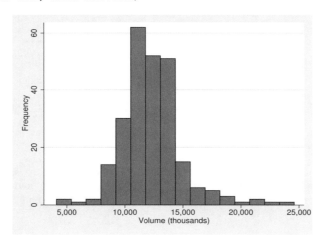

Regardless of the scale you prefer, you can specify other options to make the graph look more impressive:

```
. summarize volume
    Variable |       Obs        Mean    Std. Dev.       Min        Max
-------------+--------------------------------------------------------
      volume |       248    12320.68    2585.929       4103    23308.3
. histogram volume, freq
>         xaxis(1 2)
>         ylabel(0(10)60, grid)
>         xlabel(12321 "mean"
>              9735 "-1 s.d."
>             14907 "+1 s.d."
>              7149 "-2 s.d."
>             17493 "+2 s.d."
>             20078 "+3 s.d."
>             22664 "+4 s.d."
>                                    , axis(2) grid gmax)
>         xtitle("", axis(2))
>         subtitle("S&P 500, January 2001 - December 2001")
>         note("Source:  Yahoo! Finance and Commodity Systems, Inc.")
(bin=15, start=4103, width=1280.3533)
```

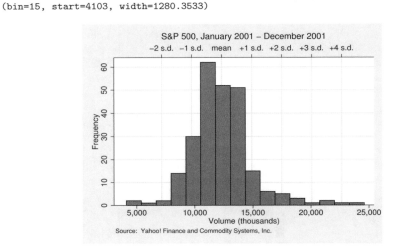

For an explanation of the xaxis() option—it created the upper and lower x axis—see [G-3] *axis_choice_options*. For an explanation of the ylabel() and xlabel() options, see [G-3] *axis_label_options*. For an explanation of the subtitle() and note() options, see [G-3] *title_options*.

Overlaying normal and kernel density estimates

Specifying normal will overlay a normal density over the histogram. It would be enough to type

```
. histogram volume, normal
```

but we will add the option to our more impressive rendition:

```
. summarize volume
    Variable |       Obs        Mean    Std. Dev.       Min        Max
-------------+--------------------------------------------------------
      volume |       248    12320.68    2585.929       4103    23308.3
```

```
. histogram volume, freq normal
>           xaxis(1 2)
>           ylabel(0(10)60, grid)
>           xlabel(12321 "mean"
>              9735 "-1 s.d."
>             14907 "+1 s.d."
>              7149 "-2 s.d."
>             17493 "+2 s.d."
>             20078 "+3 s.d."
>             22664 "+4 s.d."
>                              , axis(2) grid gmax)
>           xtitle("", axis(2))
>           subtitle("S&P 500, January 2001 - December 2001")
>           note("Source:  Yahoo! Finance and Commodity Systems, Inc.")
(bin=15, start=4103, width=1280.3533)
```

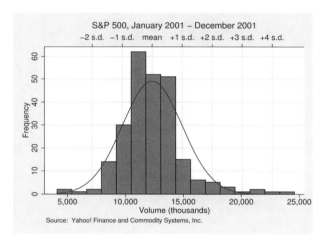

If we instead wanted to overlay a kernel density estimate, we could specify kdensity in place of normal.

Histograms of discrete variables

Specify histogram's discrete option when you wish to treat the data as discrete—when you wish each unique value of the variable to be assigned its own bin. For instance, in the automobile data, mpg is a continuous variable, but the mileage ratings have been measured to integer precision. If we were to type

```
. use http://www.stata-press.com/data/r12/auto
(1978 Automobile Data)

. histogram mpg
(bin=8, start=12, width=3.625)
```

mpg would be treated as continuous and categorized into eight bins by the default number-of-bins calculation, which is based on the number of observations, 74.

Adding the `discrete` option makes a histogram with a bin for each of the 21 unique values.

```
. histogram mpg, discrete
(start=12, width=1)
```

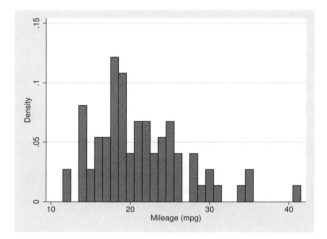

Just as in the continuous case, the y axis was reported in density, and we could specify the `fraction` or `frequency` options if we wanted it to be reported differently. Below we specify `frequency`, we specify `addlabels` to add a report of frequencies printed above the bars, we specify `ylabel(,grid)` to add horizontal grid lines, and we specify `xlabel(12(2)42)` to label the values 12, 14, ..., 42 on the x axis:

```
. histogram mpg, discrete freq addlabels ylabel(,grid) xlabel(12(2)42)
(start=12, width=1)
```

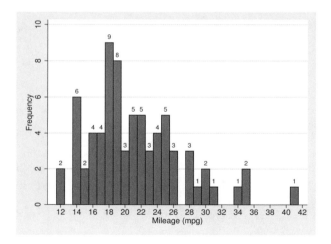

Use with by()

histogram may be used with graph twoway's by(); for example,

```
. use http://www.stata-press.com/data/r12/auto
(1978 Automobile Data)
. histogram mpg, discrete by(foreign)
```

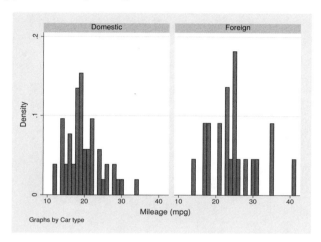

Here results would be easier to compare if the graphs were presented in one column:

```
. histogram mpg, discrete by(foreign, col(1))
```

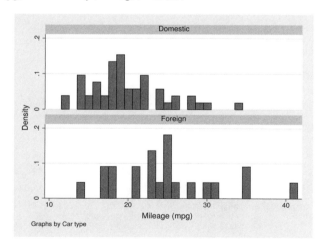

col(1) is a by() suboption—see [G-3] *by_option*—and there are other useful suboptions, such as total, which will add an overall total histogram. total is a suboption of by(), not an option of histogram, so you would type

```
. histogram mpg, discrete by(foreign, total)
```

and not histogram mpg, discrete by(foreign) total.

As another example, Lipset (1993) reprinted data from the *New York Times* (November 5, 1992) collected by the Voter Research and Surveys based on questionnaires completed by 15,490 U.S. presidential voters from 300 polling places on election day in 1992.

```
. use http://www.stata-press.com/data/r12/voter
. histogram candi [freq=pop], discrete fraction by(inc, total)
> gap(40) xlabel(2 3 4, valuelabel)
```

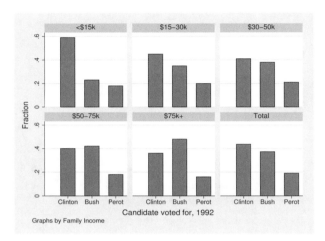

We specified gap(40) to reduce the width of the bars by 40%. We also used xlabel()'s valuelabel suboption, which caused our bars to be labeled "Clinton", "Bush", and "Perot", rather than 2, 3, and 4; see [G-3] *axis_label_options*.

Methods and formulas

histogram is implemented as an ado-file.

References

Cox, N. J. 2004. Speaking Stata: Graphing distributions. *Stata Journal* 4: 66–88.

———. 2005. Speaking Stata: Density probability plots. *Stata Journal* 5: 259–273.

Harrison, D. A. 2005. Stata tip 20: Generating histogram bin variables. *Stata Journal* 5: 280–281.

Lipset, S. M. 1993. The significance of the 1992 election. *PS: Political Science and Politics* 26: 7–16.

Pollock, P. H., III. 2011. *A Stata Companion to Political Analysis*. 2nd ed. Washington, DC: CQ Press.

Also see

[R] **kdensity** — Univariate kernel density estimation

[R] **spikeplot** — Spike plots and rootograms

[G-2] **graph twoway histogram** — Histogram plots

hsearch — Search help files

yntax

hsearch *word(s)*

hsearch *word(s)*, build

hsearch, build

escription

hsearch *word(s)* searches the help files for *word(s)* and presents a clickable list in the Viewer.

hsearch *word(s)*, build does the same thing but builds a new index first.

hsearch, build rebuilds the index but performs no search.

ption

build forces the index that hsearch uses to be built or rebuilt.

The index is automatically built the first time you use hsearch, and it is automatically rebuilt if you have recently installed an ado-file update by using update; see [R] **update**. Thus the build option is rarely specified.

You should specify build if you have recently installed user-written ado-files by using net install (see [R] **net**) or ssc (see [R] **ssc**), or if you have recently updated any of your own help files.

emarks

Remarks are presented under the following headings:

> *Using hsearch*
> *Alternatives to hsearch*
> *Recommendations*
> *How hsearch works*

sing hsearch

You use hsearch to find help for commands and features installed on your computer. If you wanted to find commands related to Mills' ratio, you would type

. hsearch Mills' ratio

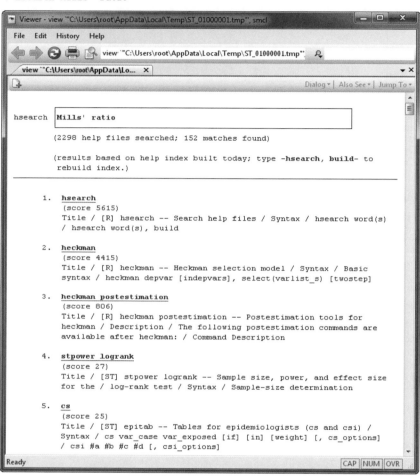

You could just as well type

. hsearch Mill's ratio

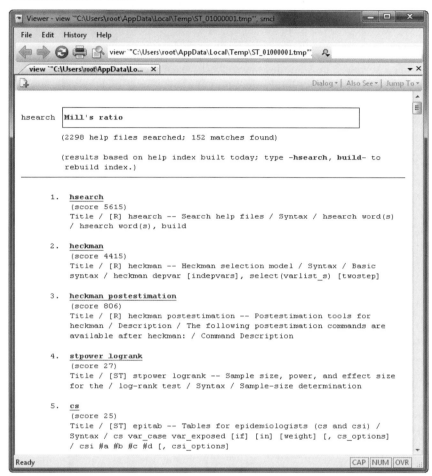

or type any of

. hsearch Mills ratio
. hsearch mills ratio

or even

. hsearch ratio mills

because word order, capitalization, and punctuation do not matter.

Alternatives to hsearch

Alternatives to hsearch are search and findit:

```
. search mills ratio
. findit mills ratio
```

search, like hsearch, searches commands already installed on your computer. search searches the keywords; hsearch searches the help files themselves. Hence, hsearch usually finds everything that search finds and more. The fewer things that search finds should be more relevant.

findit searches keywords just as search does, but findit searches the web as well as your computer and so may find commands that you might wish to install.

Recommendations

- In general, hsearch is better than search. hsearch finds more and better organizes the list of what it finds.

- When you know that Stata can do what you are looking for but you cannot remember the command name or when you know that you installed a relevant user-written package, use hsearch.

- When you are unsure whether Stata can do a certain task, use hsearch first and then use findit.

How hsearch works

hsearch searches the .sthlp files.

Finding all those files and then looking through them would take a long time if hsearch did that every time you used it. Instead, hsearch builds an index of the .sthlp files and then searches that.

That file is called sthlpindex.idx and is stored in your PERSONAL directory.

Every so often, hsearch automatically rebuilds the index so that it accurately reflects what is installed on your computer. You can force hsearch to rebuild the index at any time by typing

```
. hsearch, build
```

Methods and formulas

hsearch is implemented as an ado-file.

Also see

[R] **net search** — Search the Internet for installable packages

[R] **search** — Search Stata documentation

Title

inequality — Inequality measures

Remarks

Stata does not have commands for inequality measures, except `roctab` has an option to report Gini and Pietra indices; see [R] **roctab**. Stata users, however, have developed an excellent suite of commands, many of which have been published in the *Stata Journal* (SJ) and in the *Stata Technical Bulletin* (STB).

Issue	Insert	Author(s)	Command	Description
STB-48	gr35	N. J. Cox	psm, qsm, pdagum, qdagum	Diagnostic plots for assessing Singh–Maddala and Dagum distributions fit by MLE
STB-23	sg31	R. Goldstein	rspread	Measures of diversity: Absolute and relative
STB-48	sg104	S. P. Jenkins	sumdist, xfrac, ineqdeco, geivars, ineqfac, povdeco	Analysis of income distributions
STB-48	sg106	S. P. Jenkins	smfit, dagumfit	Fitting Singh–Maddala and Dagum distributions by maximum likelihood
STB-48	sg107	S. P. Jenkins, P. Van Kerm	glcurve	Generalized Lorenz curves and related graphs
STB-49	sg107.1	S. P. Jenkins, P. Van Kerm	glcurve	update of sg107
SJ-1-1	gr0001	S. P. Jenkins, P. Van Kerm	glcurve7	update for Stata 7 of sg107.1
SJ-7-2	gr0001_3	S. P. Van Kerm, P. Jenkins	glcurve	update for Stata 8 of gr0001; *install this version*
STB-48	sg108	P. Van Kerm	poverty	Computing poverty indices
STB-51	sg115	D. Jolliffe, B. Krushelnytskyy	ineqerr	Bootstrap standard errors for indices of inequality
STB-51	sg117	D. Jolliffe, A. Semykina	sepov	Robust standard errors for the Foster–Greer–Thorbecke class of poverty indices
SJ-8-4	st0100_1	A. López-Feldman	descogini	Decomposing inequality and obtaining marginal effects
SJ-6-4	snp15_7	R. Newson	somersd	Gini coefficient is a special case of Somers' D
STB-23	sg30	E. Whitehouse	lorenz, inequal, atkinson, relsgini	Measures of inequality in Stata

More commands may be available; enter Stata and type `search inequality measure, all`.

> Max Otto Lorenz (1876–1959) was born in Iowa and studied at the Universities of Iowa and Wisconsin. He proposed what is now known as the Lorenz curve in 1905. Lorenz worked for the Interstate Commerce Commission between 1911 and 1944, mainly with transportation data. His hobbies included calendar reform and Interlingua, a proposed international language.

To download and install the Jenkins and Van Kerm `glcurve` command from the Internet, for instance, you could

1. Select **Help > SJ and User-written Programs**.

2. Click on *Stata Journal*.

3. Click on *sj4-4*.

4. Click on *gr0001_1*.

5. Click on *click here to install*.

or you could instead do the following:

1. Navigate to the appropriate SJ issue:

    a. Type `net from http://www.stata-journal.com/software`
       Type `net cd sj4-4`

    or

    b. Type `net from http://www.stata-journal.com/software/sj4-4`

2. Type `net describe gr0001_1`

3. Type `net install gr0001_1`

To download and install the Jenkins `sumdist` command from the Internet, for instance, you could

1. Select **Help > SJ and User-written Programs**.

2. Click on *STB*.

3. Click on *stb48*.

4. Click on *sg104*.

5. Click on *click here to install*.

or you could instead do the following:

1. Navigate to the appropriate STB issue:

    a. Type `net from http://www.stata.com`
       Type `net cd stb`
       Type `net cd stb48`

    or

    b. Type `net from http://www.stata.com/stb/stb48`

2. Type `net describe sg104`

3. Type `net install sg104`

References

Cox, N. J. 1999. gr35: Diagnostic plots for assessing Singh–Maddala and Dagum distributions fitted by MLE. *Stata Technical Bulletin* 48: 2–4. Reprinted in *Stata Technical Bulletin Reprints*, vol. 8, pp. 72–74. College Station, TX: Stata Press.

Goldstein, R. 1995. sg31: Measures of diversity: Absolute and relative. *Stata Technical Bulletin* 23: 23–26. Reprinted in *Stata Technical Bulletin Reprints*, vol. 4, pp. 150–154. College Station, TX: Stata Press.

Haughton, J., and S. R. Khandker. 2009. *Handbook on Poverty + Inequality*. Washington, DC: World Bank.

Jenkins, S. P. 1999a. sg104: Analysis of income distributions. *Stata Technical Bulletin* 48: 4–18. Reprinted in *Stata Technical Bulletin Reprints*, vol. 8, pp. 243–260. College Station, TX: Stata Press.

———. 1999b. sg106: Fitting Singh–Maddala and Dagum distributions by maximum likelihood. *Stata Technical Bulletin* 48: 19–25. Reprinted in *Stata Technical Bulletin Reprints*, vol. 8, pp. 261–268. College Station, TX: Stata Press.

Jenkins, S. P., and P. Van Kerm. 1999a. sg107: Generalized Lorenz curves and related graphs. *Stata Technical Bulletin* 48: 25–29. Reprinted in *Stata Technical Bulletin Reprints*, vol. 8, pp. 269–274. College Station, TX: Stata Press.

———. 1999b. sg107.1: Generalized Lorenz curves and related graphs. *Stata Technical Bulletin* 49: 23. Reprinted in *Stata Technical Bulletin Reprints*, vol. 9, p. 171. College Station, TX: Stata Press.

———. 2001. Generalized Lorenz curves and related graphs: An update for Stata 7. *Stata Journal* 1: 107–112.

———. 2004. gr0001_1: Software Updates: Generalized Lorenz curves and related graphs. *Stata Journal* 4: 490.

———. 2006. gr0001_2: Software Updates: Generalized Lorenz curves and related graphs. *Stata Journal* 6: 597.

———. 2007. gr0001_3: Software Updates: Generalized Lorenz curves and related graphs. *Stata Journal* 7: 280.

Jolliffe, D., and B. Krushelnytskyy. 1999. sg115: Bootstrap standard errors for indices of inequality. *Stata Technical Bulletin* 51: 28–32. Reprinted in *Stata Technical Bulletin Reprints*, vol. 9, pp. 191–196. College Station, TX: Stata Press.

Jolliffe, D., and A. Semykina. 1999. sg117: Robust standard errors for the Foster–Greer–Thorbecke class of poverty indices. *Stata Technical Bulletin* 51: 34–36. Reprinted in *Stata Technical Bulletin Reprints*, vol. 9, pp. 200–203. College Station, TX: Stata Press.

Kleiber, C., and S. Kotz. 2003. *Statistical Size Distributions in Economics and Actuarial Sciences*. Hoboken, NJ: Wiley.

López-Feldman, A. 2006. Decomposing inequality and obtaining marginal effects. *Stata Journal* 6: 106–111.

———. 2008. Software Updates: Decomposing inequality and obtaining marginal effects. *Stata Journal* 8: 594.

Lorenz, M. O. 1905. Methods of measuring the concentration of wealth. *American Statistical Association* 9: 209–219.

Newson, R. 2006. Confidence intervals for rank statistics: Percentile slopes, differences, and ratios. *Stata Journal* 6: 497–520.

Van Kerm, P. 1999. sg108: Computing poverty indices. *Stata Technical Bulletin* 48: 29–33. Reprinted in *Stata Technical Bulletin Reprints*, vol. 8, pp. 274–278. College Station, TX: Stata Press.

Whitehouse, E. 1995. sg30: Measures of inequality in Stata. *Stata Technical Bulletin* 23: 20–23. Reprinted in *Stata Technical Bulletin Reprints*, vol. 4, pp. 146–150. College Station, TX: Stata Press.

Title

intreg — Interval regression

Syntax

    intreg *depvar₁* *depvar₂* [*indepvars*] [*if*] [*in*] [*weight*] [, *options*]

options	Description
Model	
<u>nocon</u>stant	suppress constant term
<u>het</u>(*varlist*[, <u>nocons</u>tant])	independent variables to model the variance; use noconstant to suppress constant term
<u>off</u>set(*varname*)	include *varname* in model with coefficient constrained to 1
<u>constr</u>aints(*constraints*)	apply specified linear constraints
<u>coll</u>inear	keep collinear variables
SE/Robust	
vce(*vcetype*)	*vcetype* may be oim, <u>r</u>obust, <u>cl</u>uster *clustvar*, opg, <u>boot</u>strap, or <u>jack</u>knife
Reporting	
<u>level</u>(#)	set confidence level; default is level(95)
<u>nocns</u>report	do not display constraints
display_options	control column formats, row spacing, line width, and display of omitted variables and base and empty cells
Maximization	
maximize_options	control the maximization process; seldom used
<u>coefl</u>egend	display legend instead of statistics

indepvars and *varlist* may contain factor variables; see [U] **11.4.3 Factor variables**.
depvar₁, *depvar₂*, *indepvars*, and *varlist* may contain time-series operators; see [U] **11.4.4 Time-series varlists**.
bootstrap, by, fracpoly, jackknife, mfp, nestreg, rolling, statsby, stepwise, and svy are allowed; see [U] **11.1.10 Prefix commands**.
Weights are not allowed with the bootstrap prefix; see [R] **bootstrap**.
aweights are not allowed with the jackknife prefix; see [R] **jackknife**.
vce() and weights are not allowed with the svy prefix; see [SVY] **svy**.
aweights, fweights, iweights, and pweights are allowed; see [U] **11.1.6 weight**.
coeflegend does not appear in the dialog box.
See [U] **20 Estimation and postestimation commands** for more capabilities of estimation commands.

Menu

Statistics > Linear models and related > Censored regression > Interval regression

Description

intreg fits a model of $y = [\,depvar_1,\ depvar_2\,]$ on *indepvars*, where y for each observation is point data, interval data, left-censored data, or right-censored data.

depvar₁ and *depvar₂* should have the following form:

Type of data		*depvar₁*	*depvar₂*
point data	$a = [\,a, a\,]$	a	a
interval data	$[\,a, b\,]$	a	b
left-censored data	$(-\infty, b\,]$	.	b
right-censored data	$[\,a, +\infty)$	a	.

Options

_____ ⌐ Model ⌐ _____

noconstant; see [R] **estimation options**.

het(*varlist* [, noconstant]) specifies that *varlist* be included in the specification of the conditional variance. This *varlist* enters the variance specification collectively as multiplicative heteroskedasticity.

offset(*varname*), constraints(*constraints*), collinear; see [R] **estimation options**.

_____ ⌐ SE/Robust ⌐ _____

vce(*vcetype*) specifies the type of standard error reported, which includes types that are derived from asymptotic theory, that are robust to some kinds of misspecification, that allow for intragroup correlation, and that use bootstrap or jackknife methods; see [R] **vce_option**.

_____ ⌐ Reporting ⌐ _____

level(*#*); see [R] **estimation options**.

nocnsreport; see [R] **estimation options**.

display_options: noomitted, vsquish, noemptycells, baselevels, allbaselevels, cformat(%*fmt*), pformat(%*fmt*), sformat(%*fmt*), and nolstretch; see [R] **estimation options**.

_____ ⌐ Maximization ⌐ _____

maximize_options: difficult, technique(*algorithm_spec*), iterate(*#*), [no]log, trace, gradient, showstep, hessian, showtolerance, tolerance(*#*), ltolerance(*#*), nrtolerance(*#*), nonrtolerance, and from(*init_specs*); see [R] **maximize**. These options are seldom used.

Setting the optimization type to technique(bhhh) resets the default *vcetype* to vce(opg).

The following option is available with intreg but is not shown in the dialog box:

coeflegend; see [R] **estimation options**.

Remarks

intreg is a generalization of the models fit by tobit. Cameron and Trivedi (2010, 548–550) discuss the differences among censored, truncated, and interval data. If you know that the value for the jth individual is somewhere in the interval $[\,y_{1j},\,y_{2j}\,]$, then the likelihood contribution from this individual is simply $\Pr(y_{1j} \le Y_j \le y_{2j})$. For censored data, their likelihoods contain terms of the form $\Pr(Y_j \le y_j)$ for left-censored data and $\Pr(Y_j \ge y_j)$ for right-censored data, where y_j is the observed censoring value and Y_j denotes the random variable representing the dependent variable in the model.

Hence, intreg can fit models for data where each observation represents interval data, left-censored data, right-censored data, or point data. Regardless of the type of observation, the data should be stored in the dataset as interval data; that is, two dependent variables, $depvar_1$ and $depvar_2$, are used to hold the endpoints of the interval. If the data are left-censored, the lower endpoint is $-\infty$ and is represented by a missing value, '.', or an extended missing value, '.a, .b, ..., .z', in $depvar_1$. If the data are right-censored, the upper endpoint is $+\infty$ and is represented by a missing value, '.' (or an extended missing value), in $depvar_2$. Point data are represented by the two endpoints being equal.

Type of data		$depvar_1$	$depvar_2$
point data	$a = [\,a, a\,]$	a	a
interval data	$[\,a, b\,]$	a	b
left-censored data	$(-\infty, b\,]$	.	b
right-censored data	$[\,a, +\infty\,)$	a	.

Truly missing values of the dependent variable must be represented by missing values in both $depvar_1$ and $depvar_2$.

Interval data arise naturally in many contexts, such as wage data. Often you know only that, for example, a person's salary is between \$30,000 and \$40,000. Below we give an example for wage data and show how to set up $depvar_1$ and $depvar_2$.

▷ Example 1

We have a dataset that contains the yearly wages of working women. Women were asked via a questionnaire to indicate a category for their yearly income from employment. The categories were less than 5,000, 5,001–10,000, ..., 25,001–30,000, 30,001–40,000, 40,001–50,000, and more than 50,000. The wage categories are stored in the wagecat variable.

```
. use http://www.stata-press.com/data/r12/womenwage
(Wages of women)
. tab wagecat
```

Wage category ($1000s)	Freq.	Percent	Cum.
5	14	2.87	2.87
10	83	17.01	19.88
15	158	32.38	52.25
20	107	21.93	74.18
25	57	11.68	85.86
30	30	6.15	92.01
40	19	3.89	95.90
50	14	2.87	98.77
51	6	1.23	100.00
Total	488	100.00	

A value of 5 for wagecat represents the category less than 5,000, a value of 10 represents 5,001–10,000, ..., and a value of 51 represents greater than 50,000.

To use intreg, we must create two variables, wage1 and wage2, containing the lower and upper endpoints of the wage categories. Here is one way to do it. We first create a dataset containing the nine wage categories, lag the wage categories into wage1, and match-merge this dataset with nine observations back into the main one.

```
. by wagecat: keep if _n==1
(479 observations deleted)
. generate wage1 = wagecat[_n-1]
(1 missing value generated)
. keep wagecat wage1
. save lagwage
file lagwage.dta saved
. use http://www.stata-press.com/data/r12/womenwage
(Wages of women)
. merge m:1 wagecat using lagwage
```

Result	# of obs.	
not matched	0	
matched	488	(_merge==3)

Now we create the upper endpoint and list the new variables:

```
. generate wage2 = wagecat
. replace wage2 = . if wagecat == 51
(6 real changes made, 6 to missing)
. sort age, stable
```

```
. list wage1 wage2 in 1/10
```

	wage1	wage2
1.	.	5
2.	5	10
3.	5	10
4.	10	15
5.	.	5
6.	.	5
7.	.	5
8.	5	10
9.	5	10
10.	5	10

We can now run intreg:

```
. intreg wage1 wage2 age c.age#c.age nev_mar rural school tenure

Fitting constant-only model:

Iteration 0:   log likelihood = -967.24956
Iteration 1:   log likelihood =  -967.1368
Iteration 2:   log likelihood =  -967.1368

Fitting full model:

Iteration 0:   log likelihood = -856.65324
Iteration 1:   log likelihood = -856.33294
Iteration 2:   log likelihood = -856.33293
```

Interval regression

Log likelihood = -856.33293

Number of obs = 488
LR chi2(6) = 221.61
Prob > chi2 = 0.0000

	Coef.	Std. Err.	z	P>|z|	[95% Conf. Interval]
age	.7914438	.4433604	1.79	0.074	-.0775265 1.660414
c.age#c.age	-.0132624	.0073028	-1.82	0.069	-.0275757 .0010509
nev_mar	-.2075022	.8119581	-0.26	0.798	-1.798911 1.383906
rural	-3.043044	.7757324	-3.92	0.000	-4.563452 -1.522637
school	1.334721	.1357873	9.83	0.000	1.068583 1.600859
tenure	.8000664	.1045077	7.66	0.000	.5952351 1.004898
_cons	-12.70238	6.367117	-1.99	0.046	-25.1817 -.2230583
/lnsigma	1.987823	.0346543	57.36	0.000	1.919902 2.055744
sigma	7.299626	.2529634			6.82029 7.81265

```
Observation summary:    14  left-censored observations
                         0         uncensored observations
                         6  right-censored observations
                       468          interval observations
```

We could also model these data by using an ordered probit model with oprobit (see [R] **oprobit**):

```
. oprobit wagecat age c.age#c.age nev_mar rural school tenure
Iteration 0:   log likelihood =  -881.1491
Iteration 1:   log likelihood = -764.31729
Iteration 2:   log likelihood = -763.31191
Iteration 3:   log likelihood = -763.31049
Iteration 4:   log likelihood = -763.31049
```

Ordered probit regression

Log likelihood = -763.31049

					Number of obs	=	488
					LR chi2(6)	=	235.68
					Prob > chi2	=	0.0000
					Pseudo R2	=	0.1337

wagecat	Coef.	Std. Err.	z	P>|z|	[95% Conf.	Interval]
age	.1674519	.0620333	2.70	0.007	.0458689	.289035
c.age#c.age	-.0027983	.0010214	-2.74	0.006	-.0048001	-.0007964
nev_mar	-.0046417	.1126737	-0.04	0.967	-.225478	.2161946
rural	-.5270036	.1100449	-4.79	0.000	-.7426875	-.3113196
school	.2010587	.0201189	9.99	0.000	.1616263	.2404911
tenure	.0989916	.0147887	6.69	0.000	.0700063	.127977
/cut1	2.650637	.8957245			.8950495	4.406225
/cut2	3.941018	.8979167			2.181134	5.700903
/cut3	5.085205	.9056582			3.310148	6.860263
/cut4	5.875534	.9120933			4.087864	7.663204
/cut5	6.468723	.918117			4.669247	8.268199
/cut6	6.922726	.9215455			5.11653	8.728922
/cut7	7.34471	.9237628			5.534168	9.155252
/cut8	7.963441	.9338881			6.133054	9.793828

We can directly compare the log likelihoods for the intreg and oprobit models because both likelihoods are discrete. If we had point data in our intreg estimation, the likelihood would be a mixture of discrete and continuous terms, and we could not compare it directly with the oprobit likelihood.

Here the oprobit log likelihood is significantly larger (that is, less negative), so it fits better than the intreg model. The intreg model assumes normality, but the distribution of wages is skewed and definitely nonnormal. Normality is more closely approximated if we model the log of wages.

```
. generate logwage1 = log(wage1)
(14 missing values generated)
. generate logwage2 = log(wage2)
(6 missing values generated)
. intreg logwage1 logwage2 age c.age#c.age nev_mar rural school tenure

Fitting constant-only model:

Iteration 0:   log likelihood = -889.23647
Iteration 1:   log likelihood = -889.06346
Iteration 2:   log likelihood = -889.06346

Fitting full model:

Iteration 0:   log likelihood = -773.81968
Iteration 1:   log likelihood = -773.36566
Iteration 2:   log likelihood = -773.36563
```

Interval regression				Number of obs	=	488
				LR chi2(6)	=	231.40
Log likelihood = -773.36563				Prob > chi2	=	0.0000

	Coef.	Std. Err.	z	P>|z|	[95% Conf. Interval]	
age	.0645589	.0249954	2.58	0.010	.0155689	.1135489
c.age#c.age	-.0010812	.0004115	-2.63	0.009	-.0018878	-.0002746
nev_mar	-.0058151	.0454867	-0.13	0.898	-.0949674	.0833371
rural	-.2098361	.0439454	-4.77	0.000	-.2959675	-.1237047
school	.0804832	.0076783	10.48	0.000	.0654341	.0955323
tenure	.0397144	.0058001	6.85	0.000	.0283464	.0510825
_cons	.7084023	.3593193	1.97	0.049	.0041495	1.412655
/lnsigma	-.906989	.0356265	-25.46	0.000	-.9768157	-.8371623
sigma	.4037381	.0143838			.3765081	.4329373

```
Observation summary:      14  left-censored observations
                           0     uncensored observations
                           6 right-censored observations
                         468       interval observations
```

The log likelihood of this intreg model is close to the oprobit log likelihood, and the z statistics for both models are similar.   ◁

❑ Technical note

intreg has two parameterizations for the log-likelihood function: the transformed parameterization $(\beta/\sigma, 1/\sigma)$ and the untransformed parameterization $(\beta, \ln(\sigma))$. By default, the log likelihood for intreg is parameterized in the transformed parameter space. This parameterization tends to be more convergent, but it requires that any starting values and constraints have the same parameterization, and it prevents the estimation with multiplicative heteroskedasticity. Therefore, when the het() option is specified, intreg switches to the untransformed log likelihood for the fit of the conditional-variance model. Similarly, specifying from() or constraints() causes the optimization in the untransformed parameter space to allow constraints on (and starting values for) the coefficients on the covariates without reference to σ.

The estimation results are all saved in the $(\beta, \ln(\sigma))$ metric.   ❑

Saved results

intreg saves the following in e():

Scalars
e(N)	number of observations
e(N_unc)	number of uncensored observations
e(N_lc)	number of left-censored observations
e(N_rc)	number of right-censored observations
e(N_int)	number of interval observations
e(k)	number of parameters
e(k_aux)	number of auxiliary parameters
e(k_eq)	number of equations in e(b)
e(k_eq_model)	number of equations in overall model test
e(k_dv)	number of dependent variables
e(df_m)	model degrees of freedom
e(ll)	log likelihood
e(ll_0)	log likelihood, constant-only model
e(ll_c)	log likelihood, comparison model
e(N_clust)	number of clusters
e(chi2)	χ^2
e(p)	p-value for model χ^2 test
e(sigma)	sigma
e(se_sigma)	standard error of sigma
e(rank)	rank of e(V)
e(rank0)	rank of e(V) for constant-only model
e(ic)	number of iterations
e(rc)	return code
e(converged)	1 if converged, 0 otherwise

Macros
e(cmd)	intreg
e(cmdline)	command as typed
e(depvar)	names of dependent variables
e(wtype)	weight type
e(wexp)	weight expression
e(title)	title in estimation output
e(clustvar)	name of cluster variable
e(offset)	linear offset variable
e(chi2type)	Wald or LR; type of model χ^2 test
e(vce)	*vcetype* specified in vce()
e(vcetype)	title used to label Std. Err.
e(het)	heteroskedasticity, if het() specified
e(ml_score)	program used to implement scores
e(opt)	type of optimization
e(which)	max or min; whether optimizer is to perform maximization or minimization
e(ml_method)	type of ml method
e(user)	name of likelihood-evaluator program
e(technique)	maximization technique
e(properties)	b V
e(predict)	program used to implement predict
e(footnote)	program and arguments to display footnote
e(asbalanced)	factor variables fvset as asbalanced
e(asobserved)	factor variables fvset as asobserved

Matrices
e(b)	coefficient vector
e(Cns)	constraints matrix
e(ilog)	iteration log (up to 20 iterations)
e(gradient)	gradient vector
e(V)	variance–covariance matrix of the estimators
e(V_modelbased)	model-based variance

Functions
e(sample)	marks estimation sample

Methods and formulas

`intreg` is implemented as an ado-file.

See Wooldridge (2009, sec. 17.4) or Davidson and MacKinnon (2004, sec. 11.6) for an introduction to censored and truncated regression models.

The likelihood for `intreg` subsumes that of the `tobit` models.

Let $\mathbf{y} = \mathbf{X}\boldsymbol{\beta} + \boldsymbol{\epsilon}$ be the model. $\mathbf{y}$ represents continuous outcomes—either observed or not observed. Our model assumes $\boldsymbol{\epsilon} \sim N(\mathbf{0}, \sigma^2\mathbf{I})$.

For observations $j \in \mathcal{C}$, we observe y_j, that is, point data. Observations $j \in \mathcal{L}$ are left-censored; we know only that the unobserved y_j is less than or equal to $y_{\mathcal{L}j}$, a censoring value that we do know. Similarly, observations $j \in \mathcal{R}$ are right-censored; we know only that the unobserved y_j is greater than or equal to $y_{\mathcal{R}j}$. Observations $j \in \mathcal{I}$ are intervals; we know only that the unobserved y_j is in the interval $[y_{1j}, y_{2j}]$.

The log likelihood is

$$
\ln L = -\frac{1}{2} \sum_{j \in \mathcal{C}} w_j \left\{ \left(\frac{y_j - \mathbf{x}\boldsymbol{\beta}}{\sigma} \right)^2 + \log 2\pi\sigma^2 \right\}
$$
$$
+ \sum_{j \in \mathcal{L}} w_j \log \Phi \left(\frac{y_{\mathcal{L}j} - \mathbf{x}\boldsymbol{\beta}}{\sigma} \right)
$$
$$
+ \sum_{j \in \mathcal{R}} w_j \log \left\{ 1 - \Phi \left(\frac{y_{\mathcal{R}j} - \mathbf{x}\boldsymbol{\beta}}{\sigma} \right) \right\}
$$
$$
+ \sum_{j \in \mathcal{I}} w_j \log \left\{ \Phi \left(\frac{y_{2j} - \mathbf{x}\boldsymbol{\beta}}{\sigma} \right) - \Phi \left(\frac{y_{1j} - \mathbf{x}\boldsymbol{\beta}}{\sigma} \right) \right\}
$$

where $\Phi()$ is the standard cumulative normal and w_j is the weight for the jth observation. If no weights are specified, $w_j = 1$. If `aweights` are specified, $w_j = 1$, and σ is replaced by $\sigma/\sqrt{a_j}$ in the above, where a_j are the `aweights` normalized to sum to N.

Maximization is as described in [R] **maximize**; the estimate reported as **_sigma** is $\widehat{\sigma}$.

See Amemiya (1973) for a generalization of the tobit model to variable, but known, cutoffs.

This command supports the Huber/White/sandwich estimator of the variance and its clustered version using `vce(robust)` and `vce(cluster clustvar)`, respectively. See [P] **_robust**, particularly *Maximum likelihood estimators* and *Methods and formulas*.

`intreg` also supports estimation with survey data. For details on VCEs with survey data, see [SVY] **variance estimation**.

References

Amemiya, T. 1973. Regression analysis when the dependent variable is truncated normal. *Econometrica* 41: 997–1016.

Cameron, A. C., and P. K. Trivedi. 2010. *Microeconometrics Using Stata*. Rev. ed. College Station, TX: Stata Press.

Conroy, R. M. 2005. Stings in the tails: Detecting and dealing with censored data. *Stata Journal* 5: 395–404.

Davidson, R., and J. G. MacKinnon. 1993. *Estimation and Inference in Econometrics*. New York: Oxford University Press.

———. 2004. *Econometric Theory and Methods*. New York: Oxford University Press.

Goldberger, A. S. 1983. Abnormal selection bias. In *Studies in Econometrics, Time Series, and Multivariate Statistics*, ed. S. Karlin, T. Amemiya, and L. A. Goodman, 67–84. New York: Academic Press.

Hurd, M. 1979. Estimation in truncated samples when there is heteroscedasticity. *Journal of Econometrics* 11: 247–258.

Long, J. S. 1997. *Regression Models for Categorical and Limited Dependent Variables.* Thousand Oaks, CA: Sage.

Stewart, M. B. 1983. On least squares estimation when the dependent variable is grouped. *Review of Economic Studies* 50: 737–753.

Wooldridge, J. M. 2009. *Introductory Econometrics: A Modern Approach.* 4th ed. Cincinnati, OH: South-Western.

Also see

[R] **intreg postestimation** — Postestimation tools for intreg

[R] **tobit** — Tobit regression

[R] **regress** — Linear regression

[SVY] **svy estimation** — Estimation commands for survey data

[XT] **xtintreg** — Random-effects interval-data regression models

[XT] **xttobit** — Random-effects tobit models

[U] **20 Estimation and postestimation commands**

Title

intreg postestimation — Postestimation tools for intreg

Description

The following postestimation commands are available after `intreg`:

Command	Description
contrast	contrasts and ANOVA-style joint tests of estimates
estat	AIC, BIC, VCE, and estimation sample summary
estat (svy)	postestimation statistics for survey data
estimates	cataloging estimation results
lincom	point estimates, standard errors, testing, and inference for linear combinations of coefficients
lrtest[1]	likelihood-ratio test
margins	marginal means, predictive margins, marginal effects, and average marginal effects
marginsplot	graph the results from margins (profile plots, interaction plots, etc.)
nlcom	point estimates, standard errors, testing, and inference for nonlinear combinations of coefficients
predict	predictions, residuals, influence statistics, and other diagnostic measures
predictnl	point estimates, standard errors, testing, and inference for generalized predictions
pwcompare	pairwise comparisons of estimates
suest	seemingly unrelated estimation
test	Wald tests of simple and composite linear hypotheses
testnl	Wald tests of nonlinear hypotheses

[1] `lrtest` is not appropriate with `svy` estimation results.

See the corresponding entries in the *Base Reference Manual* for details, but see [SVY] **estat** for details about `estat` (svy).

Syntax for predict

predict [*type*] *newvar* [*if*] [*in*] [, *statistic* <u>nooff</u>set]

predict [*type*] { *stub** | *newvar*~reg~ *newvar*~lnsigma~ } [*if*] [*in*] , <u>sc</u>ores

statistic	Description
Main	
xb	linear prediction; the default
stdp	standard error of the prediction
stdf	standard error of the forecast
<u>pr</u>(*a*,*b*)	$\Pr(a < y_j < b)$
e(*a*,*b*)	$E(y_j \mid a < y_j < b)$
<u>ystar</u>(*a*,*b*)	$E(y_j^*)$, $y_j^* = \max\{a, \min(y_j, b)\}$

These statistics are available both in and out of sample; type predict ... if e(sample) ... if wanted only for the estimation sample.

stdf is not allowed with svy postestimation results.

where *a* and *b* may be numbers or variables; *a* missing ($a \geq .$) means $-\infty$, and *b* missing ($b \geq .$) means $+\infty$; see [U] **12.2.1 Missing values**.

Menu

Statistics > Postestimation > Predictions, residuals, etc.

Options for predict

    ⌐ Main ⌐

xb, the default, calculates the linear prediction.

stdp calculates the standard error of the prediction, which can be thought of as the standard error of the predicted expected value or mean for the observation's covariate pattern. The standard error of the prediction is also referred to as the standard error of the fitted value.

stdf calculates the standard error of the forecast, which is the standard error of the point prediction for 1 observation. It is commonly referred to as the standard error of the future or forecast value. By construction, the standard errors produced by stdf are always larger than those produced by stdp; see *Methods and formulas* in [R] **regress postestimation**.

pr(*a*,*b*) calculates $\Pr(a < \mathbf{x}_j\mathbf{b} + u_j < b)$, the probability that $y_j \mid \mathbf{x}_j$ would be observed in the interval (a, b).

    a and *b* may be specified as numbers or variable names; *lb* and *ub* are variable names;
    pr(20,30) calculates $\Pr(20 < \mathbf{x}_j\mathbf{b} + u_j < 30)$;
    pr(*lb*,*ub*) calculates $\Pr(lb < \mathbf{x}_j\mathbf{b} + u_j < ub)$; and
    pr(20,*ub*) calculates $\Pr(20 < \mathbf{x}_j\mathbf{b} + u_j < ub)$.

    a missing ($a \geq .$) means $-\infty$; pr(.,30) calculates $\Pr(-\infty < \mathbf{x}_j\mathbf{b} + u_j < 30)$;
    pr(*lb*,30) calculates $\Pr(-\infty < \mathbf{x}_j\mathbf{b} + u_j < 30)$ in observations for which $lb \geq .$
    and calculates $\Pr(lb < \mathbf{x}_j\mathbf{b} + u_j < 30)$ elsewhere.

b missing ($b \geq .$) means $+\infty$; $\mathtt{pr(20,.)}$ calculates $\Pr(+\infty > \mathbf{x}_j\mathbf{b} + u_j > 20)$; $\mathtt{pr(20,}ub\mathtt{)}$ calculates $\Pr(+\infty > \mathbf{x}_j\mathbf{b} + u_j > 20)$ in observations for which $ub \geq .$ and calculates $\Pr(20 < \mathbf{x}_j\mathbf{b} + u_j < ub)$ elsewhere.

$\mathtt{e(}a\mathtt{,}b\mathtt{)}$ calculates $E(\mathbf{x}_j\mathbf{b} + u_j \mid a < \mathbf{x}_j\mathbf{b} + u_j < b)$, the expected value of $y_j|\mathbf{x}_j$ conditional on $y_j|\mathbf{x}_j$ being in the interval (a, b), meaning that $y_j|\mathbf{x}_j$ is truncated. a and b are specified as they are for $\mathtt{pr()}$.

$\mathtt{ystar(}a\mathtt{,}b\mathtt{)}$ calculates $E(y_j^*)$, where $y_j^* = a$ if $\mathbf{x}_j\mathbf{b} + u_j \leq a$, $y_j^* = b$ if $\mathbf{x}_j\mathbf{b} + u_j \geq b$, and $y_j^* = \mathbf{x}_j\mathbf{b} + u_j$ otherwise, meaning that y_j^* is censored. a and b are specified as they are for $\mathtt{pr()}$.

$\mathtt{nooffset}$ is relevant only if you specified $\mathtt{offset(}varname\mathtt{)}$. It modifies the calculations made by $\mathtt{predict}$ so that they ignore the offset variable; the linear prediction is treated as $\mathbf{x}_j\mathbf{b}$ rather than as $\mathbf{x}_j\mathbf{b} + \mathrm{offset}_j$.

$\mathtt{scores}$ calculates equation-level score variables.

The first new variable will contain $\partial \ln L / \partial(\mathbf{x}_j\boldsymbol{\beta})$.

The second new variable will contain $\partial \ln L / \partial \ln \sigma$.

Methods and formulas

All postestimation commands listed above are implemented as ado-files.

Also see

[R] **intreg** — Interval regression

[U] **20 Estimation and postestimation commands**

Title

> **ivprobit** — Probit model with continuous endogenous regressors

Syntax

Maximum likelihood estimator

> ivprobit *depvar* $[varlist_1]$ (*varlist₂* = *varlist*ᵢᵥ) $[if]$ $[in]$ $[weight]$ $[, mle\_options]$

Two-step estimator

> ivprobit *depvar* $[varlist_1]$ (*varlist₂* = *varlist*ᵢᵥ) $[if]$ $[in]$ $[weight]$, <u>two</u>step
>
> $[tse\_options]$

mle_options	Description
Model	
<u>m</u>le	use conditional maximum-likelihood estimator; the default
asis	retain perfect predictor variables
<u>constraints</u>(*constraints*)	apply specified linear constraints
SE/Robust	
vce(*vcetype*)	*vcetype* may be oim, <u>r</u>obust, <u>cl</u>uster *clustvar*, opg, <u>boot</u>strap, or <u>jack</u>knife
Reporting	
<u>l</u>evel(*#*)	set confidence level; default is level(95)
first	report first-stage regression
<u>nocns</u>report	do not display constraints
display_options	control column formats, row spacing, line width, and display of omitted variables and base and empty cells
Maximization	
maximize_options	control the maximization process
<u>coefl</u>egend	display legend instead of statistics

tse_options	Description
Model	
* <u>two</u>step	use Newey's two-step estimator; the default is mle
asis	retain perfect predictor variables
SE	
vce(*vcetype*)	*vcetype* may be twostep, <u>boot</u>strap, or <u>jack</u>knife
Reporting	
<u>l</u>evel(#)	set confidence level; default is level(95)
first	report first-stage regression
display_options	control column formats, row spacing, line width, and display of omitted variables and base and empty cells
<u>coefl</u>egend	display legend instead of statistics

*twostep is required.

*varlist*₁ and *varlist*ᵢᵥ may contain factor variables; see [U] **11.4.3 Factor variables**.

depvar, *varlist*₁, *varlist*₂, and *varlist*ᵢᵥ may contain time-series operators; see [U] **11.4.4 Time-series varlists**.

bootstrap, by, jackknife, rolling, statsby, and svy are allowed; see [U] **11.1.10 Prefix commands**.

Weights are not allowed with the bootstrap prefix; see [R] **bootstrap**.

vce(), first, twostep, and weights are not allowed with the svy prefix; see [SVY] **svy**.

fweights, iweights, and pweights are allowed with the maximum likelihood estimator. fweights are allowed with Newey's two-step estimator. See [U] **11.1.6 weight**.

coeflegend does not appear in the dialog box.

See [U] **20 Estimation and postestimation commands** for more capabilities of estimation commands.

Menu

Statistics > Endogenous covariates > Probit model with endogenous covariates

Description

ivprobit fits probit models where one or more of the regressors are endogenously determined. By default, ivprobit uses maximum likelihood estimation. Alternatively, Newey's (1987) minimum chi-squared estimator can be invoked with the twostep option. Both estimators assume that the endogenous regressors are continuous and are not appropriate for use with discrete endogenous regressors. See [R] **ivtobit** for tobit estimation with endogenous regressors and [R] **probit** for probit estimation when the model contains no endogenous regressors.

Options for ML estimator

⌐ Model ⌐

mle requests that the conditional maximum-likelihood estimator be used. This is the default.

asis requests that all specified variables and observations be retained in the maximization process. This option is typically not used and may introduce numerical instability. Normally, ivprobit drops any endogenous or exogenous variables that perfectly predict success or failure in the dependent variable. The associated observations are also dropped. For more information, see *Model identification* in [R] **probit**.

constraints(*constraints*); see [R] **estimation options**.

SE/Robust

vce(*vcetype*) specifies the type of standard error reported, which includes types that are derived from asymptotic theory, that are robust to some kinds of misspecification, that allow for intragroup correlation, and that use bootstrap or jackknife methods; see [R] **vce_option**.

Reporting

level(*#*); see [R] **estimation options**.

first requests that the parameters for the reduced-form equations showing the relationships between the endogenous variables and instruments be displayed. For the two-step estimator, first shows the first-stage regressions. For the maximum likelihood estimator, these parameters are estimated jointly with the parameters of the probit equation. The default is not to show these parameter estimates.

nocnsreport; see [R] **estimation options**.

display_options: noomitted, vsquish, noemptycells, baselevels, allbaselevels, cformat(*%fmt*), pformat(*%fmt*), sformat(*%fmt*), and nolstretch; see [R] **estimation options**.

Maximization

maximize_options: difficult, technique(*algorithm_spec*), iterate(*#*), [no]log, trace, gradient, showstep, hessian, showtolerance, tolerance(*#*), ltolerance(*#*), nrtolerance(*#*), nonrtolerance, and from(*init_specs*); see [R] **maximize**. This model's likelihood function can be difficult to maximize, especially with multiple endogenous variables. The difficult and technique(bfgs) options may be helpful in achieving convergence.

Setting the optimization type to technique(bhhh) resets the default *vcetype* to vce(opg).

The following option is available with ivprobit but is not shown in the dialog box:

coeflegend; see [R] **estimation options**.

Options for two-step estimator

Model

twostep is required and requests that Newey's (1987) efficient two-step estimator be used to obtain the coefficient estimates.

asis requests that all specified variables and observations be retained in the maximization process. This option is typically not used and may introduce numerical instability. Normally, ivprobit drops any endogenous or exogenous variables that perfectly predict success or failure in the dependent variable. The associated observations are also dropped. For more information, see *Model identification* in [R] **probit**.

SE

vce(*vcetype*) specifies the type of standard error reported, which includes types that are derived from asymptotic theory and that use bootstrap or jackknife methods; see [R] **vce_option**.

⌐ Reporting ⌐

`level(#)`; see [R] **estimation options**.

`first` requests that the parameters for the reduced-form equations showing the relationships between the endogenous variables and instruments be displayed. For the two-step estimator, `first` shows the first-stage regressions. For the maximum likelihood estimator, these parameters are estimated jointly with the parameters of the probit equation. The default is not to show these parameter estimates.

display_options: <u>noomit</u>ted, vsquish, <u>noempty</u>cells, <u>base</u>levels, <u>allbase</u>levels, cformat(%*fmt*), pformat(%*fmt*), sformat(%*fmt*), and nolstretch; see [R] **estimation options**.

The following option is available with `ivprobit` but is not shown in the dialog box:

`coeflegend`; see [R] **estimation options**.

Remarks

Remarks are presented under the following headings:

> *Model setup*
> *Model identification*

Model setup

`ivprobit` fits models with dichotomous dependent variables and endogenous regressors. You can use it to fit a probit model when you suspect that one or more of the regressors are correlated with the error term. `ivprobit` is to probit modeling what `ivregress` is to linear regression analysis; see [R] **ivregress** for more information.

Formally, the model is

$$y_{1i}^* = \boldsymbol{y}_{2i}\boldsymbol{\beta} + \boldsymbol{x}_{1i}\boldsymbol{\gamma} + u_i$$
$$\boldsymbol{y}_{2i} = \boldsymbol{x}_{1i}\boldsymbol{\Pi}_1 + \boldsymbol{x}_{2i}\boldsymbol{\Pi}_2 + \boldsymbol{v}_i$$

where $i = 1, \ldots, N$, $\boldsymbol{y}_{2i}$ is a $1 \times p$ vector of endogenous variables, $\boldsymbol{x}_{1i}$ is a $1 \times k_1$ vector of exogenous variables, $\boldsymbol{x}_{2i}$ is a $1 \times k_2$ vector of additional instruments, and the equation for $\boldsymbol{y}_{2i}$ is written in reduced form. By assumption, $(u_i, \boldsymbol{v}_i) \sim \mathrm{N}(\boldsymbol{0}, \boldsymbol{\Sigma})$, where σ_{11} is normalized to one to identify the model. $\boldsymbol{\beta}$ and $\boldsymbol{\gamma}$ are vectors of structural parameters, and $\boldsymbol{\Pi}_1$ and $\boldsymbol{\Pi}_2$ are matrices of reduced-form parameters. This is a recursive model: $\boldsymbol{y}_{2i}$ appears in the equation for y_{1i}^*, but y_{1i}^* does not appear in the equation for $\boldsymbol{y}_{2i}$. We do not observe y_{1i}^*; instead, we observe

$$y_{1i} = \begin{cases} 0 & y_{1i}^* < 0 \\ 1 & y_{1i}^* \geq 0 \end{cases}$$

The order condition for identification of the structural parameters requires that $k_2 \geq p$. Presumably, $\boldsymbol{\Sigma}$ is not block diagonal between u_i and $\boldsymbol{v}_i$; otherwise, $\boldsymbol{y}_{2i}$ would not be endogenous.

❑ Technical note

This model is derived under the assumption that $(u_i, \boldsymbol{v}_i)$ is independent and identically distributed multivariate normal for all i. The vce(cluster *clustvar*) option can be used to control for a lack of independence. As with most probit models, if u_i is heteroskedastic, point estimates will be inconsistent.

❑

Example 1

We have hypothetical data on 500 two-parent households, and we wish to model whether the woman is employed. We have a variable, fem_work, that is equal to one if she has a job and zero otherwise. Her decision to work is a function of the number of children at home (kids), number of years of schooling completed (fem_educ), and other household income measured in thousands of dollars (other_inc). We suspect that unobservable shocks affecting the woman's decision to hold a job also affect the household's other income. Therefore, we treat other_inc as endogenous. As an instrument, we use the number of years of schooling completed by the man (male_educ).

The syntax for specifying the exogenous, endogenous, and instrumental variables is identical to that used in ivregress; see [R] **ivregress** for details.

```
. use http://www.stata-press.com/data/r12/laborsup
. ivprobit fem_work fem_educ kids (other_inc = male_educ)
Fitting exogenous probit model
Iteration 0:   log likelihood = -344.63508
Iteration 1:   log likelihood = -255.36855
Iteration 2:   log likelihood = -255.31444
Iteration 3:   log likelihood = -255.31444
Fitting full model
Iteration 0:   log likelihood = -2371.4753
Iteration 1:   log likelihood = -2369.3178
Iteration 2:   log likelihood = -2368.2198
Iteration 3:   log likelihood = -2368.2062
Iteration 4:   log likelihood = -2368.2062
```

Probit model with endogenous regressors

Number of obs = 500
Wald chi2(3) = 163.88
Prob > chi2 = 0.0000

Log likelihood = -2368.2062

	Coef.	Std. Err.	z	P>|z|	[95% Conf. Interval]	
other_inc	-.0542756	.0060854	-8.92	0.000	-.0662027	-.0423485
fem_educ	.211111	.0268648	7.86	0.000	.1584569	.2637651
kids	-.1820929	.0478267	-3.81	0.000	-.2758316	-.0883543
_cons	.3672083	.4480724	0.82	0.412	-.5109975	1.245414
/athrho	.3907858	.1509443	2.59	0.010	.0949403	.6866313
/lnsigma	2.813383	.0316228	88.97	0.000	2.751404	2.875363
rho	.3720374	.1300519			.0946561	.5958135
sigma	16.66621	.5270318			15.66461	17.73186

```
Instrumented:  other_inc
Instruments:   fem_educ kids male_educ
```

Wald test of exogeneity (/athrho = 0): chi2(1) =   6.70 Prob > chi2 = 0.0096

Because we did not specify mle or twostep, ivprobit used the maximum likelihood estimator by default. At the top of the output, we see the iteration log. ivprobit fits a probit model ignoring endogeneity to obtain starting values for the endogenous model. The header of the output contains the sample size as well as a Wald statistic and p-value for the test of the hypothesis that all the slope coefficients are jointly zero. Below the table of coefficients, Stata reminds us that the endogenous variable is other_inc and that fem_educ, kids, and male_educ were used as instruments.

At the bottom of the output is a Wald test of the exogeneity of the instrumented variables. If the test statistic is not significant, there is not sufficient information in the sample to reject the null that there is no endogeneity. Then a regular probit regression may be appropriate; the point estimates from ivprobit are consistent, though those from probit (see [R] **probit**) are likely to have smaller standard errors.

◁

Various two-step estimators have also been proposed for the endogenous probit model, and Newey's (1987) minimum chi-squared estimator is available with the twostep option.

▷ Example 2

Refitting our labor-supply model with the two-step estimator yields

```
. ivprobit fem_work fem_educ kids (other_inc = male_educ), twostep
Checking reduced-form model...
Two-step probit with endogenous regressors       Number of obs   =      500
                                                 Wald chi2(3)    =    93.97
                                                 Prob > chi2     =   0.0000
```

	Coef.	Std. Err.	z	P>|z|	[95% Conf. Interval]	
other_inc	-.058473	.0093364	-6.26	0.000	-.0767719	-.040174
fem_educ	.227437	.0281628	8.08	0.000	.1722389	.282635
kids	-.1961748	.0496323	-3.95	0.000	-.2934522	-.0988973
_cons	.3956061	.4982649	0.79	0.427	-.5809752	1.372187

```
Instrumented:  other_inc
Instruments:   fem_educ kids male_educ
```

```
Wald test of exogeneity:       chi2(1) =     6.50          Prob > chi2 = 0.0108
```

All the coefficients have the same signs as their counterparts in the maximum likelihood model. The Wald test at the bottom of the output confirms our earlier finding of endogeneity.

◁

❏ Technical note

In a standard probit model, the error term is assumed to have a variance of one. In the probit model with endogenous regressors, we assume that (u_i, v_i) is multivariate normal with covariance matrix

$$\operatorname{Var}(u_i, v_i) = \Sigma = \begin{bmatrix} 1 & \Sigma'_{21} \\ \Sigma_{21} & \Sigma_{22} \end{bmatrix}$$

With the properties of the multivariate normal distribution, $\operatorname{Var}(u_i|v_i) = 1 - \Sigma'_{21}\Sigma_{22}^{-1}\Sigma_{21}$. As a result, Newey's estimator and other two-step probit estimators do not yield estimates of β and γ but rather β/σ and γ/σ, where σ is the square root of $\operatorname{Var}(u_i|v_i)$. Hence, we cannot directly compare the estimates obtained from Newey's estimator with those obtained via maximum likelihood or with those obtained from probit. See Wooldridge (2010, 585–594) for a discussion of Rivers and Vuong's (1988) two-step estimator. The issues raised pertaining to the interpretation of the coefficients of that estimator are identical to those that arise with Newey's estimator. Wooldridge also discusses ways to obtain marginal effects from two-step estimators.

❏

Despite the coefficients not being directly comparable to their maximum likelihood counterparts, the two-step estimator is nevertheless useful. The maximum likelihood estimator may have difficulty converging, especially with multiple endogenous variables. The two-step estimator, consisting of nothing more complicated than a probit regression, will almost certainly converge. Moreover, although the coefficients from the two models are not directly comparable, the two-step estimates can still be used to test for statistically significant relationships.

Model identification

As in the linear simultaneous-equation model, the order condition for identification requires that the number of excluded exogenous variables (that is, the additional instruments) be at least as great as the number of included endogenous variables. ivprobit checks this for you and issues an error message if the order condition is not met.

Like probit, logit, and logistic, ivprobit checks the exogenous and endogenous variables to see if any of them predict the outcome variable perfectly. It will then drop offending variables and observations and fit the model on the remaining data. Instruments that are perfect predictors do not affect estimation, so they are not checked. See *Model identification* in [R] **probit** for more information.

ivprobit will also occasionally display messages such as

> Note: 4 failures and 0 successes completely determined.

For an explanation of this message, see [R] **logit**.

Saved results

ivprobit, mle saves the following in e():

Scalars

e(N)	number of observations
e(N_cds)	number of completely determined successes
e(N_cdf)	number of completely determined failures
e(k)	number of parameters
e(k_eq)	number of equations in e(b)
e(k_eq_model)	number of equations in overall model test
e(k_aux)	number of auxiliary parameters
e(k_dv)	number of dependent variables
e(df_m)	model degrees of freedom
e(ll)	log likelihood
e(N_clust)	number of clusters
e(endog_ct)	number of endogenous regressors
e(p)	model Wald p-value
e(p_exog)	exogeneity test Wald p-value
e(chi2)	model Wald χ^2
e(chi2_exog)	Wald χ^2 test of exogeneity
e(rank)	rank of e(V)
e(ic)	number of iterations
e(rc)	return code
e(converged)	1 if converged, 0 otherwise

Macros
e(cmd)	ivprobit
e(cmdline)	command as typed
e(depvar)	name of dependent variable
e(instd)	instrumented variables
e(insts)	instruments
e(wtype)	weight type
e(wexp)	weight expression
e(title)	title in estimation output
e(clustvar)	name of cluster variable
e(chi2type)	Wald; type of model χ^2 test
e(vce)	*vcetype* specified in vce()
e(vcetype)	title used to label Std. Err.
e(asis)	asis, if specified
e(method)	ml
e(opt)	type of optimization
e(which)	max or min; whether optimizer is to perform maximization or minimization
e(ml_method)	type of ml method
e(user)	name of likelihood-evaluator program
e(technique)	maximization technique
e(properties)	b V
e(estat_cmd)	program used to implement estat
e(predict)	program used to implement predict
e(footnote)	program used to implement the footnote display
e(marginsok)	predictions allowed by margins
e(asbalanced)	factor variables fvset as asbalanced
e(asobserved)	factor variables fvset as asobserved

Matrices
e(b)	coefficient vector
e(Cns)	constraints matrix
e(rules)	information about perfect predictors
e(ilog)	iteration log (up to 20 iterations)
e(gradient)	gradient vector
e(Sigma)	$\widehat{\Sigma}$
e(V)	variance–covariance matrix of the estimators
e(V_modelbased)	model-based variance

Functions
e(sample)	marks estimation sample

`ivprobit, twostep` saves the following in `e()`:

Scalars

e(N)	number of observations
e(N_cds)	number of completely determined successes
e(N_cdf)	number of completely determined failures
e(df_m)	model degrees of freedom
e(df_exog)	degrees of freedom for χ^2 test of exogeneity
e(p)	model Wald p-value
e(p_exog)	exogeneity test Wald p-value
e(chi2)	model Wald χ^2
e(chi2_exog)	Wald χ^2 test of exogeneity
e(rank)	rank of e(V)

Macros

e(cmd)	ivprobit
e(cmdline)	command as typed
e(depvar)	name of dependent variable
e(instd)	instrumented variables
e(insts)	instruments
e(wtype)	weight type
e(wexp)	weight expression
e(chi2type)	Wald; type of model χ^2 test
e(vce)	*vcetype* specified in vce()
e(vcetype)	title used to label Std. Err.
e(asis)	asis, if specified
e(method)	twostep
e(properties)	b V
e(estat_cmd)	program used to implement estat
e(predict)	program used to implement predict
e(footnote)	program used to implement the footnote display
e(marginsok)	predictions allowed by margins
e(asbalanced)	factor variables fvset as asbalanced
e(asobserved)	factor variables fvset as asobserved

Matrices

e(b)	coefficient vector
e(Cns)	constraints matrix
e(rules)	information about perfect predictors
e(V)	variance–covariance matrix of the estimators

Functions

e(sample)	marks estimation sample

Methods and formulas

`ivprobit` is implemented as an ado-file.

Fitting limited-dependent variable models with endogenous regressors has received considerable attention in the econometrics literature. Building on the results of Amemiya (1978, 1979), Newey (1987) developed an efficient method of estimation that encompasses both Rivers and Vuong's (1988) simultaneous-equations probit model and Smith and Blundell's (1986) simultaneous-equations tobit model. With modern computers, maximum likelihood estimation is feasible as well. For compactness, we write the model as

$$y_{1i}^* = z_i\delta + u_i \tag{1a}$$

$$y_{2i} = x_i\Pi + v_i \tag{1b}$$

where $z_i = (y_{2i}, x_{1i})$, $x_i = (x_{1i}, x_{2i})$, $\delta = (\beta', \gamma')'$, and $\Pi = (\Pi'_1, \Pi'_2)'$.

Deriving the likelihood function is straightforward because we can write the joint density $f(y_{1i}, \boldsymbol{y}_{2i} | \boldsymbol{x}_i)$ as $f(y_{1i} | \boldsymbol{y}_{2i}, \boldsymbol{x}_i) f(\boldsymbol{y}_{2i} | \boldsymbol{x}_i)$. When there is an endogenous regressor, the log likelihood for observation i is

$$\ln L_i = w_i \left[y_{1i} \ln \Phi(m_i) + (1 - y_{1i}) \ln \{1 - \Phi(m_i)\} + \ln \phi \left(\frac{y_{2i} - \boldsymbol{x}_i \boldsymbol{\Pi}}{\sigma} \right) - \ln \sigma \right]$$

where

$$m_i = \frac{\boldsymbol{z}_i \boldsymbol{\delta} + \rho (y_{2i} - \boldsymbol{x}_i \boldsymbol{\Pi}) / \sigma}{(1 - \rho^2)^{\frac{1}{2}}}$$

$\Phi(\cdot)$ and $\phi(\cdot)$ are the standard normal distribution and density functions, respectively; σ is the standard deviation of v_i; ρ is the correlation coefficient between u_i and v_i; and w_i is the weight for observation i or one if no weights were specified. Instead of estimating σ and ρ, we estimate $\ln \sigma$ and atanh ρ, where

$$\text{atanh } \rho = \frac{1}{2} \ln \left(\frac{1 + \rho}{1 - \rho} \right)$$

For multiple endogenous regressors, let

$$\text{Var}(u_i, \boldsymbol{v}_i) = \boldsymbol{\Sigma} = \begin{bmatrix} 1 & \boldsymbol{\Sigma}_{21}' \\ \boldsymbol{\Sigma}_{21} & \boldsymbol{\Sigma}_{22} \end{bmatrix}$$

As in any probit model, we have imposed the normalization $\text{Var}(u_i) = 1$ to identify the model. The log likelihood for observation i is

$$\ln L_i = w_i \left[y_{1i} \ln \Phi(m_i) + (1 - y_{1i}) \ln \{1 - \Phi(m_i)\} + \ln f(\boldsymbol{y}_{2i} | \boldsymbol{x}_i) \right]$$

where

$$\ln f(\boldsymbol{y}_{2i} | \boldsymbol{x}_i) = -\frac{p}{2} \ln 2\pi - \frac{1}{2} \ln |\boldsymbol{\Sigma}_{22}| - \frac{1}{2} (\boldsymbol{y}_{2i} - \boldsymbol{x}_i \boldsymbol{\Pi}) \boldsymbol{\Sigma}_{22}^{-1} (\boldsymbol{y}_{2i} - \boldsymbol{x}_i \boldsymbol{\Pi})'$$

and

$$m_i = \left(1 - \boldsymbol{\Sigma}_{21}' \boldsymbol{\Sigma}_{22}^{-1} \boldsymbol{\Sigma}_{21} \right)^{-\frac{1}{2}} \left\{ \boldsymbol{z}_i \boldsymbol{\delta} + (\boldsymbol{y}_{2i} - \boldsymbol{x}_i \boldsymbol{\Pi}) \boldsymbol{\Sigma}_{22}^{-1} \boldsymbol{\Sigma}_{21} \right\}$$

Instead of maximizing the log-likelihood function with respect to $\boldsymbol{\Sigma}$, we maximize with respect to the Cholesky decomposition $\boldsymbol{S}$ of $\boldsymbol{\Sigma}$; that is, there exists a lower triangular matrix, $\boldsymbol{S}$, such that $\boldsymbol{SS}' = \boldsymbol{\Sigma}$. This maximization ensures that $\boldsymbol{\Sigma}$ is positive definite, as a covariance matrix must be. Let

$$\boldsymbol{S} = \begin{bmatrix} 1 & 0 & 0 & \cdots & 0 \\ s_{21} & s_{22} & 0 & \cdots & 0 \\ s_{31} & s_{32} & s_{33} & \cdots & 0 \\ \vdots & \vdots & \vdots & \ddots & \vdots \\ s_{p+1,1} & s_{p+1,2} & s_{p+1,3} & \cdots & s_{p+1,p+1} \end{bmatrix}$$

With maximum likelihood estimation, this command supports the Huber/White/sandwich estimator of the variance and its clustered version using `vce(robust)` and `vce(cluster clustvar)`, respectively. See [P] _robust, particularly *Maximum likelihood estimators* and *Methods and formulas*.

The maximum likelihood version of `heckman` also supports estimation with survey data. For details on VCEs with survey data, see [SVY] **variance estimation**.

The two-step estimates are obtained using Newey's (1987) minimum chi-squared estimator. The reduced-form equation for y_{1i}^* is

$$y_{1i}^* = (\boldsymbol{x}_i \boldsymbol{\Pi} + \boldsymbol{v}_i)\boldsymbol{\beta} + \boldsymbol{x}_{1i}\boldsymbol{\gamma} + u_i$$
$$= \boldsymbol{x}_i\boldsymbol{\alpha} + \boldsymbol{v}_i\boldsymbol{\beta} + u_i$$
$$= \boldsymbol{x}_i\boldsymbol{\alpha} + \nu_i$$

where $\nu_i = \boldsymbol{v}_i\boldsymbol{\beta} + u_i$. Because u_i and $\boldsymbol{v}_i$ are jointly normal, ν_i is also normal. Note that

$$\boldsymbol{\alpha} = \begin{bmatrix} \boldsymbol{\Pi}_1 \\ \boldsymbol{\Pi}_2 \end{bmatrix}\boldsymbol{\beta} + \begin{bmatrix} \boldsymbol{I} \\ \boldsymbol{0} \end{bmatrix}\boldsymbol{\gamma} = D(\boldsymbol{\Pi})\boldsymbol{\delta}$$

where $D(\boldsymbol{\Pi}) = (\boldsymbol{\Pi}, \boldsymbol{I}_1)$ and $\boldsymbol{I}_1$ is defined such that $\boldsymbol{x}_i \boldsymbol{I}_1 = \boldsymbol{x}_{1i}$. Letting $\widehat{\boldsymbol{z}}_i = (\boldsymbol{x}_i\widehat{\boldsymbol{\Pi}}, \boldsymbol{x}_{1i})$, $\widehat{\boldsymbol{z}}_i\boldsymbol{\delta} = \boldsymbol{x}_i D(\widehat{\boldsymbol{\Pi}})\boldsymbol{\delta}$, where $D(\widehat{\boldsymbol{\Pi}}) = (\widehat{\boldsymbol{\Pi}}, \boldsymbol{I}_1)$. Thus one estimator of $\boldsymbol{\alpha}$ is $D(\widehat{\boldsymbol{\Pi}})\boldsymbol{\delta}$; denote this estimator by $\widehat{\boldsymbol{D}}\boldsymbol{\delta}$.

$\boldsymbol{\alpha}$ could also be estimated directly as the solution to

$$\max_{\boldsymbol{\alpha},\boldsymbol{\lambda}} \sum_{i=1}^{N} l(y_{1i}, \boldsymbol{x}_i\boldsymbol{\alpha} + \widehat{\boldsymbol{v}}_i\boldsymbol{\lambda}) \tag{2}$$

where $l(\cdot)$ is the log likelihood for probit. Denote this estimator by $\widetilde{\boldsymbol{\alpha}}$. The inclusion of the $\widehat{\boldsymbol{v}}_i\boldsymbol{\lambda}$ term follows because the multivariate normality of $(u_i, \boldsymbol{v}_i)$ implies that, conditional on $\boldsymbol{y}_{2i}$, the expected value of u_i is nonzero. Because $\boldsymbol{v}_i$ is unobservable, the least-squares residuals from fitting (1b) are used.

Amemiya (1978) shows that the estimator of $\boldsymbol{\delta}$ defined by

$$\max_{\boldsymbol{\delta}} (\widetilde{\boldsymbol{\alpha}} - \widehat{\boldsymbol{D}}\boldsymbol{\delta})'\widehat{\boldsymbol{\Omega}}^{-1}(\widetilde{\boldsymbol{\alpha}} - \widehat{\boldsymbol{D}}\boldsymbol{\delta})$$

where $\widehat{\boldsymbol{\Omega}}$ is a consistent estimator of the covariance of $\sqrt{N}(\widetilde{\boldsymbol{\alpha}} - \widehat{\boldsymbol{D}}\boldsymbol{\delta})$, is asymptotically efficient relative to all other estimators that minimize the distance between $\widetilde{\boldsymbol{\alpha}}$ and $D(\widehat{\boldsymbol{\Pi}})\boldsymbol{\delta}$. Thus an efficient estimator of $\boldsymbol{\delta}$ is

$$\widehat{\boldsymbol{\delta}} = (\widehat{\boldsymbol{D}}'\widehat{\boldsymbol{\Omega}}^{-1}\widehat{\boldsymbol{D}})^{-1}\widehat{\boldsymbol{D}}'\widehat{\boldsymbol{\Omega}}^{-1}\widetilde{\boldsymbol{\alpha}} \tag{3}$$

and

$$\text{Var}(\widehat{\boldsymbol{\delta}}) = (\widehat{\boldsymbol{D}}'\widehat{\boldsymbol{\Omega}}^{-1}\widehat{\boldsymbol{D}})^{-1} \tag{4}$$

To implement this estimator, we need $\widehat{\boldsymbol{\Omega}}^{-1}$.

Consider the two-step maximum likelihood estimator that results from first fitting (1b) by OLS and computing the residuals $\widehat{\boldsymbol{v}}_i = \boldsymbol{y}_{2i} - \boldsymbol{x}_i\widehat{\boldsymbol{\Pi}}$. The estimator is then obtained by solving

$$\max_{\boldsymbol{\delta},\boldsymbol{\lambda}} \sum_{i=1}^{N} l(y_{1i}, \boldsymbol{z}_i\boldsymbol{\delta} + \widehat{\boldsymbol{v}}_i\boldsymbol{\lambda})$$

This is the two-step instrumental variables (2SIV) estimator proposed by Rivers and Vuong (1988), and its role will become apparent shortly.

From Proposition 5 of Newey (1987), $\sqrt{N}(\tilde{\alpha} - \hat{D}\delta) \overset{d}{\longrightarrow} \mathrm{N}(0, \Omega)$, where

$$\Omega = J_{\alpha\alpha}^{-1} + (\lambda - \beta)' \Sigma_{22} (\lambda - \beta) Q^{-1}$$

and $\Sigma_{22} = E\{v_i' v_i\}$. $J_{\alpha\alpha}^{-1}$ is simply the covariance matrix of $\tilde{\alpha}$, ignoring that $\hat{\Pi}$ is an estimated parameter matrix. Moreover, Newey shows that the covariance matrix from an OLS regression of $y_{2i}(\hat{\lambda} - \hat{\beta})$ on x_i is a consistent estimator of the second term. $\hat{\lambda}$ can be obtained from solving (2), and the 2SIV estimator yields a consistent estimate, $\hat{\beta}$.

Mechanically, estimation proceeds in several steps.

1. Each of the endogenous right-hand-side variables is regressed on all the exogenous variables, and the fitted values and residuals are calculated. The matrix $\hat{D} = D(\hat{\Pi})$ is assembled from the estimated coefficients.

2. `probit` is used to solve (2) and obtain $\tilde{\alpha}$ and $\hat{\lambda}$. The portion of the covariance matrix corresponding to α, $J_{\alpha\alpha}^{-1}$, is also saved.

3. The 2SIV estimator is evaluated, and the parameters $\hat{\beta}$ corresponding to y_{2i} are collected.

4. $y_{2i}(\hat{\lambda} - \hat{\beta})$ is regressed on x_i. The covariance matrix of the parameters from this regression is added to $J_{\alpha\alpha}^{-1}$, yielding $\hat{\Omega}$.

5. Evaluating (3) and (4) yields the estimates $\hat{\delta}$ and $\mathrm{Var}(\hat{\delta})$.

6. A Wald test of the null hypothesis $H_0 : \lambda = 0$, using the 2SIV estimates, serves as our test of exogeneity.

The two-step estimates are not directly comparable to those obtained from the maximum likelihood estimator or from `probit`. The argument is the same for Newey's efficient estimator as for Rivers and Vuong's (1988) 2SIV estimator, so we consider the simpler 2SIV estimator. From the properties of the normal distribution,

$$E(u_i | v_i) = v_i \Sigma_{22}^{-1} \Sigma_{21} \qquad \text{and} \qquad \mathrm{Var}(u_i | v_i) = 1 - \Sigma_{21}' \Sigma_{22}^{-1} \Sigma_{21}$$

We write u_i as $u_i = v_i \Sigma_{22}^{-1} \Sigma_{21} + e_i = v_i \lambda + e_i$, where $e_i \sim \mathrm{N}(0, 1 - \rho^2)$, $\rho^2 = \Sigma_{21}' \Sigma_{22}^{-1} \Sigma_{21}$, and e_i is independent of v_i. In the second stage of 2SIV, we use a probit regression to estimate the parameters of

$$y_{1i} = z_i \delta + v_i \lambda + e_i$$

Because v_i is unobservable, we use the sample residuals from the first-stage regressions.

$$\Pr(y_{1i} = 1 | z_i, v_i) = \Pr(z_i \delta + v_i \lambda + e_i > 0 | z_i, v_i) = \Phi\left\{ (1 - \rho^2)^{-\frac{1}{2}} (z_i \delta + v_i \lambda) \right\}$$

Hence, as mentioned previously, 2SIV and Newey's estimator do not estimate δ and λ but rather

$$\delta_\rho = \frac{1}{(1 - \rho^2)^{\frac{1}{2}}} \delta \qquad \text{and} \qquad \lambda_\rho = \frac{1}{(1 - \rho^2)^{\frac{1}{2}}} \lambda$$

Acknowledgments

The two-step estimator is based on the `probitiv` command written by Jonah Gelbach, Department of Economics, Yale University, and the `ivprob` command written by Joe Harkness, Institute of Policy Studies, Johns Hopkins University.

References

Amemiya, T. 1978. The estimation of a simultaneous equation generalized probit model. *Econometrica* 46: 1193–1205.

———. 1979. The estimation of a simultaneous-equation tobit model. *International Economic Review* 20: 169–181.

Finlay, K., and L. M. Magnusson. 2009. Implementing weak-instrument robust tests for a general class of instrumental-variables models. *Stata Journal* 9: 398–421.

Miranda, A., and S. Rabe-Hesketh. 2006. Maximum likelihood estimation of endogenous switching and sample selection models for binary, ordinal, and count variables. *Stata Journal* 6: 285–308.

Newey, W. K. 1987. Efficient estimation of limited dependent variable models with endogenous explanatory variables. *Journal of Econometrics* 36: 231–250.

Rivers, D., and Q. H. Vuong. 1988. Limited information estimators and exogeneity tests for simultaneous probit models. *Journal of Econometrics* 39: 347–366.

Smith, R. J., and R. Blundell. 1986. An exogeneity test for the simultaneous equation tobit model with an application to labor supply. *Econometrica* 54: 679–685.

Wooldridge, J. M. 2010. *Econometric Analysis of Cross Section and Panel Data.* 2nd ed. Cambridge, MA: MIT Press.

Also see

Title

> **ivprobit postestimation** — Postestimation tools for ivprobit

Description

The following postestimation commands are of special interest after `ivprobit`:

Command	Description
estat classification	report various summary statistics, including the classification table
lroc	compute area under ROC curve and graph the curve
lsens	graph sensitivity and specificity versus probability cutoff

These commands are not appropriate after the two-step estimator or the svy prefix.

For information about these commands, see [R] **logistic postestimation**.

The following standard postestimation commands are also available:

Command	Description
contrast	contrasts and ANOVA-style joint tests of estimates
estat[1]	AIC, BIC, VCE, and estimation sample summary
estat (svy)	postestimation statistics for survey data
estimates	cataloging estimation results
hausman	Hausman's specification test
lincom	point estimates, standard errors, testing, and inference for linear combinations of coefficients
lrtest[2]	likelihood-ratio test; not available with two-step estimator
margins	marginal means, predictive margins, marginal effects, and average marginal effects
marginsplot	graph the results from margins (profile plots, interaction plots, etc.)
nlcom	point estimates, standard errors, testing, and inference for nonlinear combinations of coefficients
predict	predictions, residuals, influence statistics, and other diagnostic measures
predictnl	point estimates, standard errors, testing, and inference for generalized predictions
pwcompare	pairwise comparisons of estimates
suest[1]	seemingly unrelated estimation
test	Wald tests of simple and composite linear hypotheses
testnl	Wald tests of nonlinear hypotheses

[1] estat ic and suest are not appropriate after ivprobit, twostep.

[2] lrtest is not appropriate with svy estimation results.

See the corresponding entries in the *Base Reference Manual* for details, but see [SVY] **estat** for details about estat (svy).

Syntax for predict

After ML or twostep

    predict $\begin{bmatrix} type \end{bmatrix}$ *newvar* $\begin{bmatrix} if \end{bmatrix}$ $\begin{bmatrix} in \end{bmatrix}$ $\begin{bmatrix} ,\ statistic\ \underline{rules}\ \text{asif} \end{bmatrix}$

After ML

    predict $\begin{bmatrix} type \end{bmatrix}$ $\{\ stub*\ |\ newvarlist\ \}$ $\begin{bmatrix} if \end{bmatrix}$ $\begin{bmatrix} in \end{bmatrix}$, $\underline{sc}$ores

statistic	Description
Main	
xb	linear prediction; the default
stdp	standard error of the linear prediction
<u>pr</u>	probability of a positive outcome; not available with two-step estimator

These statistics are available both in and out of sample; type predict ... if e(sample) ... if wanted only for the estimation sample.

Menu

Statistics > Postestimation > Predictions, residuals, etc.

Options for predict

    ⌐ Main ⌐

xb, the default, calculates the linear prediction.

stdp calculates the standard error of the linear prediction.

pr calculates the probability of a positive outcome. pr is not available with the two-step estimator.

rules requests that Stata use any rules that were used to identify the model when making the prediction. By default, Stata calculates missing for excluded observations. rules is not available with the two-step estimator.

asif requests that Stata ignore the rules and the exclusion criteria and calculate predictions for all observations possible using the estimated parameters from the model. asif is not available with the two-step estimator.

scores, not available with twostep, calculates equation-level score variables.

    For models with one endogenous regressor, four new variables are created.

       The first new variable will contain $\partial \ln L/\partial(z_i\delta)$.

       The second new variable will contain $\partial \ln L/\partial(x_i\Pi)$.

       The third new variable will contain $\partial \ln L/\partial$ atanh ρ.

       The fourth new variable will contain $\partial \ln L/\partial \ln\sigma$.

    For models with p endogenous regressors, $p + \{(p+1)(p+2)\}/2$ new variables are created.

       The first new variable will contain $\partial \ln L/\partial(z_i\delta)$.

The second through $(p+1)$th new variables will contain $\partial \ln L / \partial(x_i \Pi_k)$, $k = 1, \ldots, p$, where Π_k is the kth column of Π.

The remaining score variables will contain the partial derivatives of $\ln L$ with respect to s_{21}, $s_{31}, \ldots, s_{p+1,1}, s_{22}, \ldots, s_{p+1,2}, \ldots, s_{p+1,p+1}$, where $s_{m,n}$ denotes the (m, n) element of the Cholesky decomposition of the error covariance matrix.

Remarks

Remarks are presented under the following headings:

Marginal effects
Obtaining predicted values

Marginal effects

▷ Example 1

We can obtain marginal effects by using the `margins` command after `ivprobit`. We will calculate average marginal effects by using the labor-supply model of example 1 in [R] **ivprobit**.

```
. use http://www.stata-press.com/data/r12/laborsup

. ivprobit fem_work fem_educ kids (other_inc = male_educ)
  (output omitted)

. margins, dydx(*) predict(pr)
```

| Average marginal effects | | | | Number of obs | = | 500 |
| Model VCE   : OIM | | | | | | |

Expression : Probability of positive outcome, predict(pr)
dy/dx w.r.t. : other_inc fem_educ kids male_educ

	dy/dx	Delta-method Std. Err.	z	P>|z|	[95% Conf. Interval]	
other_inc	-.014015	.0009836	-14.25	0.000	-.0159428	-.0120872
fem_educ	.0545129	.0066007	8.26	0.000	.0415758	.06745
kids	-.0470199	.0123397	-3.81	0.000	-.0712052	-.0228346
male_educ	0	(omitted)				

Here we see that a \$1,000 increase in `other_inc` leads to an average decrease of 0.014 in the probability that the woman has a job. `male_edu` has no effect because it appears only as an instrument.

◁

Obtaining predicted values

After fitting your model with `ivprobit`, you can obtain the linear prediction and its standard error for both the estimation sample and other samples by using the `predict` command; see [U] **20 Estimation and postestimation commands** and [R] **predict**. If you had used the maximum likelihood estimator, you could also obtain the probability of a positive outcome.

predict's pr option calculates the probability of a positive outcome, remembering any rules used to identify the model, and calculates missing for excluded observations. predict's rules option uses the rules in predicting probabilities, whereas predict's asif option ignores both the rules and the exclusion criteria and calculates probabilities for all possible observations by using the estimated parameters from the model. See *Obtaining predicted values* in [R] **probit postestimation** for an example.

Methods and formulas

All postestimation commands listed above are implemented as ado-files.

The linear prediction is calculated as $z_i\widehat{\delta}$, where $\widehat{\delta}$ is the estimated value of δ, and z_i and δ are defined in (1a) of [R] **ivprobit**. The probability of a positive outcome is $\Phi(z_i\widehat{\delta})$, where $\Phi(\cdot)$ is the standard normal distribution function.

Also see

[R] **ivprobit** — Probit model with continuous endogenous regressors

[U] **20 Estimation and postestimation commands**

Title

> **ivregress** — Single-equation instrumental-variables regression

Syntax

ivregress *estimator* *depvar* [*varlist*$_1$] (*varlist*$_2$ = *varlist*$_{iv}$) [*if*] [*in*] [*weight*]

[, *options*]

estimator	Description
2sls	two-stage least squares (2SLS)
liml	limited-information maximum likelihood (LIML)
gmm	generalized method of moments (GMM)

options	Description
Model	
noconstant	suppress constant term
hascons	has user-supplied constant
GMM[1]	
wmatrix(*wmtype*)	*wmtype* may be robust, cluster *clustvar*, hac *kernel*, or unadjusted
center	center moments in weight matrix computation
igmm	use iterative instead of two-step GMM estimator
eps(#)[2]	specify # for parameter convergence criterion; default is eps(1e-6)
weps(#)[2]	specify # for weight matrix convergence criterion; default is weps(1e-6)
optimization_options[2]	control the optimization process; seldom used
SE/Robust	
vce(*vcetype*)	*vcetype* may be unadjusted, robust, cluster *clustvar*, bootstrap, jackknife, or hac *kernel*
Reporting	
level(#)	set confidence level; default is level(95)
first	report first-stage regression
small	make degrees-of-freedom adjustments and report small-sample statistics
noheader	display only the coefficient table
depname(*depname*)	substitute dependent variable name
eform(*string*)	report exponentiated coefficients and use *string* to label them
display_options	control column formats, row spacing, line width, and display of omitted variables and base and empty cells
perfect	do not check for collinearity between endogenous regressors and excluded instruments
coeflegend	display legend instead of statistics

[1]These options may be specified only when gmm is specified.

[2]These options may be specified only when igmm is specified.

varlist$_1$ and varlist$_{iv}$ may contain factor variables; see [U] **11.4.3 Factor variables**.

depvar, varlist$_1$, varlist$_2$, and varlist$_{iv}$ may contain time-series operators; see [U] **11.4.4 Time-series varlists**.

bootstrap, by, jackknife, rolling, statsby, and svy are allowed; see [U] **11.1.10 Prefix commands**.

Weights are not allowed with the bootstrap prefix; see [R] **bootstrap**.

aweights are not allowed with the jackknife prefix; see [R] **jackknife**.

hascons, vce(), noheader, depname(), and weights are not allowed with the svy prefix; see [SVY] **svy**.

aweights, fweights, iweights, and pweights are allowed; see [U] **11.1.6 weight**.

perfect and coeflegend do not appear in the dialog box.

See [U] **20 Estimation and postestimation commands** for more capabilities of estimation commands.

Menu

Statistics > Endogenous covariates > Single-equation instrumental-variables regression

Description

ivregress fits a linear regression of *depvar* on *varlist*$_1$ and *varlist*$_2$, using *varlist*$_{iv}$ (along with *varlist*$_1$) as instruments for *varlist*$_2$. ivregress supports estimation via two-stage least squares (2SLS), limited-information maximum likelihood (LIML), and generalized method of moments (GMM).

In the language of instrumental variables, *varlist*$_1$ and *varlist*$_{iv}$ are the exogenous variables, and *varlist*$_2$ are the endogenous variables.

Options

‾‾‾‾| Model |‾‾‾

noconstant; see [R] **estimation options**.

hascons indicates that a user-defined constant or its equivalent is specified among the independent variables.

‾‾‾‾| GMM |‾‾‾

wmatrix(*wmtype*) specifies the type of weighting matrix to be used in conjunction with the GMM estimator.

Specifying wmatrix(robust) requests a weighting matrix that is optimal when the error term is heteroskedastic. wmatrix(robust) is the default.

Specifying wmatrix(cluster *clustvar*) requests a weighting matrix that accounts for arbitrary correlation among observations within clusters identified by *clustvar*.

Specifying wmatrix(hac *kernel* #) requests a heteroskedasticity- and autocorrelation-consistent (HAC) weighting matrix using the specified kernel (see below) with # lags. The bandwidth of a kernel is equal to # + 1.

Specifying wmatrix(hac *kernel* opt) requests an HAC weighting matrix using the specified kernel, and the lag order is selected using Newey and West's (1994) optimal lag-selection algorithm.

Specifying wmatrix(hac *kernel*) requests an HAC weighting matrix using the specified kernel and $N - 2$ lags, where N is the sample size.

There are three kernels available for HAC weighting matrices, and you may request each one by using the name used by statisticians or the name perhaps more familiar to economists:

<u>bart</u>lett or <u>nw</u>est requests the Bartlett (Newey–West) kernel;

<u>par</u>zen or <u>g</u>allant requests the Parzen (Gallant 1987) kernel; and

<u>q</u>uadraticspectral or <u>and</u>rews requests the quadratic spectral (Andrews 1991) kernel.

Specifying wmatrix(unadjusted) requests a weighting matrix that is suitable when the errors are homoskedastic. The GMM estimator with this weighting matrix is equivalent to the 2SLS estimator.

center requests that the sample moments be centered (demeaned) when computing GMM weight matrices. By default, centering is not done.

igmm requests that the iterative GMM estimator be used instead of the default two-step GMM estimator. Convergence is declared when the relative change in the parameter vector from one iteration to the next is less than eps() or the relative change in the weight matrix is less than weps().

eps(#) specifies the convergence criterion for successive parameter estimates when the iterative GMM estimator is used. The default is eps(1e-6). Convergence is declared when the relative difference between successive parameter estimates is less than eps() and the relative difference between successive estimates of the weighting matrix is less than weps().

weps(#) specifies the convergence criterion for successive estimates of the weighting matrix when the iterative GMM estimator is used. The default is weps(1e-6). Convergence is declared when the relative difference between successive parameter estimates is less than eps() and the relative difference between successive estimates of the weighting matrix is less than weps().

optimization_options: <u>iter</u>ate(#), [<u>no</u>]<u>log</u>. iterate() specifies the maximum number of iterations to perform in conjunction with the iterative GMM estimator. The default is 16,000 or the number set using set maxiter (see [R] **maximize**). log/nolog specifies whether to show the iteration log. These options are seldom used.

    ⌐ SE/Robust ⌐

vce(*vcetype*) specifies the type of standard error reported, which includes types that are robust to some kinds of misspecification, that allow for intragroup correlation, and that use bootstrap or jackknife methods; see [R] *vce_option*.

vce(unadjusted), the default for 2sls and liml, specifies that an unadjusted (nonrobust) VCE matrix be used. The default for gmm is based on the *wmtype* specified in the wmatrix() option; see wmatrix(*wmtype*) above. If wmatrix() is specified with gmm but vce() is not, then *vcetype* is set equal to *wmtype*. To override this behavior and obtain an unadjusted (nonrobust) VCE matrix, specify vce(unadjusted).

ivregress also allows the following:

vce(hac *kernel* [#|opt]) specifies that an HAC covariance matrix be used. The syntax used with vce(hac *kernel* ...) is identical to that used with wmatrix(hac *kernel* ...); see wmatrix(*wmtype*) above.

    ⌐ Reporting ⌐

level(#); see [R] **estimation options**.

first requests that the first-stage regression results be displayed.

small requests that the degrees-of-freedom adjustment $N/(N-k)$ be made to the variance–covariance matrix of parameters and that small-sample F and t statistics be reported, where N is the sample size and k is the number of parameters estimated. By default, no degrees-of-freedom adjustment is made, and Wald and z statistics are reported. Even with this option, no degrees-of-freedom adjustment is made to the weighting matrix when the GMM estimator is used.

noheader suppresses the display of the summary statistics at the top of the output, displaying only the coefficient table.

depname(*depname*) is used only in programs and ado-files that use ivregress to fit models other than instrumental-variables regression. depname() may be specified only at estimation time. *depname* is recorded as the identity of the dependent variable, even though the estimates are calculated using *depvar*. This method affects the labeling of the output—not the results calculated—but could affect later calculations made by predict, where the residual would be calculated as deviations from *depname* rather than *depvar*. depname() is most typically used when *depvar* is a temporary variable (see [P] **macro**) used as a proxy for *depname*.

eform(*string*) is used only in programs and ado-files that use ivregress to fit models other than instrumental-variables regression. eform() specifies that the coefficient table be displayed in "exponentiated form", as defined in [R] **maximize**, and that *string* be used to label the exponentiated coefficients in the table.

display_options: <u>noomit</u>ted, vsquish, <u>noempty</u>cells, <u>base</u>levels, <u>allbase</u>levels, cformat(%*fmt*), pformat(%*fmt*), sformat(%*fmt*), and nolstretch; see [R] **estimation options**.

The following options are available with ivregress but are not shown in the dialog box:

perfect requests that ivregress not check for collinearity between the endogenous regressors and excluded instruments, allowing one to specify "perfect" instruments. This option cannot be used with the LIML estimator. This option may be required when using ivregress to implement other estimators.

coeflegend; see [R] **estimation options**.

Remarks

ivregress performs instrumental-variables regression and weighted instrumental-variables regression. For a general discussion of instrumental variables, see Baum (2006), Cameron and Trivedi (2005; 2010, chap. 6) Davidson and MacKinnon (1993, 2004), Greene (2012, chap. 8), and Wooldridge (2009, 2010). See Hall (2005) for a lucid presentation of GMM estimation. Angrist and Pischke (2009, chap. 4) offer a casual yet thorough introduction to instrumental-variables estimators, including their use in estimating treatment effects. Some of the earliest work on simultaneous systems can be found in Cowles Commission monographs—Koopmans and Marschak (1950) and Koopmans and Hood (1953)—with the first developments of 2SLS appearing in Theil (1953) and Basmann (1957). However, Stock and Watson (2011, 422–424) present an example of the method of instrumental variables that was first published in 1928 by Philip Wright.

The syntax for ivregress assumes that you want to fit one equation from a system of equations or an equation for which you do not want to specify the functional form for the remaining equations of the system. To fit a full system of equations, using either 2SLS equation-by-equation or three-stage least squares, see [R] **reg3**. An advantage of ivregress is that you can fit one equation of a multiple-equation system without specifying the functional form of the remaining equations.

Formally, the model fit by ivregress is

$$y_i = \mathbf{y}_i \beta_1 + \mathbf{x}_{1i} \beta_2 + u_i \tag{1}$$

$$\mathbf{y}_i = \mathbf{x}_{1i} \mathbf{\Pi}_1 + \mathbf{x}_{2i} \mathbf{\Pi}_2 + \mathbf{v}_i \tag{2}$$

Here y_i is the dependent variable for the ith observation, $\mathbf{y}_i$ represents the endogenous regressors (*varlist*$_2$ in the syntax diagram), $\mathbf{x}_{1i}$ represents the included exogenous regressors (*varlist*$_1$ in the syntax diagram), and $\mathbf{x}_{2i}$ represents the excluded exogenous regressors (*varlist*$_{iv}$ in the syntax diagram). $\mathbf{x}_{1i}$ and $\mathbf{x}_{2i}$ are collectively called the instruments. u_i and $\mathbf{v}_i$ are zero-mean error terms, and the correlations between u_i and the elements of $\mathbf{v}_i$ are presumably nonzero.

The rest of the discussion is presented under the following headings:

> *2SLS and LIML estimators*
> *GMM estimator*

2SLS and LIML estimators

The most common instrumental-variables estimator is 2SLS.

▷ Example 1: 2SLS estimator

We have state data from the 1980 census on the median dollar value of owner-occupied housing (hsngval) and the median monthly gross rent (rent). We want to model rent as a function of hsngval and the percentage of the population living in urban areas (pcturban):

$$\texttt{rent}_i = \beta_0 + \beta_1 \texttt{hsngval}_i + \beta_2 \texttt{pcturban}_i + u_i$$

where i indexes states and u_i is an error term.

Because random shocks that affect rental rates in a state probably also affect housing values, we treat hsngval as endogenous. We believe that the correlation between hsngval and u is not equal to zero. On the other hand, we have no reason to believe that the correlation between pcturban and u is nonzero, so we assume that pcturban is exogenous.

Because we are treating hsngval as an endogenous regressor, we must have one or more additional variables available that are correlated with hsngval but uncorrelated with u. Moreover, these excluded exogenous variables must not affect rent directly, because if they do then they should be included in the regression equation we specified above. In our dataset, we have a variable for family income (faminc) and for region of the country (region) that we believe are correlated with hsngval but not the error term. Together, pcturban, faminc, and factor variables 2.region, 3.region, and 4.region constitute our set of instruments.

To fit the equation in Stata, we specify the dependent variable and the list of included exogenous variables. In parentheses, we specify the endogenous regressors, an equal sign, and the excluded exogenous variables. Only the additional exogenous variables must be specified to the right of the equal sign; the exogenous variables that appear in the regression equation are automatically included as instruments.

Here we fit our model with the 2SLS estimator:

```
. use http://www.stata-press.com/data/r12/hsng
(1980 Census housing data)
. ivregress 2sls rent pcturban (hsngval = faminc i.region)
Instrumental variables (2SLS) regression        Number of obs  =        50
                                                 Wald chi2(2)   =     90.76
                                                 Prob > chi2    =    0.0000
                                                 R-squared      =    0.5989
                                                 Root MSE       =    22.166
```

rent	Coef.	Std. Err.	z	P>|z|	[95% Conf. Interval]	
hsngval	.0022398	.0003284	6.82	0.000	.0015961	.0028836
pcturban	.081516	.2987652	0.27	0.785	−.504053	.667085
_cons	120.7065	15.22839	7.93	0.000	90.85942	150.5536

```
Instrumented:  hsngval
Instruments:   pcturban faminc 2.region 3.region 4.region
```

As we would expect, states with higher housing values have higher rental rates. The proportion of a state's population that is urban does not have a significant effect on rents.

◁

❑ Technical note

In a simultaneous-equations framework, we could write the model we just fit as

$$\texttt{hsngval}_i = \pi_0 + \pi_1 \texttt{faminc}_i + \pi_2 \texttt{2.region}_i + \pi_3 \texttt{3.region}_i + \pi_4 \texttt{4.region}_i + v_i$$

$$\texttt{rent}_i = \beta_0 + \beta_1 \texttt{hsngval}_i + \beta_2 \texttt{pcturban}_i + u_i$$

which here happens to be recursive (triangular), because hsngval appears in the equation for rent but rent does not appear in the equation for hsngval. In general, however, systems of simultaneous equations are not recursive. Because this system is recursive, we could fit the two equations individually via OLS if we were willing to assume that u and v were independent. For a more detailed discussion of triangular systems, see Kmenta (1997, 719–720).

Historically, instrumental-variables estimation and systems of simultaneous equations were taught concurrently, and older textbooks describe instrumental-variables estimation solely in the context of simultaneous equations. However, in recent decades, the treatment of endogeneity and instrumental-variables estimation has taken on a much broader scope, while interest in the specification of complete systems of simultaneous equations has waned. Most recent textbooks, such as Cameron and Trivedi (2005), Davidson and MacKinnon (1993, 2004), and Wooldridge (2009, 2010), treat instrumental-variables estimation as an integral part of the modern economists' toolkit and introduce it long before shorter discussions on simultaneous equations.

❑

In addition to the 2SLS member of the κ-class estimators, ivregress implements the LIML estimator. Both theoretical and Monte Carlo exercises indicate that the LIML estimator may yield less bias and confidence intervals with better coverage rates than the 2SLS estimator. See Poi (2006) and Stock, Wright, and Yogo (2002) (and the papers cited therein) for Monte Carlo evidence.

▷ Example 2: LIML estimator

Here we refit our model with the LIML estimator:

```
. ivregress liml rent pcturban (hsngval = faminc i.region)
Instrumental variables (LIML) regression          Number of obs =        50
                                                   Wald chi2(2)  =     75.71
                                                   Prob > chi2   =    0.0000
                                                   R-squared     =    0.4901
                                                   Root MSE      =    24.992
```

rent	Coef.	Std. Err.	z	P>|z|	[95% Conf. Interval]	
hsngval	.0026686	.0004173	6.39	0.000	.0018507	.0034865
pcturban	-.1827391	.3571132	-0.51	0.609	-.8826681	.5171899
_cons	117.6087	17.22625	6.83	0.000	83.84587	151.3715

```
Instrumented:  hsngval
Instruments:   pcturban faminc 2.region 3.region 4.region
```

These results are qualitatively similar to the 2SLS results, although the coefficient on hsngval is about 19% higher.

◁

GMM estimator

Since the celebrated paper of Hansen (1982), the GMM has been a popular method of estimation in economics and finance, and it lends itself well to instrumental-variables estimation. The basic principle is that we have some *moment* or *orthogonality* conditions of the form

$$E(\mathbf{z}_i u_i) = \mathbf{0} \tag{3}$$

From (1), we have $u_i = y_i - \mathbf{y}_i\boldsymbol{\beta}_1 - \mathbf{x}_{1i}\boldsymbol{\beta}_2$. What are the elements of the instrument vector $\mathbf{z}_i$? By assumption, $\mathbf{x}_{1i}$ is uncorrelated with u_i, as are the excluded exogenous variables $\mathbf{x}_{2i}$, and so we use $\mathbf{z}_i = [\mathbf{x}_{1i} \ \mathbf{x}_{2i}]$. The moment conditions are simply the mathematical representation of the assumption that the instruments are exogenous—that is, the instruments are orthogonal to (uncorrelated with) u_i.

If the number of elements in $\mathbf{z}_i$ is just equal to the number of unknown parameters, then we can apply the analogy principle to (3) and solve

$$\frac{1}{N}\sum_i \mathbf{z}_i u_i = \frac{1}{N}\sum_i \mathbf{z}_i \left(y_i - \mathbf{y}_i\boldsymbol{\beta}_1 - \mathbf{x}_{1i}\boldsymbol{\beta}_2\right) = \mathbf{0} \tag{4}$$

This equation is known as the method of moments estimator. Here where the number of instruments equals the number of parameters, the method of moments estimator coincides with the 2SLS estimator, which also coincides with what has historically been called the indirect least-squares estimator (Judge et al. 1985, 595).

The "generalized" in GMM addresses the case in which the number of instruments (columns of $\mathbf{z}_i$) exceeds the number of parameters to be estimated. Here there is no unique solution to the population moment conditions defined in (3), so we cannot use (4). Instead, we define the objective function

$$Q(\boldsymbol{\beta}_1, \boldsymbol{\beta}_2) = \left(\frac{1}{N}\sum_i \mathbf{z}_i u_i\right)' \mathbf{W} \left(\frac{1}{N}\sum_i \mathbf{z}_i u_i\right) \tag{5}$$

where $\mathbf{W}$ is a positive-definite matrix with the same number of rows and columns as the number of columns of $\mathbf{z}_i$. $\mathbf{W}$ is known as the weighting matrix, and we specify its structure with the wmatrix() option. The GMM estimator of (β_1, β_2) minimizes $Q(\beta_1, \beta_2)$; that is, the GMM estimator chooses β_1 and β_2 to make the moment conditions as close to zero as possible for a given $\mathbf{W}$. For a more general GMM estimator, see [R] **gmm**. gmm does not restrict you to fitting a single linear equation, though the syntax is more complex.

A well-known result is that if we define the matrix $\mathbf{S}_0$ to be the covariance of $\mathbf{z}_i u_i$ and set $\mathbf{W} = \mathbf{S}_0^{-1}$, then we obtain the optimal two-step GMM estimator, where by optimal estimator we mean the one that results in the smallest variance given the moment conditions defined in (3).

Suppose that the errors u_i are heteroskedastic but independent among observations. Then

$$\mathbf{S}_0 = E(\mathbf{z}_i u_i u_i \mathbf{z}_i') = E(u_i^2 \mathbf{z}_i \mathbf{z}_i')$$

and the sample analogue is

$$\widehat{\mathbf{S}} = \frac{1}{N} \sum_i \widehat{u}_i^2 \mathbf{z}_i \mathbf{z}_i' \tag{6}$$

To implement this estimator, we need estimates of the sample residuals $\widehat{u}_i$. ivregress gmm obtains the residuals by estimating β_1 and β_2 by 2SLS and then evaluates (6) and sets $\mathbf{W} = \widehat{\mathbf{S}}^{-1}$. Equation (6) is the same as the center term of the "sandwich" robust covariance matrix available from most Stata estimation commands through the vce(robust) option.

▷ Example 3: GMM estimator

Here we refit our model of rents by using the GMM estimator, allowing for heteroskedasticity in u_i:

```
. ivregress gmm rent pcturban (hsngval = faminc i.region), wmatrix(robust)
Instrumental variables (GMM) regression          Number of obs   =        50
                                                  Wald chi2(2)    =    112.09
                                                  Prob > chi2     =    0.0000
                                                  R-squared       =    0.6616
GMM weight matrix: Robust                         Root MSE        =    20.358

                 |               Robust
          rent |      Coef.   Std. Err.      z    P>|z|     [95% Conf. Interval]
---------------+----------------------------------------------------------------
       hsngval |   .0014643   .0004473     3.27   0.001     .0005877    .002341
      pcturban |   .7615482   .2895105     2.63   0.009     .1941181   1.328978
         _cons |   112.1227   10.80234    10.38   0.000     90.95052   133.2949

Instrumented:  hsngval
Instruments:   pcturban faminc 2.region 3.region 4.region
```

Because we requested that a heteroskedasticity-consistent weighting matrix be used during estimation but did not specify the vce() option, ivregress reported standard errors that are robust to heteroskedasticity. Had we specified vce(unadjusted), we would have obtained standard errors that would be correct only if the weighting matrix $\mathbf{W}$ does in fact converge to $\mathbf{S}_0^{-1}$.

◁

❑ Technical note

Many software packages that implement GMM estimation use the same heteroskedasticity-consistent weighting matrix we used in the previous example to obtain the optimal two-step estimates but do not use a heteroskedasticity-consistent VCE, even though they may label the standard errors as being "robust". To replicate results obtained from other packages, you may have to use the vce(unadjusted) option. See *Methods and formulas* below for a discussion of robust covariance matrix estimation in the GMM framework.

❑

By changing our definition of S_0, we can obtain GMM estimators suitable for use with other types of data that violate the assumption that the errors are independent and identically distributed. For example, you may have a dataset that consists of multiple observations for each person in a sample. The observations that correspond to the same person are likely to be correlated, and the estimation technique should account for that lack of independence. Say that in your dataset, people are identified by the variable personid and you type

```
. ivregress gmm ..., wmatrix(cluster personid)
```

Here ivregress estimates S_0 as

$$\widehat{\mathbf{S}} = \frac{1}{N} \sum_{c \in C} \mathbf{q}_c \mathbf{q}_c'$$

where C denotes the set of clusters and

$$\mathbf{q}_c = \sum_{i \in c_j} \widehat{u}_i \mathbf{z}_i$$

where c_j denotes the jth cluster. This weighting matrix accounts for the within-person correlation among observations, so the GMM estimator that uses this version of S_0 will be more efficient than the estimator that ignores this correlation.

▷ Example 4: GMM estimator with clustering

We have data from the National Longitudinal Survey on young women's wages as reported in a series of interviews from 1968 through 1988, and we want to fit a model of wages as a function of each woman's age and age squared, job tenure, birth year, and level of education. We believe that random shocks that affect a woman's wage also affect her job tenure, so we treat tenure as endogenous. As additional instruments, we use her union status, number of weeks worked in the past year, and a dummy indicating whether she lives in a metropolitan area. Because we have several observations for each woman (corresponding to interviews done over several years), we want to control for clustering on each person.

```
. use http://www.stata-press.com/data/r12/nlswork
(National Longitudinal Survey.  Young Women 14-26 years of age in 1968)
. ivregress gmm ln_wage age c.age#c.age birth_yr grade
> (tenure = union wks_work msp), wmatrix(cluster idcode)
Instrumental variables (GMM) regression         Number of obs =    18625
                                                Wald chi2(5)  =  1807.17
                                                Prob > chi2   =   0.0000
                                                R-squared     =        .
GMM weight matrix: Cluster (idcode)             Root MSE      =  .46951
                        (Std. Err. adjusted for 4110 clusters in idcode)
```

		Robust					
ln_wage	Coef.	Std. Err.	z	P>|z|	[95% Conf.	Interval]
tenure	.099221	.0037764	26.27	0.000	.0918194	.1066227
age	.0171146	.0066895	2.56	0.011	.0040034	.0302259
c.age#c.age	-.0005191	.000111	-4.68	0.000	-.0007366	-.0003016
birth_yr	-.0085994	.0021932	-3.92	0.000	-.012898	-.0043008
grade	.071574	.0029938	23.91	0.000	.0657062	.0774417
_cons	.8575071	.1616274	5.31	0.000	.5407231	1.174291

```
Instrumented:  tenure
Instruments:   age c.age#c.age birth_yr grade union wks_work msp
```

Both job tenure and years of schooling have significant positive effects on wages.

◁

Time-series data are often plagued by serial correlation. In these cases, we can construct a weighting matrix to account for the fact that the error in period t is probably correlated with the errors in periods $t-1$, $t-2$, etc. An HAC weighting matrix can be used to account for both serial correlation and potential heteroskedasticity.

To request an HAC weighting matrix, you specify the wmatrix(hac *kernel* [# | opt]) option. *kernel* specifies which of three kernels to use: bartlett, parzen, or quadraticspectral. *kernel* determines the amount of weight given to lagged values when computing the HAC matrix, and # denotes the maximum number of lags to use. Many texts refer to the bandwidth of the kernel instead of the number of lags; the bandwidth is equal to the number of lags plus one. If neither opt nor # is specified, then $N-2$ lags are used, where N is the sample size.

If you specify wmatrix(hac *kernel* opt), then ivregress uses Newey and West's (1994) algorithm for automatically selecting the number of lags to use. Although the authors' Monte Carlo simulations do show that the procedure may result in size distortions of hypothesis tests, the procedure is still useful when little other information is available to help choose the number of lags.

For more on GMM estimation, see Baum (2006); Baum, Schaffer, and Stillman (2003, 2007); Cameron and Trivedi (2005); Davidson and MacKinnon (1993, 2004); Hayashi (2000); or Wooldridge (2010). See Newey and West (1987) for an introduction to HAC covariance matrix estimation.

Saved results

ivregress saves the following in e():

Scalars

e(N)	number of observations
e(mss)	model sum of squares
e(df_m)	model degrees of freedom
e(rss)	residual sum of squares
e(df_r)	residual degrees of freedom
e(r2)	R^2
e(r2_a)	adjusted R^2
e(F)	F statistic
e(rmse)	root mean squared error
e(N_clust)	number of clusters
e(chi2)	χ^2
e(kappa)	κ used in LIML estimator
e(J)	value of GMM objective function
e(wlagopt)	lags used in HAC weight matrix (if Newey–West algorithm used)
e(vcelagopt)	lags used in HAC VCE matrix (if Newey–West algorithm used)
e(rank)	rank of e(V)
e(iterations)	number of GMM iterations (0 if not applicable)

Macros

e(cmd)	ivregress
e(cmdline)	command as typed
e(depvar)	name of dependent variable
e(instd)	instrumented variable
e(insts)	instruments
e(constant)	noconstant or hasconstant if specified
e(wtype)	weight type
e(wexp)	weight expression
e(title)	title in estimation output
e(clustvar)	name of cluster variable
e(hac_kernel)	HAC kernel
e(hac_lag)	HAC lag
e(vce)	*vcetype* specified in vce()
e(vcetype)	title used to label Std. Err.
e(estimator)	2sls, liml, or gmm
e(exogr)	exogenous regressors
e(wmatrix)	*wmtype* specified in wmatrix()
e(moments)	centered if center specified
e(small)	small if small-sample statistics
e(depname)	*depname* if depname(*depname*) specified; otherwise same as e(depvar)
e(properties)	b V
e(estat_cmd)	program used to implement estat
e(predict)	program used to implement predict
e(footnote)	program used to implement footnote display
e(marginsok)	predictions allowed by margins
e(marginsnotok)	predictions disallowed by margins
e(asbalanced)	factor variables fvset as asbalanced
e(asobserved)	factor variables fvset as asobserved

Matrices

e(b)	coefficient vector
e(Cns)	constraints matrix
e(W)	weight matrix used to compute GMM estimates
e(S)	moment covariance matrix used to compute GMM variance–covariance matrix
e(V)	variance–covariance matrix of the estimators
e(V_modelbased)	model-based variance

Functions

e(sample)	marks estimation sample

Methods and formulas

ivregress is implemented as an ado-file.

Methods and formulas are presented under the following headings:

> *Notation*
> *2SLS and LIML estimators*
> *GMM estimator*

Notation

Items printed in lowercase and italicized (for example, x) are scalars. Items printed in lowercase and boldfaced (for example, $\mathbf{x}$) are vectors. Items printed in uppercase and boldfaced (for example, $\mathbf{X}$) are matrices.

The model is

$$\mathbf{y} = \mathbf{Y}\boldsymbol{\beta}_1 + \mathbf{X}_1\boldsymbol{\beta}_2 + \mathbf{u} = \mathbf{X}\boldsymbol{\beta} + \mathbf{u}$$

$$\mathbf{Y} = \mathbf{X}_1\boldsymbol{\Pi}_1 + \mathbf{X}_2\boldsymbol{\Pi}_2 + \mathbf{v} = \mathbf{Z}\boldsymbol{\Pi} + \mathbf{V}$$

where $\mathbf{y}$ is an $N \times 1$ vector of the left-hand-side variable; N is the sample size; $\mathbf{Y}$ is an $N \times p$ matrix of p endogenous regressors; $\mathbf{X}_1$ is an $N \times k_1$ matrix of k_1 included exogenous regressors; $\mathbf{X}_2$ is an $N \times k_2$ matrix of k_2 excluded exogenous variables, $\mathbf{X} = [\mathbf{Y}\ \mathbf{X}_1]$, $\mathbf{Z} = [\mathbf{X}_1\ \mathbf{X}_2]$; $\mathbf{u}$ is an $N \times 1$ vector of errors; $\mathbf{V}$ is an $N \times p$ matrix of errors; $\boldsymbol{\beta} = [\boldsymbol{\beta}_1\ \boldsymbol{\beta}_2]$ is a $k = (p + k_1) \times 1$ vector of parameters; and $\boldsymbol{\Pi}$ is a $(k_1 + k_2) \times p$ vector of parameters. If a constant term is included in the model, then one column of $\mathbf{X}_1$ contains all ones.

Let $\mathbf{v}$ be a column vector of weights specified by the user. If no weights are specified, $\mathbf{v} = \mathbf{1}$. Let $\mathbf{w}$ be a column vector of normalized weights. If no weights are specified or if the user specified fweights or iweights, $\mathbf{w} = \mathbf{v}$; otherwise, $\mathbf{w} = \left\{\mathbf{v}/(\mathbf{1}'\mathbf{v})\right\}(\mathbf{1}'\mathbf{1})$. Let $\mathbf{D}$ denote the $N \times N$ matrix with $\mathbf{w}$ on the main diagonal and zeros elsewhere. If no weights are specified, $\mathbf{D}$ is the identity matrix.

The weighted number of observations n is defined as $\mathbf{1}'\mathbf{w}$. For iweights, this is truncated to an integer. The *sum of the weights* is $\mathbf{1}'\mathbf{v}$. Define $c = 1$ if there is a constant in the regression and zero otherwise.

The order condition for identification requires that $k_2 \geq p$: the number of excluded exogenous variables must be at least as great as the number of endogenous regressors.

In the following formulas, if weights are specified, $\mathbf{X}_1'\mathbf{X}_1$, $\mathbf{X}'\mathbf{X}$, $\mathbf{X}'\mathbf{y}$, $\mathbf{y}'\mathbf{y}$, $\mathbf{Z}'\mathbf{Z}$, $\mathbf{Z}'\mathbf{X}$, and $\mathbf{Z}'\mathbf{y}$ are replaced with $\mathbf{X}_1'\mathbf{D}\mathbf{X}_1$, $\mathbf{X}'\mathbf{D}\mathbf{X}$, $\mathbf{X}'\mathbf{D}\mathbf{y}$, $\mathbf{y}'\mathbf{D}\mathbf{y}$, $\mathbf{Z}'\mathbf{D}\mathbf{Z}$, $\mathbf{Z}'\mathbf{D}\mathbf{X}$, and $\mathbf{Z}'\mathbf{D}\mathbf{y}$, respectively. We suppress the $\mathbf{D}$ below to simplify the notation.

2SLS and LIML estimators

Define the κ-class estimator of β as

$$\mathbf{b} = \left\{\mathbf{X}'(\mathbf{I} - \kappa\mathbf{M_Z})^{-1}\mathbf{X}\right\}^{-1}\mathbf{X}'(\mathbf{I} - \kappa\mathbf{M_Z})^{-1}\mathbf{y}$$

where $\mathbf{M_Z} = \mathbf{I} - \mathbf{Z}(\mathbf{Z}'\mathbf{Z})^{-1}\mathbf{Z}'$. The 2SLS estimator results from setting $\kappa = 1$. The LIML estimator results from selecting κ to be the minimum eigenvalue of $(\mathbf{Y}'\mathbf{M_Z}\mathbf{Y})^{-1/2}\mathbf{Y}'\mathbf{M_{X_1}}\mathbf{Y}(\mathbf{Y}'\mathbf{M_Z}\mathbf{Y})^{-1/2}$, where $\mathbf{M_{X_1}} = \mathbf{I} - \mathbf{X_1}(\mathbf{X_1}'\mathbf{X_1})^{-1}\mathbf{X_1}'$.

The total sum of squares (TSS) equals $\mathbf{y}'\mathbf{y}$ if there is no intercept and $\mathbf{y}'\mathbf{y} - \left\{(\mathbf{1}'\mathbf{y})^2/n\right\}$ otherwise. The degrees of freedom are $n-c$. The error sum of squares (ESS) is defined as $\mathbf{y}'\mathbf{y} - 2\mathbf{b}\mathbf{X}'\mathbf{y} + \mathbf{b}'\mathbf{X}'\mathbf{X}\mathbf{b}$. The model sum of squares (MSS) equals TSS $-$ ESS. The degrees of freedom are $k - c$.

The mean squared error, s^2, is defined as ESS$/(n - k)$ if small is specified and ESS$/n$ otherwise. The root mean squared error is s, its square root.

If $c = 1$ and small is not specified, a Wald statistic, W, of the joint significance of the $k - 1$ parameters of β except the constant term is calculated; $W \sim \chi^2(k - 1)$. If $c = 1$ and small is specified, then an F statistic is calculated as $F = W/(k - 1)$; $F \sim F(k - 1, n - k)$.

The R-squared is defined as $R^2 = 1 - $ ESS/TSS.

The adjusted R-squared is $R_a^2 = 1 - (1 - R^2)(n - c)/(n - k)$.

If robust is not specified, then $\mathrm{Var}(\mathbf{b}) = s^2\left\{\mathbf{X}'(\mathbf{I} - \kappa\mathbf{M_Z})^{-1}\mathbf{X}\right\}^{-1}$. For a discussion of robust variance estimates in regression and regression with instrumental variables, see *Methods and formulas* in [R] **regress**. If small is not specified, then $k = 0$ in the formulas given there.

This command also supports estimation with survey data. For details on VCEs with survey data, see [SVY] **variance estimation**.

GMM estimator

We obtain an initial consistent estimate of β by using the 2SLS estimator; see above. Using this estimate of β, we compute the weighting matrix $\mathbf{W}$ and calculate the GMM estimator

$$\mathbf{b}_{\mathrm{GMM}} = \left\{\mathbf{X}'\mathbf{Z}\mathbf{W}\mathbf{Z}'\mathbf{X}\right\}^{-1}\mathbf{X}'\mathbf{Z}\mathbf{W}\mathbf{Z}'\mathbf{y}$$

The variance of $\mathbf{b}_{\mathrm{GMM}}$ is

$$\mathrm{Var}(\mathbf{b}_{\mathrm{GMM}}) = n\left\{\mathbf{X}'\mathbf{Z}\mathbf{W}\mathbf{Z}'\mathbf{X}\right\}^{-1}\mathbf{X}'\mathbf{Z}\mathbf{W}\widehat{\mathbf{S}}\mathbf{W}\mathbf{Z}'\mathbf{X}\left\{\mathbf{X}'\mathbf{Z}\mathbf{W}\mathbf{Z}'\mathbf{X}\right\}^{-1}$$

$\mathrm{Var}(\mathbf{b}_{\mathrm{GMM}})$ is of the sandwich form $\mathbf{DMD}$; see [P] **_robust**. If the user specifies the small option, ivregress implements a small-sample adjustment by multiplying the VCE by $N/(N - k)$.

If vce(unadjusted) is specified, then we set $\widehat{\mathbf{S}} = \mathbf{W}^{-1}$ and the VCE reduces to the "optimal" GMM variance estimator

$$\mathrm{Var}(\beta_{\mathrm{GMM}}) = n\left\{\mathbf{X}'\mathbf{Z}\mathbf{W}\mathbf{Z}'\mathbf{X}\right\}^{-1}$$

However, if $\mathbf{W}^{-1}$ is not a good estimator of $E(\mathbf{z}_i u_i u_i \mathbf{z}_i')$, then the optimal GMM estimator is inefficient, and inference based on the optimal variance estimator could be misleading.

$\mathbf{W}$ is calculated using the residuals from the initial 2SLS estimates, whereas $\mathbf{S}$ is estimated using the residuals based on $\mathbf{b}_{\text{GMM}}$. The wmatrix() option affects the form of $\mathbf{W}$, whereas the vce() option affects the form of $\mathbf{S}$. Except for different residuals being used, the formulas for $\mathbf{W}^{-1}$ and $\mathbf{S}$ are identical, so we focus on estimating $\mathbf{W}^{-1}$.

If wmatrix(unadjusted) is specified, then

$$\mathbf{W}^{-1} = \frac{s^2}{n} \sum_i \mathbf{z}_i \mathbf{z}_i'$$

where $s^2 = \sum_i u_i^2/n$. This weight matrix is appropriate if the errors are homoskedastic.

If wmatrix(robust) is specified, then

$$\mathbf{W}^{-1} = \frac{1}{n} \sum_i u_i^2 \mathbf{z}_i \mathbf{z}_i'$$

which is appropriate if the errors are heteroskedastic.

If wmatrix(cluster *clustvar*) is specified, then

$$\mathbf{W}^{-1} = \frac{1}{n} \sum_c \mathbf{q}_c \mathbf{q}_c'$$

where c indexes clusters,

$$\mathbf{q}_c = \sum_{i \in c_j} u_i \mathbf{z}_i$$

and c_j denotes the jth cluster.

If wmatrix(hac *kernel* $\left[\#\right]$) is specified, then

$$\mathbf{W}^{-1} = \frac{1}{n} \sum_i u_i^2 \mathbf{z}_i \mathbf{z}_i' + \frac{1}{n} \sum_{l=1}^{l=n-1} \sum_{i=l+1}^{i=n} K(l,m) u_i u_{i-l} \left(\mathbf{z}_i \mathbf{z}_{i-l}' + \mathbf{z}_{i-l} \mathbf{z}_i' \right)$$

where $m = \#$ if $\#$ is specified and $m = n - 2$ otherwise. Define $z = l/(m+1)$. If *kernel* is nwest, then

$$K(l,m) = \begin{cases} 1 - z & 0 \le z \le 1 \\ 0 & \text{otherwise} \end{cases}$$

If *kernel* is gallant, then

$$K(l,m) = \begin{cases} 1 - 6z^2 + 6z^3 & 0 \le z \le 0.5 \\ 2(1-z)^3 & 0.5 < z \le 1 \\ 0 & \text{otherwise} \end{cases}$$

If *kernel* is quadraticspectral, then

$$K(l,m) = \begin{cases} 1 & z = 0 \\ 3 \left\{ \sin(\theta)/\theta - \cos(\theta) \right\} /\theta^2 & \text{otherwise} \end{cases}$$

where $\theta = 6\pi z/5$.

If wmatrix(hac *kernel* opt) is specified, then ivregress uses Newey and West's (1994) automatic lag-selection algorithm, which proceeds as follows. Define $\mathbf{h}$ to be a $(k_1 + k_2) \times 1$ vector containing ones in all rows except for the row corresponding to the constant term (if present); that row contains a zero. Define

$$f_i = (u_i \mathbf{z}_i) \mathbf{h}$$

$$\widehat{\sigma}_j = \frac{1}{n} \sum_{i=j+1}^{n} f_i f_{i-j} \qquad j = 0, \dots, m^*$$

$$\widehat{s}^{(q)} = 2 \sum_{j=1}^{m^*} \widehat{\sigma}_j j^q$$

$$\widehat{s}^{(0)} = \widehat{\sigma}_0 + 2 \sum_{j=1}^{m^*} \widehat{\sigma}_j$$

$$\widehat{\gamma} = c_\gamma \left\{ \left(\frac{\widehat{s}^{(q)}}{\widehat{s}^{(0)}} \right)^2 \right\}^{1/2q+1}$$

$$m = \widehat{\gamma} n^{1/(2q+1)}$$

where q, m^*, and c_γ depend on the kernel specified:

Kernel	q	m^*	c_γ
Bartlett	1	int $\left\{ 20(T/100)^{2/9} \right\}$	1.1447
Parzen	2	int $\left\{ 20(T/100)^{4/25} \right\}$	2.6614
Quadratic spectral	2	int $\left\{ 20(T/100)^{2/25} \right\}$	1.3221

where $\text{int}(x)$ denotes the integer obtained by truncating x toward zero. For the Bartlett and Parzen kernels, the optimal lag is $\min\{\text{int}(m), m^*\}$. For the quadratic spectral, the optimal lag is $\min\{m, m^*\}$.

If center is specified, when computing weighting matrices ivregress replaces the term $u_i z_i$ in the formulas above with $u_i \mathbf{z}_i - \overline{uz}$, where $\overline{uz} = \sum_i u_i \mathbf{z}_i / N$.

References

Andrews, D. W. K. 1991. Heteroskedasticity and autocorrelation consistent covariance matrix estimation. *Econometrica* 59: 817–858.

Angrist, J. D., and J.-S. Pischke. 2009. *Mostly Harmless Econometrics: An Empiricist's Companion*. Princeton, NJ: Princeton University Press.

Basmann, R. L. 1957. A generalized classical method of linear estimation of coefficients in a structural equation. *Econometrica* 25: 77–83.

Baum, C. F. 2006. *An Introduction to Modern Econometrics Using Stata*. College Station, TX: Stata Press.

Baum, C. F., M. E. Schaffer, and S. Stillman. 2003. Instrumental variables and GMM: Estimation and testing. *Stata Journal* 3: 1–31.

——. 2007. Enhanced routines for instrumental variables/generalized method of moments estimation and testing. *Stata Journal* 7: 465–506.

Cameron, A. C., and P. K. Trivedi. 2005. *Microeconometrics: Methods and Applications*. New York: Cambridge University Press.

——. 2010. *Microeconometrics Using Stata*. Rev. ed. College Station, TX: Stata Press.

Davidson, R., and J. G. MacKinnon. 1993. *Estimation and Inference in Econometrics*. New York: Oxford University Press.

———. 2004. *Econometric Theory and Methods*. New York: Oxford University Press.

Finlay, K., and L. M. Magnusson. 2009. Implementing weak-instrument robust tests for a general class of instrumental-variables models. *Stata Journal* 9: 398–421.

Gallant, A. R. 1987. *Nonlinear Statistical Models*. New York: Wiley.

Greene, W. H. 2012. *Econometric Analysis*. 7th ed. Upper Saddle River, NJ: Prentice Hall.

Hall, A. R. 2005. *Generalized Method of Moments*. Oxford: Oxford University Press.

Hansen, L. P. 1982. Large sample properties of generalized method of moments estimators. *Econometrica* 50: 1029–1054.

Hayashi, F. 2000. *Econometrics*. Princeton, NJ: Princeton University Press.

Judge, G. G., W. E. Griffiths, R. C. Hill, H. Lütkepohl, and T.-C. Lee. 1985. *The Theory and Practice of Econometrics*. 2nd ed. New York: Wiley.

Kmenta, J. 1997. *Elements of Econometrics*. 2nd ed. Ann Arbor: University of Michigan Press.

Koopmans, T. C., and W. C. Hood. 1953. *Studies in Econometric Method*. New York: Wiley.

Koopmans, T. C., and J. Marschak. 1950. *Statistical Inference in Dynamic Economic Models*. New York: Wiley.

Newey, W. K., and K. D. West. 1987. A simple, positive semi-definite, heteroskedasticity and autocorrelation consistent covariance matrix. *Econometrica* 55: 703–708.

———. 1994. Automatic lag selection in covariance matrix estimation. *Review of Economic Studies* 61: 631–653.

Nichols, A. 2007. Causal inference with observational data. *Stata Journal* 7: 507–541.

Poi, B. P. 2006. Jackknife instrumental variables estimation in Stata. *Stata Journal* 6: 364–376.

Stock, J. H., and M. W. Watson. 2011. *Introduction to Econometrics*. 3rd ed. Boston: Addison–Wesley.

Stock, J. H., J. H. Wright, and M. Yogo. 2002. A survey of weak instruments and weak identification in generalized method of moments. *Journal of Business and Economic Statistics* 20: 518–529.

Theil, H. 1953. *Repeated Least Squares Applied to Complete Equation Systems*. Mimeograph from the Central Planning Bureau, The Hague.

Wooldridge, J. M. 2009. *Introductory Econometrics: A Modern Approach*. 4th ed. Cincinnati, OH: South-Western.

———. 2010. *Econometric Analysis of Cross Section and Panel Data*. 2nd ed. Cambridge, MA: MIT Press.

Wright, P. G. 1928. *The Tariff on Animal and Vegetable Oils*. New York: Macmillan.

Also see

[R] **ivregress postestimation** — Postestimation tools for ivregress

[R] **gmm** — Generalized method of moments estimation

[R] **ivprobit** — Probit model with continuous endogenous regressors

[R] **ivtobit** — Tobit model with continuous endogenous regressors

[R] **reg3** — Three-stage estimation for systems of simultaneous equations

[R] **regress** — Linear regression

[SVY] **svy estimation** — Estimation commands for survey data

[XT] **xtivreg** — Instrumental variables and two-stage least squares for panel-data models

Stata Structural Equation Modeling Reference Manual

[U] **20 Estimation and postestimation commands**

Title

ivregress postestimation — Postestimation tools for ivregress

Description

The following postestimation commands are of special interest after `ivregress`:

Command	Description
estat endogenous	perform tests of endogeneity
estat firststage	report "first-stage" regression statistics
estat overid	perform tests of overidentifying restrictions

These commands are not appropriate after the `svy` prefix.

For information about these commands, see below.

The following postestimation commands are also available:

Command	Description
contrast	contrasts and ANOVA-style joint tests of estimates
estat	VCE and estimation sample summary
estat (svy)	postestimation statistics for survey data
estimates	cataloging estimation results
hausman	Hausman's specification test
lincom	point estimates, standard errors, testing, and inference for linear combinations of coefficients
margins	marginal means, predictive margins, marginal effects, and average marginal effects
marginsplot	graph the results from margins (profile plots, interaction plots, etc.)
nlcom	point estimates, standard errors, testing, and inference for nonlinear combinations of coefficients
predict	predictions, residuals, influence statistics, and other diagnostic measures
predictnl	point estimates, standard errors, testing, and inference for generalized predictions
pwcompare	pairwise comparisons of estimates
test	Wald tests of simple and composite linear hypotheses
testnl	Wald tests of nonlinear hypotheses

See the corresponding entries in the *Base Reference Manual* for details, but see [SVY] **estat** for details about estat (svy).

Special-interest postestimation commands

`estat endogenous` performs tests to determine whether endogenous regressors in the model are in fact exogenous. After GMM estimation, the C (difference-in-Sargan) statistic is reported. After 2SLS estimation with an unadjusted VCE, the Durbin (1954) and Wu–Hausman (Wu 1974; Hausman 1978) statistics are reported. After 2SLS estimation with a robust VCE, Wooldridge's (1995) robust score test and a robust regression-based test are reported. In all cases, if the test statistic is significant, then the variables being tested must be treated as endogenous. `estat endogenous` is not available after LIML estimation.

estat firststage reports various statistics that measure the relevance of the excluded exogenous variables. By default, whether the equation has one or more than one endogenous regressor determines what statistics are reported.

estat overid performs tests of overidentifying restrictions. If the 2SLS estimator was used, Sargan's (1958) and Basmann's (1960) χ^2 tests are reported, as is Wooldridge's (1995) robust score test; if the LIML estimator was used, Anderson and Rubin's (1950) χ^2 test and Basmann's F test are reported; and if the GMM estimator was used, Hansen's (1982) J statistic χ^2 test is reported. A statistically significant test statistic always indicates that the instruments may not be valid.

Syntax for predict

predict $\begin{bmatrix} type \end{bmatrix}$ newvar $\begin{bmatrix} if \end{bmatrix}$ $\begin{bmatrix} in \end{bmatrix}$ $\begin{bmatrix} , statistic \end{bmatrix}$

predict $\begin{bmatrix} type \end{bmatrix}$ $\{ stub* | newvarlist \}$ $\begin{bmatrix} if \end{bmatrix}$ $\begin{bmatrix} in \end{bmatrix}$, scores

statistic	Description
Main	
xb	linear prediction; the default
residuals	residuals
stdp	standard error of the prediction
stdf	standard error of the forecast
pr(a,b)	$\Pr(a < y_j < b)$
e(a,b)	$E(y_j \mid a < y_j < b)$
ystar(a,b)	$E(y_j^*)$, $y_j^* = \max\{a, \min(y_j, b)\}$

These statistics are available both in and out of sample; type predict ... if e(sample) ... if wanted only for the estimation sample.

stdf is not allowed with svy estimation results.

where a and b may be numbers or variables; a missing ($a \geq .$) means $-\infty$, and b missing ($b \geq .$) means $+\infty$; see [U] **12.2.1 Missing values**.

Menu

Statistics > Postestimation > Predictions, residuals, etc.

Options for predict

⌐ Main ⌐

xb, the default, calculates the linear prediction.

residuals calculates the residuals, that is, $y_j - x_j \mathbf{b}$. These are based on the estimated equation when the observed values of the endogenous variables are used—not the projections of the instruments onto the endogenous variables.

stdp calculates the standard error of the prediction, which can be thought of as the standard error of the predicted expected value or mean for the observation's covariate pattern. This is also referred to as the standard error of the fitted value.

stdf calculates the standard error of the forecast, which is the standard error of the point prediction for 1 observation. It is commonly referred to as the standard error of the future or forecast value. By construction, the standard errors produced by stdf are always larger than those produced by stdp; see *Methods and formulas* in [R] **regress postestimation**.

pr(a,b) calculates $\Pr(a < \mathbf{x}_j\mathbf{b} + u_j < b)$, the probability that $y_j|\mathbf{x}_j$ would be observed in the interval (a, b).

a and b may be specified as numbers or variable names; *lb* and *ub* are variable names;
pr(20,30) calculates $\Pr(20 < \mathbf{x}_j\mathbf{b} + u_j < 30)$;
pr(*lb,ub*) calculates $\Pr(lb < \mathbf{x}_j\mathbf{b} + u_j < ub)$; and
pr(20,*ub*) calculates $\Pr(20 < \mathbf{x}_j\mathbf{b} + u_j < ub)$.

a missing ($a \geq .$) means $-\infty$; pr(.,30) calculates $\Pr(-\infty < \mathbf{x}_j\mathbf{b} + u_j < 30)$;
pr(*lb*,30) calculates $\Pr(-\infty < \mathbf{x}_j\mathbf{b} + u_j < 30)$ in observations for which *lb* $\geq$.
and calculates $\Pr(lb < \mathbf{x}_j\mathbf{b} + u_j < 30)$ elsewhere.

b missing ($b \geq .$) means $+\infty$; pr(20,.) calculates $\Pr(+\infty > \mathbf{x}_j\mathbf{b} + u_j > 20)$;
pr(20,*ub*) calculates $\Pr(+\infty > \mathbf{x}_j\mathbf{b} + u_j > 20)$ in observations for which *ub* $\geq$.
and calculates $\Pr(20 < \mathbf{x}_j\mathbf{b} + u_j < ub)$ elsewhere.

e(a,b) calculates $E(\mathbf{x}_j\mathbf{b} + u_j \mid a < \mathbf{x}_j\mathbf{b} + u_j < b)$, the expected value of $y_j|\mathbf{x}_j$ conditional on $y_j|\mathbf{x}_j$ being in the interval (a, b), meaning that $y_j|\mathbf{x}_j$ is truncated.
a and b are specified as they are for pr().

ystar(a,b) calculates $E(y_j^*)$, where $y_j^* = a$ if $\mathbf{x}_j\mathbf{b} + u_j \leq a$, $y_j^* = b$ if $\mathbf{x}_j\mathbf{b} + u_j \geq b$, and $y_j^* = \mathbf{x}_j\mathbf{b} + u_j$ otherwise, meaning that y_j^* is censored. a and b are specified as they are for pr().

scores calculates the scores for the model. A new score variable is created for each endogenous regressor, as well as an equation-level score that applies to all exogenous variables and constant term (if present).

Syntax for estat endogenous

estat <u>endog</u>enous [*varlist*] [, <u>l</u>ags(#) forceweights forcenonrobust]

Menu

Statistics > Postestimation > Reports and statistics

Options for estat endogenous

lags(#) specifies the number of lags to use for prewhitening when computing the heteroskedasticity- and autocorrelation-consistent (HAC) version of the score test of endogeneity. Specifying lags(0) requests no prewhitening. This option is valid only when the model was fit via 2SLS and an HAC covariance matrix was requested when the model was fit. The default is lags(1).

forceweights requests that the tests of endogeneity be computed even though aweights, pweights, or iweights were used in the previous estimation. By default, these tests are conducted only after unweighted or frequency-weighted estimation. The reported critical values may be inappropriate for weighted data, so the user must determine whether the critical values are appropriate for a given application.

forcenonrobust requests that the Durbin and Wu–Hausman tests be performed after 2SLS estimation even though a robust VCE was used at estimation time. This option is available only if the model was fit by 2SLS.

Syntax for estat firststage

> estat firststage [, all forcenonrobust]

Menu

Statistics > Postestimation > Reports and statistics

Options for estat firststage

all requests that all first-stage goodness-of-fit statistics be reported regardless of whether the model contains one or more endogenous regressors. By default, if the model contains one endogenous regressor, then the first-stage R^2, adjusted R^2, partial R^2, and F statistics are reported, whereas if the model contains multiple endogenous regressors, then Shea's partial R^2 and adjusted partial R^2 are reported instead.

forcenonrobust requests that the minimum eigenvalue statistic and its critical values be reported even though a robust VCE was used at estimation time. The reported critical values assume that the errors are independent and identically distributed (i.i.d.) normal, so the user must determine whether the critical values are appropriate for a given application.

Syntax for estat overid

> estat overid [, lags(#) forceweights forcenonrobust]

Menu

Statistics > Postestimation > Reports and statistics

Options for estat overid

lags(#) specifies the number of lags to use for prewhitening when computing the heteroskedasticity- and autocorrelation-consistent (HAC) version of the score test of overidentifying restrictions. Specifying lags(0) requests no prewhitening. This option is valid only when the model was fit via 2SLS and an HAC covariance matrix was requested when the model was fit. The default is lags(1).

forceweights requests that the tests of overidentifying restrictions be computed even though aweights, pweights, or iweights were used in the previous estimation. By default, these tests are conducted only after unweighted or frequency-weighted estimation. The reported critical values may be inappropriate for weighted data, so the user must determine whether the critical values are appropriate for a given application.

forcenonrobust requests that the Sargan and Basmann tests of overidentifying restrictions be performed after 2SLS or LIML estimation even though a robust VCE was used at estimation time. These tests assume that the errors are i.i.d. normal, so the user must determine whether the critical values are appropriate for a given application.

Remarks

Remarks are presented under the following headings:

> *estat endogenous*
> *estat firststage*
> *estat overid*

estat endogenous

A natural question to ask is whether a variable presumed to be endogenous in the previously fit model could instead be treated as exogenous. If the endogenous regressors are in fact exogenous, then the OLS estimator is more efficient; and depending on the strength of the instruments and other factors, the sacrifice in efficiency by using an instrumental-variables estimator can be significant. Thus, unless an instrumental-variables estimator is really needed, OLS should be used instead. estat endogenous provides several tests of endogeneity after 2SLS and GMM estimation.

▷ Example 1

In example 1 of [R] **ivregress**, we fit a model of the average rental rate for housing in a state as a function of the percentage of the population living in urban areas and the average value of houses. We treated hsngval as endogenous because unanticipated shocks that affect rental rates probably affect house prices as well. We used family income and region dummies as additional instruments for hsngval. Here we test whether we could treat hsngval as exogenous.

```
. use http://www.stata-press.com/data/r12/hsng
(1980 Census housing data)
. ivregress 2sls rent pcturban (hsngval = faminc i.region)
 (output omitted )
. estat endogenous
 Tests of endogeneity
 Ho: variables are exogenous
 Durbin (score) chi2(1)          =  12.8473  (p = 0.0003)
 Wu-Hausman F(1,46)              =  15.9067  (p = 0.0002)
```

Because we did not specify any variable names after the estat endogenous command, Stata by default tested all the endogenous regressors (namely, hsngval) in our model. The null hypothesis of the Durbin and Wu–Hausman tests is that the variable under consideration can be treated as exogenous. Here both test statistics are highly significant, so we reject the null of exogeneity; we must continue to treat hsngval as endogenous.

◁

The difference between the Durbin and Wu–Hausman tests of endogeneity is that the former uses an estimate of the error term's variance based on the model assuming the variables being tested are exogenous, while the latter uses an estimate of the error variance based on the model assuming the variables being tested are endogenous. Under the null hypothesis that the variables being tested are exogenous, both estimates of the error variance are consistent. What we label the Wu–Hausman statistic is Wu's (1974) "T_2" statistic, which Hausman (1978) showed can be calculated very easily via linear regression. Baum, Schaffer, and Stillman (2003, 2007) provide a lucid discussion of these tests.

When you fit a model with multiple endogenous regressors, you can test the exogeneity of a subset of the regressors while continuing to treat the others as endogenous. For example, say you have three endogenous regressors, y1, y2, and y3, and you fit your model by typing

> . ivregress *depvar* ... (y1 y2 y3 = ...)

Suppose you are confident that y1 must be treated as endogenous, but you are undecided about y2 and y3. To test whether y2 and y3 can be treated as exogenous, you would type

> . estat endogenous y2 y3

The Durbin and Wu–Hausman tests assume that the error term is i.i.d. Therefore, if you requested a robust VCE at estimation time, estat endogenous will instead report Wooldridge's (1995) score test and a regression-based test of exogeneity. Both these tests can tolerate heteroskedastic and autocorrelated errors, while only the regression-based test is amenable to clustering.

Example 2

We refit our housing model, requesting robust standard errors, and then test the exogeneity of hsngval:

```
. use http://www.stata-press.com/data/r12/hsng
(1980 Census housing data)
. ivregress 2sls rent pcturban (hsngval = faminc i.region), vce(robust)
(output omitted)
. estat endogenous

Tests of endogeneity
Ho: variables are exogenous

Robust score chi2(1)            =  2.10428  (p = 0.1469)
Robust regression F(1,46)       =  4.31101  (p = 0.0435)
```

Wooldridge's score test does not reject the null hypothesis that hsngval is exogenous at conventional significance levels ($p = 0.1469$). However, the regression-based test does reject the null hypothesis at the 5% significance level ($p = 0.0435$). Typically, these two tests yield the same conclusion; the fact that our dataset has only 50 observations could be contributing to the discrepancy. Here we would be inclined to continue to treat hsngval as endogenous. Even if hsngval is exogenous, the 2SLS estimates are still consistent. On the other hand, if hsngval is in fact endogenous, the OLS estimates would not be consistent. Moreover, as we will see in our discussion of the estat overid command, our additional instruments may be invalid. To test whether an endogenous variable can be treated as exogenous, we must have a valid set of instruments to use to fit the model in the first place!

◁

Unlike the Durbin and Wu–Hausman tests, Wooldridge's score and the regression-based tests do not allow you to test a subset of the endogenous regressors in the model; you can test only whether all the endogenous regressors are in fact exogenous.

After GMM estimation, estat endogenous calculates what Hayashi (2000, 220) calls the C statistic, also known as the difference-in-Sargan statistic. The C statistic can be made robust to heteroskedasticity, autocorrelation, and clustering; and the version reported by estat endogenous is determined by the weight matrix requested via the wmatrix() option used when fitting the model with ivregress. Additionally, the test can be used to determine the exogeneity of a subset of the endogenous regressors, regardless of the type of weight matrix used.

If you fit your model using the LIML estimator, you can use the hausman command to carry out a traditional Hausman (1978) test between the OLS and LIML estimates.

estat firststage

For an excluded exogenous variable to be a valid instrument, it must be sufficiently correlated with the included endogenous regressors but uncorrelated with the error term. In recent decades, researchers have paid considerable attention to the issue of instruments that are only weakly correlated with the endogenous regressors. In such cases, the usual 2SLS, GMM, and LIML estimators are biased toward the OLS estimator, and inference based on the standard errors reported by, for example, `ivregress` can be severely misleading. For more information on the theory behind instrumental-variables estimation with weak instruments, see Nelson and Startz (1990); Staiger and Stock (1997); Hahn and Hausman (2003); the survey article by Stock, Wright, and Yogo (2002); and Angrist and Pischke (2009, chap. 4).

When the instruments are only weakly correlated with the endogenous regressors, some Monte Carlo evidence suggests that the LIML estimator performs better than the 2SLS and GMM estimators; see, for example, Poi (2006) and Stock, Wright, and Yogo (2002) (and the papers cited therein). On the other hand, the LIML estimator often results in confidence intervals that are somewhat larger than those from the 2SLS estimator.

Moreover, using more instruments is not a solution, because the biases of instrumental-variables estimators increase with the number of instruments. See Hahn and Hausman (2003).

`estat firststage` produces several statistics for judging the explanatory power of the instruments and is most easily explained with examples.

▷ Example 3

Again building on the model fit in example 1 of [R] **ivregress**, we now explore the degree of correlation between the additional instruments `faminc`, `2.region`, `3.region`, and `4.region` and the endogenous regressor `hsngval`:

```
. use http://www.stata-press.com/data/r12/hsng
(1980 Census housing data)
. ivregress 2sls rent pcturban (hsngval = faminc i.region)
 (output omitted)
. estat firststage
  First-stage regression summary statistics
```

Variable	R-sq.	Adjusted R-sq.	Partial R-sq.	F(4,44)	Prob > F
hsngval	0.6908	0.6557	0.5473	13.2978	0.0000

```
  Minimum eigenvalue statistic = 13.2978
  Critical Values                        # of endogenous regressors:    1
  Ho: Instruments are weak               # of excluded instruments:     4
```

	5%	10%	20%	30%
2SLS relative bias	16.85	10.27	6.71	5.34

	10%	15%	20%	25%
2SLS Size of nominal 5% Wald test	24.58	13.96	10.26	8.31
LIML Size of nominal 5% Wald test	5.44	3.87	3.30	2.98

To understand these results, recall that the first-stage regression is

$$\texttt{hsngval}_i = \pi_0 + \pi_1 \texttt{pcturban}_i + \pi_2 \texttt{faminc} + \pi_3 2.\texttt{region} + \pi_4 3.\texttt{region} + \pi_5 4.\texttt{region} + v_i$$

where v_i is an error term. The column marked "R-sq." is the simple R^2 from fitting the first-stage regression by OLS, and the column marked "Adjusted R-sq." is the adjusted R^2 from that regression. Higher values purportedly indicate stronger instruments, and instrumental-variables estimators exhibit less bias when the instruments are strongly correlated with the endogenous variable.

Looking at just the R^2 and adjusted R^2 can be misleading, however. If hsngval were strongly correlated with the included exogenous variable pcturban but only weakly correlated with the additional instruments, then these statistics could be large even though a weak-instrument problem is present.

The partial R^2 statistic measures the correlation between hsngval and the additional instruments after *partialling out* the effect of pcturban. Unlike the R^2 and adjusted R^2 statistics, the partial R^2 statistic will not be inflated because of strong correlation between hsngval and pcturban. Bound, Jaeger, and Baker (1995) and others have promoted using this statistic.

The column marked "F(4, 44)" is an F statistic for the joint significance of π_2, π_3, π_4, and π_5, the coefficients on the additional instruments. Its p-value is listed in the column marked "Prob > F". If the F statistic is not significant, then the additional instruments have no significant explanatory power for hsngval after controlling for the effect of pcturban. However, Hall, Rudebusch, and Wilcox (1996) used Monte Carlo simulation to show that simply having an F statistic that is significant at the typical 5% or 10% level is not sufficient. Stock, Wright, and Yogo (2002) suggest that the F statistic should exceed 10 for inference based on the 2SLS estimator to be reliable when there is one endogenous regressor.

estat firststage also presents the Cragg and Donald (1993) minimum eigenvalue statistic as a further test of weak instruments. Stock and Yogo (2005) discuss two characterizations of weak instruments: first, weak instruments cause instrumental-variables estimators to be biased; second, hypothesis tests of parameters estimated by instrumental-variables estimators may suffer from severe size distortions. The test statistic in our example is 13.30, which is identical to the F statistic just discussed because our model contains one endogenous regressor.

The null hypothesis of each of Stock and Yogo's tests is that the set of instruments is weak. To perform these tests, we must first choose either the largest relative bias of the 2SLS estimator we are willing to tolerate or the largest rejection rate of a nominal 5% Wald test we are willing to tolerate. If the test statistic exceeds the critical value, we can conclude that our instruments are not weak.

The row marked "2SLS relative bias" contains critical values for the test that the instruments are weak based on the bias of the 2SLS estimator *relative to* the bias of the OLS estimator. For example, from past experience we might know that the OLS estimate of a parameter β may be 50% too high. Saying that we are willing to tolerate a 10% relative bias means that we are willing to tolerate a bias of the 2SLS estimator no greater than 5% (that is, 10% of 50%). In our rental rate model, if we are willing to tolerate a 10% relative bias, then we can conclude that our instruments are not weak because the test statistic of 13.30 exceeds the critical value of 10.22. However, if we were willing to tolerate only a relative bias of 5%, we would conclude that our instruments are weak because $13.30 < 16.85$.

The rows marked "2SLS Size of nominal 5% Wald test" and "LIML Size of nominal 5% Wald test" contain critical values pertaining to Stock and Yogo's (2005) second characterization of weak instruments. This characterization defines a set of instruments to be weak if a Wald test at the 5% level can have an actual rejection rate of no more than 10%, 15%, 20%, or 25%. Using the current example, suppose that we are willing to accept a rejection rate of at most 10%. Because $13.30 < 24.58$, we cannot reject the null hypothesis of weak instruments. On the other hand, if we use the LIML estimator instead, then we can reject the null hypothesis because $13.30 > 5.44$.

◁

❏ Technical note

Stock and Yogo (2005) tabulated critical values for 2SLS relative biases of 5%, 10%, 20%, and 30% for models with 1, 2, or 3 endogenous regressors and between 3 and 30 excluded exogenous variables (instruments). They also provide critical values for worst-case rejection rates of 5%, 10%, 20%, and 25% for nominal 5% Wald tests of the endogenous regressors with 1 or 2 endogenous regressors and between 1 and 30 instruments. If the model previously fit by ivregress has more instruments or endogenous regressors than these limits, the critical values are not shown. Stock and Yogo did not consider GMM estimators.

❏

When the model being fit contains more than one endogenous regressor, the R^2 and F statistics described above can overstate the relevance of the excluded instruments. Suppose that there are two endogenous regressors, Y_1 and Y_2, and that there are two additional instruments, z_1 and z_2. Say that z_1 is highly correlated with both Y_1 and Y_2 but z_2 is not correlated with either Y_1 or Y_2. Then the first-stage regression of Y_1 on z_1 and z_2 (along with the included exogenous variables) will produce large R^2 and F statistics, as will the regression of Y_2 on z_1, z_2, and the included exogenous variables. Nevertheless, the lack of correlation between z_2 and Y_1 and Y_2 is problematic. Here, although the order condition indicates that the model is just identified (the number of excluded instruments equals the number of endogenous regressors), the irrelevance of z_2 implies that the model is in fact not identified. Even if the model is overidentified, including irrelevant instruments can adversely affect the properties of instrumental-variables estimators, because their biases increase as the number of instruments increases.

▷ Example 4

estat firststage presents different statistics when the model contains multiple endogenous regressors. For illustration, we refit our model of rental rates, assuming that both hsngval and faminc are endogenously determined. We use i.region along with popden, a measure of population density, as additional instruments.

```
. ivregress 2sls rent pcturban (hsngval faminc = i.region popden)
 (output omitted )

. estat firststage
```

Shea's partial R-squared

Variable	Shea's Partial R-sq.	Shea's Adj. Partial R-sq.
hsngval	0.3477	0.2735
faminc	0.1893	0.0972

Minimum eigenvalue statistic = 2.51666

Critical Values Ho: Instruments are weak	# of endogenous regressors: 2 # of excluded instruments: 4			
	5%	10%	20%	30%
2SLS relative bias	11.04	7.56	5.57	4.73
	10%	15%	20%	25%
2SLS Size of nominal 5% Wald test	16.87	9.93	7.54	6.28
LIML Size of nominal 5% Wald test	4.72	3.39	2.99	2.79

Consider the endogenous regressor hsngval. Part of its variation is attributable to its correlation with the other regressors pcturban and faminc. The other component of hsngval's variation is peculiar to it and orthogonal to the variation in the other regressors. Similarly, we can think of the instruments as predicting the variation in hsngval in two ways, one stemming from the fact that the predicted values of hsngval are correlated with the predicted values of the other regressors and one from the variation in the predicted values of hsngval that is orthogonal to the variation in the predicted values of the other regressors.

What really matters for instrumental-variables estimation is whether the component of hsngval that is orthogonal to the other regressors can be explained by the component of the predicted value of hsngval that is orthogonal to the predicted values of the other regressors in the model. Shea's (1997) partial R^2 statistic measures this correlation. Because the bias of instrumental-variables estimators increases as more instruments are used, Shea's adjusted partial R^2 statistic is often used instead, as it makes a degrees-of-freedom adjustment for the number of instruments, analogous to the adjusted R^2 measure used in OLS regression. Although what constitutes a "low" value for Shea's partial R^2 depends on the specifics of the model being fit and the data used, these results, taken in isolation, do not strike us as being a particular cause for concern.

However, with this specification the minimum eigenvalue statistic is low. We cannot reject the null hypothesis of weak instruments for either of the characterizations we have discussed.

◁

By default, estat firststage determines which statistics to present based on the number of endogenous regressors in the model previously fit. However, you can specify the all option to obtain all the statistics.

❑ Technical note

If the previous estimation was conducted using aweights, pweights, or iweights, then the first-stage regression summary statistics are computed using those weights. However, in these cases the minimum eigenvalue statistic and its critical values are not available.

If the previous estimation included a robust VCE, then the first-stage F statistic is based on a robust VCE as well; for example, if you fit your model with an HAC VCE using the Bartlett kernel and four lags, then the F statistic reported is based on regression results using an HAC VCE using the Bartlett kernel and four lags. By default, the minimum eigenvalue statistic and its critical values are not displayed. You can use the forcenonrobust option to obtain them in these cases; the minimum eigenvalue statistic is computed using the weights, though the critical values reported may not be appropriate.

❑

estat overid

In addition to the requirement that instrumental variables be correlated with the endogenous regressors, the instruments must also be uncorrelated with the structural error term. If the model is overidentified, meaning that the number of additional instruments exceeds the number of endogenous regressors, then we can test whether the instruments are uncorrelated with the error term. If the model is just identified, then we cannot perform a test of overidentifying restrictions.

The estimator you used to fit the model determines which tests of overidentifying restrictions estat overid reports. If you used the 2SLS estimator without a robust VCE, estat overid reports Sargan's (1958) and Basmann's (1960) χ^2 tests. If you used the 2SLS estimator and requested a robust

VCE, Wooldridge's robust score test of overidentifying restrictions is performed instead; without a robust VCE, Wooldridge's test statistic is identical to Sargan's test statistic. If you used the LIML estimator, `estat overid` reports the Anderson–Rubin (1950) likelihood-ratio test and Basmann's (1960) F test. `estat overid` reports Hansen's (1982) J statistic if you used the GMM estimator. Davidson and MacKinnon (1993, 235–236) give a particularly clear explanation of the intuition behind tests of overidentifying restrictions. Also see Judge et al. (1985, 614–616) for a summary of tests of overidentifying restrictions for the 2SLS and LIML estimators.

Tests of overidentifying restrictions actually test two different things simultaneously. One, as we have discussed, is whether the instruments are uncorrelated with the error term. The other is that the equation is misspecified and that one or more of the excluded exogenous variables should in fact be included in the structural equation. Thus a significant test statistic could represent either an invalid instrument or an incorrectly specified structural equation.

▷ Example 5

Here we refit the model that treated just `hsngval` as endogenous using 2SLS, and then we perform tests of overidentifying restrictions:

```
. ivregress 2sls rent pcturban (hsngval = faminc i.region)
(output omitted)
. estat overid
Tests of overidentifying restrictions:
Sargan (score) chi2(3)  =   11.2877  (p = 0.0103)
Basmann chi2(3)         =   12.8294  (p = 0.0050)
```

Both test statistics are significant at the 5% test level, which means that either one or more of our instruments are invalid or that our structural model is specified incorrectly.

One possibility is that the error term in our structural model is heteroskedastic. Both Sargan's and Basmann's tests assume that the errors are i.i.d.; if the errors are not i.i.d., then these tests are not valid. Here we refit the model by requesting heteroskedasticity-robust standard errors, and then we use `estat overid` to obtain Wooldridge's score test of overidentifying restrictions, which is robust to heteroskedasticity.

```
. ivregress 2sls rent pcturban (hsngval = faminc i.region), vce(robust)
(output omitted)
. estat overid
Test of overidentifying restrictions:
Score chi2(3)           =    6.8364  (p = 0.0773)
```

Here we no longer reject the null hypothesis that our instruments are valid at the 5% significance level, though we do reject the null at the 10% level. You can verify that the robust standard error on the coefficient for `hsngval` is more than twice as large as its nonrobust counterpart and that the robust standard error for `pcturban` is nearly 50% larger.

◁

❑ Technical note

The test statistic for the test of overidentifying restrictions performed after GMM estimation is simply the sample size times the value of the objective function $Q(\beta_1, \beta_2)$ defined in (5) of [R] **ivregress**, evaluated at the GMM parameter estimates. If the weighting matrix $\mathbf{W}$ is optimal, meaning that $\mathbf{W} = \text{Var}(\mathbf{z}_i u_i)$, then $Q(\beta_1, \beta_2) \overset{A}{\sim} \chi^2(q)$, where q is the number of overidentifying restrictions. However, if the estimated $\mathbf{W}$ is not optimal, then the test statistic will not have an asymptotic χ^2 distribution.

Like the Sargan and Basmann tests of overidentifying restrictions for the 2SLS estimator, the Anderson–Rubin and Basmann tests after LIML estimation are predicated on the errors' being i.i.d. If the previous LIML results were reported with robust standard errors, then estat overid by default issues an error message and refuses to report the Anderson–Rubin and Basmann test statistics. You can use the forcenonrobust option to override this behavior. You can also use forcenonrobust to obtain the Sargan and Basmann test statistics after 2SLS estimation with robust standard errors.

❏

By default, estat overid issues an error message if the previous estimation was conducted using aweights, pweights, or iweights. You can use the forceweights option to override this behavior, though the test statistics may no longer have the expected χ^2 distributions.

Saved results

After 2SLS estimation, estat endogenous saves the following in r():

Scalars
r(durbin)	Durbin χ^2 statistic
r(p_durbin)	p-value for Durbin χ^2 statistic
r(wu)	Wu–Hausman F statistic
r(p_wu)	p-value for Wu–Hausman F statistic
r(df)	degrees of freedom
r(wudf_r)	denominator degrees of freedom for Wu–Hausman F
r(r_score)	robust score statistic
r(p_r_score)	p-value for robust score statistic
r(hac_score)	HAC score statistic
r(p_hac_score)	p-value for HAC score statistic
r(lags)	lags used in prewhitening
r(regF)	regression-based F statistic
r(p_regF)	p-value for regression-based F statistic
r(regFdf_n)	regression-based F numerator degrees of freedom
r(regFdf_r)	regression-based F denominator degrees of freedom

After GMM estimation, estat endogenous saves the following in r():

Scalars
r(C)	C χ^2 statistic
r(p_C)	p-value for C χ^2 statistic
r(df)	degrees of freedom

estat firststage saves the following in r():

Scalars
r(mineig)	minimum eigenvalue statistic

Matrices
r(mineigcv)	critical values for minimum eigenvalue statistic
r(multiresults)	Shea's partial R^2 statistics
r(singleresults)	first-stage R^2 and F statistics

After 2SLS estimation, `estat overid` saves the following in `r()`:

Scalars
    `r(lags)`   lags used in prewhitening
    `r(df)` χ^2 degrees of freedom
    `r(score)`   score χ^2 statistic
    `r(p_score)` p-value for score χ^2 statistic
    `r(basmann)`   Basmann χ^2 statistic
    `r(p_basmann)` p-value for Basmann χ^2 statistic
    `r(sargan)`   Sargan χ^2 statistic
    `r(p_sargan)` p-value for Sargan χ^2 statistic

After LIML estimation, `estat overid` saves the following in `r()`:

Scalars
    `r(ar)`   Anderson–Rubin χ^2 statistic
    `r(p_ar)` p-value for Anderson–Rubin χ^2 statistic
    `r(ar_df)` χ^2 degrees of freedom
    `r(basmann)`   Basmann F statistic
    `r(p_basmann)` p-value for Basmann F statistic
    `r(basmann_df_n)` F numerator degrees of freedom
    `r(basmann_df_d)` F denominator degrees of freedom

After GMM estimation, `estat overid` saves the following in `r()`:

Scalars
    `r(HansenJ)`   Hansen's J χ^2 statistic
    `r(p_HansenJ)` p-value for Hansen's J χ^2 statistic
    `r(J_df)` χ^2 degrees of freedom

Methods and formulas

All postestimation commands listed above are implemented as ado-files.

Methods and formulas are presented under the following headings:

> *Notation*
> *estat endogenous*
> *estat firststage*
> *estat overid*

Notation

Recall from [R] **ivregress** that the model is

$$\mathbf{y} = \mathbf{Y}\boldsymbol{\beta}_1 + \mathbf{X}_1\boldsymbol{\beta}_2 + \mathbf{u} = \mathbf{X}\boldsymbol{\beta} + \mathbf{u}$$

$$\mathbf{Y} = \mathbf{X}_1\boldsymbol{\Pi}_1 + \mathbf{X}_2\boldsymbol{\Pi}_2 + \mathbf{V} = \mathbf{Z}\boldsymbol{\Pi} + \mathbf{V}$$

where $\mathbf{y}$ is an $N \times 1$ vector of the left-hand-side variable, N is the sample size, $\mathbf{Y}$ is an $N \times p$ matrix of p endogenous regressors, $\mathbf{X}_1$ is an $N \times k_1$ matrix of k_1 included exogenous regressors, $\mathbf{X}_2$ is an $N \times k_2$ matrix of k_2 excluded exogenous variables, $\mathbf{X} = [\mathbf{Y}\ \mathbf{X}_1]$, $\mathbf{Z} = [\mathbf{X}_1\ \mathbf{X}_2]$, $\mathbf{u}$ is an $N \times 1$ vector of errors, $\mathbf{V}$ is an $N \times p$ matrix of errors, $\boldsymbol{\beta} = [\boldsymbol{\beta}_1\ \boldsymbol{\beta}_2]$ is a $k = (p + k_1) \times 1$ vector of parameters, and $\boldsymbol{\Pi}$ is a $(k_1 + k_2) \times p$ vector of parameters. If a constant term is included in the model, then one column of $\mathbf{X}_1$ contains all ones.

estat endogenous

Partition $\mathbf{Y}$ as $\mathbf{Y} = [\mathbf{Y}_1 \ \mathbf{Y}_2]$, where $\mathbf{Y}_1$ represents the p_1 endogenous regressors whose endogeneity is being tested and $\mathbf{Y}_2$ represents the p_2 endogenous regressors whose endogeneity is not being tested. If the endogeneity of all endogenous regressors is being tested, $\mathbf{Y} = \mathbf{Y}_1$ and $p_2 = 0$. After GMM estimation, estat endogenous refits the model treating $\mathbf{Y}_1$ as exogenous using the same type of weight matrix as requested at estimation time with the wmatrix() option; denote the Sargan statistic from this model by J_e and the estimated weight matrix by $\mathbf{W}_e$. Let $\mathbf{S}_e = \mathbf{W}_e^{-1}$. estat endogenous removes from $\mathbf{S}_e$ the rows and columns corresponding to the variables represented by $\mathbf{Y}_1$; denote the inverse of the resulting matrix by $\mathbf{W}'_e$. Next estat endogenous fits the model treating both $\mathbf{Y}_1$ and $\mathbf{Y}_2$ as endogenous, using the weight matrix $\mathbf{W}'_e$; denote the Sargan statistic from this model by J_c. Then $C = (J_e - J_c) \sim \chi^2(p_1)$. If one simply used the J statistic from the original model fit by ivregress in place of J_c, then in finite samples $J_e - J$ might be negative. The procedure used by estat endogenous is guaranteed to yield $C \geq 0$; see Hayashi (2000, 220).

Let $\widehat{\mathbf{u}}_c$ denote the residuals from the model treating both $\mathbf{Y}_1$ and $\mathbf{Y}_2$ as endogenous, and let $\widehat{\mathbf{u}}_e$ denote the residuals from the model treating only $\mathbf{Y}_2$ as endogenous. Then Durbin's (1954) statistic is

$$D = \frac{\widehat{\mathbf{u}}'_e \mathbf{P}_{ZY_1} \widehat{\mathbf{u}}_e - \widehat{\mathbf{u}}'_c \mathbf{P}_Z \widehat{\mathbf{u}}_c}{\widehat{\mathbf{u}}'_e \widehat{\mathbf{u}}_e / N}$$

where $\mathbf{P}_Z = \mathbf{Z}(\mathbf{Z}'\mathbf{Z})^{-1}\mathbf{Z}'$ and $\mathbf{P}_{ZY_1} = [\mathbf{Z} \ \mathbf{Y}_1]([\mathbf{Z} \ \mathbf{Y}_1]'[\mathbf{Z} \ \mathbf{Y}_1])^{-1}[\mathbf{Z} \ \mathbf{Y}_1]'$ $D \sim \chi^2(p_1)$. The Wu–Hausman (Wu 1974; Hausman 1978) statistic is

$$WH = \frac{(\widehat{\mathbf{u}}'_e \mathbf{P}_{ZY_1} \widehat{\mathbf{u}}_e - \widehat{\mathbf{u}}'_c \mathbf{P}_Z \widehat{\mathbf{u}}_c)/p_1}{\{\widehat{\mathbf{u}}'_e \widehat{\mathbf{u}}_e - (\widehat{\mathbf{u}}'_e \mathbf{P}_{ZY_1} \widehat{\mathbf{u}}_e - \widehat{\mathbf{u}}'_c \mathbf{P}_Z \widehat{\mathbf{u}}_c)\}/(N - k_1 - p - p_1)}$$

$WH \sim F(p_1, N - k_1 - p - p_1)$. Baum, Schaffer, and Stillman (2003, 2007) discuss these tests in more detail.

Next we describe Wooldridge's (1995) score test. The nonrobust version of Wooldridge's test is identical to Durbin's test. Suppose a robust covariance matrix was used at estimation time. Let $\widehat{\mathbf{e}}$ denote the sample residuals obtained by fitting the model via OLS, treating $\mathbf{Y}$ as exogenous. We then regress each variable represented in $\mathbf{Y}$ on $\mathbf{Z}$; call the residuals for the jth regression $\widehat{\mathbf{r}}_j$, $j = 1, \ldots, p$. Define $\widehat{k}_{ij} = \widehat{e}_i \widehat{r}_{ij}$, $i = 1, \ldots, N$. We then run the regression

$$\mathbf{1} = \theta_1 \widehat{\mathbf{k}}_1 + \cdots + \theta_p \widehat{\mathbf{k}}_p + \epsilon$$

where $\mathbf{1}$ is an $N \times 1$ vector of ones and ϵ is a regression error term. $N - \text{RSS} \sim \chi^2(p)$, where RSS is the residual sum of squares from the regression just described. If instead an HAC VCE was used at estimation time, then before running the final regression we prewhiten the $\widehat{\mathbf{k}}_j$ series by using a VAR(q) model, where q is the number of lags specified with the lags() option.

The regression-based test proceeds as follows. Following Hausman (1978, 1259), we regress $\mathbf{Y}$ on $\mathbf{Z}$ and obtain the residuals $\widehat{\mathbf{V}}$. Next we fit the augmented regression

$$\mathbf{y} = \mathbf{Y}\beta_1 + \mathbf{X}_1\beta_2 + \widehat{\mathbf{V}}\gamma + \epsilon$$

by OLS regression, where ϵ is a regression error term. A test of the exogeneity of $\mathbf{Y}$ is equivalent to a test of $\gamma = \mathbf{0}$. As Cameron and Trivedi (2005, 276) suggest, this test can be made robust to heteroskedasticity, autocorrelation, or clustering by using the appropriate robust VCE when testing $\gamma = \mathbf{0}$. When a nonrobust VCE is used, this test is equivalent to the Wu–Hausman test described earlier. One cannot simply fit this augmented regression via 2SLS to test the endogeneity of a subset of the endogenous regressors; Davidson and MacKinnon (1993, 229–231) discuss a test of $\gamma = \mathbf{0}$ for the homoskedastic version of the augmented regression fit by 2SLS, but an appropriate robust test is not apparent.

estat firststage

When the structural equation includes one endogenous regressor, `estat firststage` fits the regression

$$\mathbf{Y} = \mathbf{X}_1\boldsymbol{\pi}_1 + \mathbf{X}_2\boldsymbol{\pi}_2 + \mathbf{v}$$

via OLS. The R^2 and adjusted R^2 from that regression are reported in the output, as well as the F statistic from the Wald test of H_0: $\boldsymbol{\pi}_2 = \mathbf{0}$. To obtain the partial R^2 statistic, `estat firststage` fits the regression

$$\mathbf{M}_{\mathbf{X}_1}\mathbf{y} = \mathbf{M}_{\mathbf{X}_1}\mathbf{X}_2\boldsymbol{\xi} + \boldsymbol{\epsilon}$$

by OLS, where $\boldsymbol{\epsilon}$ is a regression error term, $\boldsymbol{\xi}$ is a $k_2 \times 1$ parameter vector, and $\mathbf{M}_{\mathbf{X}_1} = \mathbf{I} - \mathbf{X}_1(\mathbf{X}_1'\mathbf{X}_1)^{-1}\mathbf{X}_1'$; that is, the partial R^2 is the R^2 between $\mathbf{y}$ and $\mathbf{X}_2$ after eliminating the effects of $\mathbf{X}_1$. If the model contains multiple endogenous regressors and the `all` option is specified, these statistics are calculated for each endogenous regressor in turn.

To calculate Shea's partial R^2, let $\mathbf{y}_1$ denote the endogenous regressor whose statistic is being calculated and $\mathbf{Y}_0$ denote the other endogenous regressors. Define $\widetilde{\mathbf{y}}_1$ as the residuals obtained from regressing $\mathbf{y}_1$ on $\mathbf{Y}_0$ and $\mathbf{X}_1$. Let $\widehat{\mathbf{y}}_1$ denote the fitted values obtained from regressing $\mathbf{y}_1$ on $\mathbf{X}_1$ and $\mathbf{X}_2$; that is, $\widehat{\mathbf{y}}_1$ are the fitted values from the first-stage regression for $\mathbf{y}_1$, and define the columns of $\widehat{\mathbf{Y}}_0$ analogously. Finally, let $\widetilde{\widehat{\mathbf{y}}}_1$ denote the residuals from regressing $\widehat{\mathbf{y}}_1$ on $\widehat{\mathbf{Y}}_0$ and $\mathbf{X}_1$. Shea's partial R^2 is the simple R^2 from the regression of $\widetilde{\mathbf{y}}_1$ on $\widetilde{\widehat{\mathbf{y}}}_1$; denote this as R_S^2. Shea's adjusted partial R^2 is equal to $1 - (1 - R_S^2)(N - 1)/(N - k_Z + 1)$ if a constant term is included and $1 - (1 - R_S^2)(N - 1)/(N - k_Z)$ if there is no constant term included in the model, where $k_Z = k_1 + k_2$. For one endogenous regressor, one instrument, no exogenous regressors, and a constant term, R_S^2 equals the adjusted R_S^2.

The Stock and Yogo minimum eigenvalue statistic, first proposed by Cragg and Donald (1993) as a test for underidentification, is the minimum eigenvalue of the matrix

$$\mathbf{G} = \frac{1}{k_Z}\widehat{\boldsymbol{\Sigma}}_{\mathbf{VV}}^{-1/2}\mathbf{Y}'\mathbf{M}_{\mathbf{X}_1}'\mathbf{X}_2(\mathbf{X}_2'\mathbf{M}_{\mathbf{X}_1}\mathbf{X}_2)^{-1}\mathbf{X}_2'\mathbf{M}_{\mathbf{X}_1}\mathbf{Y}\widehat{\boldsymbol{\Sigma}}_{\mathbf{VV}}^{-1/2}$$

where

$$\widehat{\boldsymbol{\Sigma}}_{\mathbf{VV}} = \frac{1}{N - k_Z}\mathbf{Y}'\mathbf{M}_{\mathbf{Z}}\mathbf{Y}$$

$\mathbf{M}_{\mathbf{Z}} = \mathbf{I} - \mathbf{Z}(\mathbf{Z}'\mathbf{Z})^{-1}\mathbf{Z}'$, and $\mathbf{Z} = [\mathbf{X}_1 \ \mathbf{X}_2]$. Critical values are obtained from the tables in Stock and Yogo (2005).

estat overid

The Sargan (1958) and Basmann (1960) χ^2 statistics are calculated by running the auxiliary regression

$$\widehat{\mathbf{u}} = \mathbf{Z}\boldsymbol{\delta} + \mathbf{e}$$

where $\widehat{\mathbf{u}}$ are the sample residuals from the model and $\mathbf{e}$ is an error term. Then Sargan's statistic is

$$S = N\left(1 - \frac{\widehat{\mathbf{e}}'\widehat{\mathbf{e}}}{\widehat{\mathbf{u}}'\widehat{\mathbf{u}}}\right)$$

where $\widehat{\mathbf{e}}$ are the residuals from that auxiliary regression. Basmann's statistic is calculated as

$$B = S\frac{N - k_Z}{N - S}$$

Both S and B are distributed $\chi^2(m)$, where m, the number of overidentifying restrictions, is equal to $k_Z - k$, where k is the number of endogenous regressors.

Wooldridge's (1995) score test of overidentifying restrictions is identical to Sargan's (1958) statistic under the assumption of i.i.d. and therefore is not recomputed unless a robust VCE was used at estimation time. If a heteroskedasticity-robust VCE was used, Wooldridge's test proceeds as follows. Let $\widehat{\mathbf{Y}}$ denote the $N \times k$ matrix of fitted values obtained by regressing the endogenous regressors on $\mathbf{X}_1$ and $\mathbf{X}_2$. Let $\mathbf{Q}$ denote an $N \times m$ matrix of excluded exogenous variables; the test statistic to be calculated is invariant to whichever m of the k_2 excluded exogenous variables is chosen. Define the ith element of $\widehat{\mathbf{k}}_j$, $i = 1, \ldots, N$, $j = 1, \ldots, m$, as

$$k_{ij} = \widehat{q}_{ij} u_i$$

where $\widehat{q}_{ij}$ is the ith element of $\widehat{\mathbf{q}}_j$, the fitted values from regressing the jth column of $\mathbf{Q}$ on $\widehat{\mathbf{Y}}$ and $\mathbf{X}_1$. Finally, fit the regression

$$\mathbf{1} = \theta_1 \widehat{\mathbf{k}}_1 + \cdots + \theta_m \widehat{\mathbf{k}}_m + \epsilon$$

where $\mathbf{1}$ is an $N \times 1$ vector of ones and ϵ is a regression error term, and calculate the residual sum of squares, RSS. Then the test statistic is $W = N - \text{RSS}$. $W \sim \chi^2(m)$. If an HAC VCE was used at estimation, then the $\widehat{\mathbf{k}}_j$ are prewhitened using a VAR(p) model, where p is specified using the `lags()` option.

The Anderson–Rubin (1950), AR, test of overidentifying restrictions for use after the LIML estimator is calculated as $\text{AR} = N(\kappa - 1)$, where κ is the minimal eigenvalue of a certain matrix defined in *Methods and formulas* of [R] **ivregress**. $\text{AR} \sim \chi^2(m)$. (Some texts define this statistic as $N \ln(\kappa)$ because $\ln(x) \approx (x - 1)$ for x near one.) Basmann's F statistic for use after the LIML estimator is calculated as $B_F = (\kappa - 1)(N - k_Z)/m$. $B_F \sim F(m, N - k_Z)$.

Hansen's J statistic is simply the sample size times the value of the GMM objective function defined in (5) of [R] **ivregress**, evaluated at the estimated parameter values. Under the null hypothesis that the overidentifying restrictions are valid, $J \sim \chi^2(m)$.

References

Anderson, T. W., and H. Rubin. 1950. The asymptotic properties of estimates of the parameters of a single equation in a complete system of stochastic equations. *Annals of Mathematical Statistics* 21: 570–582.

Angrist, J. D., and J.-S. Pischke. 2009. *Mostly Harmless Econometrics: An Empiricist's Companion*. Princeton, NJ: Princeton University Press.

Basmann, R. L. 1960. On finite sample distributions of generalized classical linear identifiability test statistics. *Journal of the American Statistical Association* 55: 650–659.

Baum, C. F., M. E. Schaffer, and S. Stillman. 2003. Instrumental variables and GMM: Estimation and testing. *Stata Journal* 3: 1–31.

———. 2007. Enhanced routines for instrumental variables/generalized method of moments estimation and testing. *Stata Journal* 7: 465–506.

Bound, J., D. A. Jaeger, and R. M. Baker. 1995. Problems with instrumental variables estimation when the correlation between the instruments and the endogenous explanatory variable is weak. *Journal of the American Statistical Association* 90: 443–450.

Cameron, A. C., and P. K. Trivedi. 2005. *Microeconometrics: Methods and Applications*. New York: Cambridge University Press.

Cragg, J. G., and S. G. Donald. 1993. Testing identifiability and specification in instrumental variable models. *Econometric Theory* 9: 222–240.

Davidson, R., and J. G. MacKinnon. 1993. *Estimation and Inference in Econometrics*. New York: Oxford University Press.

Durbin, J. 1954. Errors in variables. *Review of the International Statistical Institute* 22: 23–32.

Hahn, J., and J. A. Hausman. 2003. Weak instruments: Diagnosis and cures in empirical econometrics. *American Economic Review Papers and Proceedings* 93: 118–125.

Hall, A. R., G. D. Rudebusch, and D. W. Wilcox. 1996. Judging instrument relevance in instrumental variables estimation. *International Economic Review* 37: 283–298.

Hansen, L. P. 1982. Large sample properties of generalized method of moments estimators. *Econometrica* 50: 1029–1054.

Hausman, J. A. 1978. Specification tests in econometrics. *Econometrica* 46: 1251–1271.

Hayashi, F. 2000. *Econometrics*. Princeton, NJ: Princeton University Press.

Judge, G. G., W. E. Griffiths, R. C. Hill, H. Lütkepohl, and T.-C. Lee. 1985. *The Theory and Practice of Econometrics*. 2nd ed. New York: Wiley.

Nelson, C. R., and R. Startz. 1990. The distribution of the instrumental variable estimator and its t ratio when the instrument is a poor one. *Journal of Business* 63: S125–S140.

Poi, B. P. 2006. Jackknife instrumental variables estimation in Stata. *Stata Journal* 6: 364–376.

Sargan, J. D. 1958. The estimation of economic relationships using instrumental variables. *Econometrica* 26: 393–415.

Shea, J. 1997. Instrument relevance in multivariate linear models: A simple measure. *Review of Economics and Statistics* 79: 348–352.

Staiger, D., and J. H. Stock. 1997. Instrumental variables regression with weak instruments. *Econometrica* 65: 557–586.

Stock, J. H., J. H. Wright, and M. Yogo. 2002. A survey of weak instruments and weak identification in generalized method of moments. *Journal of Business and Economic Statistics* 20: 518–529.

Stock, J. H., and M. Yogo. 2005. Testing for weak instruments in linear IV regression. In *Identification and Inference for Econometric Models: Essays in Honor of Thomas Rothenberg*, ed. D. W. K. Andrews and J. H. Stock, 80–108. New York: Cambridge University Press.

Wooldridge, J. M. 1995. Score diagnostics for linear models estimated by two stage least squares. In *Advances in Econometrics and Quantitative Economics: Essays in Honor of Professor C. R. Rao*, ed. G. S. Maddala, P. C. B. Phillips, and T. N. Srinivasan, 66–87. Oxford: Blackwell.

Wu, D.-M. 1974. Alternative tests of independence between stochastic regressors and disturbances: Finite sample results. *Econometrica* 42: 529–546.

Also see

[R] **ivregress** — Single-equation instrumental-variables regression

[U] **20 Estimation and postestimation commands**

Title

> **ivtobit** — Tobit model with continuous endogenous regressors

Syntax

Maximum likelihood estimator

> ivtobit *depvar* [*varlist*₁] (*varlist*₂ = *varlist*ᵢᵥ) [*if*] [*in*] [*weight*],
>
> ll[(#)] ul[(#)] [*mle_options*]

Two-step estimator

> ivtobit *depvar* [*varlist*₁] (*varlist*₂ = *varlist*ᵢᵥ) [*if*] [*in*] [*weight*], <u>two</u>step
>
> ll[(#)] ul[(#)] [*tse_options*]

mle_options	Description
Model	
* ll[(#)]	lower limit for left censoring
* ul[(#)]	upper limit for right censoring
<u>mle</u>	use conditional maximum-likelihood estimator; the default
<u>constraints</u>(*constraints*)	apply specified linear constraints
SE/Robust	
vce(*vcetype*)	*vcetype* may be oim, <u>r</u>obust, <u>cl</u>uster *clustvar*, opg, <u>boot</u>strap, or <u>jackk</u>nife
Reporting	
<u>l</u>evel(#)	set confidence level; default is level(95)
first	report first-stage regression
<u>nocns</u>report	do not display constraints
display_options	control column formats, row spacing, line width, and display of omitted variables and base and empty cells
Maximization	
maximize_options	control the maximization process
<u>coefl</u>egend	display legend instead of statistics

*You must specify at least one of ll[(#)] and ul[(#)].

tse_options	Description
Model	
* <u>twostep</u>	use Newey's two-step estimator; the default is `mle`
* `ll` $\left[(\#)\right]$	lower limit for left censoring
* `ul` $\left[(\#)\right]$	upper limit for right censoring
SE	
`vce(`*vcetype*`)`	*vcetype* may be `twostep`, <u>boot</u>`strap`, or <u>jack</u>`knife`
Reporting	
<u>l</u>`evel(#)`	set confidence level; default is `level(95)`
`first`	report first-stage regression
display_options	control column formats, row spacing, line width, and display of omitted variables and base and empty cells
<u>coefl</u>`egend`	display legend instead of statistics

*`twostep` is required. You must specify at least one of `ll`$\left[(\#)\right]$ and `ul`$\left[(\#)\right]$.

varlist$_1$ and *varlist*$_{iv}$ may contain factor variables; see [U] **11.4.3 Factor variables**.
depvar, *varlist*$_1$, *varlist*$_2$, and *varlist*$_{iv}$ may contain time-series operators; see [U] **11.4.4 Time-series varlists**.
`bootstrap`, `by`, `jackknife`, `rolling`, `statsby`, and `svy` are allowed; see [U] **11.1.10 Prefix commands**.
Weights are not allowed with the `bootstrap` prefix; see [R] **bootstrap**.
`vce()`, `first`, `twostep`, and weights are not allowed with the `svy` prefix; see [SVY] **svy**.
`fweights`, `iweights`, and `pweights` are allowed with the maximum likelihood estimator. `fweights` are
   allowed with Newey's two-step estimator. See [U] **11.1.6 weight**.
`coeflegend` does not appear in the dialog box.
See [U] **20 Estimation and postestimation commands** for more capabilities of estimation commands.

Menu

Statistics > Endogenous covariates > Tobit model with endogenous covariates

Description

ivtobit fits tobit models where one or more of the regressors is endogenously determined. By default, `ivtobit` uses maximum likelihood estimation. Alternatively, Newey's (1987) minimum chi-squared estimator can be invoked with the `twostep` option. Both estimators assume that the endogenous regressors are continuous and so are not appropriate for use with discrete endogenous regressors. See [R] **ivprobit** for probit estimation with endogenous regressors and [R] **tobit** for tobit estimation when the model contains no endogenous regressors.

Options for ML estimator

⌐ Model ⌐

`ll(#)` and `ul(#)` indicate the lower and upper limits for censoring, respectively. You may specify one or both. Observations with *depvar* $\leq$ `ll()` are left-censored; observations with *depvar* $\geq$ `ul()` are right-censored; and remaining observations are not censored. You do not have to specify the censoring values at all. It is enough to type `ll`, `ul`, or both. When you do not specify a censoring value, `ivtobit` assumes that the lower limit is the minimum observed in the data (if `ll` is specified) and that the upper limit is the maximum (if `ul` is specified).

mle requests that the conditional maximum-likelihood estimator be used. This is the default.

constraints(*constraints*); see [R] **estimation options**.

SE/Robust

vce(*vcetype*) specifies the type of standard error reported, which includes types that are derived from asymptotic theory, that are robust to some kinds of misspecification, that allow for intragroup correlation, and that use bootstrap or jackknife methods; see [R] *vce_option*.

Reporting

level(*#*); see [R] **estimation options**.

first requests that the parameters for the reduced-form equations showing the relationships between the endogenous variables and instruments be displayed. For the two-step estimator, first shows the first-stage regressions. For the maximum likelihood estimator, these parameters are estimated jointly with the parameters of the tobit equation. The default is not to show these parameter estimates.

nocnsreport; see [R] **estimation options**.

display_options: noomitted, vsquish, noemptycells, baselevels, allbaselevels, cformat(%*fmt*), pformat(%*fmt*), sformat(%*fmt*), and nolstretch; see [R] **estimation options**.

Maximization

maximize_options: difficult, technique(*algorithm_spec*), iterate(*#*), [no]log, trace, gradient, showstep, hessian, showtolerance, tolerance(*#*), ltolerance(*#*), nrtolerance(*#*), nonrtolerance, and from(*init_specs*); see [R] **maximize**. This model's likelihood function can be difficult to maximize, especially with multiple endogenous variables. The difficult and technique(bfgs) options may be helpful in achieving convergence.

Setting the optimization type to technique(bhhh) resets the default *vcetype* to vce(opg).

The following option is available with ivtobit but is not shown in the dialog box:

coeflegend; see [R] **estimation options**.

Options for two-step estimator

Model

twostep is required and requests that Newey's (1987) efficient two-step estimator be used to obtain the coefficient estimates.

ll(*#*) and ul(*#*) indicate the lower and upper limits for censoring, respectively. You may specify one or both. Observations with *depvar* $\leq$ ll() are left-censored; observations with *depvar* $\geq$ ul() are right-censored; and remaining observations are not censored. You do not have to specify the censoring values at all. It is enough to type ll, ul, or both. When you do not specify a censoring value, ivtobit assumes that the lower limit is the minimum observed in the data (if ll is specified) and that the upper limit is the maximum (if ul is specified).

SE

vce(*vcetype*) specifies the type of standard error reported, which includes types that are derived from asymptotic theory and that use bootstrap or jackknife methods; see [R] *vce_option*.

⌐ Reporting ⌐

level(#); see [R] **estimation options**.

first requests that the parameters for the reduced-form equations showing the relationships between the endogenous variables and instruments be displayed. For the two-step estimator, first shows the first-stage regressions. For the maximum likelihood estimator, these parameters are estimated jointly with the parameters of the tobit equation. The default is not to show these parameter estimates.

display_options: noomitted, vsquish, noemptycells, baselevels, allbaselevels, cformat(%*fmt*), pformat(%*fmt*), sformat(%*fmt*), and nolstretch; see [R] **estimation options**.

The following option is available with ivtobit but is not shown in the dialog box:

coeflegend; see [R] **estimation options**.

Remarks

ivtobit fits models with censored dependent variables and endogenous regressors. You can use it to fit a tobit model when you suspect that one or more of the regressors is correlated with the error term. ivtobit is to tobit what ivregress is to linear regression analysis; see [R] **ivregress** for more information.

Formally, the model is

$$y^*_{1i} = \boldsymbol{y}_{2i}\boldsymbol{\beta} + \boldsymbol{x}_{1i}\boldsymbol{\gamma} + u_i$$
$$\boldsymbol{y}_{2i} = \boldsymbol{x}_{1i}\boldsymbol{\Pi}_1 + \boldsymbol{x}_{2i}\boldsymbol{\Pi}_2 + \boldsymbol{v}_i$$

where $i = 1, \ldots, N$; $\boldsymbol{y}_{2i}$ is a $1 \times p$ vector of endogenous variables; $\boldsymbol{x}_{1i}$ is a $1 \times k_1$ vector of exogenous variables; $\boldsymbol{x}_{2i}$ is a $1 \times k_2$ vector of additional instruments; and the equation for $\boldsymbol{y}_{2i}$ is written in reduced form. By assumption $(u_i, \boldsymbol{v}_i) \sim \mathrm{N}(\boldsymbol{0})$. $\boldsymbol{\beta}$ and $\boldsymbol{\gamma}$ are vectors of structural parameters, and $\boldsymbol{\Pi}_1$ and $\boldsymbol{\Pi}_2$ are matrices of reduced-form parameters. We do not observe y^*_{1i}; instead, we observe

$$y_{1i} = \begin{cases} a & y^*_{1i} < a \\ y^*_{1i} & a \leq y^*_{1i} \leq b \\ b & y^*_{1i} > b \end{cases}$$

The order condition for identification of the structural parameters is that $k_2 \geq p$. Presumably, $\boldsymbol{\Sigma}$ is not block diagonal between u_i and $\boldsymbol{v}_i$; otherwise, $\boldsymbol{y}_{2i}$ would not be endogenous.

❑ Technical note

This model is derived under the assumption that $(u_i, \boldsymbol{v}_i)$ is independent and identically distributed multivariate normal for all i. The vce(cluster *clustvar*) option can be used to control for a lack of independence. As with the standard tobit model without endogeneity, if u_i is heteroskedastic, point estimates will be inconsistent.

❑

▶ Example 1

Using the same dataset as in [R] **ivprobit**, we now want to estimate women's incomes. In our hypothetical world, all women who choose not to work receive $10,000 in welfare and child-support payments. Therefore, we never observe incomes under $10,000: a woman offered a job with an annual wage less than that would not accept and instead would collect the welfare payment. We model income as a function of the number of years of schooling completed, the number of children at home, and other household income. We again believe that other_inc is endogenous, so we use male_educ as an instrument.

```
. use http://www.stata-press.com/data/r12/laborsup

. ivtobit fem_inc fem_educ kids (other_inc = male_educ), ll

Fitting exogenous tobit model

Fitting full model

Iteration 0:   log likelihood = -3228.4224
Iteration 1:   log likelihood = -3226.2882
Iteration 2:   log likelihood =  -3226.085
Iteration 3:   log likelihood = -3226.0845
Iteration 4:   log likelihood = -3226.0845
```

Tobit model with endogenous regressors			Number of obs	=	500
			Wald chi2(3)	=	117.42
Log likelihood = -3226.0845			Prob > chi2	=	0.0000

	Coef.	Std. Err.	z	P>|z|	[95% Conf. Interval]	
other_inc	-.9045399	.1329762	-6.80	0.000	-1.165168	-.6439114
fem_educ	3.272391	.3968708	8.25	0.000	2.494538	4.050243
kids	-3.312357	.7218628	-4.59	0.000	-4.727182	-1.897532
_cons	19.24735	7.372391	2.61	0.009	4.797725	33.69697
/alpha	.2907654	.1379965	2.11	0.035	.0202972	.5612336
/lns	2.874031	.0506672	56.72	0.000	2.774725	2.973337
/lnv	2.813383	.0316228	88.97	0.000	2.751404	2.875363
s	17.70826	.897228			16.03422	19.55707
v	16.66621	.5270318			15.66461	17.73186

```
Instrumented:  other_inc
Instruments:   fem_educ kids male_educ
```

Wald test of exogeneity (/alpha = 0): chi2(1) =   4.44 Prob > chi2 = 0.0351

```
    Obs. summary:         272  left-censored observations at fem_inc<=10
                          228      uncensored observations
                            0 right-censored observations
```

Because we did not specify mle or twostep, ivtobit used the maximum likelihood estimator by default. ivtobit fits a tobit model, ignoring endogeneity, to get starting values for the full model. The header of the output contains the maximized log likelihood, the number of observations, and a Wald statistic and p-value for the test of the hypothesis that all the slope coefficients are jointly zero. At the end of the output, we see a count of the censored and uncensored observations.

Near the bottom of the output is a Wald test of the exogeneity of the instrumented variables. If the test statistic is not significant, there is not sufficient information in the sample to reject the null hypothesis of no endogeneity. Then the point estimates from ivtobit are consistent, although those from tobit are likely to have smaller standard errors.

◁

Various two-step estimators have also been proposed for the endogenous tobit model, and Newey's (1987) minimum chi-squared estimator is available with the `twostep` option.

▷ Example 2

Refitting our labor-supply model with the two-step estimator yields

```
. ivtobit fem_inc fem_educ kids (other_inc = male_educ), ll twostep
```

Two-step tobit with endogenous regressors   Number of obs   =   500
                                                   Wald chi2(3)   =   117.38
                                                   Prob > chi2   =   0.0000

	Coef.	Std. Err.	z	P>|z|	[95% Conf.	Interval]
other_inc	-.9045397	.1330015	-6.80	0.000	-1.165218	-.6438616
fem_educ	3.27239	.3969399	8.24	0.000	2.494402	4.050378
kids	-3.312356	.7220066	-4.59	0.000	-4.727463	-1.897249
_cons	19.24735	7.37392	2.61	0.009	4.794728	33.69997

Instrumented: other_inc
Instruments: fem_educ kids male_educ

Wald test of exogeneity: chi2(1) =   4.64   Prob > chi2 = 0.0312

Obs. summary: 272 left-censored observations at fem_inc<=10
                       228   uncensored observations
                         0 right-censored observations

All the coefficients have the same signs as their counterparts in the maximum likelihood model. The Wald test at the bottom of the output confirms our earlier finding of endogeneity.

◁

❑ Technical note

In the tobit model with endogenous regressors, we assume that $(u_i, \boldsymbol{v}_i)$ is multivariate normal with covariance matrix

$$\text{Var}(u_i, \boldsymbol{v}_i) = \boldsymbol{\Sigma} = \begin{bmatrix} \sigma_u^2 & \boldsymbol{\Sigma}_{21}' \\ \boldsymbol{\Sigma}_{21} & \boldsymbol{\Sigma}_{22} \end{bmatrix}$$

Using the properties of the multivariate normal distribution, $\text{Var}(u_i|\boldsymbol{v}_i) \equiv \sigma_{u|v}^2 = \sigma_u^2 - \boldsymbol{\Sigma}_{21}' \boldsymbol{\Sigma}_{22}^{-1} \boldsymbol{\Sigma}_{21}$. Calculating the marginal effects on the conditional expected values of the observed and latent dependent variables and on the probability of censoring requires an estimate of σ_u^2. The two-step estimator identifies only $\sigma_{u|v}^2$, not σ_u^2, so only the linear prediction and its standard error are available after you have used the `twostep` option. However, unlike the two-step probit estimator described in [R] **ivprobit**, the two-step tobit estimator does identify $\boldsymbol{\beta}$ and $\boldsymbol{\gamma}$. See Wooldridge (2010, 683–684) for more information.

❑

Saved results

ivtobit, mle saves the following in e():

Scalars

e(N)	number of observations
e(N_unc)	number of uncensored observations
e(N_lc)	number of left-censored observations
e(N_rc)	number of right-censored observations
e(llopt)	contents of ll()
e(ulopt)	contents of ul()
e(k)	number of parameters
e(k_eq)	number of equations in e(b)
e(k_eq_model)	number of equations in overall model test
e(k_aux)	number of auxiliary parameters
e(k_dv)	number of dependent variables
e(df_m)	model degrees of freedom
e(ll)	log likelihood
e(N_clust)	number of clusters
e(endog_ct)	number of endogenous regressors
e(p)	model Wald p-value
e(p_exog)	exogeneity test Wald p-value
e(chi2)	model Wald χ^2
e(chi2_exog)	Wald χ^2 test of exogeneity
e(rank)	rank of e(V)
e(ic)	number of iterations
e(rc)	return code
e(converged)	1 if converged, 0 otherwise

Macros

e(cmd)	ivtobit
e(cmdline)	command as typed
e(depvar)	name of dependent variable
e(instd)	instrumented variables
e(insts)	instruments
e(wtype)	weight type
e(wexp)	weight expression
e(title)	title in estimation output
e(clustvar)	name of cluster variable
e(chi2type)	Wald; type of model χ^2 test
e(vce)	*vcetype* specified in vce()
e(vcetype)	title used to label Std. Err.
e(method)	ml
e(opt)	type of optimization
e(which)	max or min; whether optimizer is to perform maximization or minimization
e(ml_method)	type of ml method
e(user)	name of likelihood-evaluator program
e(technique)	maximization technique
e(properties)	b V
e(predict)	program used to implement predict
e(footnote)	program used to implement the footnote display
e(marginsok)	predictions allowed by margins
e(asbalanced)	factor variables fvset as asbalanced
e(asobserved)	factor variables fvset as asobserved

Matrices
    e(b)   coefficient vector
    e(Cns)   constraints matrix
    e(ilog)   iteration log (up to 20 iterations)
    e(gradient)   gradient vector
    e(Sigma) $\widehat{\Sigma}$
    e(V)   variance–covariance matrix of the estimators
    e(V_modelbased)   model-based variance

Functions
    e(sample)   marks estimation sample

ivtobit, twostep saves the following in e():

Scalars
    e(N)   number of observations
    e(N_unc)   number of uncensored observations
    e(N_lc)   number of left-censored observations
    e(N_rc)   number of right-censored observations
    e(llopt)   contents of ll()
    e(ulopt)   contents of ul()
    e(df_m)   model degrees of freedom
    e(df_exog)   degrees of freedom for χ^2 test of exogeneity
    e(p)   model Wald p-value
    e(p_exog)   exogeneity test Wald p-value
    e(chi2)   model Wald χ^2
    e(chi2_exog)   Wald χ^2 test of exogeneity
    e(rank)   rank of e(V)

Macros
    e(cmd)   ivtobit
    e(cmdline)   command as typed
    e(depvar)   name of dependent variable
    e(instd)   instrumented variables
    e(insts)   instruments
    e(wtype)   weight type
    e(wexp)   weight expression
    e(chi2type)   Wald; type of model χ^2 test
    e(vce)   *vcetype* specified in vce()
    e(vcetype)   title used to label Std. Err.
    e(method)   twostep
    e(properties)   b V
    e(predict)   program used to implement predict
    e(footnote)   program used to implement the footnote display
    e(marginsok)   predictions allowed by margins
    e(asbalanced)   factor variables fvset as asbalanced
    e(asobserved)   factor variables fvset as asobserved

Matrices
    e(b)   coefficient vector
    e(Cns)   constraints matrix
    e(V)   variance–covariance matrix of the estimators
    e(V_modelbased)   model-based variance

Functions
    e(sample)   marks estimation sample

Methods and formulas

ivtobit is implemented as an ado-file.

The estimation procedure used by ivtobit is similar to that used by ivprobit. For compactness, we write the model as

$$y_{1i}^* = z_i \delta + u_i \tag{1a}$$

$$y_{2i} = x_i \Pi + v_i \tag{1b}$$

where $z_i = (y_{2i}, x_{1i})$, $x_i = (x_{1i}, x_{2i})$, $\delta = (\beta', \gamma')'$, and $\Pi = (\Pi_1', \Pi_2')'$. We do not observe y_{1i}^*; instead, we observe

$$y_{1i} = \begin{cases} a & y_{1i}^* < a \\ y_{1i}^* & a \le y_{1i}^* \le b \\ b & y_{1i}^* > b \end{cases}$$

(u_i, v_i) is distributed multivariate normal with mean zero and covariance matrix

$$\Sigma = \begin{bmatrix} \sigma_u^2 & \Sigma_{21}' \\ \Sigma_{21} & \Sigma_{22} \end{bmatrix}$$

Using the properties of the multivariate normal distribution, we can write $u_i = v_i' \alpha + \epsilon_i$, where $\alpha = \Sigma_{22}^{-1} \Sigma_{21}$; $\epsilon_i \sim N(0; \sigma_{u|v}^2)$, where $\sigma_{u|v}^2 = \sigma_u^2 - \Sigma_{21}' \Sigma_{22}^{-1} \Sigma_{21}$; and ϵ_i is independent of v_i, z_i, and x_i.

The likelihood function is straightforward to derive because we can write the joint density $f(y_{1i}, y_{2i} | x_i)$ as $f(y_{1i} | y_{2i}, x_i) f(y_{2i} | x_i)$. With one endogenous regressor,

$$\ln f(y_{2i} | x_i) = -\frac{1}{2} \left\{ \ln 2\pi + \ln \sigma_v^2 + \frac{(y_{2i} - x_i \Pi)^2}{\sigma_v^2} \right\}$$

and

$$\ln f(y_{1i} | y_{2i}, x_i) = \begin{cases} \ln \left\{ 1 - \Phi \left(\frac{m_i - a}{\sigma_{u|v}} \right) \right\} & y_{1i} = a \\ -\frac{1}{2} \left\{ \ln 2\pi + \ln \sigma_{u|v}^2 + \frac{(y_{1i} - m_i)^2}{\sigma_{u|v}^2} \right\} & a < y_{1i} < b \\ \ln \Phi \left(\frac{m_i - b}{\sigma_{u|v}} \right) & y_{1i} = b \end{cases}$$

where

$$m_i = z_i \delta + \alpha (y_{2i} - x_i \Pi)$$

and $\Phi(\cdot)$ is the normal distribution function so that the log likelihood for observation i is

$$\ln L_i = w_i \left\{ \ln f(y_{1i} | y_{2i}, x_i) + \ln f(y_{2i} | x_i) \right\}$$

where w_i is the weight for observation i or one if no weights were specified. Instead of estimating $\sigma_{u|v}$ and σ_v directly, we estimate $\ln \sigma_{u|v}$ and $\ln \sigma_v$.

For multiple endogenous regressors, we have

$$\ln f(y_{2i} | x_i) = -\frac{1}{2} \left(\ln 2\pi + \ln |\Sigma_{22}| + v_i' \Sigma_{22}^{-1} v_i \right)$$

and $\ln f(y_{1i} | y_{2i}, x_i)$ is the same as before, except that now

$$m_i = z_i \delta + (y_{2i} - x_i \Pi) \Sigma_{22}^{-1} \Sigma_{21}$$

Instead of maximizing the log-likelihood function with respect to Σ, we maximize with respect to the Cholesky decomposition S of Σ; that is, there exists a lower triangular matrix S such that $SS' = \Sigma$. This maximization ensures that Σ is positive definite, as a covariance matrix must be. Let

$$
S = \begin{bmatrix}
s_{11} & 0 & 0 & \cdots & 0 \\
s_{21} & s_{22} & 0 & \cdots & 0 \\
s_{31} & s_{32} & s_{33} & \cdots & 0 \\
\vdots & \vdots & \vdots & \ddots & \vdots \\
s_{p+1,1} & s_{p+1,2} & s_{p+1,3} & \cdots & s_{p+1,p+1}
\end{bmatrix}
$$

With maximum likelihood estimation, this command supports the Huber/White/sandwich estimator of the variance and its clustered version using vce(robust) and vce(cluster *clustvar*), respectively. See [P] **_robust**, particularly *Maximum likelihood estimators* and *Methods and formulas*.

The maximum likelihood version of ivtobit also supports estimation with survey data. For details on VCEs with survey data, see [SVY] **variance estimation**.

The two-step estimates are obtained using Newey's (1987) minimum chi-squared estimator. The procedure is identical to the one described in [R] **ivprobit**, except that tobit is used instead of probit.

Acknowledgments

The two-step estimator is based on the tobitiv command written by Jonah Gelbach, Department of Economics, Yale University, and the ivtobit command written by Joe Harkness, Institute of Policy Studies, Johns Hopkins University.

References

Finlay, K., and L. M. Magnusson. 2009. Implementing weak-instrument robust tests for a general class of instrumental-variables models. *Stata Journal* 9: 398–421.

Miranda, A., and S. Rabe-Hesketh. 2006. Maximum likelihood estimation of endogenous switching and sample selection models for binary, ordinal, and count variables. *Stata Journal* 6: 285–308.

Newey, W. K. 1987. Efficient estimation of limited dependent variable models with endogenous explanatory variables. *Journal of Econometrics* 36: 231–250.

Wooldridge, J. M. 2010. *Econometric Analysis of Cross Section and Panel Data*. 2nd ed. Cambridge, MA: MIT Press.

Also see

[R] **ivtobit postestimation** — Postestimation tools for ivtobit

[R] **gmm** — Generalized method of moments estimation

[R] **ivprobit** — Probit model with continuous endogenous regressors

[R] **ivregress** — Single-equation instrumental-variables regression

[R] **regress** — Linear regression

[R] **tobit** — Tobit regression

[SVY] **svy estimation** — Estimation commands for survey data

[XT] **xtintreg** — Random-effects interval-data regression models

[XT] **xttobit** — Random-effects tobit models

[U] **20 Estimation and postestimation commands**

Title

> **ivtobit postestimation** — Postestimation tools for ivtobit

Description

The following postestimation commands are available after `ivtobit`:

Command	Description
contrast	contrasts and ANOVA-style joint tests of estimates
estat[1]	AIC, BIC, VCE, and estimation sample summary
estat (svy)	postestimation statistics for survey data
estimates	cataloging estimation results
hausman	Hausman's specification test
lincom	point estimates, standard errors, testing, and inference for linear combinations of coefficients
lrtest[2]	likelihood-ratio test; not available with two-step estimator
margins	marginal means, predictive margins, marginal effects, and average marginal effects
marginsplot	graph the results from margins (profile plots, interaction plots, etc.)
nlcom	point estimates, standard errors, testing, and inference for nonlinear combinations of coefficients
predict	predictions, residuals, influence statistics, and other diagnostic measures
predictnl	point estimates, standard errors, testing, and inference for generalized predictions
pwcompare	pairwise comparisons of estimates
suest[1]	seemingly unrelated estimation
test	Wald tests of simple and composite linear hypotheses
testnl	Wald tests of nonlinear hypotheses

[1] `estat ic` and `suest` are not appropriate after `ivtobit, twostep`.
[2] `lrtest` is not appropriate with svy estimation results.

See the corresponding entries in the *Base Reference Manual* for details, but see [SVY] **estat** for details about `estat` (svy).

Syntax for predict

After ML or twostep

predict [*type*] *newvar* [*if*] [*in*] [, *statistic*]

After ML

predict [*type*] { *stub** | *newvarlist* } [*if*] [*in*] , <u>sc</u>ores

statistic	Description
Main	
xb	linear prediction; the default
stdp	standard error of the linear prediction
stdf	standard error of the forecast; not available with two-step estimator
<u>pr</u>(*a,b*)	$\Pr(a < y_j < b)$; not available with two-step estimator
e(*a,b*)	$E(y_j | a < y_j < b)$; not available with two-step estimator
<u>y</u>star(*a,b*)	$E(y_j^*)$, $y_j = \max\{a, \min(y_j, b)\}$; not available with two-step estimator

These statistics are available both in and out of sample; type predict ... if e(sample) ... if wanted only for the estimation sample.

stdf is not allowed with svy estimation results.

where *a* and *b* may be numbers or variables; *a* missing ($a \geq .$) means $-\infty$, and *b* missing ($b \geq .$) means $+\infty$; see [U] **12.2.1 Missing values**.

Menu

Statistics > Postestimation > Predictions, residuals, etc.

Options for predict

    ⌐ Main ⌐

xb, the default, calculates the linear prediction.

stdp calculates the standard error of the linear prediction. It can be thought of as the standard error of the predicted expected value or mean for the observation's covariate pattern. The standard error of the prediction is also referred to as the standard error of the fitted value.

stdf calculates the standard error of the forecast, which is the standard error of the point prediction for 1 observation. It is commonly referred to as the standard error of the future or forecast value. By construction, the standard errors produced by stdf are always larger than those produced by stdp; see *Methods and formulas* in [R] **regress postestimation**. stdf is not available with the two-step estimator.

pr(*a,b*) calculates $\Pr(a < \mathbf{x}_j\mathbf{b} + u_j < b)$, the probability that $y_j | \mathbf{x}_j$ would be observed in the interval (a, b).

    a and *b* may be specified as numbers or variable names; *lb* and *ub* are variable names;
    pr(20,30) calculates $\Pr(20 < \mathbf{x}_j\mathbf{b} + u_j < 30)$;
    pr(*lb,ub*) calculates $\Pr(lb < \mathbf{x}_j\mathbf{b} + u_j < ub)$; and
    pr(20,*ub*) calculates $\Pr(20 < \mathbf{x}_j\mathbf{b} + u_j < ub)$.

a missing ($a \geq .$) means $-\infty$; $\mathtt{pr(.,30)}$ calculates $\Pr(-\infty < \mathbf{x}_j\mathbf{b} + u_j < 30)$; $\mathtt{pr}(lb,30)$ calculates $\Pr(-\infty < \mathbf{x}_j\mathbf{b} + u_j < 30)$ in observations for which $lb \geq .$ and calculates $\Pr(lb < \mathbf{x}_j\mathbf{b} + u_j < 30)$ elsewhere.

b missing ($b \geq .$) means $+\infty$; $\mathtt{pr(20,.)}$ calculates $\Pr(+\infty > \mathbf{x}_j\mathbf{b} + u_j > 20)$; $\mathtt{pr}(20,ub)$ calculates $\Pr(+\infty > \mathbf{x}_j\mathbf{b} + u_j > 20)$ in observations for which $ub \geq .$ and calculates $\Pr(20 < \mathbf{x}_j\mathbf{b} + u_j < ub)$ elsewhere.

$\mathtt{pr}(a,b)$ is not available with the two-step estimator.

$\mathtt{e}(a,b)$ calculates $E(\mathbf{x}_j\mathbf{b} + u_j \mid a < \mathbf{x}_j\mathbf{b} + u_j < b)$, the expected value of $y_j | \mathbf{x}_j$ conditional on $y_j | \mathbf{x}_j$ being in the interval (a, b), meaning that $y_j | \mathbf{x}_j$ is truncated. a and b are specified as they are for $\mathtt{pr()}$. $\mathtt{e}(a,b)$ is not available with the two-step estimator.

$\mathtt{ystar}(a,b)$ calculates $E(y_j^*)$, where $y_j^* = a$ if $\mathbf{x}_j\mathbf{b} + u_j \leq a$, $y_j^* = b$ if $\mathbf{x}_j\mathbf{b} + u_j \geq b$, and $y_j^* = \mathbf{x}_j\mathbf{b} + u_j$ otherwise, meaning that y_j^* is censored. a and b are specified as they are for $\mathtt{pr()}$. $\mathtt{ystar}(a,b)$ is not available with the two-step estimator.

$\mathtt{scores}$, not available with $\mathtt{twostep}$, calculates equation-level score variables.

For models with one endogenous regressor, five new variables are created.

The first new variable will contain $\partial \ln L / \partial(\mathbf{z}_i\boldsymbol{\delta})$.

The second new variable will contain $\partial \ln L / \partial(\mathbf{x}_i\boldsymbol{\Pi})$.

The third new variable will contain $\partial \ln L / \partial\alpha$.

The fourth new variable will contain $\partial \ln L / \partial \ln\sigma_{u|v}$.

The fifth new variable will contain $\partial \ln L / \partial \ln\sigma_v$.

For models with p endogenous regressors, $p + \{(p+1)(p+2)\}/2 + 1$ new variables are created.

The first new variable will contain $\partial \ln L / \partial(\mathbf{z}_i\boldsymbol{\delta})$.

The second through $(p+1)$th new score variables will contain $\partial \ln L / \partial(\mathbf{x}_i\boldsymbol{\Pi}_k)$, $k = 1, \ldots, p$, where $\boldsymbol{\Pi}_k$ is the kth column of $\boldsymbol{\Pi}$.

The remaining score variables will contain the partial derivatives of $\ln L$ with respect to s_{11}, $s_{21}, \ldots, s_{p+1,1}, s_{22}, \ldots, s_{p+1,2}, \ldots, s_{p+1,p+1}$, where $s_{m,n}$ denotes the (m, n) element of the Cholesky decomposition of the error covariance matrix.

Remarks

Remarks are presented under the following headings:

> *Marginal effects*
> *Obtaining predicted values*

Marginal effects

▶ Example 1

We can obtain average marginal effects by using the $\mathtt{margins}$ command after $\mathtt{ivtobit}$. For the labor-supply model of example 1 in [R] **ivtobit**, suppose that we wanted to know the average marginal effects on the woman's expected income, conditional on her income being greater than \$10,000.

```
. use http://www.stata-press.com/data/r12/laborsup
. ivtobit fem_inc fem_educ kids (other_inc = male_educ), ll
  (output omitted)
. margins, dydx(*) predict(e(10, .))
```

Average marginal effects   Number of obs   =   500
Model VCE   : OIM

Expression   : E(fem_inc|fem_inc>10), predict(e(10, .))
dy/dx w.r.t. : other_inc fem_educ kids male_educ

	dy/dx	Delta-method Std. Err.	z	P>|z|	[95% Conf. Interval]	
other_inc	-.3420189	.0553591	-6.18	0.000	-.4505208	-.233517
fem_educ	1.237336	.1534025	8.07	0.000	.9366723	1.537999
kids	-1.252447	.2725166	-4.60	0.000	-1.78657	-.7183246
male_educ	0	(omitted)				

In our sample, increasing the number of children in the family by one decreases the expected wage by \$1,252 on average (wages in our dataset are measured in thousands of dollars). male_edu has no effect because it appears only as an instrument.

◁

Obtaining predicted values

After fitting your model using ivtobit, you can obtain the linear prediction and its standard error for both the estimation sample and other samples using the predict command. If you used the maximum likelihood estimator, you can also obtain conditional expected values of the observed and latent dependent variables, the standard error of the forecast, and the probability of observing the dependent variable in a specified interval. See [U] **20 Estimation and postestimation commands** and [R] **predict**.

Methods and formulas

All postestimation commands listed above are implemented as ado-files.

The linear prediction is calculated as $z_i\widehat{\delta}$, where $\widehat{\delta}$ is the estimated value of δ, and z_i and δ are defined in (1a) of [R] **ivtobit**. Expected values and probabilities are calculated using the same formulas as those used by the standard exogenous tobit model.

Also see

[R] **ivtobit** — Tobit model with continuous endogenous regressors

[U] **20 Estimation and postestimation commands**

Title

> **jackknife** — Jackknife estimation

Syntax

> jackknife *exp_list* $\left[\ ,\ options\ eform\_option\right]$: *command*

options	Description
Main	
<u>e</u>class	number of observations used is stored in e(N)
<u>r</u>class	number of observations used is stored in r(N)
n(*exp*)	specify *exp* that evaluates to the number of observations used
Options	
<u>cl</u>uster(*varlist*)	variables identifying sample clusters
<u>id</u>cluster(*newvar*)	create new cluster ID variable
<u>sav</u>ing(*filename*, ...)	save results to *filename*; save statistics in double precision; save results to *filename* every # replications
keep	keep pseudovalues
mse	use MSE formula for variance estimation
Reporting	
<u>l</u>evel(#)	set confidence level; default is level(95)
<u>nota</u>ble	suppress table of results
<u>noh</u>eader	suppress table header
<u>nol</u>egend	suppress table legend
<u>v</u>erbose	display the full table legend
nodots	suppress replication dots
<u>noi</u>sily	display any output from *command*
<u>t</u>race	trace *command*
<u>ti</u>tle(*text*)	use *text* as title for jackknife results
display_options	control column formats, row spacing, line width, and display of omitted variables and base and empty cells
Advanced	
nodrop	do not drop observations
reject(*exp*)	identify invalid results
eform_option	display coefficient table in exponentiated form
<u>coefl</u>egend	display legend instead of statistics

svy is allowed; see [SVY] **svy jackknife**.

All weight types supported by *command* are allowed except aweights; see [U] **11.1.6 weight**.

eform_option and coeflegend do not appear in the dialog box.

See [U] **20 Estimation and postestimation commands** for more capabilities of estimation commands.

exp_list contains	(*name*: *elist*)
	elist
	eexp
elist contains	*newvar* = (*exp*)
	(*exp*)
eexp is	*specname*
	[*eqno*]*specname*
specname is	_b
	_b[]
	_se
	_se[]
eqno is	# #
	name

exp is a standard Stata expression; see [U] **13 Functions and expressions**.

Distinguish between [], which are to be typed, and [ ], which indicate optional arguments.

Menu

Statistics > Resampling > Jackknife estimation

Description

jackknife performs jackknife estimation. Typing

  . jackknife *exp_list*: *command*

executes *command* once for each observation in the dataset, leaving the associated observation out of the calculations that make up *exp_list*.

command defines the statistical command to be executed. Most Stata commands and user-written programs can be used with jackknife, as long as they follow standard Stata syntax and allow the if qualifier; see [U] **11 Language syntax**. The by prefix may not be part of *command*.

exp_list specifies the statistics to be collected from the execution of *command*. If *command* changes the contents in e(b), *exp_list* is optional and defaults to _b.

Many estimation commands allow the vce(jackknife) option. For those commands, we recommend using vce(jackknife) over jackknife because the estimation command already handles clustering and other model-specific details for you. The jackknife prefix command is intended for use with nonestimation commands, such as summarize, user-written commands, or functions of coefficients.

jknife is a synonym for jackknife.

Options

  ⌐ Main ⌐

eclass, rclass, and n(*exp*) specify where *command* saves the number of observations on which it based the calculated results. We strongly advise you to specify one of these options.

eclass specifies that *command* save the number of observations in e(N).

rclass specifies that *command* save the number of observations in r(N).

n(*exp*) specifies an expression that evaluates to the number of observations used. Specifying n(r(N)) is equivalent to specifying the rclass option. Specifying n(e(N)) is equivalent to specifying the eclass option. If *command* saves the number of observations in r(N1), specify n(r(N1)).

If you specify no options, jackknife will assume eclass or rclass, depending on which of e(N) and r(N) is not missing (in that order). If both e(N) and r(N) are missing, jackknife assumes that all observations in the dataset contribute to the calculated result. If that assumption is incorrect, the reported standard errors will be incorrect. For instance, say that you specify

        . jackknife coef=_b[x2]: myreg y x1 x2 x3

where myreg uses e(n) instead of e(N) to identify the number of observations used in calculations. Further assume that observation 42 in the dataset has x3 equal to missing. The 42nd observation plays no role in obtaining the estimates, but jackknife has no way of knowing that and will use the wrong N. If, on the other hand, you specify

        . jackknife coef=_b[x2], n(e(n)): myreg y x1 x2 x3

jackknife will notice that observation 42 plays no role. The n(e(n)) option is specified because myreg is an estimation command but it saves the number of observations used in e(n) (instead of the standard e(N)). When jackknife runs the regression omitting the 42nd observation, jackknife will observe that e(n) has the same value as when jackknife previously ran the regression using all the observations. Thus jackknife will know that myreg did not use the observation.

⌐ Options ⌐

cluster(*varlist*) specifies the variables identifying sample clusters. If cluster() is specified, one cluster is left out of each call to *command*, instead of 1 observation.

idcluster(*newvar*) creates a new variable containing a unique integer identifier for each resampled cluster, starting at 1 and leading up to the number of clusters. This option may be specified only when the cluster() option is specified. idcluster() helps identify the cluster to which a pseudovalue belongs.

saving(*filename*[, *suboptions*]) creates a Stata data file (.dta file) consisting of (for each statistic in *exp_list*) a variable containing the replicates.

  double specifies that the results for each replication be stored as doubles, meaning 8-byte reals. By default, they are stored as floats, meaning 4-byte reals. This option may be used without the saving() option to compute the variance estimates by using double precision.

  every(*#*) specifies that results be written to disk every *#*th replication. every() should be specified only in conjunction with saving() when *command* takes a long time for each replication. This option will allow recovery of partial results should some other software crash your computer. See [P] **postfile**.

  replace specifies that *filename* be overwritten if it exists. This option does not appear in the dialog box.

keep specifies that new variables be added to the dataset containing the pseudovalues of the requested statistics. For instance, if you typed

        . jackknife coef=_b[x2], eclass keep: regress y x1 x2 x3

new variable coef would be added to the dataset containing the pseudovalues for _b[x2]. Let b be the value of _b[x2] when all observations are used to fit the model, and let $b(j)$ be the value when the jth observation is omitted. The pseudovalues are defined as

$$\text{pseudovalue}_j = N\{b - b(j)\} + b(j)$$

where N is the number of observations used to produce b.

When the cluster() option is specified, each cluster is given at most one nonmissing pseudovalue. The keep option implies the nodrop option.

mse specifies that jackknife compute the variance by using deviations of the replicates from the observed value of the statistics based on the entire dataset. By default, jackknife computes the variance by using deviations of the pseudovalues from their mean.

⌐ Reporting ⌐

level(#); see [R] **estimation options**.

notable suppresses the display of the table of results.

noheader suppresses the display of the table header. This option implies nolegend.

nolegend suppresses the display of the table legend. The table legend identifies the rows of the table with the expressions they represent.

verbose specifies that the full table legend be displayed. By default, coefficients and standard errors are not displayed.

nodots suppresses display of the replication dots. By default, one dot character is displayed for each successful replication. A red 'x' is displayed if *command* returns an error or if one of the values in *exp_list* is missing.

noisily specifies that any output from *command* be displayed. This option implies the nodots option.

trace causes a trace of the execution of *command* to be displayed. This option implies the noisily option.

title(*text*) specifies a title to be displayed above the table of jackknife results; the default title is Jackknife results or what is produced in e(title) by an estimation command.

display_options: noomitted, vsquish, noemptycells, baselevels, allbaselevels, cformat(%*fmt*), pformat(%*fmt*), sformat(%*fmt*), and nolstretch; see [R] **estimation options**.

⌐ Advanced ⌐

nodrop prevents observations outside e(sample) and the if and in qualifiers from being dropped before the data are resampled.

reject(*exp*) identifies an expression that indicates when results should be rejected. When *exp* is true, the resulting values are reset to missing values.

The following options are available with jackknife but are not shown in the dialog box:

eform_option causes the coefficient table to be displayed in exponentiated form; see [R] *eform_option*. *command* determines which *eform_option* is allowed (eform(*string*) and eform are always allowed).

command determines which of the following are allowed (eform(*string*) and eform are always allowed):

eform_option	Description
eform(*string*)	use *string* for the column title
eform	exponentiated coefficient, *string* is exp(b)
hr	hazard ratio, *string* is Haz. Ratio
shr	subhazard ratio, *string* is SHR
irr	incidence-rate ratio, *string* is IRR
or	odds ratio, *string* is Odds Ratio
rrr	relative-risk ratio, *string* is RRR

coeflegend; see [R] **estimation options**.

Remarks

Remarks are presented under the following headings:

Introduction
Jackknifed standard deviation
Collecting multiple statistics
Collecting coefficients

Introduction

Although the jackknife—developed in the late 1940s and early 1950s—is of largely historical interest today, it is still useful in searching for overly influential observations. This feature is often forgotten. In any case, the jackknife is

- an alternative, first-order unbiased estimator for a statistic;

- a data-dependent way to calculate the standard error of the statistic and to obtain significance levels and confidence intervals; and

- a way of producing measures called pseudovalues for each observation, reflecting the observation's influence on the overall statistic.

The idea behind the simplest form of the jackknife—the one implemented here—is to repeatedly calculate the statistic in question, each time omitting just one of the dataset's observations. Assume that our statistic of interest is the sample mean. Let y_j be the jth observation of our data on some measurement y, where $j = 1, \ldots, N$ and N is the sample size. If $\overline{y}$ is the sample mean of y using the entire dataset and $\overline{y}_{(j)}$ is the mean when the jth observation is omitted, then

$$\overline{y} = \frac{(N-1)\,\overline{y}_{(j)} + y_j}{N}$$

Solving for y_j, we obtain

$$y_j = N\,\overline{y} - (N-1)\,\overline{y}_{(j)}$$

These are the pseudovalues that jackknife calculates. To move this discussion beyond the sample mean, let $\widehat{\theta}$ be the value of our statistic (not necessarily the sample mean) using the entire dataset, and let $\widehat{\theta}_{(j)}$ be the computed value of our statistic with the jth observation omitted. The pseudovalue for the jth observation is

$$\widehat{\theta}_j^* = N\,\widehat{\theta} - (N-1)\,\widehat{\theta}_{(j)}$$

The mean of the pseudovalues is the alternative, first-order unbiased estimator mentioned above, and the standard error of the mean of the pseudovalues is an estimator for the standard error of $\widehat{\theta}$ (Tukey 1958).

When the cluster() option is given, clusters are omitted instead of observations, and N is the number of clusters instead of the sample size.

The jackknife estimate of variance has been largely replaced by the bootstrap estimate (see [R] **bootstrap**), which is widely viewed as more efficient and robust. The use of jackknife pseudovalues to detect outliers is too often forgotten and is something the bootstrap does not provide. See Mosteller and Tukey (1977, 133–163) and Mooney and Duval (1993, 22–27) for more information.

▷ Example 1

As our first example, we will show that the jackknife standard error of the sample mean is equivalent to the standard error of the sample mean computed using the classical formula in the ci command. We use the double option to compute the standard errors with the same precision as the ci command.

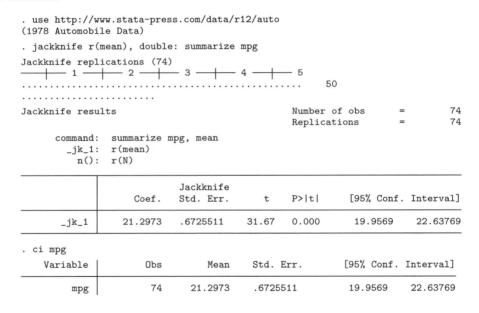

```
. use http://www.stata-press.com/data/r12/auto
(1978 Automobile Data)

. jackknife r(mean), double: summarize mpg

Jackknife replications (74)
───────── 1 ───── 2 ───── 3 ───── 4 ───── 5
..................................................  50
......................

Jackknife results                          Number of obs   =        74
                                           Replications    =        74

        command:  summarize mpg, mean
         _jk_1:  r(mean)
          n():  r(N)
```

	Coef.	Jackknife Std. Err.	t	P>|t|	[95% Conf. Interval]	
_jk_1	21.2973	.6725511	31.67	0.000	19.9569	22.63769

```
. ci mpg
```

Variable	Obs	Mean	Std. Err.	[95% Conf. Interval]	
mpg	74	21.2973	.6725511	19.9569	22.63769

◁

Jackknifed standard deviation

▷ Example 2

Mosteller and Tukey (1977, 139–140) request a 95% confidence interval for the standard deviation of the 11 values:

$$0.1, \quad 0.1, \quad 0.1, \quad 0.4, \quad 0.5, \quad 1.0, \quad 1.1, \quad 1.3, \quad 1.9, \quad 1.9, \quad 4.7$$

Stata's summarize command calculates the mean and standard deviation and saves them as r(mean) and r(sd). To obtain the jackknifed standard deviation of the 11 values and save the pseudovalues as a new variable, sd, we would type

```
. clear
. input x

              x
  1. 0.1
  2. 0.1
  3. 0.1
  4. 0.4
  5. 0.5
  6. 1.0
  7. 1.1
  8. 1.3
  9. 1.9
 10. 1.9
 11. 4.7
 12. end

. jackknife sd=r(sd), rclass keep: summarize x
(running summarize on estimation sample)
Jackknife replications (11)
────┼─── 1 ───┼─── 2 ───┼─── 3 ───┼─── 4 ───┼─── 5
. . . . . . . . . .
Jackknife results                          Number of obs    =        11
                                           Replications     =        11

        command:  summarize x
             sd:  r(sd)
            n():  r(N)
```

	Coef.	Jackknife Std. Err.	t	P>|t|	[95% Conf. Interval]
sd	1.343469	.624405	2.15	0.057	-.047792	2.73473

Interpreting the output, the standard deviation reported by summarize mpg is 1.34. The jackknife standard error is 0.62. The 95% confidence interval for the standard deviation is −0.048 to 2.73.

By specifying keep, jackknife creates in our dataset a new variable, sd, for the pseudovalues.

```
. list, sep(4)
```

	x	sd
1.	.1	1.139977
2.	.1	1.139977
3.	.1	1.139977
4.	.4	.8893147
5.	.5	.824267
6.	1	.632489
7.	1.1	.6203189
8.	1.3	.6218889
9.	1.9	.835419
10.	1.9	.835419
11.	4.7	7.703949

The jackknife estimate is the average of the sd variable, so sd contains the individual values of our statistic. We can see that the last observation is substantially larger than the others. The last observation is certainly an outlier, but whether that reflects the considerable information it contains or indicates that it should be excluded from analysis depends on the context of the problem. Here Mosteller and Tukey created the dataset by sampling from an exponential distribution, so the observation is informative.   ◁

▷ Example 3

Let's repeat the example above using the automobile dataset, obtaining the standard error of the standard deviation of mpg.

```
. use http://www.stata-press.com/data/r12/auto, clear
(1978 Automobile Data)
. jackknife sd=r(sd), rclass keep: summarize mpg
(running summarize on estimation sample)
Jackknife replications (74)
```
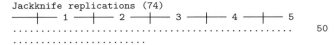
```
Jackknife results                         Number of obs      =        74
                                          Replications       =        74

        command:  summarize mpg
             sd:  r(sd)
            n():  r(N)
```

	Coef.	Jackknife Std. Err.	t	P>|t|	[95% Conf. Interval]
sd	5.785503	.6072509	9.53	0.000	4.575254 6.995753

Let's look at sd more carefully:

```
. summarize sd, detail
                     pseudovalues: r(sd)
```

	Percentiles	Smallest		
1%	2.870471	2.870471		
5%	2.870471	2.870471		
10%	2.906255	2.870471	Obs	74
25%	3.328489	2.870471	Sum of Wgt.	74
50%	3.948335		Mean	5.817374
		Largest	Std. Dev.	5.22377
75%	6.844418	17.34316		
90%	9.597018	19.7617	Variance	27.28777
95%	17.34316	19.7617	Skewness	4.07202
99%	38.60905	38.60905	Kurtosis	23.37823

```
. list make mpg sd if sd > 30
```

	make	mpg	sd
71.	VW Diesel	41	38.60905

Here the VW Diesel is the only diesel car in our dataset.   ◁

Collecting multiple statistics

▷ Example 4

jackknife is not limited to collecting just one statistic. For instance, we can use summarize, detail and then obtain the jackknife estimate of the standard deviation and skewness. summarize, detail saves the standard deviation in r(sd) and the skewness in r(skewness), so we might type

```
. use http://www.stata-press.com/data/r12/auto, clear
(1978 Automobile Data)
. jackknife sd=r(sd) skew=r(skewness), rclass: summarize mpg, detail
(running summarize on estimation sample)
Jackknife replications (74)
────┼─── 1 ───┼─── 2 ───┼─── 3 ───┼─── 4 ───┼─── 5
.................................................  50
......................
Jackknife results                        Number of obs    =        74
                                         Replications     =        74
      command:  summarize mpg, detail
           sd:  r(sd)
         skew:  r(skewness)
          n():  r(N)
```

	Coef.	Jackknife Std. Err.	t	P>|t|	[95% Conf. Interval]
sd	5.785503	.6072509	9.53	0.000	4.575254	6.995753
skew	.9487176	.3367242	2.82	0.006	.2776272	1.619808

◁

Collecting coefficients

▷ Example 5

jackknife can also collect coefficients from estimation commands. For instance, using auto.dta, we might wish to obtain the jackknife standard errors of the coefficients from a regression in which we model the mileage of a car by its weight and trunk space. To do this, we could refer to the coefficients as _b[weight], _b[trunk], _se[weight], and _se[trunk] in the exp_list, or we could simply use the extended expressions _b. In fact, jackknife assumes _b by default when used with estimation commands.

```
. use http://www.stata-press.com/data/r12/auto
(1978 Automobile Data)

. jackknife: regress mpg weight trunk
(running regress on estimation sample)

Jackknife replications (74)
──────┼─── 1 ──┼─── 2 ──┼─── 3 ──┼─── 4 ──┼─── 5
..................................................   50
......................
```

```
Linear regression                          Number of obs   =        74
                                           Replications    =        74
                                           F(   2,     73) =     78.10
                                           Prob > F        =    0.0000
                                           R-squared       =    0.6543
                                           Adj R-squared   =    0.6446
                                           Root MSE        =    3.4492
```

mpg	Coef.	Jackknife Std. Err.	t	P>|t|	[95% Conf. Interval]	
weight	-.0056527	.0010216	-5.53	0.000	-.0076887	-.0036167
trunk	-.096229	.1486236	-0.65	0.519	-.3924354	.1999773
_cons	39.68913	1.873324	21.19	0.000	35.9556	43.42266

If you are going to use `jackknife` to estimate standard errors of model coefficients, we recommend using the `vce(jackknife)` option when it is allowed with the estimation command; see [R] *vce_option*.

```
. regress mpg weight trunk, vce(jackknife, nodots)
Linear regression                          Number of obs   =        74
                                           Replications    =        74
                                           F(   2,     73) =     78.10
                                           Prob > F        =    0.0000
                                           R-squared       =    0.6543
                                           Adj R-squared   =    0.6446
                                           Root MSE        =    3.4492
```

mpg	Coef.	Jackknife Std. Err.	t	P>|t|	[95% Conf. Interval]	
weight	-.0056527	.0010216	-5.53	0.000	-.0076887	-.0036167
trunk	-.096229	.1486236	-0.65	0.519	-.3924354	.1999773
_cons	39.68913	1.873324	21.19	0.000	35.9556	43.42266

◁

John Wilder Tukey (1915–2000) was born in Massachusetts. He studied chemistry at Brown and mathematics at Princeton and afterward worked at both Princeton and Bell Labs, as well as being involved in a great many government projects, consultancies, and committees. He made outstanding contributions to several areas of statistics, including time series, multiple comparisons, robust statistics, and exploratory data analysis. Tukey was extraordinarily energetic and inventive, not least in his use of terminology: he is credited with inventing the terms bit and software, in addition to ANOVA, boxplot, data analysis, hat matrix, jackknife, stem-and-leaf plot, trimming, and winsorizing, among many others. Tukey's direct and indirect impacts mark him as one of the greatest statisticians of all time.

Saved results

jknife saves the following in e():

Scalars
e(N)	sample size
e(N_reps)	number of complete replications
e(N_misreps)	number of incomplete replications
e(N_clust)	number of clusters
e(k_eq)	number of equations in e(b)
e(k_extra)	number of extra equations
e(k_exp)	number of expressions
e(k_eexp)	number of extended expressions (_b or _se)
e(df_r)	degrees of freedom

Macros
e(cmdname)	command name from *command*
e(cmd)	same as e(cmdname) or jackknife
e(command)	*command*
e(cmdline)	command as typed
e(prefix)	jackknife
e(wtype)	weight type
e(wexp)	weight expression
e(title)	title in estimation output
e(cluster)	cluster variables
e(pseudo)	new variables containing pseudovalues
e(nfunction)	e(N), r(N), n() option, or empty
e(exp#)	expression for the #th statistic
e(mse)	from mse option
e(vce)	jackknife
e(vcetype)	title used to label Std. Err.
e(properties)	b V

Matrices
e(b)	observed statistics
e(b_jk)	jackknife estimates
e(V)	jackknife variance–covariance matrix
e(V_modelbased)	model-based variance

When *exp_list* is _b, jackknife will also carry forward most of the results already in e() from *command*.

Methods and formulas

jackknife is implemented as an ado-file.

Let $\widehat{\theta}$ be the observed value of the statistic, that is, the value of the statistic calculated using the original dataset. Let $\widehat{\theta}_{(j)}$ be the value of the statistic computed by leaving out the jth observation (or cluster); thus $j = 1, 2, \ldots, N$ identifies an individual observation (or cluster), and N is the total number of observations (or clusters). The jth pseudovalue is given by

$$\widehat{\theta}_j^* = \widehat{\theta}_{(j)} + N\{\widehat{\theta} - \widehat{\theta}_{(j)}\}$$

When the mse option is specified, the standard error is estimated as

$$\widehat{se} = \left\{ \frac{N-1}{N} \sum_{j=1}^{N} (\widehat{\theta}_{(j)} - \widehat{\theta})^2 \right\}^{1/2}$$

and the jackknife estimate is

$$\bar{\theta}_{(.)} = \frac{1}{N} \sum_{j=1}^{N} \widehat{\theta}_{(j)}$$

Otherwise, the standard error is estimated as

$$\widehat{se} = \left\{ \frac{1}{N(N-1)} \sum_{j=1}^{N} (\widehat{\theta}_j^* - \bar{\theta}^*)^2 \right\}^{1/2} \qquad \bar{\theta}^* = \frac{1}{N} \sum_{j=1}^{N} \widehat{\theta}_j^*$$

where $\bar{\theta}^*$ is the jackknife estimate. The variance–covariance matrix is similarly computed.

References

Brillinger, D. R. 2002. John W. Tukey: His life and professional contributions. *Annals of Statistics* 30: 1535–1575.

Gould, W. W. 1995. sg34: Jackknife estimation. *Stata Technical Bulletin* 24: 25–29. Reprinted in *Stata Technical Bulletin Reprints*, vol. 4, pp. 165–170. College Station, TX: Stata Press.

Mooney, C. Z., and R. D. Duval. 1993. *Bootstrapping: A Nonparametric Approach to Statistical Inference*. Newbury Park, CA: Sage.

Mosteller, F., and J. W. Tukey. 1977. *Data Analysis and Regression: A Second Course in Statistics*. Reading, MA: Addison–Wesley.

Tukey, J. W. 1958. Bias and confidence in not-quite large samples. Abstract in *Annals of Mathematical Statistics* 29: 614.

Also see

[R] **jackknife postestimation** — Postestimation tools for jackknife

[R] **bootstrap** — Bootstrap sampling and estimation

[R] **permute** — Monte Carlo permutation tests

[R] **simulate** — Monte Carlo simulations

[SVY] **svy jackknife** — Jackknife estimation for survey data

[U] **13.5 Accessing coefficients and standard errors**

[U] **13.6 Accessing results from Stata commands**

[U] **20 Estimation and postestimation commands**

---
jackknife postestimation — Postestimation tools for jackknife
---

Description

The following postestimation commands are available after `jackknife`:

Command	Description
*`contrast`	contrasts and ANOVA-style joint tests of estimates
`estat`	AIC, BIC, VCE, and estimation sample summary
`estimates`	cataloging estimation results
`lincom`	point estimates, standard errors, testing, and inference for linear combinations of coefficients
*`margins`	marginal means, predictive margins, marginal effects, and average marginal effects
*`marginsplot`	graph the results from margins (profile plots, interaction plots, etc.)
`nlcom`	point estimates, standard errors, testing, and inference for nonlinear combinations of coefficients
*`predict`	predictions, residuals, influence statistics, and other diagnostic measures
*`predictnl`	point estimates, standard errors, testing, and inference for nonlinear combinations of coefficients
*`pwcompare`	pairwise comparisons of estimates
`test`	Wald tests of simple and composite linear hypotheses
`testnl`	Wald tests of nonlinear hypotheses

*This postestimation command is allowed only if it may be used after *command*.

See the corresponding entries in the *Base Reference Manual* for details.

Syntax for predict

The syntax of `predict` (and whether `predict` is even allowed) following `jackknife` depends on the *command* used with `jackknife`.

Methods and formulas

All postestimation commands listed above are implemented as ado-files.

Also see

[R] **jackknife** — Jackknife estimation

[U] **20 Estimation and postestimation commands**

Title

> **kappa** — Interrater agreement

Syntax

Interrater agreement, two unique raters

> kap *varname*$_1$ *varname*$_2$ $[$ *if* $]$ $[$ *in* $]$ $[$ *weight* $]$ $[$, *options* $]$

Weights for weighting disagreements

> kapwgt *wgtid* $[$ 1 \ # 1 $[$ \ # # 1 ... $]$ $]$

Interrater agreement, nonunique raters, variables record ratings for each rater

> kap *varname*$_1$ *varname*$_2$ *varname*$_3$ $[$ ... $]$ $[$ *if* $]$ $[$ *in* $]$ $[$ *weight* $]$

Interrater agreement, nonunique raters, variables record frequency of ratings

> kappa *varlist* $[$ *if* $]$ $[$ *in* $]$

options	Description
Main	
<u>tab</u>	display table of assessments
<u>wgt</u>(*wgtid*)	specify how to weight disagreements; see *Options* for alternatives
<u>ab</u>solute	treat rating categories as absolute

fweights are allowed; see [U] **11.1.6 weight**.

Menu

kap: two unique raters

Statistics > Epidemiology and related > Other > Interrater agreement, two unique raters

kapwgt

Statistics > Epidemiology and related > Other > Define weights for the above (kap)

kap: nonunique raters

Statistics > Epidemiology and related > Other > Interrater agreement, nonunique raters

kappa

Statistics > Epidemiology and related > Other > Interrater agreement, nonunique raters with frequencies

Description

kap (first syntax) calculates the kappa-statistic measure of interrater agreement when there are two unique raters and two or more ratings.

kapwgt defines weights for use by kap in measuring the importance of disagreements.

kap (second syntax) and kappa calculate the kappa-statistic measure when there are two or more (nonunique) raters and two outcomes, more than two outcomes when the number of raters is fixed, and more than two outcomes when the number of raters varies. kap (second syntax) and kappa produce the same results; they merely differ in how they expect the data to be organized.

kap assumes that each observation is a subject. $varname_1$ contains the ratings by the first rater, $varname_2$ by the second rater, and so on.

kappa also assumes that each observation is a subject. The variables, however, record the frequencies with which ratings were assigned. The first variable records the number of times the first rating was assigned, the second variable records the number of times the second rating was assigned, and so on.

Options

___Main___

tab displays a tabulation of the assessments by the two raters.

wgt(*wgtid*) specifies that *wgtid* be used to weight disagreements. You can define your own weights by using kapwgt; wgt() then specifies the name of the user-defined matrix. For instance, you might define

    . kapwgt mine 1 \ .8 1 \ 0 .8 1 \ 0 0 .8 1

and then

    . kap rata ratb, wgt(mine)

Also, two prerecorded weights are available.

wgt(w) specifies weights $1 - |i - j|/(k - 1)$, where i and j index the rows and columns of the ratings by the two raters and k is the maximum number of possible ratings.

wgt(w2) specifies weights $1 - \{(i - j)/(k - 1)\}^2$.

absolute is relevant only if wgt() is also specified. The absolute option modifies how i, j, and k are defined and how corresponding entries are found in a user-defined weighting matrix. When absolute is not specified, i and j refer to the row and column index, not to the ratings themselves. Say that the ratings are recorded as $\{0, 1, 1.5, 2\}$. There are four ratings; $k = 4$, and i and j are still 1, 2, 3, and 4 in the formulas above. Index 3, for instance, corresponds to rating $= 1.5$. This system is convenient but can, with some data, lead to difficulties.

When absolute is specified, all ratings must be integers, and they must be coded from the set $\{1, 2, 3, \ldots\}$. Not all values need be used; integer values that do not occur are simply assumed to be unobserved.

Remarks

Remarks are presented under the following headings:

> *Two raters*
> *More than two raters*

The kappa-statistic measure of agreement is scaled to be 0 when the amount of agreement is what would be expected to be observed by chance and 1 when there is perfect agreement. For intermediate values, Landis and Koch (1977a, 165) suggest the following interpretations:

below 0.0	Poor
0.00–0.20	Slight
0.21–0.40	Fair
0.41–0.60	Moderate
0.61–0.80	Substantial
0.81–1.00	Almost perfect

Two raters

▷ Example 1

Consider the classification by two radiologists of 85 xeromammograms as normal, benign disease, suspicion of cancer, or cancer (a subset of the data from Boyd et al. [1982] and discussed in the context of kappa in Altman [1991, 403–405]).

```
. use http://www.stata-press.com/data/r12/rate2
(Altman p. 403)
. tabulate rada radb
```

Radiologist A's assessment	Radiologist B's assessment				Total
	normal	benign	suspect	cancer	
normal	21	12	0	0	33
benign	4	17	1	0	22
suspect	3	9	15	2	29
cancer	0	0	0	1	1
Total	28	38	16	3	85

Our dataset contains two variables: `rada`, radiologist A's assessment, and `radb`, radiologist B's assessment. Each observation is a patient.

We can obtain the kappa measure of interrater agreement by typing

```
. kap rada radb
```

Agreement	Expected Agreement	Kappa	Std. Err.	Z	Prob>Z
63.53%	30.82%	0.4728	0.0694	6.81	0.0000

If each radiologist had made his determination randomly (but with probabilities equal to the overall proportions), we would expect the two radiologists to agree on 30.8% of the patients. In fact, they agreed on 63.5% of the patients, or 47.3% of the way between random agreement and perfect agreement. The amount of agreement indicates that we can reject the hypothesis that they are making their determinations randomly.

◁

▷ Example 2: Weighted kappa, prerecorded weight w

There is a difference between two radiologists disagreeing about whether a xeromammogram indicates cancer or the suspicion of cancer and disagreeing about whether it indicates cancer or is normal. The weighted kappa attempts to deal with this. kap provides two "prerecorded" weights, w and w2:

```
. kap rada radb, wgt(w)
Ratings weighted by:
    1.0000    0.6667    0.3333    0.0000
    0.6667    1.0000    0.6667    0.3333
    0.3333    0.6667    1.0000    0.6667
    0.0000    0.3333    0.6667    1.0000

                 Expected
Agreement       Agreement      Kappa    Std. Err.          Z    Prob>Z

  86.67%          69.11%       0.5684     0.0788         7.22   0.0000
```

The w weights are given by $1 - |i - j|/(k - 1)$, where i and j index the rows of columns of the ratings by the two raters and k is the maximum number of possible ratings. The weighting matrix is printed above the table. Here the rows and columns of the 4×4 matrix correspond to the ratings normal, benign, suspicious, and cancerous.

A weight of 1 indicates that an observation should count as perfect agreement. The matrix has 1s down the diagonals—when both radiologists make the same assessment, they are in agreement. A weight of, say, 0.6667 means that they are in two-thirds agreement. In our matrix, they get that score if they are "one apart"—one radiologist assesses cancer and the other is merely suspicious, or one is suspicious and the other says benign, and so on. An entry of 0.3333 means that they are in one-third agreement, or, if you prefer, two-thirds disagreement. That is the score attached when they are "two apart". Finally, they are in complete disagreement when the weight is zero, which happens only when they are three apart—one says cancer and the other says normal.   ◁

▷ Example 3: Weighted kappa, prerecorded weight w2

The other prerecorded weight is w2, where the weights are given by $1 - \{(i - j)/(k - 1)\}^2$:

```
. kap rada radb, wgt(w2)
Ratings weighted by:
    1.0000    0.8889    0.5556    0.0000
    0.8889    1.0000    0.8889    0.5556
    0.5556    0.8889    1.0000    0.8889
    0.0000    0.5556    0.8889    1.0000

                 Expected
Agreement       Agreement      Kappa    Std. Err.          Z    Prob>Z

  94.77%          84.09%       0.6714     0.1079         6.22   0.0000
```

The w2 weight makes the categories even more alike and is probably inappropriate here.   ◁

▷ Example 4: Weighted kappa, user-defined weights

In addition to using prerecorded weights, we can define our own weights with the kapwgt command. For instance, we might feel that suspicious and cancerous are reasonably similar, that benign and normal are reasonably similar, but that the suspicious/cancerous group is nothing like the benign/normal group:

```
. kapwgt xm 1 \ .8 1 \ 0 0 1 \ 0 0 .8 1
. kapwgt xm
1.0000
0.8000 1.0000
0.0000 0.0000 1.0000
0.0000 0.0000 0.8000 1.0000
```

We name the weights xm, and after the weight name, we enter the lower triangle of the weighting matrix, using \ to separate rows. We have four outcomes, so we continued entering numbers until we had defined the fourth row of the weighting matrix. If we type kapwgt followed by a name and nothing else, it shows us the weights recorded under that name. Satisfied that we have entered them correctly, we now use the weights to recalculate kappa:

```
. kap rada radb, wgt(xm)
Ratings weighted by:
    1.0000    0.8000    0.0000    0.0000
    0.8000    1.0000    0.0000    0.0000
    0.0000    0.0000    1.0000    0.8000
    0.0000    0.0000    0.8000    1.0000
```

| | Expected | | | | |
Agreement	Agreement	Kappa	Std. Err.	Z	Prob>Z
80.47%	52.67%	0.5874	0.0865	6.79	0.0000

◁

❑ Technical note

In addition to using weights for weighting the differences in categories, you can specify Stata's traditional weights for weighting the data. In the examples above, we have 85 observations in our dataset—one for each patient. If we only knew the table of outcomes—that there were 21 patients rated normal by both radiologists, etc.—it would be easier to enter the table into Stata and work from it. The easiest way to enter the data is with tabi; see [R] **tabulate twoway**.

```
. tabi 21 12 0 0 \ 4 17 1 0 \ 3 9 15 2 \ 0 0 0 1, replace
```

| | | | col | | |
row	1	2	3	4	Total
1	21	12	0	0	33
2	4	17	1	0	22
3	3	9	15	2	29
4	0	0	0	1	1
Total	28	38	16	3	85

Pearson chi2(9) =   77.8111   Pr = 0.000

tabi reported the Pearson χ^2 for this table, but we do not care about it. The important thing is that, with the replace option, tabi left the table in memory:

```
. list in 1/5
```

	row	col	pop
1.	1	1	21
2.	1	2	12
3.	1	3	0
4.	1	4	0
5.	2	1	4

The variable `row` is radiologist A's assessment, `col` is radiologist B's assessment, and `pop` is the number so assessed by both. Thus

```
. kap row col [freq=pop]
```

Agreement	Expected Agreement	Kappa	Std. Err.	Z	Prob>Z
63.53%	30.82%	0.4728	0.0694	6.81	0.0000

If we are going to keep these data, the names `row` and `col` are not indicative of what the data reflect. We could type (see [U] **12.6 Dataset, variable, and value labels**)

```
. rename row rada
. rename col radb
. label var rada "Radiologist A's assessment"
. label var radb "Radiologist B's assessment"
. label define assess 1 normal 2 benign 3 suspect 4 cancer
. label values rada assess
. label values radb assess
. label data "Altman p. 403"
```

`kap`'s `tab` option, which can be used with or without weighted data, shows the table of assessments:

```
. kap rada radb [freq=pop], tab
```

Radiologist A's assessment	Radiologist B's assessment				Total
	normal	benign	suspect	cancer	
normal	21	12	0	0	33
benign	4	17	1	0	22
suspect	3	9	15	2	29
cancer	0	0	0	1	1
Total	28	38	16	3	85

Agreement	Expected Agreement	Kappa	Std. Err.	Z	Prob>Z
63.53%	30.82%	0.4728	0.0694	6.81	0.0000

❑

❏ Technical note

You have data on individual patients. There are two raters, and the possible ratings are 1, 2, 3, and 4, but neither rater ever used rating 3:

```
. use http://www.stata-press.com/data/r12/rate2no3, clear
. tabulate ratera raterb
```

ratera	raterb 1	2	4	Total
1	6	4	3	13
2	5	3	3	11
4	1	1	26	28
Total	12	8	32	52

Here `kap` would determine that the ratings are from the set $\{1, 2, 4\}$ because those were the only values observed. `kap` would expect a user-defined weighting matrix to be 3×3, and if it were not, `kap` would issue an error message. In the formula-based weights, the calculation would be based on $i, j = 1, 2, 3$ corresponding to the three observed ratings $\{1, 2, 4\}$.

Specifying the `absolute` option would clarify that the ratings are 1, 2, 3, and 4; it just so happens that rating 3 was never assigned. If a user-defined weighting matrix were also specified, `kap` would expect it to be 4×4 or larger (larger because we can think of the ratings being 1, 2, 3, 4, 5, ... and it just so happens that ratings 5, 6, ... were never observed, just as rating 3 was not observed). In the formula-based weights, the calculation would be based on $i, j = 1, 2, 4$.

```
. kap ratera raterb, wgt(w)
Ratings weighted by:
   1.0000   0.5000   0.0000
   0.5000   1.0000   0.5000
   0.0000   0.5000   1.0000
```

Agreement	Expected Agreement	Kappa	Std. Err.	Z	Prob>Z
79.81%	57.17%	0.5285	0.1169	4.52	0.0000

```
. kap ratera raterb, wgt(w) absolute
Ratings weighted by:
   1.0000   0.6667   0.0000
   0.6667   1.0000   0.3333
   0.0000   0.3333   1.0000
```

Agreement	Expected Agreement	Kappa	Std. Err.	Z	Prob>Z
81.41%	55.08%	0.5862	0.1209	4.85	0.0000

If all conceivable ratings are observed in the data, specifying `absolute` makes no difference. For instance, if rater A assigns ratings $\{1, 2, 4\}$ and rater B assigns $\{1, 2, 3, 4\}$, the complete set of assigned ratings is $\{1, 2, 3, 4\}$, the same that `absolute` would specify. Without `absolute`, it makes no difference whether the ratings are coded $\{1, 2, 3, 4\}$, $\{0, 1, 2, 3\}$, $\{1, 7, 9, 100\}$, $\{0, 1, 1.5, 2.0\}$, or otherwise.

❏

More than two raters

For more than two raters, the mathematics are such that the two raters are not considered unique. For instance, if there are three raters, there is no assumption that the three raters who rate the first subject are the same as the three raters who rate the second. Although we call this the "more than two raters" case, it can be used with two raters when the raters' identities vary.

The nonunique rater case can be usefully broken down into three subcases: 1) there are two possible ratings, which we will call positive and negative; 2) there are more than two possible ratings, but the number of raters per subject is the same for all subjects; and 3) there are more than two possible ratings, and the number of raters per subject varies. kappa handles all these cases. To emphasize that there is no assumption of constant identity of raters across subjects, the variables specified contain counts of the number of raters rating the subject into a particular category.

> Jacob Cohen (1923–1998) was born in New York City. After studying psychology at City College of New York and New York University, he worked as a medical psychologist until 1959 when he became a full professor in the Department of Psychology at New York University. He made many contributions to research methods, including the kappa measure. He persistently emphasized the value of multiple regression and the importance of power and of measuring effects rather than testing significance.

Example 5: Two ratings

Fleiss, Levin, and Paik (2003, 612) offers the following hypothetical ratings by different sets of raters on 25 subjects:

Subject	No. of raters	No. of pos. ratings	Subject	No. of raters	No. of pos. ratings
1	2	2	14	4	3
2	2	0	15	2	0
3	3	2	16	2	2
4	4	3	17	3	1
5	3	3	18	2	1
6	4	1	19	4	1
7	3	0	20	5	4
8	5	0	21	3	2
9	2	0	22	4	0
10	4	4	23	3	0
11	5	5	24	3	3
12	3	3	25	2	2
13	4	4			

We have entered these data into Stata, and the variables are called subject, raters, and pos. kappa, however, requires that we specify variables containing the number of positive ratings and negative ratings, that is, pos and raters-pos:

```
. use http://www.stata-press.com/data/r12/p612
. gen neg = raters-pos
. kappa pos neg
```

Two-outcomes, multiple raters:

Kappa	Z	Prob>Z
0.5415	5.28	0.0000

We would have obtained the same results if we had typed kappa neg pos.

◁

▷ Example 6: More than two ratings, constant number of raters, kappa

Each of 10 subjects is rated into one of three categories by five raters (Fleiss, Levin, and Paik 2003, 615):

```
. use http://www.stata-press.com/data/r12/p615, clear
. list
```

	subject	cat1	cat2	cat3
1.	1	1	4	0
2.	2	2	0	3
3.	3	0	0	5
4.	4	4	0	1
5.	5	3	0	2
6.	6	1	4	0
7.	7	5	0	0
8.	8	0	4	1
9.	9	1	0	4
10.	10	3	0	2

We obtain the kappa statistic:

```
. kappa cat1-cat3
```

Outcome	Kappa	Z	Prob>Z
cat1	0.2917	2.92	0.0018
cat2	0.6711	6.71	0.0000
cat3	0.3490	3.49	0.0002
combined	0.4179	5.83	0.0000

The first part of the output shows the results of calculating kappa for each of the categories separately against an amalgam of the remaining categories. For instance, the cat1 line is the two-rating kappa, where positive is cat1 and negative is cat2 or cat3. The test statistic, however, is calculated differently (see *Methods and formulas*). The combined kappa is the appropriately weighted average of the individual kappas. There is considerably less agreement about the rating of subjects into the first category than there is for the second.

◁

▷ Example 7: More than two ratings, constant number of raters, kap

Now suppose that we have the same data as in the previous example but that the data are organized differently:

```
. use http://www.stata-press.com/data/r12/p615b
. list
```

	subject	rater1	rater2	rater3	rater4	rater5
1.	1	1	2	2	2	2
2.	2	1	1	3	3	3
3.	3	3	3	3	3	3
4.	4	1	1	1	1	3
5.	5	1	1	1	3	3
6.	6	1	2	2	2	2
7.	7	1	1	1	1	1
8.	8	2	2	2	2	3
9.	9	1	3	3	3	3
10.	10	1	1	1	3	3

Here we would use kap rather than kappa because the variables record ratings for each rater.

```
. kap rater1 rater2 rater3 rater4 rater5
There are 5 raters per subject:
```

Outcome	Kappa	Z	Prob>Z
1	0.2917	2.92	0.0018
2	0.6711	6.71	0.0000
3	0.3490	3.49	0.0002
combined	0.4179	5.83	0.0000

It does not matter which rater is which when there are more than two raters.

◁

Example 8: More than two ratings, varying number of raters, kappa

In this unfortunate case, kappa can be calculated, but there is no test statistic for testing against $\kappa > 0$. We do nothing differently—kappa calculates the total number of raters for each subject, and, if it is not a constant, kappa suppresses the calculation of test statistics.

```
. use http://www.stata-press.com/data/r12/rvary
. list
```

	subject	cat1	cat2	cat3
1.	1	1	3	0
2.	2	2	0	3
3.	3	0	0	5
4.	4	4	0	1
5.	5	3	0	2
6.	6	1	4	0
7.	7	5	0	0
8.	8	0	4	1
9.	9	1	0	2
10.	10	3	0	2

```
. kappa cat1-cat3
```

Outcome	Kappa	Z	Prob>Z
cat1	0.2685	.	.
cat2	0.6457	.	.
cat3	0.2938	.	.
combined	0.3816	.	.

Note: number of ratings per subject vary; cannot calculate test
        statistics.

◁

▷ Example 9: More than two ratings, varying number of raters, kap

This case is similar to the previous example, but the data are organized differently:

```
. use http://www.stata-press.com/data/r12/rvary2
. list
```

	subject	rater1	rater2	rater3	rater4	rater5
1.	1	1	2	2	.	2
2.	2	1	1	3	3	3
3.	3	3	3	3	3	3
4.	4	1	1	1	1	3
5.	5	1	1	1	3	3
6.	6	1	2	2	2	2
7.	7	1	1	1	1	1
8.	8	2	2	2	2	3
9.	9	1	3	.	.	3
10.	10	1	1	1	3	3

Here we specify kap instead of kappa because the variables record ratings for each rater.

```
. kap rater1-rater5
```
There are between 3 and 5 (median = 5.00) raters per subject:

Outcome	Kappa	Z	Prob>Z
1	0.2685	.	.
2	0.6457	.	.
3	0.2938	.	.
combined	0.3816	.	.

Note: number of ratings per subject vary; cannot calculate test
        statistics.

◁

Saved results

kap and kappa save the following in r():

Scalars

r(N)	number of subjects (kap only)	r(kappa)	kappa
r(prop_o)	observed proportion of agreement (kap only)	r(z)	z statistic
r(prop_e)	expected proportion of agreement (kap only)	r(se)	standard error for kappa statistic

Methods and formulas

kap, kapwgt, and kappa are implemented as ado-files.

The kappa statistic was first proposed by Cohen (1960). The generalization for weights reflecting the relative seriousness of each possible disagreement is due to Cohen (1968). The analysis-of-variance approach for $k = 2$ and $m \geq 2$ is due to Landis and Koch (1977b). See Altman (1991, 403–409) or Dunn (2000, chap. 2) for an introductory treatment and Fleiss, Levin, and Paik (2003, chap. 18) for a more detailed treatment. All formulas below are as presented in Fleiss, Levin, and Paik (2003). Let m be the number of raters, and let k be the number of rating outcomes.

Methods and formulas are presented under the following headings:

kap: m = 2
kappa: m > 2, k = 2
kappa: m > 2, k > 2

kap: m = 2

Define w_{ij} ($i = 1, \ldots, k$ and $j = 1, \ldots, k$) as the weights for agreement and disagreement (wgt()), or, if the data are not weighted, define $w_{ii} = 1$ and $w_{ij} = 0$ for $i \neq j$. If wgt(w) is specified, $w_{ij} = 1 - |i - j|/(k - 1)$. If wgt(w2) is specified, $w_{ij} = 1 - \{(i - j)/(k - 1)\}^2$.

The observed proportion of agreement is

$$p_o = \sum_{i=1}^{k} \sum_{j=1}^{k} w_{ij} p_{ij}$$

where p_{ij} is the fraction of ratings i by the first rater and j by the second. The expected proportion of agreement is

$$p_e = \sum_{i=1}^{k} \sum_{j=1}^{k} w_{ij} p_{i.} p_{.j}$$

where $p_{i.} = \sum_j p_{ij}$ and $p_{.j} = \sum_i p_{ij}$.

Kappa is given by $\widehat{\kappa} = (p_o - p_e)/(1 - p_e)$.

The standard error of $\widehat{\kappa}$ for testing against 0 is

$$\widehat{s}_0 = \frac{1}{(1 - p_e)\sqrt{n}} \left(\left[\sum_i \sum_j p_{i.} p_{.j} \{w_{ij} - (\overline{w}_{i.} + \overline{w}_{.j})\}^2 \right] - p_e^2 \right)^{1/2}$$

where n is the number of subjects being rated, $\overline{w}_{i.} = \sum_j p_{.j} w_{ij}$, and $\overline{w}_{.j} = \sum_i p_{i.} w_{ij}$. The test statistic $Z = \widehat{\kappa}/\widehat{s}_0$ is assumed to be distributed $N(0, 1)$.

kappa: m > 2, k = 2

Each subject i, $i = 1, \ldots, n$, is found by x_i of m_i raters to be positive (the choice as to what is labeled positive is arbitrary).

The overall proportion of positive ratings is $\overline{p} = \sum_i x_i/(n\overline{m})$, where $\overline{m} = \sum_i m_i/n$. The between-subjects mean square is (approximately)

$$B = \frac{1}{n} \sum_i \frac{(x_i - m_i\overline{p})^2}{m_i}$$

and the within-subject mean square is

$$W = \frac{1}{n(\overline{m} - 1)} \sum_i \frac{x_i(m_i - x_i)}{m_i}$$

Kappa is then defined as

$$\widehat{\kappa} = \frac{B - W}{B + (\overline{m} - 1)W}$$

The standard error for testing against 0 (Fleiss and Cuzick 1979) is approximately equal to and is calculated as

$$\widehat{s}_0 = \frac{1}{(\overline{m} - 1)\sqrt{n\overline{m}_H}} \left\{ 2(\overline{m}_H - 1) + \frac{(\overline{m} - \overline{m}_H)(1 - 4\overline{p}\overline{q})}{\overline{m}\,\overline{p}\,\overline{q}} \right\}^{1/2}$$

where $\overline{m}_H$ is the harmonic mean of m_i and $\overline{q} = 1 - \overline{p}$.

The test statistic $Z = \widehat{\kappa}/\widehat{s}_0$ is assumed to be distributed $N(0, 1)$.

kappa: m > 2, k > 2

Let x_{ij} be the number of ratings on subject i, $i = 1, \ldots, n$, into category j, $j = 1, \ldots, k$. Define $\overline{p}_j$ as the overall proportion of ratings in category j, $\overline{q}_j = 1 - \overline{p}_j$, and let $\widehat{\kappa}_j$ be the kappa statistic given above for $k = 2$ when category j is compared with the amalgam of all other categories. Kappa is

$$\overline{\kappa} = \frac{\sum_j \overline{p}_j \overline{q}_j \widehat{\kappa}_j}{\sum_j \overline{p}_j \overline{q}_j}$$

(Landis and Koch 1977b). In the case where the number of raters per subject, $\sum_j x_{ij}$, is a constant m for all i, Fleiss, Nee, and Landis (1979) derived the following formulas for the approximate standard errors. The standard error for testing $\widehat{\kappa}_j$ against 0 is

$$\widehat{s}_j = \left\{ \frac{2}{nm(m - 1)} \right\}^{1/2}$$

and the standard error for testing $\overline{\kappa}$ is

$$\overline{s} = \frac{\sqrt{2}}{\sum_j \overline{p}_j \overline{q}_j \sqrt{nm(m-1)}} \left\{ \left(\sum_j \overline{p}_j \overline{q}_j \right)^2 - \sum_j \overline{p}_j \overline{q}_j (\overline{q}_j - \overline{p}_j) \right\}^{1/2}$$

References

Abramson, J. H., and Z. H. Abramson. 2001. *Making Sense of Data: A Self-Instruction Manual on the Interpretation of Epidemiological Data.* 3rd ed. New York: Oxford University Press.

Altman, D. G. 1991. *Practical Statistics for Medical Research.* London: Chapman & Hall/CRC.

Boyd, N. F., C. Wolfson, M. Moskowitz, T. Carlile, M. Petitclerc, H. A. Ferri, E. Fishell, A. Gregoire, M. Kiernan, J. D. Longley, I. S. Simor, and A. B. Miller. 1982. Observer variation in the interpretation of xeromammograms. *Journal of the National Cancer Institute* 68: 357–363.

Campbell, M. J., D. Machin, and S. J. Walters. 2007. *Medical Statistics: A Textbook for the Health Sciences.* 4th ed. Chichester, UK: Wiley.

Cohen, J. 1960. A coefficient of agreement for nominal scales. *Educational and Psychological Measurement* 20: 37–46.

———. 1968. Weighted kappa: Nominal scale agreement with provision for scaled disagreement or partial credit. *Psychological Bulletin* 70: 213–220.

Cox, N. J. 2006. Assessing agreement of measurements and predictions in geomorphology. *Geomorphology* 76: 332–346.

Dunn, G. 2000. *Statistics in Psychiatry.* London: Arnold.

Fleiss, J. L., and J. Cuzick. 1979. The reliability of dichotomous judgments: Unequal numbers of judges per subject. *Applied Psychological Measurement* 3: 537–542.

Fleiss, J. L., B. Levin, and M. C. Paik. 2003. *Statistical Methods for Rates and Proportions.* 3rd ed. New York: Wiley.

Fleiss, J. L., J. C. M. Nee, and J. R. Landis. 1979. Large sample variance of kappa in the case of different sets of raters. *Psychological Bulletin* 86: 974–977.

Gould, W. W. 1997. stata49: Interrater agreement. *Stata Technical Bulletin* 40: 2–8. Reprinted in *Stata Technical Bulletin Reprints*, vol. 7, pp. 20–28. College Station, TX: Stata Press.

Landis, J. R., and G. G. Koch. 1977a. The measurement of observer agreement for categorical data. *Biometrics* 33: 159–174.

———. 1977b. A one-way components of variance model for categorical data. *Biometrics* 33: 671–679.

Reichenheim, M. E. 2000. sxd3: Sample size for the kappa-statistic of interrater agreement. *Stata Technical Bulletin* 58: 41–45. Reprinted in *Stata Technical Bulletin Reprints*, vol. 10, pp. 382–387. College Station, TX: Stata Press.

———. 2004. Confidence intervals for the kappa statistic. *Stata Journal* 4: 421–428.

Shrout, P. E. 2001. Jacob Cohen (1923–1998). *American Psychologist* 56: 166.

Steichen, T. J., and N. J. Cox. 1998a. sg84: Concordance correlation coefficient. *Stata Technical Bulletin* 43: 35–39. Reprinted in *Stata Technical Bulletin Reprints*, vol. 8, pp. 137–143. College Station, TX: Stata Press.

———. 1998b. sg84.1: Concordance correlation coefficient, revisited. *Stata Technical Bulletin* 45: 21–23. Reprinted in *Stata Technical Bulletin Reprints*, vol. 8, pp. 143–145. College Station, TX: Stata Press.

———. 2000a. sg84.3: Concordance correlation coefficient: Minor corrections. *Stata Technical Bulletin* 58: 9. Reprinted in *Stata Technical Bulletin Reprints*, vol. 10, p. 137. College Station, TX: Stata Press.

———. 2000b. sg84.2: Concordance correlation coefficient: Update for Stata 6. *Stata Technical Bulletin* 54: 25–26. Reprinted in *Stata Technical Bulletin Reprints*, vol. 9, pp. 169–170. College Station, TX: Stata Press.

———. 2002. A note on the concordance correlation coefficient. *Stata Journal* 2: 183–189.

Title

kdensity — Univariate kernel density estimation

Syntax

kdensity *varname* [*if*] [*in*] [*weight*] [, *options*]

options	Description
Main	
kernel(*kernel*)	specify kernel function; default is kernel(epanechnikov)
bwidth(*#*)	half-width of kernel
generate(*newvar_x newvar_d*)	store the estimation points in *newvar_x* and the density estimate in *newvar_d*
n(*#*)	estimate density using *#* points; default is min(N, 50)
at(*var_x*)	estimate density using the values specified by *var_x*
nograph	suppress graph
Kernel plot	
cline_options	affect rendition of the plotted kernel density estimate
Density plots	
normal	add normal density to the graph
normopts(*cline_options*)	affect rendition of normal density
student(*#*)	add Student's t density with *#* degrees of freedom to the graph
stopts(*cline_options*)	affect rendition of the Student's t density
Add plots	
addplot(*plot*)	add other plots to the generated graph
Y axis, X axis, Titles, Legend, Overall	
twoway_options	any options other than by() documented in [G-3] ***twoway_options***

kernel	Description
epanechnikov	Epanechnikov kernel function; the default
epan2	alternative Epanechnikov kernel function
biweight	biweight kernel function
cosine	cosine trace kernel function
gaussian	Gaussian kernel function
parzen	Parzen kernel function
rectangle	rectangle kernel function
triangle	triangle kernel function

fweights, aweights, and iweights are allowed; see [U] **11.1.6 weight**.

884

Menu

Statistics > Nonparametric analysis > Kernel density estimation

Description

kdensity produces kernel density estimates and graphs the result.

Options

```
┌─ Main ─────────────────────────────────────────────────────────────────
```

kernel(*kernel*) specifies the kernel function for use in calculating the kernel density estimate. The default kernel is the Epanechnikov kernel (epanechnikov).

bwidth(#) specifies the half-width of the kernel, the width of the density window around each point. If bwidth() is not specified, the "optimal" width is calculated and used. The optimal width is the width that would minimize the mean integrated squared error if the data were Gaussian and a Gaussian kernel were used, so it is not optimal in any global sense. In fact, for multimodal and highly skewed densities, this width is usually too wide and oversmooths the density (Silverman 1992).

generate(*newvar$_x$ newvar$_d$*) stores the results of the estimation. *newvar$_x$* will contain the points at which the density is estimated. *newvar$_d$* will contain the density estimate.

n(#) specifies the number of points at which the density estimate is to be evaluated. The default is $\min(N, 50)$, where N is the number of observations in memory.

at(*var$_x$*) specifies a variable that contains the values at which the density should be estimated. This option allows you to more easily obtain density estimates for different variables or different subsamples of a variable and then overlay the estimated densities for comparison.

nograph suppresses the graph. This option is often used with the generate() option.

```
┌─ Kernel plot ──────────────────────────────────────────────────────────
```

cline_options affect the rendition of the plotted kernel density estimate. See [G-3] ***cline_options***.

```
┌─ Density plots ────────────────────────────────────────────────────────
```

normal requests that a normal density be overlaid on the density estimate for comparison.

normopts(*cline_options*) specifies details about the rendition of the normal curve, such as the color and style of line used. See [G-3] ***cline_options***.

student(#) specifies that a Student's t density with # degrees of freedom be overlaid on the density estimate for comparison.

stopts(*cline_options*) affects the rendition of the Student's t density. See [G-3] ***cline_options***.

```
┌─ Add plots ────────────────────────────────────────────────────────────
```

addplot(*plot*) provides a way to add other plots to the generated graph. See [G-3] ***addplot_option***.

twoway_options are any of the options documented in [G-3] *twoway_options*, excluding by(). These include options for titling the graph (see [G-3] *title_options*) and for saving the graph to disk (see [G-3] *saving_option*).

Remarks

Kernel density estimators approximate the density $f(x)$ from observations on x. Histograms do this, too, and the histogram itself is a kind of kernel density estimate. The data are divided into nonoverlapping intervals, and counts are made of the number of data points within each interval. Histograms are bar graphs that depict these frequency counts—the bar is centered at the midpoint of each interval—and its height reflects the average number of data points in the interval.

In more general kernel density estimates, the range is still divided into intervals, and estimates of the density at the center of intervals are produced. One difference is that the intervals are allowed to overlap. We can think of sliding the interval—called a window—along the range of the data and collecting the center-point density estimates. The second difference is that, rather than merely counting the number of observations in a window, a kernel density estimator assigns a weight between 0 and 1—based on the distance from the center of the window—and sums the weighted values. The function that determines these weights is called the kernel.

Kernel density estimates have the advantages of being smooth and of being independent of the choice of origin (corresponding to the location of the bins in a histogram).

See Salgado-Ugarte, Shimizu, and Taniuchi (1993) and Fox (1990) for discussions of kernel density estimators that stress their use as exploratory data-analysis tools.

Cox (2007) gives a lucid introductory tutorial on kernel density estimation with several Stata produced examples. He provides tips and tricks for working with skewed or bounded distributions and applying the same techniques to estimate the intensity function of a point process.

▷ Example 1: Histogram and kernel density estimate

Goeden (1978) reports data consisting of 316 length observations of coral trout. We wish to investigate the underlying density of the lengths. To begin on familiar ground, we might draw a histogram. In [R] **histogram**, we suggest setting the bins to $\min(\sqrt{n}, 10 \cdot \log_{10} n)$, which for $n = 316$ is roughly 18:

```
. use http://www.stata-press.com/data/r12/trocolen
. histogram length, bin(18)
(bin=18, start=226, width=19.777778)
```

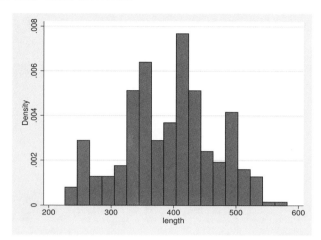

The kernel density estimate, on the other hand, is smooth.

```
. kdensity length
```

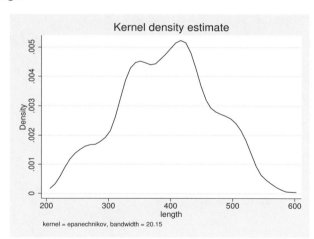

Kernel density estimators are, however, sensitive to an assumption, just as are histograms. In histograms, we specify a number of bins. For kernel density estimators, we specify a width. In the graph above, we used the default width. `kdensity` is smarter than `twoway histogram` in that its default width is not a fixed constant. Even so, the default width is not necessarily best.

`kdensity` saves the width in the returned scalar `bwidth`, so typing `display r(bwidth)` reveals it. Doing this, we discover that the width is approximately 20.

Widths are similar to the inverse of the number of bins in a histogram in that smaller widths provide more detail. The units of the width are the units of x, the variable being analyzed. The width is specified as a half-width, meaning that the kernel density estimator with half-width 20 corresponds to sliding a window of size 40 across the data.

We can specify half-widths for ourselves by using the `bwidth()` option. Smaller widths do not smooth the density as much:

. kdensity length, bwidth(10)

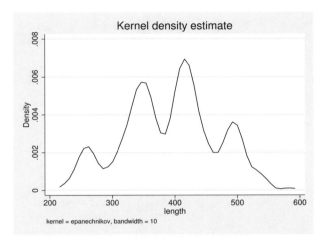

. kdensity length, bwidth(15)

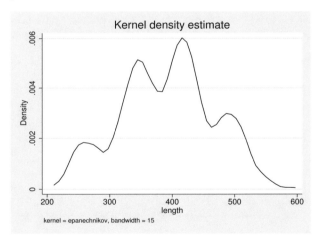

◁

▷ Example 2: Different kernels can produce different results

When widths are held constant, different kernels can produce surprisingly different results. This is really an attribute of the kernel and width combination; for a given width, some kernels are more sensitive than others at identifying peaks in the density estimate.

We can see this when using a dataset with lots of peaks. In the automobile dataset, we characterize the density of `weight`, the weight of the vehicles. Below we compare the Epanechnikov and Parzen kernels.

```
. use http://www.stata-press.com/data/r12/auto
(1978 Automobile Data)
. kdensity weight, kernel(epanechnikov) nograph generate(x epan)
. kdensity weight, kernel(parzen) nograph generate(x2 parzen)
. label var epan "Epanechnikov density estimate"
. label var parzen "Parzen density estimate"
. line epan parzen x, sort ytitle(Density) legend(cols(1))
```

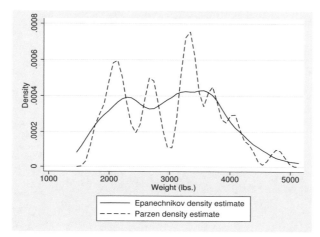

We did not specify a width, so we obtained the default width. That width is not a function of the selected kernel, but of the data. See *Methods and formulas* for the calculation of the optimal width.

◁

Example 3: Density with overlaid normal density

In examining the density estimates, we may wish to overlay a normal density or a Student's t density for comparison. Using automobile weights, we can get an idea of the distance from normality by using the normal option.

```
. kdensity weight, kernel(epanechnikov) normal
```

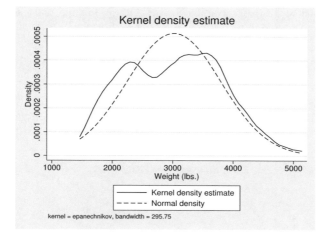

◁

▷ Example 4: Compare two densities

We also may want to compare two or more densities. In this example, we will compare the density estimates of the weights for the foreign and domestic cars.

```
. use http://www.stata-press.com/data/r12/auto, clear
(1978 Automobile Data)
. kdensity weight, nograph generate(x fx)
. kdensity weight if foreign==0, nograph generate(fx0) at(x)
. kdensity weight if foreign==1, nograph generate(fx1) at(x)
. label var fx0 "Domestic cars"
. label var fx1 "Foreign cars"
. line fx0 fx1 x, sort ytitle(Density)
```

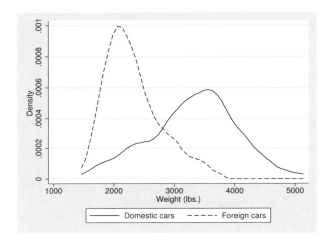

◁

❑ Technical note

Although all the examples we included had densities of less than 1, the density may exceed 1.

The probability density $f(x)$ of a continuous variable, x, has the units and dimensions of the reciprocal of x. If x is measured in meters, $f(x)$ has units 1/meter. Thus the density is not measured on a probability scale, so it is possible for $f(x)$ to exceed 1.

To see this, think of a uniform density on the interval 0 to 1. The area under the density curve is 1: this is the product of the density, which is constant at 1, and the range, which is 1. If the variable is then transformed by doubling, the area under the curve remains 1 and is the product of the density, constant at 0.5, and the range, which is 2. Conversely, if the variable is transformed by halving, the area under the curve also remains at 1 and is the product of the density, constant at 2, and the range, which is 0.5. (Strictly, the range is measured in certain units, and the density is measured in the reciprocal of those units, so the units cancel on multiplication.)

❑

Saved results

kdensity saves the following in r():

Scalars

r(bwidth)	kernel bandwidth
r(n)	number of points at which the estimate was evaluated
r(scale)	density bin width

Macros

r(kernel)	name of kernel

Methods and formulas

kdensity is implemented as an ado-file.

A kernel density estimate is formed by summing the weighted values calculated with the kernel function K, as in

$$\widehat{f}_K = \frac{1}{qh} \sum_{i=1}^{n} w_i K \left(\frac{x - X_i}{h} \right)$$

where $q = \sum_i w_i$ if weights are frequency weights (fweight) or analytic weights (aweight), and $q = 1$ if weights are importance weights (iweights). Analytic weights are rescaled so that $\sum_i w_i = n$ (see [U] **11 Language syntax**). If weights are not used, then $w_i = 1$, for $i = 1, \ldots, n$. kdensity includes seven different kernel functions. The Epanechnikov is the default function if no other kernel is specified and is the most efficient in minimizing the mean integrated squared error.

Kernel	Formula	
Biweight	$K[z] = \begin{cases} \frac{15}{16}(1 - z^2)^2 \\ 0 \end{cases}$	if $\|z\| < 1$ otherwise
Cosine	$K[z] = \begin{cases} 1 + \cos(2\pi z) \\ 0 \end{cases}$	if $\|z\| < 1/2$ otherwise
Epanechnikov	$K[z] = \begin{cases} \frac{3}{4}(1 - \frac{1}{5}z^2)/\sqrt{5} \\ 0 \end{cases}$	if $\|z\| < \sqrt{5}$ otherwise
Epan2	$K[z] = \begin{cases} \frac{3}{4}(1 - z^2) \\ 0 \end{cases}$	if $\|z\| < 1$ otherwise
Gaussian	$K[z] = \frac{1}{\sqrt{2\pi}} e^{-z^2/2}$	
Parzen	$K[z] = \begin{cases} \frac{4}{3} - 8z^2 + 8\|z\|^3 \\ 8(1 - \|z\|)^3/3 \\ 0 \end{cases}$	if $\|z\| \leq 1/2$ if $1/2 < \|z\| \leq 1$ otherwise
Rectangular	$K[z] = \begin{cases} 1/2 \\ 0 \end{cases}$	if $\|z\| < 1$ otherwise
Triangular	$K[z] = \begin{cases} 1 - \|z\| \\ 0 \end{cases}$	if $\|z\| < 1$ otherwise

From the definitions given in the table, we can see that the choice of h will drive how many values are included in estimating the density at each point. This value is called the *window width* or *bandwidth*. If the window width is not specified, it is determined as

$$m = \min\left(\sqrt{\text{variance}_x}, \quad \frac{\text{interquartile range}_x}{1.349}\right)$$

$$h = \frac{0.9m}{n^{1/5}}$$

where x is the variable for which we wish to estimate the kernel and n is the number of observations.

Most researchers agree that the choice of kernel is not as important as the choice of bandwidth. There is a great deal of literature on choosing bandwidths under various conditions; see, for example, Parzen (1962) or Tapia and Thompson (1978). Also see Newton (1988) for a comparison with sample spectral density estimation in time-series applications.

Acknowledgments

We gratefully acknowledge the previous work by Isaías H. Salgado-Ugarte of Universidad Nacional Autónoma de México, and Makoto Shimizu and Toru Taniuchi of the University of Tokyo; see Salgado-Ugarte, Shimizu, and Taniuchi (1993). Their article provides a good overview of the subject of univariate kernel density estimation and presents arguments for its use in exploratory data analysis.

References

Cox, N. J. 2005. Speaking Stata: Density probability plots. *Stata Journal* 5: 259–273.

———. 2007. Kernel estimation as a basic tool for geomorphological data analysis. *Earth Surface Processes and Landforms* 32: 1902–1912.

Fiorio, C. V. 2004. Confidence intervals for kernel density estimation. *Stata Journal* 4: 168–179.

Fox, J. 1990. Describing univariate distributions. In *Modern Methods of Data Analysis*, ed. J. Fox and J. S. Long, 58–125. Newbury Park, CA: Sage.

Goeden, G. B. 1978. A monograph of the coral trout, *Plectropomus leopardus* (Lacépède). *Queensland Fisheries Services Research Bulletin* 1: 1–42.

Kohler, U., and F. Kreuter. 2009. *Data Analysis Using Stata*. 2nd ed. College Station, TX: Stata Press.

Newton, H. J. 1988. *TIMESLAB: A Time Series Analysis Laboratory*. Belmont, CA: Wadsworth.

Parzen, E. 1962. On estimation of a probability density function and mode. *Annals of Mathematical Statistics* 33: 1065–1076.

Royston, P., and N. J. Cox. 2005. A multivariable scatterplot smoother. *Stata Journal* 5: 405–412.

Salgado-Ugarte, I. H., and M. A. Pérez-Hernández. 2003. Exploring the use of variable bandwidth kernel density estimators. *Stata Journal* 3: 133–147.

Salgado-Ugarte, I. H., M. Shimizu, and T. Taniuchi. 1993. snp6: Exploring the shape of univariate data using kernel density estimators. *Stata Technical Bulletin* 16: 8–19. Reprinted in *Stata Technical Bulletin Reprints*, vol. 3, pp. 155–173. College Station, TX: Stata Press.

———. 1995a. snp6.1: ASH, WARPing, and kernel density estimation for univariate data. *Stata Technical Bulletin* 26: 23–31. Reprinted in *Stata Technical Bulletin Reprints*, vol. 5, pp. 161–172. College Station, TX: Stata Press.

———. 1995b. snp6.2: Practical rules for bandwidth selection in univariate density estimation. *Stata Technical Bulletin* 27: 5–19. Reprinted in *Stata Technical Bulletin Reprints*, vol. 5, pp. 172–190. College Station, TX: Stata Press.

———. 1997. snp13: Nonparametric assessment of multimodality for univariate data. *Stata Technical Bulletin* 38: 27–35. Reprinted in *Stata Technical Bulletin Reprints*, vol. 7, pp. 232–243. College Station, TX: Stata Press.

Scott, D. W. 1992. *Multivariate Density Estimation: Theory, Practice, and Visualization*. New York: Wiley.

Silverman, B. W. 1992. *Density Estimation for Statistics and Data Analysis*. London: Chapman & Hall.

Simonoff, J. S. 1996. *Smoothing Methods in Statistics*. New York: Springer.

Steichen, T. J. 1998. gr33: Violin plots. *Stata Technical Bulletin* 46: 13–18. Reprinted in *Stata Technical Bulletin Reprints*, vol. 8, pp. 57–65. College Station, TX: Stata Press.

Tapia, R. A., and J. R. Thompson. 1978. *Nonparametric Probability Density Estimation*. Baltimore: Johns Hopkins University Press.

Van Kerm, P. 2003. Adaptive kernel density estimation. *Stata Journal* 3: 148–156.

Wand, M. P., and M. C. Jones. 1995. *Kernel Smoothing*. London: Chapman & Hall.

Also see

[R] **histogram** — Histograms for continuous and categorical variables

Title

> **ksmirnov** — Kolmogorov–Smirnov equality-of-distributions test

Syntax

One-sample Kolmogorov–Smirnov test

> ksmirnov *varname* = *exp* $\begin{bmatrix} if \end{bmatrix}$ $\begin{bmatrix} in \end{bmatrix}$

Two-sample Kolmogorov–Smirnov test

> ksmirnov *varname* $\begin{bmatrix} if \end{bmatrix}$ $\begin{bmatrix} in \end{bmatrix}$, by(*groupvar*) $\begin{bmatrix} \underline{e}xact \end{bmatrix}$

Menu

one-sample

Statistics > Nonparametric analysis > Tests of hypotheses > One-sample Kolmogorov-Smirnov test

two-sample

Statistics > Nonparametric analysis > Tests of hypotheses > Two-sample Kolmogorov-Smirnov test

Description

ksmirnov performs one- and two-sample Kolmogorov–Smirnov tests of the equality of distributions. In the first syntax, *varname* is the variable whose distribution is being tested, and *exp* must evaluate to the corresponding (theoretical) cumulative. In the second syntax, *groupvar* must take on two distinct values. The distribution of *varname* for the first value of *groupvar* is compared with that of the second value.

When testing for normality, please see [R] **sktest** and [R] **swilk**.

Options for two-sample test

‾‾‾‾⌐ Main ⌐‾‾

by(*groupvar*) is required. It specifies a binary variable that identifies the two groups.

exact specifies that the exact p-value be computed. This may take a long time if $n > 50$.

Remarks

▷ Example 1: Two-sample test

Say that we have data on x that resulted from two different experiments, labeled as group==1 and group==2. Our data contain

894

```
. use http://www.stata-press.com/data/r12/ksxmpl
. list
```

	group	x
1.	2	2
2.	1	0
3.	2	3
4.	1	4
5.	1	5
6.	2	8
7.	2	10

We wish to use the two-sample Kolmogorov–Smirnov test to determine if there are any differences in the distribution of x for these two groups:

```
. ksmirnov x, by(group)
Two-sample Kolmogorov-Smirnov test for equality of distribution functions
  Smaller group       D       P-value  Corrected

  1:                0.5000    0.424
  2:               -0.1667    0.909
  Combined K-S:     0.5000    0.785        0.735
```

The first line tests the hypothesis that x for group 1 contains *smaller* values than for group 2. The largest difference between the distribution functions is 0.5. The approximate p-value for this is 0.424, which is not significant.

The second line tests the hypothesis that x for group 1 contains *larger* values than for group 2. The largest difference between the distribution functions in this direction is 0.1667. The approximate p-value for this small difference is 0.909.

Finally, the approximate p-value for the combined test is 0.785, corrected to 0.735. The p-values ksmirnov calculates are based on the asymptotic distributions derived by Smirnov (1933). These approximations are not good for small samples ($n < 50$). They are too conservative—real p-values tend to be substantially smaller. We have also included a less conservative approximation for the nondirectional hypothesis based on an empirical continuity correction—the 0.735 reported in the third column.

That number, too, is only an approximation. An exact value can be calculated using the exact option:

```
. ksmirnov x, by(group) exact
Two-sample Kolmogorov-Smirnov test for equality of distribution functions
  Smaller group       D       P-value     Exact

  1:                0.5000    0.424
  2:               -0.1667    0.909
  Combined K-S:     0.5000    0.785      0.657
```

◁

Example 2: One-sample test

Let's now test whether x in the example above is distributed normally. Kolmogorov–Smirnov is not a particularly powerful test in testing for normality, and we do not endorse such use of it; see [R] **sktest** and [R] **swilk** for better tests.

In any case, we will test against a normal distribution with the same mean and standard deviation:

```
. summarize x
    Variable │      Obs        Mean    Std. Dev.        Min         Max
─────────────┼─────────────────────────────────────────────────────────
           x │        7    4.571429    3.457222           0          10
. ksmirnov x = normal((x-4.571429)/3.457222)
```

One-sample Kolmogorov-Smirnov test against theoretical distribution
            normal((x-4.571429)/3.457222)

Smaller group	D	P-value	Corrected
x:	0.1650	0.683	
Cumulative:	-0.1250	0.803	
Combined K-S:	0.1650	0.991	0.978

Because Stata has no way of knowing that we based this calculation on the calculated mean and standard deviation of x, the test statistics will be slightly conservative in addition to being approximations. Nevertheless, they clearly indicate that the data cannot be distinguished from normally distributed data.

◁

Saved results

ksmirnov saves the following in r():

Scalars

r(D_1)	D from line 1	r(D)	combined D
r(p_1)	p-value from line 1	r(p)	combined p-value
r(D_2)	D from line 2	r(p_cor)	corrected combined p-value
r(p_2)	p-value from line 2	r(p_exact)	exact combined p-value

Macros

r(group1)	name of group from line 1	r(group2)	name of group from line 2

Methods and formulas

ksmirnov is implemented as an ado-file.

In general, the Kolmogorov–Smirnov test (Kolmogorov 1933; Smirnov 1933; also see Conover [1999], 428–465) is not very powerful against differences in the tails of distributions. In return for this, it is fairly powerful for alternative hypotheses that involve lumpiness or clustering in the data.

The directional hypotheses are evaluated with the statistics

$$D^+ = \max_x \left\{ F(x) - G(x) \right\}$$
$$D^- = \min_x \left\{ F(x) - G(x) \right\}$$

where $F(x)$ and $G(x)$ are the empirical distribution functions for the sample being compared. The combined statistic is

$$D = \max \left(|D^+|, |D^-| \right)$$

The p-value for this statistic may be obtained by evaluating the asymptotic limiting distribution. Let m be the sample size for the first sample, and let n be the sample size for the second sample. Smirnov (1933) shows that

$$\lim_{m,n\to\infty} \Pr\left\{ \sqrt{mn/(m+n)}D_{m,n} \leq z \right\} = 1 - 2\sum_{i=1}^{\infty} (-1)^{i-1} \exp\left(-2i^2 z^2\right)$$

The first five terms form the approximation P_a used by Stata. The exact p-value is calculated by a counting algorithm; see Gibbons and Chakraborti (2011, 236–238). A corrected p-value was obtained by modifying the asymptotic p-value by using a numerical approximation technique:

$$Z = \Phi^{-1}(P_a) + 1.04/\min(m,n) + 2.09/\max(m,n) - 1.35/\sqrt{mn/(m+n)}$$
$$p\text{-value} = \Phi(Z)$$

where $\Phi(\cdot)$ is the cumulative normal distribution.

Andrei Nikolayevich Kolmogorov (1903–1987), of Russia, was one of the great mathematicians of the twentieth century, making outstanding contributions in many different branches, including set theory, measure theory, probability and statistics, approximation theory, functional analysis, classical dynamics, and theory of turbulence. He was a faculty member at Moscow State University for more than 60 years.

Nikolai Vasilyevich Smirnov (1900–1966) was a Russian statistician whose work included contributions in nonparametric statistics, order statistics, and goodness of fit. After army service and the study of philosophy and philology, he turned to mathematics and eventually rose to be head of mathematical statistics at the Steklov Mathematical Institute in Moscow.

References

Aivazian, S. A. 1997. Smirnov, Nikolai Vasil'yevich. In *Leading Personalities in Statistical Sciences: From the Seventeenth Century to the Present*, ed. N. L. Johnson and S. Kotz, 208–210. New York: Wiley.

Conover, W. J. 1999. *Practical Nonparametric Statistics*. 3rd ed. New York: Wiley.

Gibbons, J. D., and S. Chakraborti. 2011. *Nonparametric Statistical Inference*. 5th ed. Boca Raton, FL: Chapman & Hall/CRC.

Goerg, S. J., and J. Kaiser. 2009. Nonparametric testing of distributions—the Epss–Singleton two-sample test using the empirical characteristic function. *Stata Journal* 9: 454–465.

Jann, B. 2008. Multinomial goodness-of-fit: Large-sample tests with survey design correction and exact tests for small samples. *Stata Journal* 8: 147–169.

Johnson, N. L., and S. Kotz. 1997. Kolmogorov, Andrei Nikolayevich. In *Leading Personalities in Statistical Sciences: From the Seventeenth Century to the Present*, ed. N. L. Johnson and S. Kotz, 255–256. New York: Wiley.

Kolmogorov, A. N. 1933. Sulla determinazione empirica di una legge di distribuzione. *Giornale dell' Istituto Italiano degli Attuari* 4: 83–91.

Riffenburgh, R. H. 2005. *Statistics in Medicine*. 2nd ed. New York: Elsevier.

Smirnov, N. V. 1933. Estimate of deviation between empirical distribution functions in two independent samples. *Bulletin Moscow University* 2: 3–16.

Also see

[R] **runtest** — Test for random order

[R] **sktest** — Skewness and kurtosis test for normality

[R] **swilk** — Shapiro–Wilk and Shapiro–Francia tests for normality

Title

> **kwallis** — Kruskal–Wallis equality-of-populations rank test

Syntax

kwallis *varname* [*if*] [*in*] , by(*groupvar*)

Menu

Statistics > Nonparametric analysis > Tests of hypotheses > Kruskal-Wallis rank test

Description

kwallis tests the hypothesis that several samples are from the same population. In the syntax diagram above, *varname* refers to the variable recording the outcome, and *groupvar* refers to the variable denoting the population. by() is required.

Option

by(*groupvar*) is required. It specifies a variable that identifies the groups.

Remarks

Example 1

We have data on the 50 states. The data contain the median age of the population, medage, and the region of the country, region, for each state. We wish to test for the equality of the median age distribution across all four regions simultaneously:

```
. use http://www.stata-press.com/data/r12/census
(1980 Census data by state)
. kwallis medage, by(region)
Kruskal-Wallis equality-of-populations rank test
```

region	Obs	Rank Sum
NE	9	376.50
N Cntrl	12	294.00
South	16	398.00
West	13	206.50

```
chi-squared =      17.041 with 3 d.f.
probability =       0.0007
chi-squared with ties =      17.062 with 3 d.f.
probability =       0.0007
```

From the output, we see that we can reject the hypothesis that the populations are the same at any level below 0.07%.

◁

899

Saved results

kwallis saves the following in r():

Scalars
    r(df)   degrees of freedom
    r(chi2) χ^2
    r(chi2_adj) χ^2 adjusted for ties

Methods and formulas

kwallis is implemented as an ado-file.

The Kruskal–Wallis test (Kruskal and Wallis 1952, 1953; also see Altman [1991, 213–215]; Conover [1999, 288–297]; and Riffenburgh [2005, 287–291]) is a multiple-sample generalization of the two-sample Wilcoxon (also called Mann–Whitney) rank sum test (Wilcoxon 1945; Mann and Whitney 1947). Samples of sizes n_j, $j = 1, \ldots, m$, are combined and ranked in ascending order of magnitude. Tied values are assigned the average ranks. Let n denote the overall sample size, and let $R_j = \sum_{i=1}^{n_j} R(X_{ji})$ denote the sum of the ranks for the jth sample. The Kruskal–Wallis one-way analysis-of-variance test, H, is defined as

$$H = \frac{1}{S^2} \left\{ \sum_{j=1}^{m} \frac{R_j^2}{n_j} - \frac{n(n+1)^2}{4} \right\}$$

where

$$S^2 = \frac{1}{n-1} \left\{ \sum_{\text{all ranks}} R(X_{ji})^2 - \frac{n(n+1)^2}{4} \right\}$$

If there are no ties, this equation simplifies to

$$H = \frac{12}{n(n+1)} \sum_{j=1}^{m} \frac{R_j^2}{n_j} - 3(n+1)$$

The sampling distribution of H is approximately χ^2 with $m - 1$ degrees of freedom.

William Henry Kruskal (1919–2005) was born in New York City. He studied mathematics and statistics at Antioch College, Harvard, and Columbia, and joined the University of Chicago in 1951. He made many outstanding contributions to linear models, nonparametric statistics, government statistics, and the history and methodology of statistics.

Wilson Allen Wallis (1912–1998) was born in Philadelphia. He studied psychology and economics at the Universities of Minnesota and Chicago and at Columbia. He taught at Yale, Stanford, and Chicago, before moving as president (later chancellor) to the University of Rochester in 1962. He also served in several Republican administrations. Wallis served as editor of the *Journal of the American Statistical Association*, coauthored a popular introduction to statistics, and contributed to nonparametric statistics.

References

Altman, D. G. 1991. *Practical Statistics for Medical Research*. London: Chapman & Hall/CRC.

Conover, W. J. 1999. *Practical Nonparametric Statistics*. 3rd ed. New York: Wiley.

Fienberg, S. E., S. M. Stigler, and J. M. Tanur. 2007. The William Kruskal Legacy: 1919–2005. *Statistical Science* 22: 255–261.

Kruskal, W. H., and W. A. Wallis. 1952. Use of ranks in one-criterion variance analysis. *Journal of the American Statistical Association* 47: 583–621.

———. 1953. Errata: Use of ranks in one-criterion variance analysis. *Journal of the American Statistical Association* 48: 907–911.

Mann, H. B., and D. R. Whitney. 1947. On a test of whether one of two random variables is stochastically larger than the other. *Annals of Mathematical Statistics* 18: 50–60.

Newson, R. 2006. Confidence intervals for rank statistics: Somers' D and extensions. *Stata Journal* 6: 309–334.

Olkin, I. 1991. A conversation with W. Allen Wallis. *Statistical Science* 6: 121–140.

Riffenburgh, R. H. 2005. *Statistics in Medicine*. 2nd ed. New York: Elsevier.

Wilcoxon, F. 1945. Individual comparisons by ranking methods. *Biometrics* 1: 80–83.

Zabell, S. 1994. A conversation with William Kruskal. *Statistical Science* 9: 285–303.

Also see

[R] **nptrend** — Test for trend across ordered groups

[R] **oneway** — One-way analysis of variance

[R] **sdtest** — Variance-comparison tests

[R] **signrank** — Equality tests on matched data

Title

> **ladder** — Ladder of powers

Syntax

Ladder of powers

>   ladder *varname* [*if*] [*in*] [, <u>g</u>enerate(*newvar*) <u>noa</u>djust]

Ladder-of-powers histograms

>   gladder *varname* [*if*] [*in*] [, *histogram_options combine_options*]

Ladder-of-powers quantile–normal plots

>   qladder *varname* [*if*] [*in*] [, *qnorm_options combine_options*]

> by is allowed with ladder; see [D] **by**.

Menu

ladder

Statistics > Summaries, tables, and tests > Distributional plots and tests > Ladder of powers

gladder

Statistics > Summaries, tables, and tests > Distributional plots and tests > Ladder-of-powers histograms

qladder

Statistics > Summaries, tables, and tests > Distributional plots and tests > Ladder-of-powers quantile-normal plots

Description

ladder searches a subset of the ladder of powers (Tukey 1977) for a transform that converts *varname* into a normally distributed variable. sktest tests for normality; see [R] **sktest**. Also see [R] **boxcox**.

gladder displays nine histograms of transforms of *varname* according to the ladder of powers. gladder is useful pedagogically, but we do not advise looking at histograms for research work; ladder or qnorm (see [R] **diagnostic plots**) is preferred.

qladder displays the quantiles of transforms of *varname* according to the ladder of powers against the quantiles of a normal distribution.

Options for ladder

generate(*newvar*) saves the transformed values corresponding to the minimum chi-squared value from the table. We do not recommend using generate() because it is literal in interpreting the minimum, thus ignoring nearly equal but perhaps more interpretable transforms.

noadjust is the noadjust option to sktest; see [R] **sktest**.

Options for gladder

histogram_options affect the rendition of the histograms across all relevant transformations; see [R] **histogram**. Here the normal option is assumed, so you must supply the nonormal option to suppress the overlaid normal density. Also, gladder does not allow the width(#) option of histogram.

combine_options are any of the options documented in [G-2] **graph combine**. These include options for titling the graph (see [G-3] *title_options*) and for saving the graph to disk (see [G-3] *saving_option*).

Options for qladder

qnorm_options affect the rendition of the quantile–normal plots across all relevant transformations. See [R] **diagnostic plots**.

combine_options are any of the options documented in [G-2] **graph combine**. These include options for titling the graph (see [G-3] *title_options*) and for saving the graph to disk (see [G-3] *saving_option*).

Remarks

Example 1: ladder

We have data on the mileage rating of 74 automobiles and wish to find a transform that makes the variable normally distributed:

```
. use http://www.stata-press.com/data/r12/auto
(1978 Automobile Data)
. ladder mpg
```

Transformation	formula	chi2(2)	P(chi2)
cubic	mpg^3	43.59	0.000
square	mpg^2	27.03	0.000
identity	mpg	10.95	0.004
square root	sqrt(mpg)	4.94	0.084
log	log(mpg)	0.87	0.647
1/(square root)	1/sqrt(mpg)	0.20	0.905
inverse	1/mpg	2.36	0.307
1/square	1/(mpg^2)	11.99	0.002
1/cubic	1/(mpg^3)	24.30	0.000

If we had typed ladder mpg, gen(mpgx), the variable mpgx containing $1/\sqrt{mpg}$ would have been automatically generated for us. This is the perfect example of why you should not, in general, specify the generate() option. We also cannot reject the hypothesis that the inverse of mpg is normally distributed and that 1/mpg—gallons per mile—has a better interpretation. It is a measure of energy consumption.

◁

▷ Example 2: gladder

gladder explores the same transforms as ladder but presents results graphically:

. gladder mpg, fraction

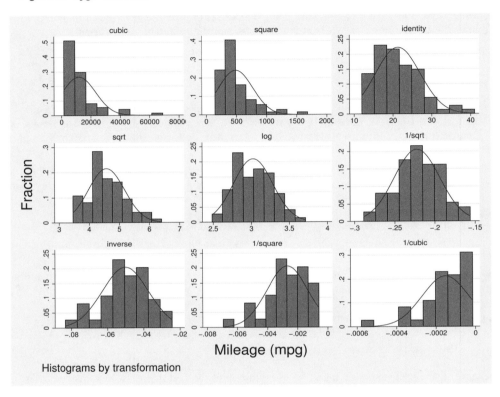

Histograms by transformation

◁

❑ Technical note

gladder is useful pedagogically, but be careful when using it for research work, especially with many observations. For instance, consider the following data on the average July temperature in degrees Fahrenheit for 954 U.S. cities:

```
. use http://www.stata-press.com/data/r12/citytemp
(City Temperature Data)
. ladder tempjuly
```

Transformation	formula	chi2(2)	P(chi2)
cubic	tempjuly^3	47.49	0.000
square	tempjuly^2	19.70	0.000
identity	tempjuly	3.83	0.147
square root	sqrt(tempjuly)	1.83	0.400
log	log(tempjuly)	5.40	0.067
1/(square root)	1/sqrt(tempjuly)	13.72	0.001
inverse	1/tempjuly	26.36	0.000
1/square	1/(tempjuly^2)	64.43	0.000
1/cubic	1/(tempjuly^3)	.	0.000

The period in the last line indicates that the χ^2 is very large; see [R] **sktest**.

From the table, we see that there is certainly a difference in normality between the square and square-root transform. If, however, you can see the difference between the transforms in the diagram below, you have better eyes than we do:

```
. gladder tempjuly, l1title("") ylabel(none) xlabel(none)
```

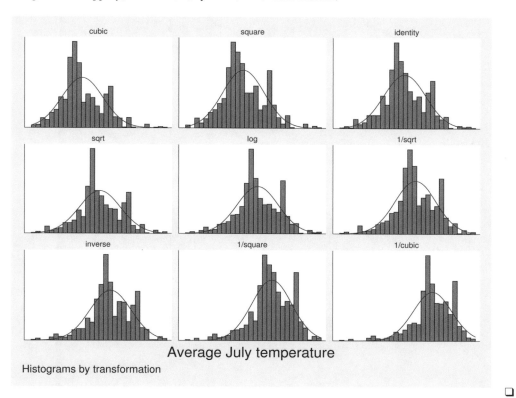

Average July temperature

Histograms by transformation

❏

▷ Example 3: qladder

A better graph for seeing normality is the quantile–normal graph, which can be produced by `qladder`.

```
. qladder tempjuly, ylabel(none) xlabel(none)
```

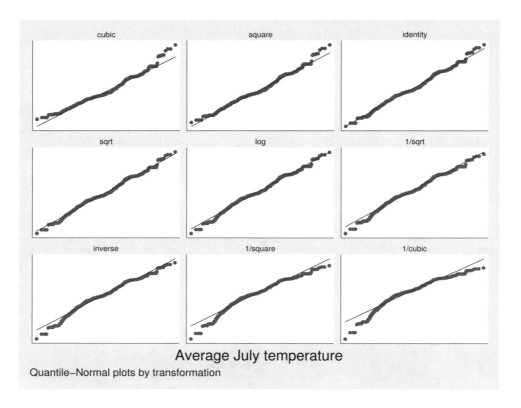

Average July temperature

Quantile–Normal plots by transformation

This graph shows that for the square transform, the upper tail—and only the upper tail—diverges from what would be expected. This divergence is detected by sktest (see [R] **sktest**) as a problem with skewness, as we would learn from using sktest to examine tempjuly squared and square rooted.

◁

Saved results

ladder saves the following in r():

Scalars

r(N)	number of observations
r(invcube)	χ^2 for inverse-cubic transformation
r(P_invcube)	significance level for inverse-cubic transformation
r(invsq)	χ^2 for inverse-square transformation
r(P_invsq)	significance level for inverse-square transformation
r(inv)	χ^2 for inverse transformation
r(P_inv)	significance level for inverse transformation
r(invsqrt)	χ^2 for inverse-root transformation
r(P_invsqrt)	significance level for inverse-root transformation
r(log)	χ^2 for log transformation
r(P_log)	significance level for log transformation
r(sqrt)	χ^2 for square-root transformation
r(P_sqrt)	significance level for square-root transformation
r(ident)	χ^2 for untransformed data
r(P_ident)	significance level for untransformed data
r(square)	χ^2 for square transformation
r(P_square)	significance level for square transformation
r(cube)	χ^2 for cubic transformation
r(P_cube)	significance level for cubic transformation

Methods and formulas

ladder, gladder, and qladder are implemented as ado-files.

For ladder, results are as reported by sktest; see [R] **sktest**. If generate() is specified, the transform with the minimum χ^2 value is chosen.

gladder sets the number of bins to $\min(\sqrt{n}, 10 \log_{10} n)$, rounded to the closest integer, where n is the number of unique values of *varname*. See [R] **histogram** for a discussion of the optimal number of bins.

Also see Findley (1990) for a ladder-of-powers variable transformation program that produces one-way graphs with overlaid box plots, in addition to histograms with overlaid normals. Buchner and Findley (1990) discuss ladder-of-powers transformations as one aspect of preliminary data analysis. Also see Hamilton (1992, 18–23) and Hamilton (2009, 142–144).

Acknowledgment

qladder was written by Jeroen Weesie, Utrecht University, Utrecht, The Netherlands.

References

Buchner, D. M., and T. W. Findley. 1990. Research in physical medicine and rehabilitation: VIII. Preliminary data analysis. *American Journal of Physical Medicine and Rehabilitation* 69: 154–169.

Cox, N. J. 2005. Speaking Stata: Density probability plots. *Stata Journal* 5: 259–273.

Findley, T. W. 1990. sed3: Variable transformation and evaluation. *Stata Technical Bulletin* 2: 15. Reprinted in *Stata Technical Bulletin Reprints*, vol. 1, pp. 85–86. College Station, TX: Stata Press.

Hamilton, L. C. 1992. *Regression with Graphics: A Second Course in Applied Statistics*. Belmont, CA: Duxbury.

———. 2009. *Statistics with Stata (Updated for Version 10)*. Belmont, CA: Brooks/Cole.

Tukey, J. W. 1977. *Exploratory Data Analysis*. Reading, MA: Addison–Wesley.

Also see

[R] **diagnostic plots** — Distributional diagnostic plots

[R] **lnskew0** — Find zero-skewness log or Box–Cox transform

[R] **lv** — Letter-value displays

[R] **sktest** — Skewness and kurtosis test for normality

Title

> **level** — Set default confidence level

Syntax

set level # [, permanently]

Description

set level specifies the default confidence level for confidence intervals for all commands that report confidence intervals. The initial value is 95, meaning 95% confidence intervals. # may be between 10.00 and 99.99, and # can have at most two digits after the decimal point.

Option

permanently specifies that, in addition to making the change right now, the level setting be remembered and become the default setting when you invoke Stata.

Remarks

To change the level of confidence intervals reported by a particular command, you need not reset the default confidence level. All commands that report confidence intervals have a level(#) option. When you do not specify the option, the confidence intervals are calculated for the default level set by set level, or for 95% if you have not reset set level.

▷ Example 1

We use the ci command to obtain the confidence interval for the mean of mpg:

```
. use http://www.stata-press.com/data/r12/auto
(1978 Automobile Data)
. ci mpg
```

Variable	Obs	Mean	Std. Err.	[95% Conf. Interval]	
mpg	74	21.2973	.6725511	19.9569	22.63769

To obtain 90% confidence intervals, we would type

```
. ci mpg, level(90)
```

Variable	Obs	Mean	Std. Err.	[90% Conf. Interval]	
mpg	74	21.2973	.6725511	20.17683	22.41776

or

```
. set level 90
. ci mpg
```

Variable	Obs	Mean	Std. Err.	[90% Conf. Interval]	
mpg	74	21.2973	.6725511	20.17683	22.41776

If we opt for the second alternative, the next time that we fit a model (say, with `regress`), 90% confidence intervals will be reported. If we wanted 95% confidence intervals, we could specify `level(95)` on the estimation command, or we could reset the default by typing `set level 95`.

The current setting of `level()` is stored as the c-class value `c(level)`; see [P] **creturn**.

◁

Also see

[R] **query** — Display system parameters

[P] **creturn** — Return c-class values

[U] **20 Estimation and postestimation commands**

[U] **20.7 Specifying the width of confidence intervals**

Title

```
lincom — Linear combinations of estimators
```

Syntax

lincom *exp* [, *options*]

options	Description
<u>ef</u>orm	generic label; exp(b); the default
or	odds ratio
hr	hazard ratio
shr	subhazard ratio
<u>irr</u>	incidence-rate ratio
<u>rrr</u>	relative-risk ratio
<u>level</u>(#)	set confidence level; default is level(95)
display_options	control column formats

where *exp* is any linear combination of coefficients that is a valid syntax for test; see [R] **test**. *exp* must not contain an equal sign.

Menu

Statistics > Postestimation > Linear combinations of estimates

Description

lincom computes point estimates, standard errors, t or z statistics, p-values, and confidence intervals for linear combinations of coefficients after any estimation command. Results can optionally be displayed as odds ratios, hazard ratios, incidence-rate ratios, or relative-risk ratios.

lincom can be used with svy estimation results; see [SVY] **svy postestimation**.

Options

eform, or, hr, shr, irr, and rrr all report coefficient estimates as $\exp(\widehat{\beta})$ rather than $\widehat{\beta}$. Standard errors and confidence intervals are similarly transformed. or is the default after logistic. The only difference in these options is how the output is labeled.

Option	Label	Explanation	Example commands
eform	exp(b)	Generic label	cloglog
or	Odds Ratio	Odds ratio	logistic, logit
hr	Haz. Ratio	Hazard ratio	stcox, streg
shr	SHR	Subhazard ratio	stcrreg
irr	IRR	Incidence-rate ratio	poisson
rrr	RRR	Relative-risk ratio	mlogit

exp may not contain any additive constants when you use the eform, or, hr, irr, or rrr option.

level(*#*) specifies the confidence level, as a percentage, for confidence intervals. The default is
   level(95) or as set by set level; see [U] **20.7 Specifying the width of confidence intervals**.

display_options: cformat(% *fmt*), pformat(% *fmt*), and sformat(% *fmt*); see [R] **estimation options**.

Remarks

Remarks are presented under the following headings:

> *Using lincom*
> *Odds ratios and incidence-rate ratios*
> *Multiple-equation models*

Using lincom

After fitting a model and obtaining estimates for coefficients $\beta_1, \beta_2, \ldots, \beta_k$, you may want to
view estimates for linear combinations of the β_i, such as $\beta_1 - \beta_2$. lincom can display estimates for
any linear combination of the form $c_0 + c_1\beta_1 + c_2\beta_2 + \cdots + c_k\beta_k$.

lincom works after any estimation command for which test works. Any valid expression for
test syntax 1 (see [R] **test**) is a valid expression for lincom.

lincom is useful for viewing odds ratios, hazard ratios, etc., for one group (that is, one set of
covariates) relative to another group (that is, another set of covariates). See the examples below.

▷ Example 1

We perform a linear regression:

```
. use http://www.stata-press.com/data/r12/regress
. regress y x1 x2 x3
```

Source	SS	df	MS		Number of obs	=	148
					F(3, 144)	=	96.12
Model	3259.3561	3	1086.45203		Prob > F	=	0.0000
Residual	1627.56282	144	11.3025196		R-squared	=	0.6670
					Adj R-squared	=	0.6600
Total	4886.91892	147	33.2443464		Root MSE	=	3.3619

y	Coef.	Std. Err.	t	P>|t|	[95% Conf. Interval]	
x1	1.457113	1.07461	1.36	0.177	-.666934	3.581161
x2	2.221682	.8610358	2.58	0.011	.5197797	3.923583
x3	-.006139	.0005543	-11.08	0.000	-.0072345	-.0050435
_cons	36.10135	4.382693	8.24	0.000	27.43863	44.76407

To see the difference of the coefficients of x2 and x1, we type

```
. lincom x2 - x1
 ( 1)  - x1 + x2 = 0
```

y	Coef.	Std. Err.	t	P>|t|	[95% Conf. Interval]	
(1)	.7645682	.9950282	0.77	0.444	-1.20218	2.731316

The expression can be any linear combination.

```
. lincom 3*x1 + 500*x3
( 1)   3*x1 + 500*x3 = 0
```

y	Coef.	Std. Err.	t	P>|t|	[95% Conf. Interval]
(1)	1.301825	3.396624	0.38	0.702	-5.411858 8.015507

Nonlinear expressions are not allowed.

```
. lincom x2/x1
not possible with test
r(131);
```

For information about estimating nonlinear expressions, see [R] **nlcom**.

◁

❑ Technical note

lincom uses the same shorthands for coefficients as does test (see [R] **test**). When you type x1, for instance, lincom knows that you mean the coefficient of x1. The formal syntax for referencing this coefficient is actually _b[x1], or alternatively, _coef[x1]. So, more formally, in the last example we could have typed

```
. lincom 3*_b[x1] + 500*_b[x3]
(output omitted )
```

❑

Odds ratios and incidence-rate ratios

After logistic regression, the or option can be specified with lincom to display odds ratios for any effect. Incidence-rate ratios after commands such as poisson can be similarly obtained by specifying the irr option.

▷ Example 2

Consider the low birthweight dataset from Hosmer and Lemeshow (2000, 25). We fit a logistic regression model of low birthweight (variable low) on the following variables:

Variable	Description	Coding
age	age in years	
race	race	1 if white, 2 if black, 3 if other
smoke	smoking status	1 if smoker, 0 if nonsmoker
ht	history of hypertension	1 if yes, 0 if no
ui	uterine irritability	1 if yes, 0 if no
lwd	maternal weight before pregnancy	1 if weight < 110 lb., 0 otherwise
ptd	history of premature labor	1 if yes, 0 if no
c.age##lwd	age main effects, lwd main effects, and their interaction	
smoke##lwd	smoke main effects, lwd main effects, and their interaction	

We first fit a model without the interaction terms by using logit.

```
. use http://www.stata-press.com/data/r12/lbw3
(Hosmer & Lemeshow data)

. logit low age lwd i.race smoke ptd ht ui

Iteration 0:   log likelihood =   -117.336
Iteration 1:   log likelihood =   -99.3982
Iteration 2:   log likelihood = -98.780418
Iteration 3:   log likelihood = -98.777998
Iteration 4:   log likelihood = -98.777998

Logistic regression                             Number of obs   =        189
                                                LR chi2(8)      =      37.12
                                                Prob > chi2     =     0.0000
Log likelihood = -98.777998                     Pseudo R2       =     0.1582
```

low	Coef.	Std. Err.	z	P>|z|	[95% Conf. Interval]	
age	-.0464796	.0373888	-1.24	0.214	-.1197603	.0268011
lwd	.8420615	.4055338	2.08	0.038	.0472299	1.636893
race						
2	1.073456	.5150753	2.08	0.037	.0639273	2.082985
3	.815367	.4452979	1.83	0.067	-.0574008	1.688135
smoke	.8071996	.404446	2.00	0.046	.0145001	1.599899
ptd	1.281678	.4621157	2.77	0.006	.3759478	2.187408
ht	1.435227	.6482699	2.21	0.027	.1646414	2.705813
ui	.6576256	.4666192	1.41	0.159	-.2569313	1.572182
_cons	-1.216781	.9556797	-1.27	0.203	-3.089878	.656317

To get the odds ratio for black smokers relative to white nonsmokers (the reference group), we type

```
. lincom 2.race + smoke, or
 ( 1)  [low]2.race + [low]smoke = 0
```

low	Odds Ratio	Std. Err.	z	P>|z|	[95% Conf. Interval]	
(1)	6.557805	4.744692	2.60	0.009	1.588176	27.07811

lincom computed $\exp(\beta_{2.\text{race}} + \beta_{\text{smoke}}) = 6.56$. To see the odds ratio for white smokers relative to black nonsmokers, we type

```
. lincom smoke - 2.race, or
 ( 1)  - [low]2.race + [low]smoke = 0
```

low	Odds Ratio	Std. Err.	z	P>|z|	[95% Conf. Interval]	
(1)	.7662425	.4430176	-0.46	0.645	.2467334	2.379603

Now let's add the interaction terms to the model (Hosmer and Lemeshow 1989, table 4.10). This time, we will use logistic rather than logit. By default, logistic displays odds ratios.

```
. logistic low i.race ht ui ptd c.age##lwd smoke##lwd
```

Logistic regression

Number of obs =	189
LR chi2(10) =	42.66
Prob > chi2 =	0.0000

Log likelihood = −96.00616

Pseudo R2 = 0.1818

low	Odds Ratio	Std. Err.	z	P>|z|	[95% Conf. Interval]	
race						
2	2.95383	1.532789	2.09	0.037	1.068277	8.167465
3	2.137589	.9919138	1.64	0.102	.8608708	5.307752
ht	3.893141	2.575201	2.05	0.040	1.064768	14.2346
ui	2.071284	.9931388	1.52	0.129	.8092926	5.301192
ptd	3.426633	1.615282	2.61	0.009	1.360252	8.632089
age	.9194513	.041896	−1.84	0.065	.8408967	1.005344
1.lwd	.1772934	.3312384	−0.93	0.354	.0045539	6.902367
lwd#c.age						
1	1.15883	.09602	1.78	0.075	.9851215	1.36317
1.smoke	3.168096	1.452378	2.52	0.012	1.289956	7.78076
smoke#lwd						
1 1	.2447849	.2003996	−1.72	0.086	.0491956	1.217988
_cons	.599443	.6519163	−0.47	0.638	.0711271	5.051971

Hosmer and Lemeshow (1989, table 4.13) consider the effects of smoking (smoke = 1) and low maternal weight before pregnancy (lwd = 1). The effect of smoking among non–low-weight mothers (lwd = 0) is given by the odds ratio 3.17 for smoke in the logistic output. The effect of smoking among low-weight mothers is given by

```
. lincom 1.smoke + 1.smoke#1.lwd
```
(1) [low]1.smoke + [low]1.smoke#1.lwd = 0

low	Odds Ratio	Std. Err.	z	P>|z|	[95% Conf. Interval]	
(1)	.7755022	.574951	−0.34	0.732	.1813465	3.316323

We did not have to specify the or option. After logistic, lincom assumes or by default.

The effect of low weight (lwd = 1) is more complicated because we fit an age × lwd interaction. We must specify the age of mothers for the effect. The effect among 30-year-old nonsmokers is given by

```
. lincom 1.lwd + 30*1.lwd#c.age
```
(1) [low]1.lwd + 30*[low]1.lwd#c.age = 0

low	Odds Ratio	Std. Err.	z	P>|z|	[95% Conf. Interval]	
(1)	14.7669	13.5669	2.93	0.003	2.439264	89.39633

lincom computed $\exp(\beta_{1wd} + 30\beta_{agelwd}) = 14.8$. It may seem odd that we entered it as 1.lwd + 30*1.lwd#c.age, but remember that these terms are just lincom's (and test's) shorthands for _b[1.lwd] and _b[1.lwd#c.age]. We could have typed

```
. lincom _b[1.lwd] + 30*_b[1.lwd#c.age]
 ( 1)   [low]1.lwd + 30*[low]1.lwd#c.age = 0
```

low	Odds Ratio	Std. Err.	z	P>|z|	[95% Conf. Interval]	
(1)	14.7669	13.5669	2.93	0.003	2.439264	89.39633

◁

Multiple-equation models

lincom also works with multiple-equation models. The only difference is how you refer to the coefficients. Recall that for multiple-equation models, coefficients are referenced using the syntax

[*eqno*] *varname*

where *eqno* is the equation number or equation name and *varname* is the corresponding variable name for the coefficient; see [U] **13.5 Accessing coefficients and standard errors** and [R] **test** for details.

▷ Example 3

Let's consider example 4 from [R] **mlogit** (Tarlov et al. 1989; Wells et al. 1989).

```
. use http://www.stata-press.com/data/r12/sysdsn1
(Health insurance data)
. mlogit insure age male nonwhite i.site, nolog
```

Multinomial logistic regression				Number of obs	=	615
				LR chi2(10)	=	42.99
				Prob > chi2	=	0.0000
Log likelihood = -534.36165				Pseudo R2	=	0.0387

insure	Coef.	Std. Err.	z	P>|z|	[95% Conf. Interval]	
Indemnity	(base outcome)					
Prepaid						
age	-.011745	.0061946	-1.90	0.058	-.0238862	.0003962
male	.5616934	.2027465	2.77	0.006	.1643175	.9590693
nonwhite	.9747768	.2363213	4.12	0.000	.5115955	1.437958
site						
2	.1130359	.2101903	0.54	0.591	-.2989296	.5250013
3	-.5879879	.2279351	-2.58	0.010	-1.034733	-.1412433
_cons	.2697127	.3284422	0.82	0.412	-.3740222	.9134476
Uninsure						
age	-.0077961	.0114418	-0.68	0.496	-.0302217	.0146294
male	.4518496	.3674867	1.23	0.219	-.268411	1.17211
nonwhite	.2170589	.4256361	0.51	0.610	-.6171725	1.05129
site						
2	-1.211563	.4705127	-2.57	0.010	-2.133751	-.2893747
3	-.2078123	.3662926	-0.57	0.570	-.9257327	.510108
_cons	-1.286943	.5923219	-2.17	0.030	-2.447872	-.1260134

To see the estimate of the sum of the coefficient of male and the coefficient of nonwhite for the Prepaid outcome, we type

```
. lincom [Prepaid]male + [Prepaid]nonwhite

( 1)  [Prepaid]male + [Prepaid]nonwhite = 0
```

insure	Coef.	Std. Err.	z	P>|z|	[95% Conf. Interval]	
(1)	1.53647	.3272489	4.70	0.000	.8950741	2.177866

To view the estimate as a ratio of relative risks (see [R] **mlogit** for the definition and interpretation), we specify the **rrr** option.

```
. lincom [Prepaid]male + [Prepaid]nonwhite, rrr

( 1)  [Prepaid]male + [Prepaid]nonwhite = 0
```

insure	RRR	Std. Err.	z	P>|z|	[95% Conf. Interval]	
(1)	4.648154	1.521103	4.70	0.000	2.447517	8.827451

◁

Saved results

lincom saves the following in r():

Scalars
    r(estimate)   point estimate
    r(se)   estimate of standard error
    r(df)   degrees of freedom

Methods and formulas

lincom is implemented as an ado-file.

References

Hosmer, D. W., Jr., and S. Lemeshow. 1989. *Applied Logistic Regression.* New York: Wiley.

———. 2000. *Applied Logistic Regression.* 2nd ed. New York: Wiley.

Tarlov, A. R., J. E. Ware, Jr., S. Greenfield, E. C. Nelson, E. Perrin, and M. Zubkoff. 1989. The medical outcomes study. An application of methods for monitoring the results of medical care. *Journal of the American Medical Association* 262: 925–930.

Wells, K. B., R. D. Hays, M. A. Burnam, W. H. Rogers, S. Greenfield, and J. E. Ware, Jr. 1989. Detection of depressive disorder for patients receiving prepaid or fee-for-service care. Results from the Medical Outcomes Survey. *Journal of the American Medical Association* 262: 3298–3302.

Also see

[R] **nlcom** — Nonlinear combinations of estimators

[R] **test** — Test linear hypotheses after estimation

[R] **testnl** — Test nonlinear hypotheses after estimation

[U] **13.5 Accessing coefficients and standard errors**

[U] **20 Estimation and postestimation commands**

Title

> **linktest** — Specification link test for single-equation models

Syntax

linktest [*if*] [*in*] [, *cmd_options*]

When if and in are not specified, the link test is performed on the same sample as the previous estimation.

Menu

Statistics > Postestimation > Tests > Specification link test for single-equation models

Description

linktest performs a link test for model specification after any single-equation estimation command, such as logistic, regress, stcox, etc.

Option

⌐ Main ⌐

cmd_options must be the same options specified with the underlying estimation command, except the *display_options* may differ.

Remarks

The form of the link test implemented here is based on an idea of Tukey (1949), which was further described by Pregibon (1980), elaborating on work in his unpublished thesis (Pregibon 1979). See *Methods and formulas* below for more details.

▷ Example 1

We want to explain the mileage ratings of cars in our automobile dataset by using the weight, engine displacement, and whether the car is manufactured outside the United States:

```
. use http://www.stata-press.com/data/r12/auto
(1978 Automobile Data)

. regress mpg weight displ foreign
```

Source	SS	df	MS
Model	1619.71935	3	539.906448
Residual	823.740114	70	11.7677159
Total	2443.45946	73	33.4720474

Number of obs =	74
F(3, 70) =	45.88
Prob > F =	0.0000
R-squared =	0.6629
Adj R-squared =	0.6484
Root MSE =	3.4304

mpg	Coef.	Std. Err.	t	P>|t|	[95% Conf. Interval]	
weight	-.0067745	.0011665	-5.81	0.000	-.0091011	-.0044479
displacement	.0019286	.0100701	0.19	0.849	-.0181556	.0220129
foreign	-1.600631	1.113648	-1.44	0.155	-3.821732	.6204699
_cons	41.84795	2.350704	17.80	0.000	37.15962	46.53628

On the basis of the R^2, we are reasonably pleased with this model.

If our model really is specified correctly, then if we were to regress mpg on the prediction and the prediction squared, the prediction squared would have no explanatory power. This is what linktest does:

```
. linktest
```

Source	SS	df	MS		
Model	1670.71514	2	835.357572	Number of obs =	74
Residual	772.744316	71	10.8837228	F(2, 71) =	76.75
				Prob > F =	0.0000
				R-squared =	0.6837
				Adj R-squared =	0.6748
Total	2443.45946	73	33.4720474	Root MSE =	3.299

mpg	Coef.	Std. Err.	t	P>|t|	[95% Conf. Interval]	
_hat	-.4127198	.6577736	-0.63	0.532	-1.724283	.8988434
_hatsq	.0338198	.015624	2.16	0.034	.0026664	.0649732
_cons	14.00705	6.713276	2.09	0.041	.6211539	27.39294

We find that the prediction squared does have explanatory power, so our specification is not as good as we thought.

Although linktest is formally a test of the specification of the dependent variable, it is often interpreted as a test that, conditional on the specification, the independent variables are specified incorrectly. We will follow that interpretation and now include weight squared in our model:

```
. regress mpg weight c.weight#c.weight displ foreign
```

Source	SS	df	MS		
Model	1699.02634	4	424.756584	Number of obs =	74
Residual	744.433124	69	10.7888859	F(4, 69) =	39.37
				Prob > F =	0.0000
				R-squared =	0.6953
				Adj R-squared =	0.6777
Total	2443.45946	73	33.4720474	Root MSE =	3.2846

mpg	Coef.	Std. Err.	t	P>|t|	[95% Conf. Interval]	
weight	-.0173257	.0040488	-4.28	0.000	-.0254028	-.0092486
c.weight#c.weight	1.87e-06	6.89e-07	2.71	0.008	4.93e-07	3.24e-06
displacement	-.0101625	.0106236	-0.96	0.342	-.031356	.011031
foreign	-2.560016	1.123506	-2.28	0.026	-4.801349	-.3186832
_cons	58.23575	6.449882	9.03	0.000	45.36859	71.10291

Now we perform the link test on our new model:

```
. linktest
```

Source	SS	df	MS					Number of obs =	74
								F(2, 71) =	81.08
Model	1699.39489	2	849.697445					Prob > F =	0.0000
Residual	744.06457	71	10.4797827					R-squared =	0.6955
								Adj R-squared =	0.6869
Total	2443.45946	73	33.4720474					Root MSE =	3.2372

mpg	Coef.	Std. Err.	t	P>|t|	[95% Conf. Interval]	
_hat	1.141987	.7612218	1.50	0.138	-.3758456	2.659821
_hatsq	-.0031916	.0170194	-0.19	0.852	-.0371272	.0307441
_cons	-1.50305	8.196444	-0.18	0.855	-17.84629	14.84019

We now pass the link test.

◁

> ## Example 2

Above we followed a standard misinterpretation of the link test—when we discovered a problem, we focused on the explanatory variables of our model. We might consider varying exactly what the link test tests. The link test told us that our dependent variable was misspecified. For those with an engineering background, mpg is indeed a strange measure. It would make more sense to model energy consumption—gallons per mile—in terms of weight and displacement:

```
. gen gpm = 1/mpg
. regress gpm weight displ foreign
```

Source	SS	df	MS					Number of obs =	74
								F(3, 70) =	76.33
Model	.009157962	3	.003052654					Prob > F =	0.0000
Residual	.002799666	70	.000039995					R-squared =	0.7659
								Adj R-squared =	0.7558
Total	.011957628	73	.000163803					Root MSE =	.00632

gpm	Coef.	Std. Err.	t	P>|t|	[95% Conf. Interval]	
weight	.0000144	2.15e-06	6.72	0.000	.0000102	.0000187
displacement	.0000186	.0000186	1.00	0.319	-.0000184	.0000557
foreign	.0066981	.0020531	3.26	0.002	.0026034	.0107928
_cons	.0008917	.0043337	0.21	0.838	-.0077515	.009535

This model looks every bit as reasonable as our original model:

```
. linktest
```

Source	SS	df	MS					Number of obs =	74
								F(2, 71) =	117.06
Model	.009175219	2	.004587609					Prob > F =	0.0000
Residual	.002782409	71	.000039189					R-squared =	0.7673
								Adj R-squared =	0.7608
Total	.011957628	73	.000163803					Root MSE =	.00626

gpm	Coef.	Std. Err.	t	P>|t|	[95% Conf. Interval]	
_hat	.6608413	.515275	1.28	0.204	-.3665877	1.68827
_hatsq	3.275857	4.936655	0.66	0.509	-6.567553	13.11927
_cons	.008365	.0130468	0.64	0.523	-.0176496	.0343795

Specifying the model in terms of gallons per mile also solves the specification problem and results in a more parsimonious specification.

◁

▷ Example 3

The link test can be used with any single-equation estimation procedure, not solely regression. Let's turn our problem around and attempt to explain whether a car is manufactured outside the United States by its mileage rating and weight. To save paper, we will specify logit's nolog option, which suppresses the iteration log:

```
. logit foreign mpg weight, nolog
```

Logistic regression

```
                                          Number of obs   =         74
                                          LR chi2(2)      =      35.72
                                          Prob > chi2     =     0.0000
Log likelihood = -27.175156              Pseudo R2       =     0.3966
```

foreign	Coef.	Std. Err.	z	P>|z|	[95% Conf. Interval]	
mpg	-.1685869	.0919175	-1.83	0.067	-.3487418	.011568
weight	-.0039067	.0010116	-3.86	0.000	-.0058894	-.001924
_cons	13.70837	4.518709	3.03	0.002	4.851859	22.56487

When we run linktest after logit, the result is another logit specification:

```
. linktest, nolog
```

Logistic regression

```
                                          Number of obs   =         74
                                          LR chi2(2)      =      36.83
                                          Prob > chi2     =     0.0000
Log likelihood = -26.615714              Pseudo R2       =     0.4090
```

foreign	Coef.	Std. Err.	z	P>|z|	[95% Conf. Interval]	
_hat	.8438531	.2738759	3.08	0.002	.3070661	1.38064
_hatsq	-.1559115	.1568642	-0.99	0.320	-.4633596	.1515366
_cons	.2630557	.4299598	0.61	0.541	-.57965	1.105761

The link test reveals no problems with our specification.

If there had been a problem, we would have been virtually forced to accept the misinterpretation of the link test—we would have reconsidered our specification of the independent variables. When using logit, we have no control over the specification of the dependent variable other than to change likelihood functions.

We admit to having seen a dataset once for which the link test rejected the logit specification. We did change the likelihood function, refitting the model using probit, and satisfied the link test. Probit has thinner tails than logit. In general, however, you will not be so lucky.

◁

❏ Technical note

You should specify the same options with `linktest` that you do with the estimation command, although you do not have to follow this advice as literally as we did in the preceding example. `logit`'s `nolog` option merely suppresses a part of the output, not what is estimated. We specified `nolog` both times to save paper.

If you are testing a tobit model, you must specify the censoring points just as you do with the `tobit` command.

If you are not sure which options are important, duplicate exactly what you specified on the estimation command.

If you do not specify `if` *exp* or `in` *range* with `linktest`, Stata will by default perform the link test on the same sample as the previous estimation. Suppose that you omitted some data when performing your estimation, but want to calculate the link test on all the data, which you might do if you believe the model is appropriate for all the data. You would type `linktest if e(sample) < .` to do this.

❏

Saved results

`linktest` saves the following in `r()`:

Scalars
`r(t)`	*t* statistic on `_hatsq`
`r(df)`	degrees of freedom

`linktest` is *not* an estimation command in the sense that it leaves previous estimation results unchanged. For instance, after running a regression and performing the link test, typing `regress` without arguments after the link test still replays the original regression.

For integrating an estimation command with `linktest`, `linktest` assumes that the name of the estimation command is stored in `e(cmd)` and that the name of the dependent variable is stored in `e(depvar)`. After estimation, it assumes that the number of degrees of freedom for the *t* test is given by `e(df_m)` if the macro is defined.

If the estimation command reports *z* statistics instead of *t* statistics, `linktest` will also report *z* statistics. The *z* statistic, however, is still returned in `r(t)`, and `r(df)` is set to a missing value.

Methods and formulas

`linktest` is implemented as an ado-file.

The link test is based on the idea that if a regression or regression-like equation is properly specified, you should be able to find no additional independent variables that are significant except by chance. One kind of specification error is called a link error. In regression, this means that the dependent variable needs a transformation or "link" function to properly relate to the independent variables. The idea of a link test is to add an independent variable to the equation that is especially likely to be significant if there is a link error.

Let

$$\mathbf{y} = f(\mathbf{X}\beta)$$

be the model and $\widehat{\beta}$ be the parameter estimates. linktest calculates

$$\_hat = \mathbf{X}\widehat{\beta}$$

and

$$\_hatsq = \_hat^2$$

The model is then refit with these two variables, and the test is based on the significance of _hatsq. This is the form suggested by Pregibon (1979) based on an idea of Tukey (1949). Pregibon (1980) suggests a slightly different method that has come to be known as "Pregibon's goodness-of-link test". We prefer the older version because it is universally applicable, straightforward, and a good second-order approximation. It can be applied to any single-equation estimation technique, whereas Pregibon's more recent tests are estimation-technique specific.

References

Pregibon, D. 1979. Data analytic methods for generalized linear models. PhD diss., University of Toronto.

——. 1980. Goodness of link tests for generalized linear models. *Applied Statistics* 29: 15–24.

Tukey, J. W. 1949. One degree of freedom for non-additivity. *Biometrics* 5: 232–242.

Also see

Title

> **lnskew0** — Find zero-skewness log or Box–Cox transform

Syntax

Zero-skewness log transform

> lnskew0 *newvar* = *exp* $\begin{bmatrix} if \end{bmatrix}$ $\begin{bmatrix} in \end{bmatrix}$ $\begin{bmatrix} , options \end{bmatrix}$

Zero-skewness Box–Cox transform

> bcskew0 *newvar* = *exp* $\begin{bmatrix} if \end{bmatrix}$ $\begin{bmatrix} in \end{bmatrix}$ $\begin{bmatrix} , options \end{bmatrix}$

options	Description
Main	
<u>delta</u>(#)	increment for derivative of skewness function; default is delta(0.02) for lnskew0 and delta(0.01) for bcskew0
<u>zero</u>(#)	value for determining convergence; default is zero(0.001)
<u>level</u>(#)	set confidence level; default is level(95)

Menu

lnskew0

Data > Create or change data > Other variable-creation commands > Zero-skewness log transform

bcskew0

Data > Create or change data > Other variable-creation commands > Box-Cox transform

Description

lnskew0 creates *newvar* $= \ln(\pm exp - k)$, choosing k and the sign of *exp* so that the skewness of *newvar* is zero.

bcskew0 creates *newvar* $= (exp^{\lambda} - 1)/\lambda$, the Box–Cox power transformation (Box and Cox 1964), choosing λ so that the skewness of *newvar* is zero. *exp* must be strictly positive. Also see [R] **boxcox** for maximum likelihood estimation of λ.

Options

_____ Main _____

delta(#) specifies the increment used for calculating the derivative of the skewness function with respect to k (lnskew0) or λ (bcskew0). The default values are 0.02 for lnskew0 and 0.01 for bcskew0.

zero(*#*) specifies a value for skewness to determine convergence that is small enough to be considered zero and is, by default, 0.001.

level(*#*) specifies the confidence level for the confidence interval for k (lnskew0) or λ (bcskew0). The confidence interval is calculated only if level() is specified. *#* is specified as an integer; 95 means 95% confidence intervals. The level() option is honored only if the number of observations exceeds 7.

Remarks

Example 1: lnskew0

Using our automobile dataset (see [U] **1.2.2 Example datasets**), we want to generate a new variable equal to $\ln(\text{mpg} - k)$ to be approximately normally distributed. mpg records the miles per gallon for each of our cars. One feature of the normal distribution is that it has skewness 0.

```
. use http://www.stata-press.com/data/r12/auto
(1978 Automobile Data)

. lnskew0 lnmpg = mpg
```

Transform	k	[95% Conf. Interval]	Skewness
ln(mpg-k)	5.383659	(not calculated)	-7.05e-06

This created the new variable $\text{lnmpg} = \ln(\text{mpg} - 5.384)$:

```
. describe lnmpg
```

variable name	storage type	display format	value label	variable label
lnmpg	float	%9.0g		ln(mpg-5.383659)

Because we did not specify the level() option, no confidence interval was calculated. At the outset, we could have typed

```
. use http://www.stata-press.com/data/r12/auto, clear
(Automobile Data)

. lnskew0 lnmpg = mpg, level(95)
```

Transform	k	[95% Conf. Interval]	Skewness
ln(mpg-k)	5.383659	-17.12339 9.892416	-7.05e-06

The confidence interval is calculated under the assumption that $\ln(\text{mpg} - k)$ really does have a normal distribution. It would be perfectly reasonable to use lnskew0, even if we did not believe that the transformed variable would have a normal distribution—if we literally wanted the zero-skewness transform—although, then the confidence interval would be an approximation of unknown quality to the true confidence interval. If we now wanted to test the believability of the confidence interval, we could also test our new variable lnmpg by using swilk (see [R] **swilk**) with the lnnormal option.

◁

❏ Technical note

lnskew0 and bcskew0 report the resulting skewness of the variable merely to reassure you of the accuracy of its results. In our example above, lnskew0 found k such that the resulting skewness was -7×10^{-6}, a number close enough to zero for all practical purposes. If we wanted to make it even smaller, we could specify the zero() option. Typing lnskew0 new=mpg, zero(1e-8) changes the estimated k to 5.383552 from 5.383659 and reduces the calculated skewness to -2×10^{-11}.

When you request a confidence interval, lnskew0 may report the lower confidence interval as '.', which should be taken as indicating the lower confidence limit $k_L = -\infty$. (This cannot happen with bcskew0.)

As an example, consider a sample of size n on x and assume that the skewness of x is positive, but not significantly so, at the desired significance level—say, 5%. Then no matter how large and negative you make k_L, there is no value extreme enough to make the skewness of $\ln(x - k_L)$ equal the corresponding percentile (97.5 for a 95% confidence interval) of the distribution of skewness in a normal distribution of the same sample size. You cannot do this because the distribution of $\ln(x - k_L)$ tends to that of x—apart from location and scale shift—as $x \to \infty$. This "problem" never applies to the upper confidence limit, k_U, because the skewness of $\ln(x - k_U)$ tends to $-\infty$ as k tends upward to the minimum value of x.

❏

▷ Example 2: bcskew0

In example 1, using lnskew0 with a variable such as mpg is probably undesirable. mpg has a natural zero, and we are shifting that zero arbitrarily. On the other hand, use of lnskew0 with a variable such as temperature measured in Fahrenheit or Celsius would be more appropriate, as the zero is indeed arbitrary.

For a variable like mpg, it makes more sense to use the Box–Cox power transform (Box and Cox 1964):

$$y^{(\lambda)} = \frac{y^\lambda - 1}{\lambda}$$

λ is free to take on any value, but $y^{(1)} = y - 1$, $y^{(0)} = \ln(y)$, and $y^{(-1)} = 1 - 1/y$.

bcskew0 works like lnskew0:

```
. bcskew0 bcmpg = mpg, level(95)
        Transform |        L    [95% Conf. Interval]     Skewness
    (mpg^L-1)/L   | -.3673283   -1.212752    .4339645     .0001898
```

The 95% confidence interval includes $\lambda = -1$ (λ is labeled L in the output), which has a rather more pleasing interpretation—gallons per mile—than $(\text{mpg}^{-0.3673} - 1)/(-0.3673)$. The confidence interval, however, is calculated assuming that the power transformed variable is normally distributed. It makes perfect sense to use bcskew0, even when you do not believe that the transformed variable will be normally distributed, but then the confidence interval is an approximation of unknown quality. If you believe that the transformed data are normally distributed, you can alternatively use boxcox to estimate λ; see [R] **boxcox**.

◁

Saved results

lnskew0 and bcskew0 save the following in r():

Scalars

r(gamma)	k (lnskew0)
r(lambda)	λ (bcskew0)
r(lb)	lower bound of confidence interval
r(ub)	upper bound of confidence interval
r(skewness)	resulting skewness of transformed variable

Methods and formulas

lnskew0 and bcskew0 are implemented as ado-files.

Skewness is as calculated by summarize; see [R] **summarize**. Newton's method with numeric, uncentered derivatives is used to estimate k (lnskew0) and λ (bcskew0). For lnskew0, the initial value is chosen so that the minimum of $x - k$ is 1, and thus $\ln(x - k)$ is 0. bcskew0 starts with $\lambda = 1$.

Acknowledgment

lnskew0 and bcskew0 were written by Patrick Royston of the MRC Clinical Trials Unit, London.

Reference

Box, G. E. P., and D. R. Cox. 1964. An analysis of transformations. *Journal of the Royal Statistical Society, Series B* 26: 211–252.

Also see

[R] **ladder** — Ladder of powers

[R] **boxcox** — Box–Cox regression models

[R] **swilk** — Shapiro–Wilk and Shapiro–Francia tests for normality

Title

> **log** — Echo copy of session to file

Syntax

Report status of log file

```
log
```

```
log query [logname | _all]
```

Open log file

```
log using filename [, append replace [text | smcl] name(logname)]
```

Close log

```
log close [logname | _all]
```

Temporarily suspend logging or resume logging

```
log {off | on} [logname]
```

Report status of command log file

```
cmdlog
```

Open command log file

```
cmdlog using filename [, append replace]
```

Close command log, temporarily suspend logging, or resume logging

```
cmdlog {close | on | off}
```

Set default format for logs

```
set logtype {text | smcl} [, permanently]
```

Specify screen width

```
set linesize #
```

In addition to using the log command, you may access the capabilities of log by selecting **File > Log** from the menu and choosing one of the options in the list.

928

Menu

File > Log

Description

log allows you to make a full record of your Stata session. A log is a file containing what you type and Stata's output. You may start multiple log files at the same time, and you may refer to them with a *logname*. If you do not specify a *logname*, Stata will use the name <unnamed>.

cmdlog allows you to make a record of what you type during your Stata session. A command log contains only what you type, so it is a subset of a full log.

You can make full logs, command logs, or both simultaneously. Neither is produced until you tell Stata to start logging.

Command logs are always text files, making them easy to convert into do-files. (In this respect, it would make more sense if the default extension of a command log file was .do because command logs are do-files. The default is .txt, not .do, however, to keep you from accidentally overwriting your important do-files.)

Full logs are recorded in one of two formats: Stata Markup and Control Language (SMCL) or plain text. The default is SMCL, but you can use set logtype to change that, or you can specify an option to state the format you wish. We recommend SMCL because it preserves fonts and colors. SMCL logs can be converted to text or to other formats by using the translate command; see [R] **translate**. You can also use translate to produce printable versions of SMCL logs. SMCL logs can be viewed and printed from the Viewer, as can any text file; see [R] **view**.

When using multiple log files, you may have up to five SMCL logs and five text logs open at the same time.

log or cmdlog, typed without arguments, reports the status of logging. log query, when passed an optional *logname*, reports the status of that log.

log using and cmdlog using open a log file. log close and cmdlog close close the file. Between times, log off and cmdlog off, and log on and cmdlog on, can temporarily suspend and resume logging.

If *filename* is specified without an extension, one of the suffixes .smcl, .log, or .txt is added. The extension .smcl or .log is added by log, depending on whether the file format is SMCL or text. The extension .txt is added by cmdlog. If *filename* contains embedded spaces, remember to enclose it in double quotes.

set logtype specifies the default format in which full logs are to be recorded. Initially, full logs are recorded in SMCL format.

set linesize specifies the maximum width, in characters, of Stata output. Most commands in Stata do not respect linesize, because it is not important for most commands. Most users never need to set linesize, because it will automatically be reset if you resize your Results window. This is also why there is no permanently option allowed with set linesize. set linesize is for use with commands such as list and display and is typically used by programmers who wish the output of those commands to be wider or narrower than the current width of the Results window.

Options for use with both log and cmdlog

append specifies that results be appended to an existing file. If the file does not already exist, a new file is created.

replace specifies that *filename*, if it already exists, be overwritten. When you do not specify either replace or append, the file is assumed to be new. If the specified file already exists, an error message is issued and logging is not started.

Options for use with log

text and smcl specify the format in which the log is to be recorded. The default is complicated to describe but is what you would expect:

If you specify the file as *filename*.smcl, the default is to write the log in SMCL format (regardless of the value of set logtype).

If you specify the file as *filename*.log, the default is to write the log in text format (regardless of the value of set logtype).

If you type *filename* without an extension and specify neither the smcl option nor the text option, the default is to write the file according to the value of set logtype. If you have not set logtype, then that default is SMCL. Also, the *filename* you specified will be fixed to read *filename*.smcl if a SMCL log is being created or *filename*.log if a text log is being created.

If you specify either the text or smcl option, then what you specify determines how the log is written. If *filename* was specified without an extension, the appropriate extension is added for you.

If you open multiple log files, you may choose a different format for each file.

name(*logname*) specifies an optional name you may use to refer to the log while it is open. You can start multiple log files, give each a different *logname*, and then close, temporarily suspend, or resume them each individually.

Option for use with set logtype

permanently specifies that, in addition to making the change right now, the logtype setting be remembered and become the default setting when you invoke Stata.

Remarks

For a detailed explanation of logs, see [U] **15 Saving and printing output—log files**.

When you open a full log, the default is to show the name of the file and a time and date stamp:

```
. log using myfile
```

```
      name:  <unnamed>
       log:  C:\data\proj1\myfile.smcl
  log type:  smcl
 opened on:  12 Jan 2011, 12:28:23
```
.

The above information will appear in the log. If you do not want this information to appear, precede the command by quietly:

```
. quietly log using myfile
```

quietly will not suppress any error messages or anything else you need to know.

Similarly, when you close a full log, the default is to show the full information,

```
. log close
       name:  <unnamed>
        log:  C:\data\proj1\myfile.smcl
   log type:  smcl
  closed on:  12 Jan 2011, 12:32:41
```

and that information will also appear in the log. If you want to suppress that, type quietly log close.

Saved results

log and cmdlog save the following in r():

Macros
r(name)	*logname*
r(filename)	name of file
r(status)	on or off
r(type)	smcl or text

log query _all saves the following in r():

Scalars
r(numlogs)	number of open log files

For each open log file, log query _all also saves

r(name#)	*logname*
r(filename#)	name of file
r(status#)	on or off
r(type#)	smcl or text

where # varies between 1 and the value of r(numlogs). Be aware that # will not necessarily represent the order in which the log files were first opened, nor will it necessarily remain constant for a given log file upon multiple calls to log query.

Also see

[R] **translate** — Print and translate logs

[R] **query** — Display system parameters

[GSM] **16 Saving and printing results by using logs**

[GSW] **16 Saving and printing results by using logs**

[GSU] **16 Saving and printing results by using logs**

[U] **15 Saving and printing output—log files**

Title

logistic — Logistic regression, reporting odds ratios

Syntax

logistic *depvar* *indepvars* $\big[$ *if* $\big]$ $\big[$ *in* $\big]$ $\big[$ *weight* $\big]$ $\big[$, *options* $\big]$

options	Description
Model	
<u>nocon</u>stant	suppress constant term
<u>off</u>set(*varname*)	include *varname* in model with coefficient constrained to 1
asis	retain perfect predictor variables
<u>constra</u>ints(*constraints*)	apply specified linear constraints
<u>col</u>linear	keep collinear variables
SE/Robust	
vce(*vcetype*)	*vcetype* may be oim, <u>r</u>obust, <u>cl</u>uster *clustvar*, <u>boot</u>strap, or <u>jack</u>knife
Reporting	
<u>l</u>evel(#)	set confidence level; default is level(95)
coef	report estimated coefficients
<u>nocns</u>report	do not display constraints
display_options	control column formats, row spacing, line width, and display of omitted variables and base and empty cells
Maximization	
maximize_options	control the maximization process; seldom used
<u>coefl</u>egend	display legend instead of statistics

indepvars may contain factor variables; see [U] **11.4.3 Factor variables**.
depvar and *indepvars* may contain time-series operators; see [U] **11.4.4 Time-series varlists**.
bootstrap, by, fracpoly, jackknife, mfp, mi estimate, nestreg, rolling, statsby, stepwise, and svy are allowed; see [U] **11.1.10 Prefix commands**.
vce(bootstrap) and vce(jackknife) are not allowed with the mi estimate prefix; see [MI] **mi estimate**.
Weights are not allowed with the bootstrap prefix; see [R] **bootstrap**.
vce() and weights are not allowed with the svy prefix; see [SVY] **svy**.
fweights, iweights, and pweights are allowed; see [U] **11.1.6 weight**.
coeflegend does not appear in the dialog box.
See [U] **20 Estimation and postestimation commands** for more capabilities of estimation commands.

Menu

Statistics > Binary outcomes > Logistic regression (reporting odds ratios)

Description

logistic fits a logistic regression model of *depvar* on *indepvars*, where *depvar* is a 0/1 variable (or, more precisely, a 0/non-0 variable). Without arguments, logistic redisplays the last logistic estimates. logistic displays estimates as odds ratios; to view coefficients, type logit after running logistic. To obtain odds ratios for any covariate pattern relative to another, see [R] **lincom**.

Options

────┌ Model ┐───

noconstant, offset(*varname*), constraints(*constraints*), collinear; see [R] **estimation options**.

asis forces retention of perfect predictor variables and their associated perfectly predicted observations and may produce instabilities in maximization; see [R] **probit**.

────┌ SE/Robust ┐───

vce(*vcetype*) specifies the type of standard error reported, which includes types that are derived from asymptotic theory, that are robust to some kinds of misspecification, that allow for intragroup correlation, and that use bootstrap or jackknife methods; see [R] *vce_option*.

────┌ Reporting ┐───

level(*#*); see [R] **estimation options**.

coef causes logistic to report the estimated coefficients rather than the odds ratios (exponentiated coefficients). coef may be specified when the model is fit or may be used later to redisplay results. coef affects only how results are displayed and not how they are estimated.

nocnsreport; see [R] **estimation options**.

display_options: noomitted, vsquish, noemptycells, baselevels, allbaselevels, cformat(%*fmt*), pformat(%*fmt*), sformat(%*fmt*), and nolstretch; see [R] **estimation options**.

────┌ Maximization ┐──

maximize_options: difficult, technique(*algorithm_spec*), iterate(*#*), [no]log, trace, gradient, showstep, hessian, showtolerance, tolerance(*#*), ltolerance(*#*), nrtolerance(*#*), nonrtolerance, and from(*init_specs*); see [R] **maximize**. These options are seldom used.

The following option is available with logistic but is not shown in the dialog box:

coeflegend; see [R] **estimation options**.

Remarks

Remarks are presented under the following headings:

> *logistic and logit*
> *Robust estimate of variance*

logistic and logit

logistic provides an alternative and preferred way to fit maximum-likelihood logit models, the other choice being logit ([R] **logit**).

First, let's dispose of some confusing terminology. We use the words logit and logistic to mean the same thing: maximum likelihood estimation. To some, one or the other of these words connotes transforming the dependent variable and using weighted least squares to fit the model, but that is not how we use either word here. Thus the logit and logistic commands produce the same results.

The logistic command is generally preferred to the logit command because logistic presents the estimates in terms of odds ratios rather than coefficients. To some people, this may seem disadvantageous, but you can type logit without arguments after logistic to see the underlying coefficients. You should be cautious when interpreting the odds ratio of the constant term. Usually, this odds ratio represents the baseline odds of the model when all predictor variables are set to zero. However, you must verify that a zero value for all predictor variables in the model actually makes sense before continuing with this interpretation.

Nevertheless, [R] **logit** is still worth reading because logistic shares the same features as logit, including omitting variables due to collinearity or one-way causation.

For an introduction to logistic regression, see Lemeshow and Hosmer (2005), Pagano and Gauvreau (2000, 470–487), or Pampel (2000); for a complete but nonmathematical treatment, see Kleinbaum and Klein (2010); and for a thorough discussion, see Hosmer and Lemeshow (2000). See Gould (2000) for a discussion of the interpretation of logistic regression. See Dupont (2009) or Hilbe (2009) for a discussion of logistic regression with examples using Stata. For a discussion using Stata with an emphasis on model specification, see Vittinghoff et al. (2005).

Stata has a variety of commands for performing estimation when the dependent variable is dichotomous or polytomous. See Long and Freese (2006) for a book devoted to fitting these models with Stata. Here is a list of some estimation commands that may be of interest. See [I] **estimation commands** for a complete list of all of Stata's estimation commands.

asclogit	[R] **asclogit**	Alternative-specific conditional logit (McFadden's choice) model
asmprobit	[R] **asmprobit**	Alternative-specific multinomial probit regression
asroprobit	[R] **asroprobit**	Alternative-specific rank-ordered probit regression
binreg	[R] **binreg**	Generalized linear models for the binomial family
biprobit	[R] **biprobit**	Bivariate probit regression
blogit	[R] **glogit**	Logit regression for grouped data
bprobit	[R] **glogit**	Probit regression for grouped data
clogit	[R] **clogit**	Conditional (fixed-effects) logistic regression
cloglog	[R] **cloglog**	Complementary log-log regression
exlogistic	[R] **exlogistic**	Exact logistic regression
glm	[R] **glm**	Generalized linear models
glogit	[R] **glogit**	Weighted least-squares logistic regression for grouped data
gprobit	[R] **glogit**	Weighted least-squares probit regression for grouped data
heckprob	[R] **heckprob**	Probit model with selection
hetprob	[R] **hetprob**	Heteroskedastic probit model
ivprobit	[R] **ivprobit**	Probit model with endogenous regressors
logit	[R] **logit**	Logistic regression, reporting coefficients
mlogit	[R] **mlogit**	Multinomial (polytomous) logistic regression
mprobit	[R] **mprobit**	Multinomial probit regression
nlogit	[R] **nlogit**	Nested logit regression (RUM-consistent and nonnormalized)
ologit	[R] **ologit**	Ordered logistic regression
oprobit	[R] **oprobit**	Ordered probit regression
probit	[R] **probit**	Probit regression
rologit	[R] **rologit**	Rank-ordered logistic regression
scobit	[R] **scobit**	Skewed logistic regression
slogit	[R] **slogit**	Stereotype logistic regression
svy: *cmd*	[SVY] **svy estimation**	Survey versions of many of these commands are available; see [SVY] **svy estimation**
xtcloglog	[XT] **xtcloglog**	Random-effects and population-averaged cloglog models
xtgee	[XT] **xtgee**	GEE population-averaged generalized linear models
xtlogit	[XT] **xtlogit**	Fixed-effects, random-effects, and population-averaged logit models
xtprobit	[XT] **xtprobit**	Random-effects and population-averaged probit models

▷ Example 1

Consider the following dataset from a study of risk factors associated with low birthweight described in Hosmer and Lemeshow (2000, 25).

```
. use http://www.stata-press.com/data/r12/lbw
(Hosmer & Lemeshow data)

. describe

Contains data from http://www.stata-press.com/data/r12/lbw.dta
  obs:              189                          Hosmer & Lemeshow data
  vars:              11                          15 Jan 2011 05:01
  size:           2,646
```

variable name	storage type	display format	value label	variable label
id	int	%8.0g		identification code
low	byte	%8.0g		birthweight<2500g
age	byte	%8.0g		age of mother
lwt	int	%8.0g		weight at last menstrual period
race	byte	%8.0g	race	race
smoke	byte	%8.0g		smoked during pregnancy
ptl	byte	%8.0g		premature labor history (count)
ht	byte	%8.0g		has history of hypertension
ui	byte	%8.0g		presence, uterine irritability
ftv	byte	%8.0g		number of visits to physician during 1st trimester
bwt	int	%8.0g		birthweight (grams)

```
Sorted by:
```

We want to investigate the causes of low birthweight. Here race is a categorical variable indicating whether a person is white (race = 1), black (race = 2), or some other race (race = 3). We want indicator (dummy) variables for race included in the regression, so we will use factor variables.

```
. logistic low age lwt i.race smoke ptl ht ui
Logistic regression                             Number of obs   =        189
                                                LR chi2(8)      =      33.22
                                                Prob > chi2     =     0.0001
Log likelihood =   -100.724                     Pseudo R2       =     0.1416
```

low	Odds Ratio	Std. Err.	z	P>|z|	[95% Conf. Interval]	
age	.9732636	.0354759	-0.74	0.457	.9061578	1.045339
lwt	.9849634	.0068217	-2.19	0.029	.9716834	.9984249
race						
2	3.534767	1.860737	2.40	0.016	1.259736	9.918406
3	2.368079	1.039949	1.96	0.050	1.001356	5.600207
smoke	2.517698	1.00916	2.30	0.021	1.147676	5.523162
ptl	1.719161	.5952579	1.56	0.118	.8721455	3.388787
ht	6.249602	4.322408	2.65	0.008	1.611152	24.24199
ui	2.1351	.9808153	1.65	0.099	.8677528	5.2534
_cons	1.586014	1.910496	0.38	0.702	.1496092	16.8134

The odds ratios are for a one-unit change in the variable. If we wanted the odds ratio for age to be in terms of 4-year intervals, we would type

```
. gen age4 = age/4
. logistic low age4 lwt i.race smoke ptl ht ui
(output omitted)
```

After `logistic`, we can type `logit` to see the model in terms of coefficients and standard errors:

```
. logit
```

Logistic regression				Number of obs	=	189
				LR chi2(8)	=	33.22
				Prob > chi2	=	0.0001
Log likelihood =	-100.724			Pseudo R2	=	0.1416

low	Coef.	Std. Err.	z	P>|z|	[95% Conf. Interval]	
age4	-.1084012	.1458017	-0.74	0.457	-.3941673	.1773649
lwt	-.0151508	.0069259	-2.19	0.029	-.0287253	-.0015763
race						
2	1.262647	.5264101	2.40	0.016	.2309024	2.294392
3	.8620792	.4391532	1.96	0.050	.0013548	1.722804
smoke	.9233448	.4008266	2.30	0.021	.137739	1.708951
ptl	.5418366	.346249	1.56	0.118	-.136799	1.220472
ht	1.832518	.6916292	2.65	0.008	.4769494	3.188086
ui	.7585135	.4593768	1.65	0.099	-.1418484	1.658875
_cons	.4612239	1.20459	0.38	0.702	-1.899729	2.822176

If we wanted to see the `logistic` output again, we would type `logistic` without arguments.

◁

▷ Example 2

We can specify the confidence interval for the odds ratios with the `level()` option, and we can do this either at estimation time or when replaying the model. For instance, to see our first model in example 1 with narrower, 90% confidence intervals, we might type

```
. logistic, level(90)
```

Logistic regression				Number of obs	=	189
				LR chi2(8)	=	33.22
				Prob > chi2	=	0.0001
Log likelihood =	-100.724			Pseudo R2	=	0.1416

low	Odds Ratio	Std. Err.	z	P>|z|	[90% Conf. Interval]	
age4	.8972675	.1308231	-0.74	0.457	.7059409	1.140448
lwt	.9849634	.0068217	-2.19	0.029	.9738063	.9962483
race						
2	3.534767	1.860737	2.40	0.016	1.487028	8.402379
3	2.368079	1.039949	1.96	0.050	1.149971	4.876471
smoke	2.517698	1.00916	2.30	0.021	1.302185	4.867819
ptl	1.719161	.5952579	1.56	0.118	.9726876	3.038505
ht	6.249602	4.322408	2.65	0.008	2.003487	19.49478
ui	2.1351	.9808153	1.65	0.099	1.00291	4.545424
_cons	1.586014	1.910496	0.38	0.702	.2186791	11.50288

◁

Robust estimate of variance

If you specify vce(robust), Stata reports the robust estimate of variance described in [U] **20.20 Obtaining robust variance estimates**. Here is the model previously fit with the robust estimate of variance:

```
. logistic low age lwt i.race smoke ptl ht ui, vce(robust)
Logistic regression                             Number of obs   =         189
                                                Wald chi2(8)    =       29.02
                                                Prob > chi2     =      0.0003
Log pseudolikelihood =    -100.724              Pseudo R2       =      0.1416

                          Robust
         low │ Odds Ratio Std. Err.      z    P>|z|     [95% Conf. Interval]
─────────────┼──────────────────────────────────────────────────────────────
         age │  .9732636   .0329376    -0.80   0.423     .9108015    1.040009
         lwt │  .9849634   .0070209    -2.13   0.034     .9712984    .9988206
             │
        race │
          2  │  3.534767   1.793616     2.49   0.013     1.307504    9.556051
          3  │  2.368079   1.026563     1.99   0.047     1.012512    5.538501
             │
       smoke │  2.517698   .9736417     2.39   0.017     1.179852    5.372537
         ptl │  1.719161   .7072902     1.32   0.188     .7675715    3.850476
          ht │  6.249602   4.102026     2.79   0.005     1.726445     22.6231
          ui │    2.1351   1.042775     1.55   0.120     .8197749    5.560858
       _cons │  1.586014   1.939482     0.38   0.706      .144345    17.42658
```

Also you can specify vce(cluster *clustvar*) and then, within cluster, relax the assumption of independence. To illustrate this, we have made some fictional additions to the low-birthweight data.

Say that these data are not a random sample of mothers but instead are a random sample of mothers from a random sample of hospitals. In fact, that may be true—we do not know the history of these data.

Hospitals specialize, and it would not be too incorrect to say that some hospitals specialize in more difficult cases. We are going to show two extremes. In one, all hospitals are alike, but we are going to estimate under the possibility that they might differ. In the other, hospitals are strikingly different. In both cases, we assume that patients are drawn from 20 hospitals.

In both examples, we will fit the same model, and we will type the same command to fit it. Below are the same data we have been using but with a new variable, hospid, that identifies from which of the 20 hospitals each patient was drawn (and which we have made up):

```
. use http://www.stata-press.com/data/r12/hospid1, clear
. logistic low age lwt i.race smoke ptl ht ui, vce(cluster hospid)
```

Logistic regression			Number of obs	=	189
			Wald chi2(8)	=	49.67
			Prob > chi2	=	0.0000
Log pseudolikelihood =	-100.724		Pseudo R2	=	0.1416

(Std. Err. adjusted for 20 clusters in hospid)

low	Odds Ratio	Robust Std. Err.	z	P>|z|	[95% Conf. Interval]	
age	.9732636	.0397476	-0.66	0.507	.898396	1.05437
lwt	.9849634	.0057101	-2.61	0.009	.9738352	.9962187
race						
2	3.534767	2.013285	2.22	0.027	1.157563	10.79386
3	2.368079	.8451325	2.42	0.016	1.176562	4.766257
smoke	2.517698	.8284259	2.81	0.005	1.321062	4.79826
ptl	1.719161	.6676221	1.40	0.163	.8030814	3.680219
ht	6.249602	4.066275	2.82	0.005	1.74591	22.37086
ui	2.1351	1.093144	1.48	0.138	.7827337	5.824014
_cons	1.586014	1.661913	0.44	0.660	.2034094	12.36639

The standard errors are similar to the standard errors we have previously obtained, whether we used the robust or conventional estimators. In this example, we invented the hospital IDs randomly.

Here are the results of the estimation with the same data but with a different set of hospital IDs:

```
. use http://www.stata-press.com/data/r12/hospid2
. logistic low age lwt i.race smoke ptl ht ui, vce(cluster hospid)
```

Logistic regression			Number of obs	=	189
			Wald chi2(8)	=	7.19
			Prob > chi2	=	0.5167
Log pseudolikelihood =	-100.724		Pseudo R2	=	0.1416

(Std. Err. adjusted for 20 clusters in hospid)

low	Odds Ratio	Robust Std. Err.	z	P>|z|	[95% Conf. Interval]	
age	.9732636	.0293064	-0.90	0.368	.9174862	1.032432
lwt	.9849634	.0106123	-1.41	0.160	.9643817	1.005984
race						
2	3.534767	3.120338	1.43	0.153	.6265521	19.9418
3	2.368079	1.297738	1.57	0.116	.8089594	6.932114
smoke	2.517698	1.570287	1.48	0.139	.7414969	8.548655
ptl	1.719161	.6799153	1.37	0.171	.7919045	3.732161
ht	6.249602	7.165454	1.60	0.110	.660558	59.12808
ui	2.1351	1.411977	1.15	0.251	.5841231	7.804266
_cons	1.586014	1.946253	0.38	0.707	.1431423	17.573

Note the strikingly larger standard errors. What happened? In these data, women most likely to have low-birthweight babies are sent to certain hospitals, and the decision on likeliness is based not just on age, smoking history, etc., but on other things that doctors can see but that are not recorded in our data. Thus merely because a woman is at one of the centers identifies her to be more likely to have a low-birthweight baby.

Saved results

logistic saves the following in e():

Scalars

e(N)	number of observations
e(N_cds)	number of completely determined successes
e(N_cdf)	number of completely determined failures
e(k)	number of parameters
e(k_eq)	number of equations in e(b)
e(k_eq_model)	number of equations in overall model test
e(k_dv)	number of dependent variables
e(df_m)	model degrees of freedom
e(r2_p)	pseudo-R-squared
e(ll)	log likelihood
e(ll_0)	log likelihood, constant-only model
e(N_clust)	number of clusters
e(chi2)	χ^2
e(p)	significance of model test
e(rank)	rank of e(V)
e(ic)	number of iterations
e(rc)	return code
e(converged)	1 if converged, 0 otherwise

Macros

e(cmd)	logistic
e(cmdline)	command as typed
e(depvar)	name of dependent variable
e(wtype)	weight type
e(wexp)	weight expression
e(title)	title in estimation output
e(clustvar)	name of cluster variable
e(offset)	linear offset variable
e(chi2type)	Wald or LR; type of model χ^2 test
e(vce)	*vcetype* specified in vce()
e(vcetype)	title used to label Std. Err.
e(opt)	type of optimization
e(which)	max or min; whether optimizer is to perform maximization or minimization
e(ml_method)	type of ml method
e(user)	name of likelihood-evaluator program
e(technique)	maximization technique
e(properties)	b V
e(estat_cmd)	program used to implement estat
e(predict)	program used to implement predict
e(marginsnotok)	predictions disallowed by margins
e(asbalanced)	factor variables fvset as asbalanced
e(asobserved)	factor variables fvset as asobserved

Matrices

e(b)	coefficient vector
e(Cns)	constraints matrix
e(ilog)	iteration log (up to 20 iterations)
e(gradient)	gradient vector
e(mns)	vector of means of the independent variables
e(rules)	information about perfect predictors
e(V)	variance–covariance matrix of the estimators
e(V_modelbased)	model-based variance

Functions

e(sample)	marks estimation sample

Methods and formulas

`logistic` is implemented as an ado-file.

Define $\mathbf{x}_j$ as the (row) vector of independent variables, augmented by 1, and $\mathbf{b}$ as the corresponding estimated parameter (column) vector. The logistic regression model is fit by `logit`; see [R] **logit** for details of estimation.

The odds ratio corresponding to the ith coefficient is $\psi_i = \exp(b_i)$. The standard error of the odds ratio is $s_i^{\psi} = \psi_i s_i$, where s_i is the standard error of b_i estimated by `logit`.

Define $I_j = \mathbf{x}_j \mathbf{b}$ as the predicted index of the jth observation. The predicted probability of a positive outcome is

$$p_j = \frac{\exp(I_j)}{1 + \exp(I_j)}$$

This command supports the Huber/White/sandwich estimator of the variance and its clustered version using `vce(robust)` and `vce(cluster clustvar)`, respectively. See [P] **_robust**, particularly *Maximum likelihood estimators* and *Methods and formulas*.

`logistic` also supports estimation with survey data. For details on VCEs with survey data, see [SVY] **variance estimation**.

References

Archer, K. J., and S. Lemeshow. 2006. Goodness-of-fit test for a logistic regression model fitted using survey sample data. *Stata Journal* 6: 97–105.

Brady, A. R. 1998. sbe21: Adjusted population attributable fractions from logistic regression. *Stata Technical Bulletin* 42: 8–12. Reprinted in *Stata Technical Bulletin Reprints*, vol. 7, pp. 137–143. College Station, TX: Stata Press.

Buis, M. L. 2010a. Direct and indirect effects in a logit model. *Stata Journal* 10: 11–29.

———. 2010b. Stata tip 87: Interpretation of interactions in nonlinear models. *Stata Journal* 10: 305–308.

Cleves, M. A., and A. Tosetto. 2000. sg139: Logistic regression when binary outcome is measured with uncertainty. *Stata Technical Bulletin* 55: 20–23. Reprinted in *Stata Technical Bulletin Reprints*, vol. 10, pp. 152–156. College Station, TX: Stata Press.

Collett, D. 2003. *Modelling Survival Data in Medical Research*. 2nd ed. London: Chapman & Hall/CRC.

de Irala-Estévez, J., and M. A. Martínez. 2000. sg125: Automatic estimation of interaction effects and their confidence intervals. *Stata Technical Bulletin* 53: 29–31. Reprinted in *Stata Technical Bulletin Reprints*, vol. 9, pp. 270–273. College Station, TX: Stata Press.

Dupont, W. D. 2009. *Statistical Modeling for Biomedical Researchers: A Simple Introduction to the Analysis of Complex Data*. 2nd ed. Cambridge: Cambridge University Press.

Freese, J. 2002. Least likely observations in regression models for categorical outcomes. *Stata Journal* 2: 296–300.

Garrett, J. M. 1997. sbe14: Odds ratios and confidence intervals for logistic regression models with effect modification. *Stata Technical Bulletin* 36: 15–22. Reprinted in *Stata Technical Bulletin Reprints*, vol. 6, pp. 104–114. College Station, TX: Stata Press.

Gould, W. W. 2000. sg124: Interpreting logistic regression in all its forms. *Stata Technical Bulletin* 53: 19–29. Reprinted in *Stata Technical Bulletin Reprints*, vol. 9, pp. 257–270. College Station, TX: Stata Press.

Hilbe, J. M. 1997. sg63: Logistic regression: Standardized coefficients and partial correlations. *Stata Technical Bulletin* 35: 21–22. Reprinted in *Stata Technical Bulletin Reprints*, vol. 6, pp. 162–163. College Station, TX: Stata Press.

———. 2009. *Logistic Regression Models*. Boca Raton, FL: Chapman & Hill/CRC.

Hosmer, D. W., Jr., and S. Lemeshow. 2000. *Applied Logistic Regression*. 2nd ed. New York: Wiley.

Kleinbaum, D. G., and M. Klein. 2010. *Logistic Regression: A Self-Learning Text*. 3rd ed. New York: Springer.

Lemeshow, S., and J.-R. L. Gall. 1994. Modeling the severity of illness of ICU patients: A systems update. *Journal of the American Medical Association* 272: 1049–1055.

Lemeshow, S., and D. W. Hosmer, Jr. 2005. Logistic regression. In Vol. 2 of *Encyclopedia of Biostatistics*, ed. P. Armitage and T. Colton, 2870–2880. Chichester, UK: Wiley.

Long, J. S., and J. Freese. 2006. *Regression Models for Categorical Dependent Variables Using Stata*. 2nd ed. College Station, TX: Stata Press.

Miranda, A., and S. Rabe-Hesketh. 2006. Maximum likelihood estimation of endogenous switching and sample selection models for binary, ordinal, and count variables. *Stata Journal* 6: 285–308.

Mitchell, M. N., and X. Chen. 2005. Visualizing main effects and interactions for binary logit models. *Stata Journal* 5: 64–82.

Pagano, M., and K. Gauvreau. 2000. *Principles of Biostatistics*. 2nd ed. Belmont, CA: Duxbury.

Pampel, F. C. 2000. *Logistic Regression: A Primer*. Thousand Oaks, CA: Sage.

Paul, C. 1998. sg92: Logistic regression for data including multiple imputations. *Stata Technical Bulletin* 45: 28–30. Reprinted in *Stata Technical Bulletin Reprints*, vol. 8, pp. 180–183. College Station, TX: Stata Press.

Pearce, M. S. 2000. sg148: Profile likelihood confidence intervals for explanatory variables in logistic regression. *Stata Technical Bulletin* 56: 45–47. Reprinted in *Stata Technical Bulletin Reprints*, vol. 10, pp. 211–214. College Station, TX: Stata Press.

Pregibon, D. 1981. Logistic regression diagnostics. *Annals of Statistics* 9: 705–724.

Reilly, M., and A. Salim. 2000. sg156: Mean score method for missing covariate data in logistic regression models. *Stata Technical Bulletin* 58: 25–27. Reprinted in *Stata Technical Bulletin Reprints*, vol. 10, pp. 256–258. College Station, TX: Stata Press.

Schonlau, M. 2005. Boosted regression (boosting): An introductory tutorial and a Stata plugin. *Stata Journal* 5: 330–354.

Vittinghoff, E., D. V. Glidden, S. C. Shiboski, and C. E. McCulloch. 2005. *Regression Methods in Biostatistics: Linear, Logistic, Survival, and Repeated Measures Models*. New York: Springer.

Xu, J., and J. S. Long. 2005. Confidence intervals for predicted outcomes in regression models for categorical outcomes. *Stata Journal* 5: 537–559.

Also see

Title

> **logistic postestimation** — Postestimation tools for logistic

Description

The following postestimation commands are of special interest after `logistic`:

Command	Description
estat classification	report various summary statistics, including the classification table
estat gof	Pearson or Hosmer–Lemeshow goodness-of-fit test
lroc	compute area under ROC curve and graph the curve
lsens	graph sensitivity and specificity versus probability cutoff

These commands are not appropriate after the svy prefix.

For information about these commands, see below.

The following standard postestimation commands are also available:

Command	Description
contrast	contrasts and ANOVA-style joint tests of estimates
estat	AIC, BIC, VCE, and estimation sample summary
estat (svy)	postestimation statistics for survey data
estimates	cataloging estimation results
lincom	point estimates, standard errors, testing, and inference for linear combinations of coefficients
linktest	link test for model specification
lrtest[1]	likelihood-ratio test
margins	marginal means, predictive margins, marginal effects, and average marginal effects
marginsplot	graph the results from margins (profile plots, interaction plots, etc.)
nlcom	point estimates, standard errors, testing, and inference for nonlinear combinations of coefficients
predict	predictions, residuals, influence statistics, and other diagnostic measures
predictnl	point estimates, standard errors, testing, and inference for generalized predictions
pwcompare	pairwise comparisons of estimates
suest	seemingly unrelated estimation
test	Wald tests of simple and composite linear hypotheses
testnl	Wald tests of nonlinear hypotheses

[1] `lrtest` is not appropriate with svy estimation results.

See the corresponding entries in the *Base Reference Manual* for details, but see [SVY] **estat** for details about estat (svy).

Special-interest postestimation commands

estat classification reports various summary statistics, including the classification table.

estat gof reports the Pearson goodness-of-fit test or the Hosmer–Lemeshow goodness-of-fit test.

lroc graphs the ROC curve and calculates the area under the curve.

lsens graphs sensitivity and specificity versus probability cutoff and optionally creates new variables containing these data.

estat classification, estat gof, lroc, and lsens produce statistics and graphs either for the estimation sample or for any set of observations. However, they always use the estimation sample by default. When weights, if, or in is used with logistic, it is not necessary to repeat the qualifier with these commands when you want statistics computed for the estimation sample. Specify if, in, or the all option only when you want statistics computed for a set of observations other than the estimation sample. Specify weights (only fweights are allowed with these commands) only when you want to use a different set of weights.

By default, estat classification, estat gof, lroc, and lsens use the last model fit by logistic. You may also directly specify the model to the lroc and lsens commands by inputting a vector of coefficients with the beta() option and passing the name of the dependent variable *depvar*.

estat classification and estat gof require that the current estimation results be from logistic, logit, or probit. lroc and lsens commands may also be used after logit or probit. estat classification, lroc, and lsens may also be used after ivprobit.

Syntax for predict

predict [*type*] *newvar* [*if*] [*in*] [, *statistic* <u>nooff</u>set <u>rules</u> asif]

statistic	Description
Main	
<u>pr</u>	probability of a positive outcome; the default
xb	linear prediction
stdp	standard error of the prediction
* <u>dbeta</u>	Pregibon (1981) $\Delta\widehat{\beta}$ influence statistic
* <u>deviance</u>	deviance residual
* <u>dx2</u>	Hosmer and Lemeshow (2000) $\Delta\chi^2$ influence statistic
* <u>ddeviance</u>	Hosmer and Lemeshow (2000) ΔD influence statistic
* <u>hat</u>	Pregibon (1981) leverage
* <u>number</u>	sequential number of the covariate pattern
* <u>residuals</u>	Pearson residuals; adjusted for number sharing covariate pattern
* <u>rstandard</u>	standardized Pearson residuals; adjusted for number sharing covariate pattern
<u>score</u>	first derivative of the log likelihood with respect to $\mathbf{x}_j\boldsymbol{\beta}$

Unstarred statistics are available both in and out of sample; type predict ... if e(sample) ... if wanted only for the estimation sample. Starred statistics are calculated only for the estimation sample, even when if e(sample) is not specified.

pr, xb, stdp, and score are the only options allowed with svy estimation results.

Menu

Statistics > Postestimation > Predictions, residuals, etc.

Options for predict

Main

pr, the default, calculates the probability of a positive outcome.

xb calculates the linear prediction.

stdp calculates the standard error of the linear prediction.

dbeta calculates the Pregibon (1981) $\Delta\widehat{\beta}$ influence statistic, a standardized measure of the difference in the coefficient vector that is due to deletion of the observation along with all others that share the same covariate pattern. In Hosmer and Lemeshow (2000, 144–145) jargon, this statistic is M-asymptotic; that is, it is adjusted for the number of observations that share the same covariate pattern.

deviance calculates the deviance residual.

dx2 calculates the Hosmer and Lemeshow (2000, 174) $\Delta\chi^2$ influence statistic, reflecting the decrease in the Pearson χ^2 that is due to the deletion of the observation and all others that share the same covariate pattern.

ddeviance calculates the Hosmer and Lemeshow (2000, 174) ΔD influence statistic, which is the change in the deviance residual that is due to deletion of the observation and all others that share the same covariate pattern.

hat calculates the Pregibon (1981) leverage or the diagonal elements of the hat matrix adjusted for the number of observations that share the same covariate pattern.

number numbers the covariate patterns—observations with the same covariate pattern have the same number. Observations not used in estimation have number set to missing. The first covariate pattern is numbered 1, the second 2, and so on.

residuals calculates the Pearson residual as given by Hosmer and Lemeshow (2000, 145) and adjusted for the number of observations that share the same covariate pattern.

rstandard calculates the standardized Pearson residual as given by Hosmer and Lemeshow (2000, 173) and adjusted for the number of observations that share the same covariate pattern.

score calculates the equation-level score, $\partial \ln L / \partial(\mathbf{x}_j \boldsymbol{\beta})$.

Options

nooffset is relevant only if you specified offset(*varname*) for logistic. It modifies the calculations made by predict so that they ignore the offset variable; the linear prediction is treated as $\mathbf{x}_j \mathbf{b}$ rather than as $\mathbf{x}_j \mathbf{b} + \text{offset}_j$.

rules requests that Stata use any rules that were used to identify the model when making the prediction. By default, Stata calculates missing for excluded observations. See example 1 in [R] **logit postestimation**.

asif requests that Stata ignore the rules and the exclusion criteria and calculate predictions for all observations possible by using the estimated parameter from the model. See example 1 in [R] **logit postestimation**.

Syntax for estat classification

estat <u>clas</u>sification [*if*] [*in*] [*weight*] [, *class_options*]

class_options	Description
Main	
all	display summary statistics for all observations in the data
<u>cut</u>off(*#*)	positive outcome threshold; default is cutoff(0.5)

fweights are allowed; see [U] **11.1.6 weight**.

Menu

Statistics > Postestimation > Reports and statistics

Options for estat classification

⌐ Main ⌐_____

all requests that the statistic be computed for all observations in the data, ignoring any if or in restrictions specified by logistic.

cutoff(*#*) specifies the value for determining whether an observation has a predicted positive outcome. An observation is classified as positive if its predicted probability is $\geq$ *#*. The default is 0.5.

Syntax for estat gof

estat gof [*if*] [*in*] [*weight*] [, *gof_options*]

gof_options	Description
Main	
<u>group</u>(*#*)	perform Hosmer–Lemeshow goodness-of-fit test using *#* quantiles
all	execute test for all observations in the data
<u>out</u>sample	adjust degrees of freedom for samples outside estimation sample
<u>table</u>	display table of groups used for test

fweights are allowed; see [U] **11.1.6 weight**.

Menu

Statistics > Postestimation > Reports and statistics

Options for estat gof

┌─ Main ───┐

group(#) specifies the number of quantiles to be used to group the data for the Hosmer–Lemeshow goodness-of-fit test. group(10) is typically specified. If this option is not given, the Pearson goodness-of-fit test is computed using the covariate patterns in the data as groups.

all requests that the statistic be computed for all observations in the data, ignoring any if or in restrictions specified with logistic.

outsample adjusts the degrees of freedom for the Pearson and Hosmer–Lemeshow goodness-of-fit tests for samples outside the estimation sample. See *Samples other than the estimation sample* later in this entry.

table displays a table of the groups used for the Hosmer–Lemeshow or Pearson goodness-of-fit test with predicted probabilities, observed and expected counts for both outcomes, and totals for each group.

Syntax for lroc

lroc [*depvar*] [*if*] [*in*] [*weight*] [, *lroc_options*]

lroc_options	Description
Main	
all	compute area under ROC curve and graph curve for all observations
<u>nog</u>raph	suppress graph
Advanced	
beta(*matname*)	row vector containing coefficients for a logistic model
Plot	
cline_options	change the look of the line
marker_options	change look of markers (color, size, etc.)
marker_label_options	add marker labels; change look or position
Reference line	
<u>rlopts</u>(*cline_options*)	affect rendition of the reference line
Add plots	
addplot(*plot*)	add other plots to the generated graph
Y axis, X axis, Titles, Legend, Overall	
twoway_options	any options other than by() documented in [G-3] *twoway_options*

fweights are allowed; see [U] **11.1.6 weight**.

Menu

Statistics > Binary outcomes > Postestimation > ROC curve after logistic/logit/probit/ivprobit

Options for lroc

all requests that the statistic be computed for all observations in the data, ignoring any if or in restrictions specified by logistic.

nograph suppresses graphical output.

beta(*matname*) specifies a row vector containing coefficients for a logistic model. The columns of the row vector must be labeled with the corresponding names of the independent variables in the data. The dependent variable *depvar* must be specified immediately after the command name. See *Models other than the last fitted model* later in this entry.

cline_options, *marker_options*, and *marker_label_options* affect the rendition of the ROC curve—the plotted points connected by lines. These options affect the size and color of markers, whether and how the markers are labeled, and whether and how the points are connected; see [G-3] *cline_options*, [G-3] *marker_options*, and [G-3] *marker_label_options*.

rlopts(*cline_options*) affects the rendition of the reference line; see [G-3] *cline_options*.

addplot(*plot*) provides a way to add other plots to the generated graph; see [G-3] *addplot_option*.

twoway_options are any of the options documented in [G-3] *twoway_options*, excluding by(). These include options for titling the graph (see [G-3] *title_options*) and for saving the graph to disk (see [G-3] *saving_option*).

Syntax for lsens

lsens [*depvar*] [*if*] [*in*] [*weight*] [, *lsens_options*]

lsens_options	Description
Main	
all	graph all observations in the data
genprob(*varname*)	create variable containing probability cutoffs
gensens(*varname*)	create variable containing sensitivity
genspec(*varname*)	create variable containing specificity
replace	overwrite existing variables
nograph	suppress the graph
Advanced	
beta(*matname*)	row vector containing coefficients for the model
Plot	
connect_options	affect rendition of the plotted points connected by lines
Add plots	
addplot(*plot*)	add other plots to the generated graph
Y axis, X axis, Titles, Legend, Overall	
twoway_options	any options other than by() documented in [G-3] ***twoway_options***

fweights are allowed; see [U] **11.1.6 weight**.

Menu

Statistics > Binary outcomes > Postestimation > Sensitivity/specificity plot

Options for lsens

Main

all requests that the statistic be computed for all observations in the data, ignoring any if or in restrictions specified with logistic.

genprob(*varname*), gensens(*varname*), and genspec(*varname*) specify the names of new variables created to contain, respectively, the probability cutoffs and the corresponding sensitivity and specificity.

replace requests that existing variables specified for genprob(), gensens(), or genspec() be overwritten.

nograph suppresses graphical output.

`beta(`*matname*`)` specifies a row vector containing coefficients for a logistic model. The columns of the row vector must be labeled with the corresponding names of the independent variables in the data. The dependent variable *depvar* must be specified immediately after the command name. See *Models other than the last fitted model* later in this entry.

Plot

connect_options affect the rendition of the plotted points connected by lines; see *connect_options* in [G-2] **graph twoway scatter**.

Add plots

`addplot(`*plot*`)` provides a way to add other plots to the generated graph. See [G-3] ***addplot_option***.

Y axis, X axis, Titles, Legend, Overall

twoway_options are any of the options documented in [G-3] ***twoway_options***, excluding `by()`. These include options for titling the graph (see [G-3] ***title_options***) and for saving the graph to disk (see [G-3] ***saving_option***).

Remarks

Remarks are presented under the following headings:

> *predict after logistic*
> > *predict without options*
> > *predict with the xb and stdp options*
> > *predict with the residuals option*
> > *predict with the number option*
> > *predict with the deviance option*
> > *predict with the rstandard option*
> > *predict with the hat option*
> > *predict with the dx2 option*
> > *predict with the ddeviance option*
> > *predict with the dbeta option*
> *estat classification*
> *estat gof*
> *lroc*
> *lsens*
> *Samples other than the estimation sample*
> *Models other than the last fitted model*

predict after logistic

`predict` is used after `logistic` to obtain predicted probabilities, residuals, and influence statistics for the estimation sample. The suggested diagnostic graphs below are from Hosmer and Lemeshow (2000), where they are more elaborately explained. Also see Collett (2003, 129–168) for a thorough discussion of model checking.

predict without options

Typing `predict` *newvar* after estimation calculates the predicted probability of a positive outcome.

In example 1 of [R] **logistic**, we ran the model `logistic low age lwt i.race smoke ptl ht ui`. We obtain the predicted probabilities of a positive outcome by typing

```
. use http://www.stata-press.com/data/r12/lbw
(Hosmer & Lemeshow data)
. logistic low age lwt i.race smoke ptl ht ui
  (output omitted )
. predict p
(option pr assumed; Pr(low))
. summarize p low
```

Variable	Obs	Mean	Std. Dev.	Min	Max
p	189	.3121693	.1913915	.0272559	.8391283
low	189	.3121693	.4646093	0	1

predict with the xb and stdp options

predict with the xb option calculates the linear combination $x_j\mathbf{b}$, where x_j are the independent variables in the jth observation and $\mathbf{b}$ is the estimated parameter vector. This is sometimes known as the index function because the cumulative distribution function indexed at this value is the probability of a positive outcome.

With the stdp option, predict calculates the standard error of the prediction, which is *not* adjusted for replicated covariate patterns in the data. The influence statistics described below are adjusted for replicated covariate patterns in the data.

predict with the residuals option

predict can calculate more than predicted probabilities. The Pearson residual is defined as the square root of the contribution of the covariate pattern to the Pearson χ^2 goodness-of-fit statistic, signed according to whether the observed number of positive responses within the covariate pattern is less than or greater than expected. For instance,

```
. predict r, residuals
. summarize r, detail
```

                          Pearson residual

	Percentiles	Smallest		
1%	-1.750923	-2.283885		
5%	-1.129907	-1.750923		
10%	-.9581174	-1.636279	Obs	189
25%	-.6545911	-1.636279	Sum of Wgt.	189
50%	-.3806923		Mean	-.0242299
		Largest	Std. Dev.	.9970949
75%	.8162894	2.23879		
90%	1.510355	2.317558	Variance	.9941981
95%	1.747948	3.002206	Skewness	.8618271
99%	3.002206	3.126763	Kurtosis	3.038448

We notice the prevalence of a few large positive residuals:

```
. sort r
. list id r low p age race in -5/l
```

	id	r	low	p	age	race
185.	33	2.224501	1	.1681123	19	white
186.	57	2.23879	1	.166329	15	white
187.	16	2.317558	1	.1569594	27	other
188.	77	3.002206	1	.0998678	26	white
189.	36	3.126763	1	.0927932	24	white

predict with the number option

Covariate patterns play an important role in logistic regression. Two observations are said to share the same covariate pattern if the independent variables for the two observations are identical. Although we might think of having individual observations, the statistical information in the sample can be summarized by the covariate patterns, the number of observations with that covariate pattern, and the number of positive outcomes within the pattern. Depending on the model, the number of covariate patterns can approach or be equal to the number of observations, or it can be considerably less.

Stata calculates all the residual and diagnostic statistics in terms of covariate patterns, not observations. That is, all observations with the same covariate pattern are given the same residual and diagnostic statistics. Hosmer and Lemeshow (2000, 145–145) argue that such "M-asymptotic" statistics are more useful than "N-asymptotic" statistics.

To understand the difference, think of an observed positive outcome with predicted probability of 0.8. Taking the observation in isolation, the residual must be positive—we expected 0.8 positive responses and observed 1. This may indeed be the correct residual, but not necessarily. Under the M-asymptotic definition, we ask how many successes we observed across all observations with this covariate pattern. If that number were, say, six, and there were a total of 10 observations with this covariate pattern, then the residual is negative for the covariate pattern—we expected eight positive outcomes but observed six. predict makes this kind of calculation and then attaches the same residual to all observations in the covariate pattern.

Occasionally, you might want to find all observations sharing a covariate pattern. number allows you to do this:

```
. predict pattern, number
. summarize pattern
```

Variable		Obs	Mean	Std. Dev.	Min	Max
pattern		189	89.2328	53.16573	1	182

We previously fit the model logistic low age lwt i.race smoke ptl ht ui over 189 observations. There are 182 covariate patterns in our data.

predict with the deviance option

The deviance residual is defined as the square root of the contribution to the likelihood-ratio test statistic of a saturated model versus the fitted model. It has slightly different properties from the Pearson residual (see Hosmer and Lemeshow [2000, 145–147]):

```
. predict d, deviance

. summarize d, detail
```

```
                       deviance residual

          Percentiles      Smallest
   1%      -1.843472      -1.911621
   5%      -1.33477       -1.843472
  10%      -1.148316      -1.843472       Obs                 189
  25%      -.8445325      -1.674869       Sum of Wgt.         189

  50%      -.5202702                      Mean          -.1228811
                            Largest       Std. Dev.      1.049237
  75%       .9129041       1.894089
  90%      1.541558        1.924457       Variance       1.100898
  95%      1.673338        2.146583       Skewness       .6598857
  99%      2.146583        2.180542       Kurtosis       2.036938
```

predict with the rstandard option

Pearson residuals do not have a standard deviation equal to 1. rstandard generates Pearson residuals normalized to have an *expected* standard deviation equal to 1.

```
. predict rs, rstandard

. summarize r rs

      Variable |       Obs        Mean    Std. Dev.       Min        Max

             r |       189   -.0242299    .9970949   -2.283885   3.126763
            rs |       189   -.0279135    1.026406    -2.4478    3.149081

. correlate r rs
(obs=189)

                |        r        rs

             r |   1.0000
            rs |   0.9998    1.0000
```

Remember that we previously created r containing the (unstandardized) Pearson residuals. In these data, whether we use standardized or unstandardized residuals does not matter much.

predict with the hat option

hat calculates the leverage of a covariate pattern—a scaled measure of distance in terms of the independent variables. Large values indicate covariate patterns far from the average covariate pattern that can have a large effect on the fitted model even if the corresponding residual is small. Consider the following graph:

```
. predict h, hat
. scatter h r, xline(0)
```

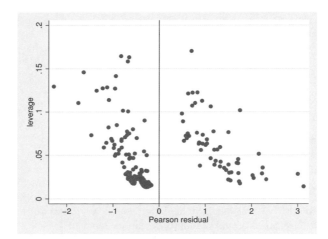

The points to the left of the vertical line are observed negative outcomes; here our data contain almost as many covariate patterns as observations, so most covariate patterns are unique. In such unique patterns, we observe either 0 or 1 success and expect p, thus forcing the sign of the residual. If we had fewer covariate patterns—if we did not have continuous variables in our model—there would be no such interpretation, and we would not have drawn the vertical line at 0.

Points on the left and right edges of the graph represent large residuals—covariate patterns that are not fit well by our model. Points at the top of our graph represent high leverage patterns. When analyzing the influence of observations on the model, we are most interested in patterns with high leverage and small residuals—patterns that might otherwise escape our attention.

predict with the dx2 option

There are many ways to measure influence, and `hat` is one example. `dx2` measures the decrease in the Pearson χ^2 goodness-of-fit statistic that would be caused by deleting an observation (and all others sharing the covariate pattern):

```
. predict dx2, dx2
. scatter dx2 p
```

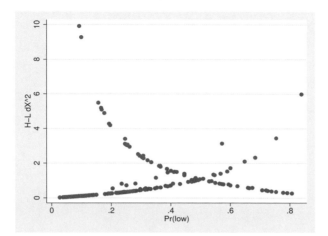

Paraphrasing Hosmer and Lemeshow (2000, 178–179), the points going from the top left to the bottom right correspond to covariate patterns with the number of positive outcomes equal to the number in the group; the points on the other curve correspond to 0 positive outcomes. In our data, most of the covariate patterns are unique, so the points tend to lie along one or the other curves; the points that are off the curves correspond to the few repeated covariate patterns in our data in which all the outcomes are not the same.

We examine this graph for large values of dx2—there are two at the top left.

predict with the ddeviance option

Another measure of influence is the change in the deviance residuals due to deletion of a covariate pattern:

```
. predict dd, ddeviance
```

As with dx2, we typically graph ddeviance against the probability of a positive outcome. We direct you to Hosmer and Lemeshow (2000, 178) for an example and for the interpretation of this graph.

predict with the dbeta option

One of the more direct measures of influence of interest to model fitters is the Pregibon (1981) dbeta measure, a measure of the change in the coefficient vector that would be caused by deleting an observation (and all others sharing the covariate pattern):

```
. predict db, dbeta
. scatter db p
```

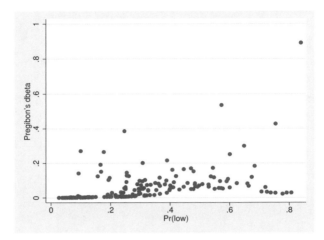

One observation has a large effect on the estimated coefficients. We can easily find this point:

```
. sort db
. list in 1
```

189.	id	low	age	lwt	race	smoke	ptl	ht	ui	ftv	bwt
	188	0	25	95	white	1	3	0	1	0	3637

p	r	pattern	d	rs	h
.8391283	-2.283885	117	-1.911621	-2.4478	.1294439

dx2	dd	db
5.991726	4.197658	.8909163

Hosmer and Lemeshow (2000, 180) suggest a graph that combines two of the influence measures:

```
. scatter dx2 p [w=db], title("Symbol size proportional to dBeta") mfcolor(none)
(analytic weights assumed)
```

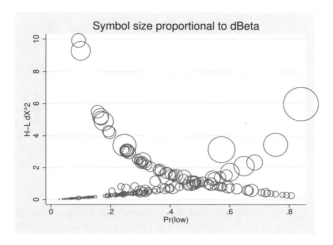

We can easily spot the most influential points by the dbeta and dx2 measures.

estat classification

▷ Example 1

estat classification presents the classification statistics and classification table after logistic.

```
. use http://www.stata-press.com/data/r12/lbw, clear
(Hosmer & Lemeshow data)
. logistic low age lwt i.race smoke ptl ht ui
  (output omitted )
. estat classification
Logistic model for low
```

Classified	— True —		Total
	D	~D	
+	21	12	33
−	38	118	156
Total	59	130	189

Classified + if predicted Pr(D) >= .5
True D defined as low != 0

Sensitivity	Pr(+| D)	35.59%
Specificity	Pr(−|~D)	90.77%
Positive predictive value	Pr(D| +)	63.64%
Negative predictive value	Pr(~D| −)	75.64%
False + rate for true ~D	Pr(+|~D)	9.23%
False − rate for true D	Pr(−| D)	64.41%
False + rate for classified +	Pr(~D| +)	36.36%
False − rate for classified −	Pr(D| −)	24.36%
Correctly classified		73.54%

By default, estat classification uses a cutoff of 0.5, although you can vary this with the cutoff() option. You can use the lsens command to review the potential cutoffs; see *lsens* below.

◁

estat gof

estat gof computes goodness-of-fit tests: either the Pearson χ^2 test or the Hosmer–Lemeshow test.

By default, estat classification, estat gof, lroc, and lsens compute statistics for the estimation sample by using the last model fit by logistic. However, samples other than the estimation sample can be specified; see *Samples other than the estimation sample* later in this entry.

▷ Example 2

estat gof, typed without options, presents the Pearson χ^2 goodness-of-fit test for the fitted model. The Pearson χ^2 goodness-of-fit test is a test of the observed against expected number of responses using cells defined by the covariate patterns; see *predict with the number option* earlier in this entry for the definition of covariate patterns.

```
. estat gof
Logistic model for low, goodness-of-fit test
        number of observations =        189
   number of covariate patterns =        182
           Pearson chi2(173) =        179.24
                Prob > chi2 =         0.3567
```

Our model fits reasonably well. However, the number of covariate patterns is close to the number of observations, making the applicability of the Pearson χ^2 test questionable but not necessarily inappropriate. Hosmer and Lemeshow (2000, 147–150) suggest regrouping the data by ordering on the predicted probabilities and then forming, say, 10 nearly equal-sized groups. estat gof with the group() option does this:

```
. estat gof, group(10)
Logistic model for low, goodness-of-fit test
  (Table collapsed on quantiles of estimated probabilities)
        number of observations =        189
            number of groups =         10
     Hosmer-Lemeshow chi2(8) =          9.65
                Prob > chi2 =         0.2904
```

Again we cannot reject our model. If we specify the table option, estat gof displays the groups along with the expected and observed number of positive responses (low-birthweight babies):

```
. estat gof, group(10) table
```

Logistic model for low, goodness-of-fit test

(Table collapsed on quantiles of estimated probabilities)

Group	Prob	Obs_1	Exp_1	Obs_0	Exp_0	Total
1	0.0827	0	1.2	19	17.8	19
2	0.1276	2	2.0	17	17.0	19
3	0.2015	6	3.2	13	15.8	19
4	0.2432	1	4.3	18	14.7	19
5	0.2792	7	4.9	12	14.1	19
6	0.3138	7	5.6	12	13.4	19
7	0.3872	6	6.5	13	12.5	19
8	0.4828	7	8.2	12	10.8	19
9	0.5941	10	10.3	9	8.7	19
10	0.8391	13	12.8	5	5.2	18

```
          number of observations =        189
                number of groups =         10
       Hosmer-Lemeshow chi2(8) =          9.65
                    Prob > chi2 =        0.2904
```

◁

❏ Technical note

estat gof with the group() option puts all observations with the same predicted probabilities into the same group. If, as in the previous example, we request 10 groups, the groups that estat gof makes are $[p_0, p_{10}], (p_{10}, p_{20}], (p_{20}, p_{30}], \ldots, (p_{90}, p_{100}]$, where p_k is the kth percentile of the predicted probabilities, with p_0 the minimum and p_{100} the maximum.

If there are many ties at the quantile boundaries, as will often happen if all independent variables are categorical and there are only a few of them, the sizes of the groups will be uneven. If the totals in some of the groups are small, the χ^2 statistic for the Hosmer–Lemeshow test may be unreliable. In this case, fewer groups should be specified, or the Pearson goodness-of-fit test may be a better choice.

❏

▷ Example 3

The table option can be used without the group() option. We would not want to specify this for our current model because there were 182 covariate patterns in the data, caused by including the two continuous variables, age and lwt, in the model. As an aside, we fit a simpler model and specify table with estat gof:

```
. logistic low i.race smoke ui
```

Logistic regression

			Number of obs	=	189
			LR chi2(4)	=	18.80
			Prob > chi2	=	0.0009
Log likelihood = -107.93404			Pseudo R2	=	0.0801

low	Odds Ratio	Std. Err.	z	P>|z|	[95% Conf. Interval]	
race						
2	3.052746	1.498087	2.27	0.023	1.166747	7.987382
3	2.922593	1.189229	2.64	0.008	1.316457	6.488285
smoke	2.945742	1.101838	2.89	0.004	1.415167	6.131715
ui	2.419131	1.047359	2.04	0.041	1.035459	5.651788
_cons	.1402209	.0512295	-5.38	0.000	.0685216	.2869447

```
. estat gof, table
```

Logistic model for low, goodness-of-fit test

Group	Prob	Obs_1	Exp_1	Obs_0	Exp_0	Total
1	0.1230	3	4.9	37	35.1	40
2	0.2533	1	1.0	3	3.0	4
3	0.2907	16	13.7	31	33.3	47
4	0.2923	15	12.6	28	30.4	43
5	0.2997	3	3.9	10	9.1	13
6	0.4978	4	4.0	4	4.0	8
7	0.4998	4	4.5	5	4.5	9
8	0.5087	2	1.5	1	1.5	3
9	0.5469	2	4.4	6	3.6	8
10	0.5577	6	5.6	4	4.4	10
11	0.7449	3	3.0	1	1.0	4

Group	Prob	race	smoke	ui
1	0.1230	white	0	0
2	0.2533	white	0	1
3	0.2907	other	0	0
4	0.2923	white	1	0
5	0.2997	black	0	0
6	0.4978	other	0	1
7	0.4998	white	1	1
8	0.5087	black	0	1
9	0.5469	other	1	0
10	0.5577	black	1	0
11	0.7449	other	1	1

```
      number of observations =        189
number of covariate patterns =         11
            Pearson chi2(6) =        5.71
               Prob > chi2 =      0.4569
```

❑ Technical note

logistic and estat gof keep track of the estimation sample. If you type logistic ... if x==1, then when you type estat gof, the statistics will be calculated on the x==1 subsample of the data automatically.

You should specify if or in with estat gof only when you wish to calculate statistics for a set of observations other than the estimation sample. See *Samples other than the estimation sample* later in this entry.

If the logistic model was fit with fweights, estat gof properly accounts for the weights in its calculations. (estat gof does not allow pweights.) You do not have to specify the weights when you run estat gof. Weights should be specified with estat gof only when you wish to use a different set of weights.

❑

lroc

Stata also has a suite of commands for performing both parametric and nonparametric receiver operating characteristic (ROC) analysis. See [R] **roc** for an overview of these commands.

lroc graphs the ROC curve—a graph of sensitivity versus one minus specificity as the cutoff c is varied—and calculates the area under it. Sensitivity is the fraction of observed positive-outcome cases that are correctly classified; specificity is the fraction of observed negative-outcome cases that are correctly classified. When the purpose of the analysis is classification, you must choose a cutoff.

The curve starts at $(0, 0)$, corresponding to $c = 1$, and continues to $(1, 1)$, corresponding to $c = 0$. A model with no predictive power would be a $45°$ line. The greater the predictive power, the more bowed the curve, and hence the area beneath the curve is often used as a measure of the predictive power. A model with no predictive power has area 0.5; a perfect model has area 1.

The ROC curve was first discussed in signal detection theory (Peterson, Birdsall, and Fox 1954) and then was quickly introduced into psychology (Tanner and Swets 1954). It has since been applied in other fields, particularly medicine (for instance, Metz [1978]). For a classic text on ROC techniques, see Green and Swets (1966).

▷ Example 4

ROC curves are typically used when the point of the analysis is classification—which it is not in our low-birthweight model. Nevertheless, the ROC curve is

```
. lroc
```

Logistic model for low

```
number of observations =        189
area under ROC curve   =     0.6658
```

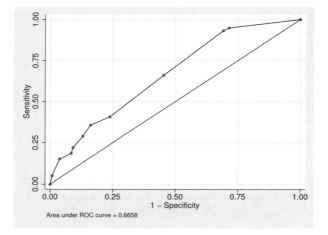

We see that the area under the curve is 0.6658.

◁

lsens

lsens also plots sensitivity and specificity; it plots both sensitivity and specificity versus probability cutoff c. The graph is equivalent to what you would get from estat classification if you varied the cutoff probability from 0 to 1.

```
. lsens
```

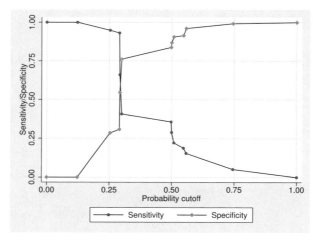

lsens optionally creates new variables containing the probability cutoff, sensitivity, and specificity.

```
. lsens, genprob(p) gensens(sens) genspec(spec) nograph
```

The variables created will have $M + 2$ distinct nonmissing values: one for each of the M covariate patterns, one for $c = 0$, and another for $c = 1$. Values are recorded for $p = 0$, for each of the observed predicted probabilities, and for $p = 1$. The total number of observations required to do this can be fewer than _N, the same as _N, or _N + 1, or _N + 2. If more observations are added, they are added at the end of the dataset and the values of the original variables are set to missing in the added observations. How the values added align with existing observations is irrelevant.

Samples other than the estimation sample

estat gof, estat classification, lroc, and lsens can be used with samples other than the estimation sample. By default, these commands remember the estimation sample used with the last logistic command. To override this, simply use an if or in restriction to select another set of observations, or specify the all option to force the command to use all the observations in the dataset.

If you use estat gof with a sample that is completely different from the estimation sample (that is, no overlap), you should also specify the outsample option so that the χ^2 statistic properly adjusts the degrees of freedom upward. For an overlapping sample, the conservative thing to do is to leave the degrees of freedom the same as they are for the estimation sample.

▷ Example 5

We want to develop a model for predicting low-birthweight babies. One approach would be to divide our data into two groups, a developmental sample and a validation sample. See Lemeshow and Gall (1994) and Tilford, Roberson, and Fiser (1995) for more information on developing prediction models and severity-scoring systems.

We will do this with the low-birthweight data that we considered previously. First, we randomly divide the data into two samples.

```
. use http://www.stata-press.com/data/r12/lbw, clear
(Hosmer & Lemeshow data)
. set seed 1
. gen r = runiform()
. sort r
. gen group = 1 if _n <= _N/2
(95 missing values generated)
. replace group = 2 if group >=.
(95 real changes made)
```

Then we fit a model using the first sample (group $= 1$), which is our developmental sample.

```
. logistic low age lwt i.race smoke ptl ht ui if group==1
Logistic regression                             Number of obs   =          94
                                                LR chi2(8)      =       29.14
                                                Prob > chi2     =      0.0003
Log likelihood = -44.293342                     Pseudo R2       =      0.2475
```

low	Odds Ratio	Std. Err.	z	P>|z|	[95% Conf. Interval]	
age	.91542	.0553937	-1.46	0.144	.8130414	1.03069
lwt	.9744276	.0112295	-2.25	0.025	.9526649	.9966874
race						
2	5.063678	3.78442	2.17	0.030	1.170327	21.90913
3	2.606209	1.657608	1.51	0.132	.7492483	9.065522
smoke	.909912	.5252898	-0.16	0.870	.2934966	2.820953
ptl	3.033543	1.507048	2.23	0.025	1.145718	8.03198
ht	21.07656	22.64788	2.84	0.005	2.565304	173.1652
ui	.988479	.6699458	-0.02	0.986	.2618557	3.731409
_cons	30.73641	56.82168	1.85	0.064	.8204589	1151.462

To test calibration in the developmental sample, we calculate the Hosmer–Lemeshow goodness-of-fit test by using `estat gof`.

```
. estat gof, group(10)
Logistic model for low, goodness-of-fit test

(Table collapsed on quantiles of estimated probabilities)
      number of observations =         94
           number of groups =         10
   Hosmer-Lemeshow chi2(8) =       6.67
               Prob > chi2 =     0.5721
```

We did not specify an `if` statement with `estat gof` because we wanted to use the estimation sample. Because the test is not significant, we are satisfied with the fit of our model.

Running `lroc` gives a measure of the discrimination:

```
. lroc, nograph
Logistic model for low
number of observations =         94
area under ROC curve   =     0.8156
```

Now we test the calibration of our model by performing a goodness-of-fit test on the validation sample. We specify the `outsample` option so that the number of degrees of freedom is 10 rather than 8.

```
. estat gof if group==2, group(10) table outsample
```
Logistic model for low, goodness-of-fit test

(Table collapsed on quantiles of estimated probabilities)

Group	Prob	Obs_1	Exp_1	Obs_0	Exp_0	Total
1	0.0725	1	0.4	9	9.6	10
2	0.1202	4	0.8	5	8.2	9
3	0.1549	3	1.3	7	8.7	10
4	0.1888	1	1.5	8	7.5	9
5	0.2609	3	2.2	7	7.8	10
6	0.3258	4	2.7	5	6.3	9
7	0.4217	2	3.7	8	6.3	10
8	0.4915	3	4.1	6	4.9	9
9	0.6265	4	5.5	6	4.5	10
10	0.9737	4	7.1	5	1.9	9

```
           number of observations =        95
              number of groups =        10
   Hosmer-Lemeshow chi2(10) =        28.03
              Prob > chi2 =         0.0018
```

We must acknowledge that our model does not fit well on the validation sample. The model's discrimination in the validation sample is appreciably lower, as well.

```
. lroc if group==2, nograph
Logistic model for low
number of observations =        95
area under ROC curve   =     0.5839
```

◁

Models other than the last fitted model

By default, estat classification, estat gof, lroc, and lsens use the last model fit by logistic. You may also directly specify the model to lroc and lsens by inputting a vector of coefficients with the beta() option and passing the name of the dependent variable *depvar* to these commands.

▷ Example 6

Suppose that someone publishes the following logistic model of low birthweight:

$$\Pr(\text{low} = 1) = F(-0.02\,\text{age} - 0.01\,\text{lwt} + 1.3\,\text{black} + 1.1\,\text{smoke} + 0.5\,\text{ptl} + 1.8\,\text{ht} + 0.8\,\text{ui} + 0.5)$$

where F is the cumulative logistic distribution. These coefficients are not odds ratios; they are the equivalent of what logit produces.

We can see whether this model fits our data. First we enter the coefficients as a row vector and label its columns with the names of the independent variables plus _cons for the constant (see [P] **matrix define** and [P] **matrix rownames**).

```
. use http://www.stata-press.com/data/r12/lbw3, clear
(Hosmer & Lemeshow data)
. matrix input b = (-.02, -.01, 1.3, 1.1, .5, 1.8, .8, .5)
. matrix colnames b = age lwt black smoke ptl ht ui _cons
```

Here we use `lroc` to examine the predictive ability of the model:

```
. lroc low, beta(b) nograph
Logistic model for low
number of observations =       189
area under ROC curve   =    0.7275
```

The area under the curve indicates that this model does have some predictive power. We could also obtain a graph of sensitivity and specificity as a function of the cutoff probability by typing

```
. lsens low, beta(b)
```

◁

Saved results

`estat classification` saves the following in `r()`:

Scalars

`r(P_corr)`	percent correctly classified
`r(P_p1)`	sensitivity
`r(P_n0)`	specificity
`r(P_p0)`	false-positive rate given true negative
`r(P_n1)`	false-negative rate given true positive
`r(P_1p)`	positive predictive value
`r(P_0n)`	negative predictive value
`r(P_0p)`	false-positive rate given classified positive
`r(P_1n)`	false-negative rate given classified negative

`estat gof` saves the following in `r()`:

Scalars

`r(N)`	number of observations
`r(m)`	number of covariate patterns or groups
`r(df)`	degrees of freedom
`r(chi2)`	χ^2

`lroc` saves the following in `r()`:

Scalars

`r(N)`	number of observations
`r(area)`	area under the ROC curve

`lsens` saves the following in `r()`:

Scalars

`r(N)`	number of observations

Methods and formulas

All postestimation commands listed above are implemented as ado-files.

Methods and formulas are presented under the following headings:

> *estat gof*
> *predict after logistic*
> *estat classification and lsens*
> *lroc*

stat gof

Let M be the total number of covariate patterns among the N observations. View the data as collapsed on covariate patterns $j = 1, 2, \ldots, M$, and define m_j as the total number of observations having covariate pattern j and y_j as the total number of positive responses among observations with covariate pattern j. Define p_j as the predicted probability of a positive outcome in covariate pattern j.

The Pearson χ^2 goodness-of-fit statistic is

$$\chi^2 = \sum_{j=1}^{M} \frac{(y_j - m_j p_j)^2}{m_j p_j (1 - p_j)}$$

This χ^2 statistic has approximately $M - k$ degrees of freedom for the estimation sample, where k is the number of independent variables, including the constant. For a sample outside the estimation sample, the statistic has M degrees of freedom.

The Hosmer–Lemeshow goodness-of-fit χ^2 (Hosmer and Lemeshow 1980; Lemeshow and Hosmer 1982; Hosmer, Lemeshow, and Klar 1988) is calculated similarly, except that rather than using the M covariate patterns as the group definition, the quantiles of the predicted probabilities are used to form groups. Let $G = \#$ be the number of quantiles requested with group(#). The smallest index $1 \leq q(i) \leq M$, such that

$$W_{q(i)} = \sum_{j=1}^{q(i)} m_j \geq \frac{N}{G}$$

gives $p_{q(i)}$ as the upper boundary of the ith quantile for $i = 1, 2, \ldots, G$. Let $q(0) = 1$ denote the first index.

The groups are then

$$[p_{q(0)}, p_{q(1)}], (p_{q(1)}, p_{q(2)}], \ldots, (p_{q(G-1)}, p_{q(G)}]$$

If the table option is given, the upper boundaries $p_{q(1)}, \ldots, p_{q(G)}$ of the groups appear next to the group number on the output.

The resulting χ^2 statistic has approximately $G - 2$ degrees of freedom for the estimation sample. For a sample outside the estimation sample, the statistic has G degrees of freedom.

predict after logistic

Index j will now be used to index observations, not covariate patterns. Define M_j for each observation as the total number of observations sharing j's covariate pattern. Define Y_j as the total number of positive responses among observations sharing j's covariate pattern.

The Pearson residual for the jth observation is defined as

$$r_j = \frac{Y_j - M_j p_j}{\sqrt{M_j p_j (1 - p_j)}}$$

For $M_j > 1$, the deviance residual d_j is defined as

$$d_j = \pm \left(2 \left[Y_j \ln \left(\frac{Y_j}{M_j p_j} \right) + (M_j - Y_j) \ln \left\{ \frac{M_j - Y_j}{M_j (1 - p_j)} \right\} \right] \right)^{1/2}$$

where the sign is the same as the sign of $(Y_j - M_j p_j)$. In the limiting cases, the deviance residual is given by

$$d_j = \begin{cases} -\sqrt{2M_j|\ln(1-p_j)|} & \text{if } Y_j = 0 \\ \sqrt{2M_j|\ln p_j|} & \text{if } Y_j = M_j \end{cases}$$

The *unadjusted* diagonal elements of the hat matrix h_{Uj} are given by $h_{Uj} = (\mathbf{XVX'})_{jj}$, where V is the estimated covariance matrix of parameters. The adjusted diagonal elements h_j created by `hat` are then $h_j = M_j p_j (1 - p_j) h_{Uj}$.

The standardized Pearson residual r_{Sj} is $r_j/\sqrt{1-h_j}$.

The Pregibon (1981) $\Delta\widehat{\beta}_j$ influence statistic is

$$\Delta\widehat{\beta}_j = \frac{r_j^2 h_j}{(1-h_j)^2}$$

The corresponding change in the Pearson χ^2 is r_{Sj}^2. The corresponding change in the deviance residual is $\Delta D_j = d_j^2/(1-h_j)$.

estat classification and lsens

Again let j index observations. Define c as the `cutoff()` specified by the user or, if not specified, as 0.5. Let p_j be the predicted probability of a positive outcome and y_j be the actual outcome, which we will treat as 0 or 1, although Stata treats it as 0 and non-0, excluding missing observations.

A prediction is classified as *positive* if $p_j \geq c$ and otherwise is classified as *negative*. The classification is *correct* if it is *positive* and $y_j = 1$ or if it is *negative* and $y_j = 0$.

Sensitivity is the fraction of $y_j = 1$ observations that are correctly classified. *Specificity* is the percentage of $y_j = 0$ observations that are correctly classified.

lroc

The ROC curve is a graph of *specificity* against $(1 - sensitivity)$. This is guaranteed to be a monotone nondecreasing function because the number of correctly predicted successes increases and the number of correctly predicted failures decreases as the classification cutoff c decreases.

The area under the ROC curve is the area on the bottom of this graph and is determined by integrating the curve. The vertices of the curve are determined by sorting the data according to the predicted index, and the integral is computed using the trapezoidal rule.

References

Archer, K. J., and S. Lemeshow. 2006. Goodness-of-fit test for a logistic regression model fitted using survey sample data. *Stata Journal* 6: 97–105.

Collett, D. 2003. *Modelling Survival Data in Medical Research.* 2nd ed. London: Chapman & Hall/CRC.

Garrett, J. M. 2000. sg157: Predicted values calculated from linear or logistic regression models. *Stata Technical Bulletin* 58: 27–30. Reprinted in *Stata Technical Bulletin Reprints*, vol. 10, pp. 258–261. College Station, TX: Stata Press.

Green, D. M., and J. A. Swets. 1966. *Signal Detection Theory and Psychophysics.* New York: Wiley.

Hosmer, D. W., Jr., and S. Lemeshow. 1980. Goodness-of-fit tests for the multiple logistic regression model. *Communications in Statistics* A9: 1043–1069.

———. 2000. *Applied Logistic Regression*. 2nd ed. New York: Wiley.

Hosmer, D. W., Jr., S. Lemeshow, and J. Klar. 1988. Goodness-of-fit testing for the logistic regression model when the estimated probabilities are small. *Biometrical Journal* 30: 911–924.

Lemeshow, S., and J.-R. L. Gall. 1994. Modeling the severity of illness of ICU patients: A systems update. *Journal of the American Medical Association* 272: 1049–1055.

Lemeshow, S., and D. W. Hosmer, Jr. 1982. A review of goodness of fit statistics for the use in the development of logistic regression models. *American Journal of Epidemiology* 115: 92–106.

Metz, C. E. 1978. Basic principles of ROC analysis. *Seminars in Nuclear Medicine* 8: 283–298.

Mitchell, M. N., and X. Chen. 2005. Visualizing main effects and interactions for binary logit models. *Stata Journal* 5: 64–82.

Peterson, W. W., T. G. Birdsall, and W. C. Fox. 1954. The theory of signal detectability. *Transactions IRE Professional Group on Information Theory* PGIT-4: 171–212.

Pregibon, D. 1981. Logistic regression diagnostics. *Annals of Statistics* 9: 705–724.

Seed, P. T., and A. Tobías. 2001. sbe36.1: Summary statistics for diagnostic tests. *Stata Technical Bulletin* 59: 25–27. Reprinted in *Stata Technical Bulletin Reprints*, vol. 10, pp. 90–93. College Station, TX: Stata Press.

Tanner, W. P., Jr., and J. A. Swets. 1954. A decision-making theory of visual detection. *Psychological Review* 61: 401–409.

Tilford, J. M., P. K. Roberson, and D. H. Fiser. 1995. sbe12: Using lfit and lroc to evaluate mortality prediction models. *Stata Technical Bulletin* 28: 14–18. Reprinted in *Stata Technical Bulletin Reprints*, vol. 5, pp. 77–81. College Station, TX: Stata Press.

Tobías, A. 2000. sbe36: Summary statistics report for diagnostic tests. *Stata Technical Bulletin* 56: 16–18. Reprinted in *Stata Technical Bulletin Reprints*, vol. 10, pp. 87–90. College Station, TX: Stata Press.

Tobías, A., and M. J. Campbell. 1998. sg90: Akaike's information criterion and Schwarz's criterion. *Stata Technical Bulletin* 45: 23–25. Reprinted in *Stata Technical Bulletin Reprints*, vol. 8, pp. 174–177. College Station, TX: Stata Press.

Wang, Z. 2007. Two postestimation commands for assessing confounding effects in epidemiological studies. *Stata Journal* 7: 183–196.

Weesie, J. 1997. sg66: Enhancements to the alpha command. *Stata Technical Bulletin* 35: 32–34. Reprinted in *Stata Technical Bulletin Reprints*, vol. 6, pp. 176–179. College Station, TX: Stata Press.

Also see

[R] **logistic** — Logistic regression, reporting odds ratios

[U] **20 Estimation and postestimation commands**

Title

> **logit** — Logistic regression, reporting coefficients

Syntax

<u>logit</u> *depvar* [*indepvars*] [*if*] [*in*] [*weight*] [, *options*]

options	Description
Model	
<u>nocon</u>stant	suppress constant term
<u>off</u>set(*varname*)	include *varname* in model with coefficient constrained to 1
asis	retain perfect predictor variables
<u>constraints</u>(*constraints*)	apply specified linear constraints
<u>coll</u>inear	keep collinear variables
SE/Robust	
vce(*vcetype*)	*vcetype* may be oim, <u>r</u>obust, <u>cl</u>uster *clustvar*, <u>boot</u>strap, or <u>jack</u>knife
Reporting	
<u>l</u>evel(#)	set confidence level; default is level(95)
or	report odds ratios
<u>nocns</u>report	do not display constraints
display_options	control column formats, row spacing, line width, and display of omitted variables and base and empty cells
Maximization	
maximize_options	control the maximization process; seldom used
<u>nocoef</u>	do not display coefficient table; seldom used
<u>coefl</u>egend	display legend instead of statistics

indepvars may contain factor variables; see [U] **11.4.3 Factor variables**.

depvar and *indepvars* may contain time-series operators; see [U] **11.4.4 Time-series varlists**.

bootstrap, by, fracpoly, jackknife, mfp, mi estimate, nestreg, rolling, statsby, stepwise, and svy are allowed; see [U] **11.1.10 Prefix commands**.

vce(bootstrap) and vce(jackknife) are not allowed with the mi estimate prefix; see [MI] **mi estimate**.

Weights are not allowed with the bootstrap prefix; see [R] **bootstrap**.

vce(), nocoef, and weights are not allowed with the svy prefix; see [SVY] **svy**.

fweights, iweights, and pweights are allowed; see [U] **11.1.6 weight**.

nocoef and coeflegend do not appear in the dialog box.

See [U] **20 Estimation and postestimation commands** for more capabilities of estimation commands.

Menu

Statistics > Binary outcomes > Logistic regression

Description

logit fits a logit model for a binary response by maximum likelihood; it models the probability of a positive outcome given a set of regressors. *depvar* equal to nonzero and nonmissing (typically *depvar* equal to one) indicates a positive outcome, whereas *depvar* equal to zero indicates a negative outcome.

Also see [R] **logistic**; logistic displays estimates as odds ratios. Many users prefer the logistic command to logit. Results are the same regardless of which you use—both are the maximum-likelihood estimator. Several auxiliary commands that can be run after logit, probit, or logistic estimation are described in [R] **logistic postestimation**. A list of related estimation commands is given in [R] **logistic**.

If estimating on grouped data, see [R] **glogit**.

Options

‾‾‾‾⌐ Model ⌐‾‾‾

noconstant, offset(*varname*), constraints(*constraints*), collinear; see [R] **estimation options**.

asis forces retention of perfect predictor variables and their associated perfectly predicted observations and may produce instabilities in maximization; see [R] **probit**.

‾‾‾‾⌐ SE/Robust ⌐‾‾‾

vce(*vcetype*) specifies the type of standard error reported, which includes types that are derived from asymptotic theory, that are robust to some kinds of misspecification, that allow for intragroup correlation, and that use bootstrap or jackknife methods; see [R] *vce_option*.

‾‾‾‾⌐ Reporting ⌐‾‾‾

level(#); see [R] **estimation options**.

or reports the estimated coefficients transformed to odds ratios, that is, e^b rather than b. Standard errors and confidence intervals are similarly transformed. This option affects how results are displayed, not how they are estimated. or may be specified at estimation or when replaying previously estimated results.

nocnsreport; see [R] **estimation options**.

display_options: noomitted, vsquish, noemptycells, baselevels, allbaselevels, cformat(%*fmt*), pformat(%*fmt*), sformat(%*fmt*), and nolstretch; see [R] **estimation options**.

‾‾‾‾⌐ Maximization ⌐‾‾‾

maximize_options: difficult, technique(*algorithm_spec*), iterate(#), [no]log, trace, gradient, showstep, hessian, showtolerance, tolerance(#), ltolerance(#), nrtolerance(#), nonrtolerance, and from(*init_specs*); see [R] **maximize**. These options are seldom used.

The following options are available with logit but are not shown in the dialog box:

nocoef specifies that the coefficient table not be displayed. This option is sometimes used by program writers but is of no use interactively.

coeflegend; see [R] **estimation options**.

Remarks

Remarks are presented under the following headings:

Basic usage
Model identification

Basic usage

logit fits maximum likelihood models with dichotomous dependent (left-hand-side) variables coded as 0/1 (or, more precisely, coded as 0 and not-0).

▷ Example 1

We have data on the make, weight, and mileage rating of 22 foreign and 52 domestic automobiles. We wish to fit a logit model explaining whether a car is foreign on the basis of its weight and mileage. Here is an overview of our data:

```
. use http://www.stata-press.com/data/r12/auto
(1978 Automobile Data)
. keep make mpg weight foreign
. describe
Contains data from http://www.stata-press.com/data/r12/auto.dta
  obs:            74                          1978 Automobile Data
  vars:            4                          13 Apr 2011 17:45
  size:         1,702                         (_dta has notes)
```

variable name	storage type	display format	value label	variable label
make	str18	%-18s		Make and Model
mpg	int	%8.0g		Mileage (mpg)
weight	int	%8.0gc		Weight (lbs.)
foreign	byte	%8.0g	origin	Car type

```
Sorted by:  foreign
     Note:  dataset has changed since last saved
. inspect foreign
foreign:  Car type
```

```
foreign:  Car type                      Number of Observations

                                    Total    Integers   Nonintegers
    |  #         Negative             -          -           -
    |  #         Zero                 52         52          -
    |  #         Positive             22         22          -
    |  #                           _____     _____     _____
    |  #    #    Total                74         74          -
    |  #    #    Missing              -
    +                              _____
    0              1                   74
      (2 unique values)

    foreign is labeled and all values are documented in the label.
```

The variable foreign takes on two unique values, 0 and 1. The value 0 denotes a domestic car, and 1 denotes a foreign car.

The model that we wish to fit is

$$\Pr(\mathtt{foreign} = 1) = F(\beta_0 + \beta_1\mathtt{weight} + \beta_2\mathtt{mpg})$$

where $F(z) = e^z/(1 + e^z)$ is the cumulative logistic distribution.

To fit this model, we type

```
. logit foreign weight mpg
Iteration 0:   log likelihood =   -45.03321
Iteration 1:   log likelihood = -29.238536
Iteration 2:   log likelihood = -27.244139
Iteration 3:   log likelihood = -27.175277
Iteration 4:   log likelihood = -27.175156
Iteration 5:   log likelihood = -27.175156
```

```
Logistic regression                             Number of obs   =          74
                                                LR chi2(2)      =       35.72
                                                Prob > chi2     =      0.0000
Log likelihood = -27.175156                     Pseudo R2       =      0.3966
```

foreign	Coef.	Std. Err.	z	P>|z|	[95% Conf. Interval]	
weight	-.0039067	.0010116	-3.86	0.000	-.0058894	-.001924
mpg	-.1685869	.0919175	-1.83	0.067	-.3487418	.011568
_cons	13.70837	4.518709	3.03	0.002	4.851859	22.56487

We find that heavier cars are less likely to be foreign and that cars yielding better gas mileage are also less likely to be foreign, at least holding the weight of the car constant.

◁

❏ Technical note

Stata interprets a value of 0 as a negative outcome (failure) and treats all other values (except missing) as positive outcomes (successes). Thus if your dependent variable takes on the values 0 and 1, then 0 is interpreted as failure and 1 as success. If your dependent variable takes on the values 0, 1, and 2, then 0 is still interpreted as failure, but both 1 and 2 are treated as successes.

If you prefer a more formal mathematical statement, when you type logit y x, Stata fits the model

$$\Pr(y_j \neq 0 \mid \mathbf{x}_j) = \frac{\exp(\mathbf{x}_j \boldsymbol{\beta})}{1 + \exp(\mathbf{x}_j \boldsymbol{\beta})}$$

❏

Model identification

The logit command has one more feature, and it is probably the most useful. logit automatically checks the model for identification and, if it is underidentified, drops whatever variables and observations are necessary for estimation to proceed. (logistic, probit, and ivprobit do this as well.)

▷ Example 2

Have you ever fit a logit model where one or more of your independent variables perfectly predicted one or the other outcome?

For instance, consider the following data:

Outcome y	Independent variable x
0	1
0	1
0	0
1	0

Say that we wish to predict the outcome on the basis of the independent variable. The outcome is always zero whenever the independent variable is one. In our data, $\Pr(y = 0 \mid x = 1) = 1$, which means that the logit coefficient on x must be minus infinity with a corresponding infinite standard error. At this point, you may suspect that we have a problem.

Unfortunately, not all such problems are so easily detected, especially if you have a lot of independent variables in your model. If you have ever had such difficulties, you have experienced one of the more unpleasant aspects of computer optimization. The computer has no idea that it is trying to solve for an infinite coefficient as it begins its iterative process. All it knows is that at each step, making the coefficient a little bigger, or a little smaller, works wonders. It continues on its merry way until either 1) the whole thing comes crashing to the ground when a numerical overflow error occurs or 2) it reaches some predetermined cutoff that stops the process. In the meantime, you have been waiting. The estimates that you finally receive, if you receive any at all, may be nothing more than numerical roundoff.

Stata watches for these sorts of problems, alerts us, fixes them, and properly fits the model.

Let's return to our automobile data. Among the variables we have in the data is one called `repair`, which takes on three values. A value of 1 indicates that the car has a poor repair record, 2 indicates an average record, and 3 indicates a better-than-average record. Here is a tabulation of our data:

```
. use http://www.stata-press.com/data/r12/repair, clear
(1978 Automobile Data)

. tabulate foreign repair
```

Car type	repair 1	2	3	Total
Domestic	10	27	9	46
Foreign	0	3	9	12
Total	10	30	18	58

All the cars with poor repair records (`repair` = 1) are domestic. If we were to attempt to predict `foreign` on the basis of the repair records, the predicted probability for the `repair` = 1 category would have to be zero. This in turn means that the logit coefficient must be minus infinity, and that would set most computer programs buzzing.

Let's try Stata on this problem.

```
. logit foreign b3.repair
note: 1.repair != 0 predicts failure perfectly
      1.repair dropped and 10 obs not used
Iteration 0:   log likelihood = -26.992087
Iteration 1:   log likelihood = -22.483187
Iteration 2:   log likelihood = -22.230498
Iteration 3:   log likelihood = -22.229139
Iteration 4:   log likelihood = -22.229138
```

| Logistic regression | | | | | Number of obs | = | 48 |

LR chi2(1) = 9.53
Prob > chi2 = 0.0020

Log likelihood = -22.229138   Pseudo R2 = 0.1765

foreign	Coef.	Std. Err.	z	P>|z|	[95% Conf. Interval]
repair						
1	0	(empty)				
2	-2.197225	.7698003	-2.85	0.004	-3.706005	-.6884436
_cons	-1.98e-16	.4714045	-0.00	1.000	-.9239359	.9239359

Remember that all the cars with poor repair records (repair = 1) are domestic, so the model cannot be fit, or at least it cannot be fit if we restrict ourselves to finite coefficients. Stata noted that fact "note: 1.repair !=0 predicts failure perfectly". This is Stata's mathematically precise way of saying what we said in English. When repair is 1, the car is domestic.

Stata then went on to say "1.repair dropped and 10 obs not used". This is Stata eliminating the problem. First 1.repair had to be removed from the model because it would have an infinite coefficient. Then the 10 observations that led to the problem had to be eliminated, as well, so as not to bias the remaining coefficients in the model. The 10 observations that are not used are the 10 domestic cars that have poor repair records.

Stata then fit what was left of the model, using the remaining observations. Because no observations remained for cars with poor repair records, Stata reports "(empty)" in the row for repair = 1.

◁

❏ Technical note

Stata is pretty smart about catching problems like this. It will catch "one-way causation by a dummy variable", as we demonstrated above.

Stata also watches for "two-way causation", that is, a variable that perfectly determines the outcome, both successes and failures. Here Stata says, "so-and-so predicts outcome perfectly" and stops. Statistics dictates that no model can be fit.

Stata also checks your data for collinear variables; it will say, "so-and-so omitted because of collinearity". No observations need to be eliminated in this case, and model fitting will proceed without the offending variable.

It will also catch a subtle problem that can arise with continuous data. For instance, if we were estimating the chances of surviving the first year after an operation, and if we included in our model age, and if all the persons over 65 died within the year, Stata would say, "age > 65 predicts failure perfectly". It would then inform us about the fix-up it takes and fit what can be fit of our model.

`logit` (and `logistic`, `probit`, and `ivprobit`) will also occasionally display messages such as

```
Note: 4 failures and 0 successes completely determined.
```

There are two causes for a message like this. The first—and most unlikely—case occurs when a continuous variable (or a combination of a continuous variable with other continuous or dummy variables) is simply a great predictor of the dependent variable. Consider Stata's `auto.dta` dataset with 6 observations removed.

```
. use http://www.stata-press.com/data/r12/auto
(1978 Automobile Data)
. drop if foreign==0 & gear_ratio > 3.1
(6 observations deleted)
. logit foreign mpg weight gear_ratio, nolog
```

Logistic regression

```
Number of obs   =       68
LR chi2(3)      =    72.64
Prob > chi2     =   0.0000
```
Log likelihood = -6.4874814
```
Pseudo R2       =   0.8484
```

foreign	Coef.	Std. Err.	z	P>|z|	[95% Conf. Interval]	
mpg	-.4944907	.2655508	-1.86	0.063	-1.014961	.0259792
weight	-.0060919	.003101	-1.96	0.049	-.0121698	-.000014
gear_ratio	15.70509	8.166234	1.92	0.054	-.300436	31.71061
_cons	-21.39527	25.41486	-0.84	0.400	-71.20747	28.41694

```
Note: 4 failures and 0 successes completely determined.
```

There are no missing standard errors in the output. If you receive the "completely determined" message and have one or more missing standard errors in your output, see the second case discussed below.

Note `gear_ratio`'s large coefficient. `logit` thought that the 4 observations with the smallest predicted probabilities were essentially predicted perfectly.

```
. predict p
(option pr assumed; Pr(foreign))
. sort p
. list p in 1/4
```

	p
1.	1.34e-10
2.	6.26e-09
3.	7.84e-09
4.	1.49e-08

If this happens to you, you do not have to do anything. Computationally, the model is sound. The second case discussed below requires careful examination.

The second case occurs when the independent terms are all dummy variables or continuous ones with repeated values (for example, age). Here one or more of the estimated coefficients will have missing standard errors. For example, consider this dataset consisting of 5 observations.

```
. use http://www.stata-press.com/data/r12/logitxmpl, clear

. list, separator(0)
```

	y	x1	x2
1.	0	0	0
2.	0	0	0
3.	0	1	0
4.	1	1	0
5.	0	0	1
6.	1	0	1

```
. logit y x1 x2

Iteration 0:    log likelihood =  -3.819085
Iteration 1:    log likelihood = -2.9527336
Iteration 2:    log likelihood = -2.8110282
Iteration 3:    log likelihood = -2.7811973
Iteration 4:    log likelihood = -2.7746107
Iteration 5:    log likelihood = -2.7730128
  (output omitted )
Iteration 15996: log likelihood = -2.7725887  (not concave)
Iteration 15997: log likelihood = -2.7725887  (not concave)
Iteration 15998: log likelihood = -2.7725887  (not concave)
Iteration 15999: log likelihood = -2.7725887  (not concave)
Iteration 16000: log likelihood = -2.7725887  (not concave)
convergence not achieved
```

```
Logistic regression                             Number of obs   =          6
                                                LR chi2(1)      =       2.09
                                                Prob > chi2     =     0.1480
Log likelihood = -2.7725887                     Pseudo R2       =     0.2740
```

y	Coef.	Std. Err.	z	P>|z|	[95% Conf. Interval]	
x1	18.3704	2	9.19	0.000	14.45047	22.29033
x2	18.3704	.	.	.	.	.
_cons	-18.3704	1.414214	-12.99	0.000	-21.14221	-15.5986

```
Note: 2 failures and 0 successes completely determined.
convergence not achieved
r(430);
```

Three things are happening here. First, `logit` iterates almost forever and then declares nonconvergence. Second, `logit` can fit the outcome ($y = 0$) for the covariate pattern $x1 = 0$ and $x2 = 0$ (that is, the first two observations) perfectly. This observation is the "2 failures and 0 successes completely determined". Third, if this observation is dropped, then x1, x2, and the constant are collinear.

This is the cause of the nonconvergence, the message "completely determined", and the missing standard errors. It happens when you have a covariate pattern (or patterns) with only one outcome and there is collinearity when the observations corresponding to this covariate pattern are dropped.

If this happens to you, confirm the causes. First, identify the covariate pattern with only one outcome. (For your data, replace x1 and x2 with the independent variables of your model.)

```
. egen pattern = group(x1 x2)

. quietly logit y x1 x2, iterate(100)

. predict p
(option pr assumed; Pr(y))
```

```
. summarize p
```

Variable	Obs	Mean	Std. Dev.	Min	Max
p	6	.3333333	.2581989	1.05e-08	.5

If successes were completely determined, that means that there are predicted probabilities that are almost 1. If failures were completely determined, that means that there are predicted probabilities that are almost 0. The latter is the case here, so we locate the corresponding value of pattern:

```
. tabulate pattern if p < 1e-7
```

group(x1 x2)	Freq.	Percent	Cum.
1	2	100.00	100.00
Total	2	100.00	

Once we omit this covariate pattern from the estimation sample, logit can deal with the collinearity:

```
. logit y x1 x2 if pattern !=1, nolog
note: x2 omitted because of collinearity
```

```
Logistic regression                             Number of obs   =          4
                                                LR chi2(1)      =       0.00
                                                Prob > chi2     =     1.0000
Log likelihood = -2.7725887                     Pseudo R2       =     0.0000
```

y	Coef.	Std. Err.	z	P>|z|	[95% Conf. Interval]	
x1	0	2	0.00	1.000	-3.919928	3.919928
x2	0	(omitted)				
_cons	0	1.414214	0.00	1.000	-2.771808	2.771808

We omit the collinear variable. Then we must decide whether to include or omit the observations with pattern = 1. We could include them,

```
. logit y x1, nolog
```

```
Logistic regression                             Number of obs   =          6
                                                LR chi2(1)      =       0.37
                                                Prob > chi2     =     0.5447
Log likelihood = -3.6356349                     Pseudo R2       =     0.0480
```

y	Coef.	Std. Err.	z	P>|z|	[95% Conf. Interval]	
x1	1.098612	1.825742	0.60	0.547	-2.479776	4.677001
_cons	-1.098612	1.154701	-0.95	0.341	-3.361784	1.164559

or exclude them,

```
. logit y x1 if pattern != 1, nolog
```

Logistic regression			Number of obs	=	4
			LR chi2(1)	=	0.00
			Prob > chi2	=	1.0000
Log likelihood = -2.7725887			Pseudo R2	=	0.0000

y	Coef.	Std. Err.	z	P>|z|	[95% Conf. Interval]
x1	0	2	0.00	1.000	-3.919928 3.919928
_cons	0	1.414214	0.00	1.000	-2.771808 2.771808

If the covariate pattern that predicts outcome perfectly is meaningful, you may want to exclude these observations from the model. Here you would report that covariate pattern such and such predicted outcome perfectly and that the best model for the rest of the data is . . . . But, more likely, the perfect prediction was simply the result of having too many predictors in the model. Then you would omit the extraneous variables from further consideration and report the best model for all the data.

❑

Saved results

logit saves the following in e():

Scalars
e(N)	number of observations
e(N_cds)	number of completely determined successes
e(N_cdf)	number of completely determined failures
e(k)	number of parameters
e(k_eq)	number of equations in e(b)
e(k_eq_model)	number of equations in overall model test
e(k_dv)	number of dependent variables
e(df_m)	model degrees of freedom
e(r2_p)	pseudo-R-squared
e(ll)	log likelihood
e(ll_0)	log likelihood, constant-only model
e(N_clust)	number of clusters
e(chi2)	χ^2
e(p)	significance of model test
e(rank)	rank of e(V)
e(ic)	number of iterations
e(rc)	return code
e(converged)	1 if converged, 0 otherwise

Macros

e(cmd)	logit
e(cmdline)	command as typed
e(depvar)	name of dependent variable
e(wtype)	weight type
e(wexp)	weight expression
e(title)	title in estimation output
e(clustvar)	name of cluster variable
e(offset)	linear offset variable
e(chi2type)	Wald or LR; type of model χ^2 test
e(vce)	*vcetype* specified in vce()
e(vcetype)	title used to label Std. Err.
e(opt)	type of optimization
e(which)	max or min; whether optimizer is to perform maximization or minimization
e(ml_method)	type of ml method
e(user)	name of likelihood-evaluator program
e(technique)	maximization technique
e(properties)	b V
e(estat_cmd)	program used to implement estat
e(predict)	program used to implement predict
e(marginsnotok)	predictions disallowed by margins
e(asbalanced)	factor variables fvset as asbalanced
e(asobserved)	factor variables fvset as asobserved

Matrices

e(b)	coefficient vector
e(Cns)	constraints matrix
e(ilog)	iteration log (up to 20 iterations)
e(gradient)	gradient vector
e(mns)	vector of means of the independent variables
e(rules)	information about perfect predictors
e(V)	variance–covariance matrix of the estimators
e(V_modelbased)	model-based variance

Functions

e(sample)	marks estimation sample

Methods and formulas

logit is implemented as an ado-file.

Cramer (2003, chap. 9) surveys the prehistory and history of the logit model. The word "logit" was coined by Berkson (1944) and is analogous to the word "probit". For an introduction to probit and logit, see, for example, Aldrich and Nelson (1984), Cameron and Trivedi (2010), Greene (2012), Jones (2007), Long (1997), Long and Freese (2006), Pampel (2000), or Powers and Xie (2008).

The likelihood function for logit is

$$\ln L = \sum_{j \in S} w_j \ln F(\mathbf{x}_j \mathbf{b}) + \sum_{j \notin S} w_j \ln\left\{1 - F(\mathbf{x}_j \mathbf{b})\right\}$$

where S is the set of all observations j, such that $y_j \neq 0$, $F(z) = e^z/(1 + e^z)$, and w_j denotes the optional weights. $\ln L$ is maximized as described in [R] **maximize**.

This command supports the Huber/White/sandwich estimator of the variance and its clustered version using vce(robust) and vce(cluster *clustvar*), respectively. See [P] **_robust**, particularly *Maximum likelihood estimators* and *Methods and formulas*. The scores are calculated as $\mathbf{u}_j = \{1 - F(\mathbf{x}_j\mathbf{b})\}\mathbf{x}_j$ for the positive outcomes and $-F(\mathbf{x}_j\mathbf{b})\mathbf{x}_j$ for the negative outcomes.

logit also supports estimation with survey data. For details on VCEs with survey data, see [SVY] **variance estimation**.

Joseph Berkson (1899–1982) was born in New York City and studied at the College of the City of New York, Columbia, and Johns Hopkins, earning both an MD and a doctorate in statistics. He then worked at Johns Hopkins before moving to the Mayo Clinic in 1931 as a biostatistician. Among many other contributions, his most influential one drew upon a long-sustained interest in the logistic function, especially his 1944 paper on bioassay, in which he introduced the term "logit". Berkson was a frequent participant in controversy—sometimes humorous, sometimes bitter—on subjects such as the evidence for links between smoking and various diseases and the relative merits of probit and logit methods and of different calculation methods.

References

Aldrich, J. H., and F. D. Nelson. 1984. *Linear Probability, Logit, and Probit Models*. Newbury Park, CA: Sage.

Archer, K. J., and S. Lemeshow. 2006. Goodness-of-fit test for a logistic regression model fitted using survey sample data. *Stata Journal* 6: 97–105.

Berkson, J. 1944. Application of the logistic function to bio-assay. *Journal of the American Statistical Association* 39: 357–365.

Buis, M. L. 2010a. Direct and indirect effects in a logit model. *Stata Journal* 10: 11–29.

———. 2010b. Stata tip 87: Interpretation of interactions in nonlinear models. *Stata Journal* 10: 305–308.

Cameron, A. C., and P. K. Trivedi. 2010. *Microeconometrics Using Stata*. Rev. ed. College Station, TX: Stata Press.

Cleves, M. A., and A. Tosetto. 2000. sg139: Logistic regression when binary outcome is measured with uncertainty. *Stata Technical Bulletin* 55: 20–23. Reprinted in *Stata Technical Bulletin Reprints*, vol. 10, pp. 152–156. College Station, TX: Stata Press.

Cramer, J. S. 2003. *Logit Models from Economics and Other Fields*. Cambridge: Cambridge University Press.

Greene, W. H. 2012. *Econometric Analysis*. 7th ed. Upper Saddle River, NJ: Prentice Hall.

Hilbe, J. M. 2009. *Logistic Regression Models*. Boca Raton, FL: Chapman & Hill/CRC.

Hosmer, D. W., Jr., and S. Lemeshow. 2000. *Applied Logistic Regression*. 2nd ed. New York: Wiley.

Jones, A. 2007. *Applied Econometrics for Health Economists: A Practical Guide*. 2nd ed. Abingdon, UK: Radcliffe.

Judge, G. G., W. E. Griffiths, R. C. Hill, H. Lütkepohl, and T.-C. Lee. 1985. *The Theory and Practice of Econometrics*. 2nd ed. New York: Wiley.

Long, J. S. 1997. *Regression Models for Categorical and Limited Dependent Variables*. Thousand Oaks, CA: Sage.

Long, J. S., and J. Freese. 2006. *Regression Models for Categorical Dependent Variables Using Stata*. 2nd ed. College Station, TX: Stata Press.

Miranda, A., and S. Rabe-Hesketh. 2006. Maximum likelihood estimation of endogenous switching and sample selection models for binary, ordinal, and count variables. *Stata Journal* 6: 285–308.

Mitchell, M. N., and X. Chen. 2005. Visualizing main effects and interactions for binary logit models. *Stata Journal* 5: 64–82.

O'Fallon, W. M. 1998. Berkson, Joseph. In Vol. 1 of *Encyclopedia of Biostatistics*, ed. P. Armitage and T. Colton, 290–295. Chichester, UK: Wiley.

Pampel, F. C. 2000. *Logistic Regression: A Primer.* Thousand Oaks, CA: Sage.

Powers, D. A., and Y. Xie. 2008. *Statistical Methods for Categorical Data Analysis.* 2nd ed. Bingley, UK: Emerald.

Pregibon, D. 1981. Logistic regression diagnostics. *Annals of Statistics* 9: 705–724.

Schonlau, M. 2005. Boosted regression (boosting): An introductory tutorial and a Stata plugin. *Stata Journal* 5: 330–354.

Xu, J., and J. S. Long. 2005. Confidence intervals for predicted outcomes in regression models for categorical outcomes. *Stata Journal* 5: 537–559.

Also see

[R] **logit postestimation** — Postestimation tools for logit

[R] **brier** — Brier score decomposition

[R] **exlogistic** — Exact logistic regression

[R] **glogit** — Logit and probit regression for grouped data

[R] **logistic** — Logistic regression, reporting odds ratios

[R] **probit** — Probit regression

[R] **roc** — Receiver operating characteristic (ROC) analysis

[MI] **estimation** — Estimation commands for use with mi estimate

[SVY] **svy estimation** — Estimation commands for survey data

[XT] **xtlogit** — Fixed-effects, random-effects, and population-averaged logit models

[U] **20 Estimation and postestimation commands**

logit postestimation — Postestimation tools for logit

Description

The following postestimation commands are of special interest after `logit`:

Command	Description
estat classification	report various summary statistics, including the classification table
estat gof	Pearson or Hosmer–Lemeshow goodness-of-fit test
lroc	compute area under ROC curve and graph the curve
lsens	graph sensitivity and specificity versus probability cutoff

These commands are not appropriate after the `svy` prefix.

For information about these commands, see [R] **logistic postestimation**.

The following standard postestimation commands are also available:

Command	Description
contrast	contrasts and ANOVA-style joint tests of estimates
estat	AIC, BIC, VCE, and estimation sample summary
estat (svy)	postestimation statistics for survey data
estimates	cataloging estimation results
lincom	point estimates, standard errors, testing, and inference for linear combinations of coefficients
linktest	link test for model specification
lrtest[1]	likelihood-ratio test
margins	marginal means, predictive margins, marginal effects, and average marginal effects
marginsplot	graph the results from margins (profile plots, interaction plots, etc.)
nlcom	point estimates, standard errors, testing, and inference for nonlinear combinations of coefficients
predict	predictions, residuals, influence statistics, and other diagnostic measures
predictnl	point estimates, standard errors, testing, and inference for generalized predictions
pwcompare	pairwise comparisons of estimates
suest	seemingly unrelated estimation
test	Wald tests of simple and composite linear hypotheses
testnl	Wald tests of nonlinear hypotheses

[1] `lrtest` is not appropriate with svy estimation results.

See the corresponding entries in the *Base Reference Manual* for details, but see [SVY] **estat** for details about `estat (svy)`.

Syntax for predict

predict [*type*] *newvar* [*if*] [*in*] [, *statistic* <u>nooff</u>set <u>rules</u> asif]

statistic	Description
Main	
<u>pr</u>	probability of a positive outcome; the default
xb	linear prediction
stdp	standard error of the prediction
*<u>dbeta</u>	Pregibon (1981) $\Delta\widehat{\beta}$ influence statistic
*<u>deviance</u>	deviance residual
*<u>dx2</u>	Hosmer and Lemeshow (2000) $\Delta\chi^2$ influence statistic
*<u>ddeviance</u>	Hosmer and Lemeshow (2000) ΔD influence statistic
*<u>hat</u>	Pregibon (1981) leverage
*<u>number</u>	sequential number of the covariate pattern
*<u>residuals</u>	Pearson residuals; adjusted for number sharing covariate pattern
*<u>rstandard</u>	standardized Pearson residuals; adjusted for number sharing covariate pattern
<u>sc</u>ore	first derivative of the log likelihood with respect to $\mathbf{x}_j\beta$

Unstarred statistics are available both in and out of sample; type predict ... if e(sample) ... if wanted only for the estimation sample. Starred statistics are calculated only for the estimation sample, even when if e(sample) is not specified.

pr, xb, stdp, and score are the only options allowed with svy estimation results.

Menu

Statistics > Postestimation > Predictions, residuals, etc.

Options for predict

⌐ Main ⌐

pr, the default, calculates the probability of a positive outcome.

xb calculates the linear prediction.

stdp calculates the standard error of the linear prediction.

dbeta calculates the Pregibon (1981) $\Delta\widehat{\beta}$ influence statistic, a standardized measure of the difference in the coefficient vector that is due to deletion of the observation along with all others that share the same covariate pattern. In Hosmer and Lemeshow (2000, 144–145) jargon, this statistic is M-asymptotic; that is, it is adjusted for the number of observations that share the same covariate pattern.

deviance calculates the deviance residual.

dx2 calculates the Hosmer and Lemeshow (2000, 174) $\Delta\chi^2$ influence statistic, reflecting the decrease in the Pearson χ^2 that is due to deletion of the observation and all others that share the same covariate pattern.

ddeviance calculates the Hosmer and Lemeshow (2000, 174) ΔD influence statistic, which is the change in the deviance residual that is due to deletion of the observation and all others that share the same covariate pattern.

hat calculates the Pregibon (1981) leverage or the diagonal elements of the hat matrix adjusted for the number of observations that share the same covariate pattern.

number numbers the covariate patterns—observations with the same covariate pattern have the same number. Observations not used in estimation have number set to missing. The first covariate pattern is numbered 1, the second 2, and so on.

residuals calculates the Pearson residual as given by Hosmer and Lemeshow (2000, 145) and adjusted for the number of observations that share the same covariate pattern.

rstandard calculates the standardized Pearson residual as given by Hosmer and Lemeshow (2000, 173) and adjusted for the number of observations that share the same covariate pattern.

score calculates the equation-level score, $\partial \ln L / \partial(\mathbf{x}_j \beta)$.

___

Options

nooffset is relevant only if you specified offset(*varname*) for logit. It modifies the calculations made by predict so that they ignore the offset variable; the linear prediction is treated as $\mathbf{x}_j \mathbf{b}$ rather than as $\mathbf{x}_j \mathbf{b} + \text{offset}_j$.

rules requests that Stata use any rules that were used to identify the model when making the prediction. By default, Stata calculates missing for excluded observations.

asif requests that Stata ignore the rules and exclusion criteria and calculate predictions for all observations possible by using the estimated parameter from the model.

Remarks

Once you have fit a logit model, you can obtain the predicted probabilities by using the predict command for both the estimation sample and other samples; see [U] **20 Estimation and postestimation commands** and [R] **predict**. Here we will make only a few more comments.

predict without arguments calculates the predicted probability of a positive outcome, that is, $\Pr(y_j = 1) = F(\mathbf{x}_j \mathbf{b})$. With the xb option, predict calculates the linear combination $\mathbf{x}_j \mathbf{b}$, where $\mathbf{x}_j$ are the independent variables in the jth observation and $\mathbf{b}$ is the estimated parameter vector. This is sometimes known as the index function because the cumulative distribution function indexed at this value is the probability of a positive outcome.

In both cases, Stata remembers any rules used to identify the model and calculates missing for excluded observations, unless rules or asif is specified. For information about the other statistics available after predict, see [R] **logistic postestimation**.

▷ Example 1

In example 2 of [R] **logit**, we fit the logit model logit foreign b3.repair. To obtain predicted probabilities, type

```
. use http://www.stata-press.com/data/r12/repair
(1978 Automobile Data)
. logit foreign b3.repair
note: 1.repair != 0 predicts failure perfectly
      1.repair dropped and 10 obs not used
  (output omitted )
. predict p
(option pr assumed; Pr(foreign))
(10 missing values generated)
```

```
. summarize foreign p
```

Variable	Obs	Mean	Std. Dev.	Min	Max
foreign	58	.2068966	.4086186	0	1
p	48	.25	.1956984	.1	.5

Stata remembers any rules used to identify the model and sets predictions to missing for any excluded observations. logit dropped the variable 1.repair from our model and excluded 10 observations. Thus when we typed predict p, those same 10 observations were again excluded, and their predictions were set to missing.

predict's rules option uses the rules in the prediction. During estimation, we were told "1.repair != 0 predicts failure perfectly", so the rule is that when 1.repair is not zero, we should predict 0 probability of success or a positive outcome:

```
. predict p2, rules
(option pr assumed; Pr(foreign))
. summarize foreign p p2
```

Variable	Obs	Mean	Std. Dev.	Min	Max
foreign	58	.2068966	.4086186	0	1
p	48	.25	.1956984	.1	.5
p2	58	.2068966	.2016268	0	.5

predict's asif option ignores the rules and exclusion criteria and calculates predictions for all observations possible by using the estimated parameters from the model:

```
. predict p3, asif
(option pr assumed; Pr(foreign))
. summarize foreign p p2 p3
```

Variable	Obs	Mean	Std. Dev.	Min	Max
foreign	58	.2068966	.4086186	0	1
p	48	.25	.1956984	.1	.5
p2	58	.2068966	.2016268	0	.5
p3	58	.2931035	.2016268	.1	.5

Which is right? What predict does by default is the most conservative approach. If many observations had been excluded because of a simple rule, we could be reasonably certain that the rules prediction is correct. The asif prediction is correct only if the exclusion is a fluke, and we would be willing to exclude the variable from the analysis anyway. Then, however, we would refit the model to include the excluded observations.

◁

▷ Example 2

We can use the command margins, contrast after logit to make comparisons on the probability scale. Let's fit a model predicting low birthweight from characteristics of the mother:

```
. use http://www.stata-press.com/data/r12/lbw, clear
(Hosmer & Lemeshow data)
```

```
. logit low age i.race i.smoke ptl i.ht i.ui
Iteration 0:   log likelihood =   -117.336
Iteration 1:   log likelihood = -103.81846
Iteration 2:   log likelihood = -103.40486
Iteration 3:   log likelihood = -103.40384
Iteration 4:   log likelihood = -103.40384
```

```
Logistic regression                              Number of obs   =        189
                                                 LR chi2(7)      =      27.86
                                                 Prob > chi2     =     0.0002
Log likelihood = -103.40384                      Pseudo R2       =     0.1187
```

low	Coef.	Std. Err.	z	P>|z|	[95% Conf. Interval]	
age	-.0403293	.0357127	-1.13	0.259	-.1103249	.0296663
race						
2	1.009436	.5025122	2.01	0.045	.0245302	1.994342
3	1.001908	.4248342	2.36	0.018	.1692485	1.834568
1.smoke	.9631876	.3904357	2.47	0.014	.1979477	1.728427
ptl	.6288678	.3399067	1.85	0.064	-.0373371	1.295073
1.ht	1.358142	.6289555	2.16	0.031	.125412	2.590872
1.ui	.8001832	.4572306	1.75	0.080	-.0959724	1.696339
_cons	-1.184127	.9187461	-1.29	0.197	-2.984837	.6165818

The coefficients are log odds-ratios: conditional on the other predictors, smoking during pregnancy is associated with an increase of 0.96 in the log odds-ratios of low birthweight. The model is linear in the log odds-scale, so the estimate of 0.96 has the same interpretation, whatever the values of the other predictors might be. We could convert 0.96 to an odds ratio by replaying the results with logit, or.

But what if we want to talk about the probability of low birthweight, and not the odds? Then we will need the command margins, contrast. We will use the r. contrast operator to compare each level of smoke with a reference level. (smoke has only two levels, so there will be only one comparison: a comparison of smokers with nonsmokers.)

```
. margins r.smoke, contrast
Contrasts of predictive margins
Model VCE    : OIM

Expression   : Pr(low), predict()
```

	df	chi2	P>chi2
smoke	1	6.32	0.0119

	Contrast	Delta-method Std. Err.	[95% Conf. Interval]	
smoke (1 vs 0)	.1832779	.0728814	.0404329	.3261229

We see that maternal smoking is associated with an 18.3% increase in the probability of low birthweight. (We received a contrast in the probability scale because predicted probabilities are the default when margins is used after logit.)

The contrast of 18.3% is a difference of margins that are computed by averaging over the predictions for observations in the estimation sample. If the values of the other predictors were different, the contrast for smoke would be different, too. Let's estimate the contrast for 25-year-old mothers:

```
. margins r.smoke, contrast at(age=25)
Contrasts of predictive margins
Model VCE    : OIM
Expression   : Pr(low), predict()
at           : age            =         25
```

	df	chi2	P>chi2
smoke	1	6.19	0.0129

	Contrast	Delta-method Std. Err.	[95% Conf. Interval]	
smoke (1 vs 0)	.1808089	.0726777	.0383632	.3232547

Specifying a maternal age of 25 changed the contrast to 18.1%. Our contrast of probabilities changed because the logit model is nonlinear in the probability scale. A contrast of log odds-ratios would not have changed.

◁

Methods and formulas

All postestimation commands listed above are implemented as ado-files.

See *Methods and formulas* of [R] **logistic postestimation** for details.

References

Archer, K. J., and S. Lemeshow. 2006. Goodness-of-fit test for a logistic regression model fitted using survey sample data. *Stata Journal* 6: 97–105.

Hosmer, D. W., Jr., and S. Lemeshow. 2000. *Applied Logistic Regression.* 2nd ed. New York: Wiley.

Pregibon, D. 1981. Logistic regression diagnostics. *Annals of Statistics* 9: 705–724.

Also see

[R] **logit** — Logistic regression, reporting coefficients

[R] **logistic postestimation** — Postestimation tools for logistic

[U] **20 Estimation and postestimation commands**

> **loneway** — Large one-way ANOVA, random effects, and reliability

loneway *response_var group_var* [*if*] [*in*] [*weight*] [, *options*]

options	Description
Main	
mean	expected value of F distribution; default is 1
median	median of F distribution; default is 1
exact	exact confidence intervals (groups must be equal with no weights)
level(#)	set confidence level; default is level(95)

by is allowed; see [D] **by**.
aweights are allowed; see [U] **11.1.6 weight**.

Menu

Statistics > Linear models and related > ANOVA/MANOVA > Large one-way ANOVA

Description

loneway fits one-way analysis-of-variance (ANOVA) models on datasets with many levels of *group_var* and presents different ancillary statistics from oneway (see [R] **oneway**):

Feature	oneway	loneway
Fit one-way model	x	x
on fewer than 376 levels	x	x
on more than 376 levels		x
Bartlett's test for equal variance	x	
Multiple-comparison tests	x	
Intragroup correlation and SE		x
Intragroup correlation confidence interval		x
Est. reliability of group-averaged score		x
Est. SD of group effect		x
Est. SD within group		x

Options

> Main

mean specifies that the expected value of the $F_{k-1,N-k}$ distribution be used as the reference point F_m in the estimation of ρ instead of the default value of 1.

median specifies that the median of the $F_{k-1,N-k}$ distribution be used as the reference point F_m in the estimation of ρ instead of the default value of 1.

exact requests that exact confidence intervals be computed, as opposed to the default asymptotic confidence intervals. This option is allowed only if the groups are equal in size and weights are not used.

level(#) specifies the confidence level, as a percentage, for confidence intervals of the coefficients. The default is level(95) or as set by set level; see [U] **20.7 Specifying the width of confidence intervals**.

Remarks

Remarks are presented under the following headings:

> *The one-way ANOVA model*
> *R-squared*
> *The random-effects ANOVA model*
> *Intraclass correlation*
> *Estimated reliability of the group-averaged score*

The one-way ANOVA model

▷ Example 1

loneway's output looks like that of oneway, except that loneway presents more information at the end. Using our automobile dataset, we have created a (numeric) variable called manufacturer_grp identifying the manufacturer of each car, and within each manufacturer we have retained a maximum of four models, selecting those with the lowest mpg. We can compute the intraclass correlation of mpg for all manufacturers with at least four models as follows:

```
. use http://www.stata-press.com/data/r12/auto7
(1978 Automobile Data)
. loneway mpg manufacturer_grp if nummake == 4
             One-way Analysis of Variance for mpg: Mileage (mpg)

                                              Number of obs =         36
                                              R-squared     =     0.5228

      Source                SS         df        MS          F      Prob > F

Between manufactur~p     621.88889      8     77.736111      3.70    0.0049
Within manufactur~p      567.75        27     21.027778

Total                   1189.6389      35     33.989683

             Intraclass       Asy.
             correlation      S.E.        [95% Conf. Interval]

               0.40270       0.18770        0.03481      0.77060

         Estimated SD of manufactur~p effect        3.765247
         Estimated SD within manufactur~p           4.585605
         Est. reliability of a manufactur~p mean     .72950
              (evaluated at n=4.00)
```

◁

In addition to the standard one-way ANOVA output, loneway produces the R-squared, the estimated standard deviation of the group effect, the estimated standard deviation within group, the intragroup correlation, the estimated reliability of the group-averaged mean, and, for unweighted data, the asymptotic standard error and confidence interval for the intragroup correlation.

R-squared

The R-squared is, of course, simply the underlying R^2 for a regression of *response_var* on the levels of *group_var*, or mpg on the various manufacturers here.

The random-effects ANOVA model

loneway assumes that we observe a variable, y_{ij}, measured for n_i elements within k groups or classes such that

$$y_{ij} = \mu + \alpha_i + \epsilon_{ij}, \quad i = 1, 2, \ldots, k, \quad j = 1, 2, \ldots, n_i$$

and α_i and ϵ_{ij} are independent zero-mean random variables with variance σ_α^2 and σ_ϵ^2, respectively. This is the random-effects ANOVA model, also known as the components-of-variance model, in which it is typically assumed that the y_{ij} are normally distributed.

The interpretation with respect to our example is that the observed value of our response variable, mpg, is created in two steps. First, the ith manufacturer is chosen, and a value, α_i, is determined—the typical mpg for that manufacturer less the overall mpg μ. Then a deviation, ϵ_{ij}, is chosen for the jth model within this manufacturer. This is how much that particular automobile differs from the typical mpg value for models from this manufacturer.

For our sample of 36 car models, the estimated standard deviations are $\sigma_\alpha = 3.8$ and $\sigma_\epsilon = 4.6$. Thus a little more than half of the variation in mpg between cars is attributable to the car model, with the rest attributable to differences between manufacturers. These standard deviations differ from those that would be produced by a (standard) fixed-effects regression in that the regression would require the sum within each manufacturer of the ϵ_{ij}, ϵ_i. for the ith manufacturer, to be zero, whereas these estimates merely impose the constraint that the sum is *expected* to be zero.

Intraclass correlation

There are various estimators of the intraclass correlation, such as the pairwise estimator, which is defined as the Pearson product-moment correlation computed over all possible pairs of observations that can be constructed within groups. For a discussion of various estimators, see Donner (1986). loneway computes what is termed the analysis of variance, or ANOVA, estimator. This intraclass correlation is the theoretical upper bound on the variation in *response_var* that is explainable by *group_var*, of which R-squared is an overestimate because of the serendipity of fitting. This correlation is comparable to an R-squared—you do not have to square it.

In our example, the intra-manu correlation, the correlation of mpg within manufacturer, is 0.40. Because aweights were not used and the default correlation was computed (that is, the mean and median options were not specified), loneway also provided the asymptotic confidence interval and standard error of the intraclass correlation estimate.

Estimated reliability of the group-averaged score

The estimated reliability of the group-averaged score or mean has an interpretation similar to that of the intragroup correlation; it is a comparable number if we average *response_var* by *group_var*, or mpg by manu in our example. It is the theoretical upper bound of a regression of manufacturer-averaged mpg on characteristics of manufacturers. Why would we want to collapse our 36-observation dataset into a 9-observation dataset of manufacturer averages? Because the 36 observations might be a mirage. When General Motors builds cars, do they sometimes put a Pontiac label and sometimes a Chevrolet label on them, so that it appears in our data as if we have two cars when we really have only one, replicated? If that were the case, and if it were the case for many other manufacturers, then we would be forced to admit that we do not have data on 36 cars; we instead have data on nine manufacturer-averaged characteristics.

Saved results

loneway saves the following in r():

Scalars

r(N)	number of observations	r(rho_t)	estimated reliability
r(rho)	intraclass correlation	r(se)	asymp. SE of intraclass correlation
r(lb)	lower bound of 95% CI for rho	r(sd_w)	estimated SD within group
r(ub)	upper bound of 95% CI for rho	r(sd_b)	estimated SD of group effect

Methods and formulas

loneway is implemented as an ado-file.

The mean squares in the loneway's ANOVA table are computed as

$$\text{MS}_\alpha = \sum_i w_{i\cdot}(\overline{y}_{i\cdot} - \overline{y}_{\cdot\cdot})^2/(k-1)$$

and

$$\text{MS}_\epsilon = \sum_i \sum_j w_{ij}(y_{ij} - \overline{y}_{i\cdot})^2/(N-k)$$

in which

$$w_{i\cdot} = \sum_j w_{ij} \quad w_{\cdot\cdot} = \sum_i w_{i\cdot} \quad \overline{y}_{i\cdot} = \sum_j w_{ij}y_{ij}/w_{i\cdot} \quad \text{and} \quad \overline{y}_{\cdot\cdot} = \sum_i w_{i\cdot}\overline{y}_{i\cdot}/w_{\cdot\cdot}$$

The corresponding expected values of these mean squares are

$$E(\text{MS}_\alpha) = \sigma_\epsilon^2 + g\sigma_\alpha^2 \quad \text{and} \quad E(\text{MS}_\epsilon) = \sigma_\epsilon^2$$

in which

$$g = \frac{w_{\cdot\cdot} - \sum_i w_{i\cdot}^2/w_{\cdot\cdot}}{k-1}$$

In the unweighted case, we get

$$g = \frac{N - \sum_i n_i^2/N}{k-1}$$

As expected, $g = m$ for the case of no weights and equal group sizes in the data, that is, $n_i = m$ for all i. Replacing the expected values with the observed values and solving yields the ANOVA estimates of σ_α^2 and σ_ϵ^2. Substituting these into the definition of the intraclass correlation

$$\rho = \frac{\sigma_\alpha^2}{\sigma_\alpha^2 + \sigma_\epsilon^2}$$

yields the ANOVA estimator of the intraclass correlation:

$$\rho_A = \frac{F_{\text{obs}} - 1}{F_{\text{obs}} - 1 + g}$$

F_{obs} is the observed value of the F statistic from the ANOVA table. For no weights and equal n_i, ρ_A = roh, which is the intragroup correlation defined by Kish (1965). Two slightly different estimators are available through the `mean` and `median` options (Gleason 1997). If either of these options is specified, the estimate of ρ becomes

$$\rho = \frac{F_{\text{obs}} - F_m}{F_{\text{obs}} + (g-1)F_m}$$

For the `mean` option, $F_m = E(F_{k-1,N-K}) = (N-k)/(N-k-2)$, that is, the expected value of the ANOVA table's F statistic. For the `median` option, F_m is simply the median of the F statistic. Setting F_m to 1 gives ρ_A, so for large samples, these different point estimators are essentially the same. Also, because the intraclass correlation of the random-effects model is by definition nonnegative, for any of the three possible point estimators, ρ is truncated to zero if F_{obs} is less than F_m.

For no weighting, interval estimators for ρ_A are computed. If the groups are equal sized (all n_i equal) and the `exact` option is specified, the following exact (assuming that the y_{ij} are normally distributed) $100(1-\alpha)\%$ confidence interval is computed:

$$\left\{ \frac{F_{\text{obs}} - F_m F_u}{F_{\text{obs}} + (g-1)F_m F_u}, \frac{F_{\text{obs}} - F_m F_l}{F_{\text{obs}} + (g-1)F_m F_l} \right\}$$

with $F_m = 1$, $F_l = F_{\alpha/2,k-1,N-k}$, and $F_u = F_{1-\alpha/2,k-1,N-k}$, $F_{\cdot,k-1,N-k}$ being the cumulative distribution function for the F distribution with $k-1$ and $N-k$ degrees of freedom. If `mean` or `median` is specified, F_m is defined as above. If the groups are equal sized and `exact` is not specified, the following asymptotic $100(1-\alpha)\%$ confidence interval for ρ_A is computed,

$$\left[\rho_A - z_{\alpha/2}\sqrt{V(\rho_A)}, \rho_A + z_{\alpha/2}\sqrt{V(\rho_A)} \right]$$

where $z_{\alpha/2}$ is the $100(1-\alpha/2)$ percentile of the standard normal distribution and $\sqrt{V(\rho_A)}$ is the asymptotic standard error of ρ defined below. This confidence interval is also available for unequal groups. It is not applicable and, therefore, not computed for the estimates of ρ provided by the `mean` and `median` options. Again, because the intraclass coefficient is nonnegative, if the lower bound is negative for either confidence interval, it is truncated to zero. As might be expected, the coverage probability of a truncated interval is higher than its nominal value.

The asymptotic standard error of ρ_A, assuming that the y_{ij} are normally distributed, is also computed when appropriate, namely, for unweighted data and when ρ_A is computed (neither the `mean` option nor the `median` option is specified):

$$V(\rho_A) = \frac{2(1-\rho)^2}{g^2}(A + B + C)$$

with

$$A = \frac{\{1 + \rho(g-1)\}^2}{N - k}$$

$$B = \frac{(1-\rho)\{1 + \rho(2g-1)\}}{k - 1}$$

$$C = \frac{\rho^2\{\sum n_i^2 - 2N^{-1}\sum n_i^3 + N^{-2}(\sum n_i^2)^2\}}{(k-1)^2}$$

and ρ_A is substituted for ρ (Donner 1986).

The estimated reliability of the group-averaged score, known as the Spearman–Brown prediction formula in the psychometric literature (Winer, Brown, and Michels 1991, 1014), is

$$\rho_t = \frac{t\rho}{1 + (t-1)\rho}$$

for group size t. loneway computes ρ_t for $t = g$.

The estimated standard deviation of the group effect is $\sigma_\alpha = \sqrt{(\text{MS}_\alpha - \text{MS}_\epsilon)/g}$. This deviation comes from the assumption that an observation is derived by adding a group effect to a within-group effect.

The estimated standard deviation within group is the square root of the mean square due to error, or $\sqrt{\text{MS}_\epsilon}$.

Acknowledgment

We thank John Gleason of Syracuse University (retired) for his contributions to improving loneway.

References

Donner, A. 1986. A review of inference procedures for the intraclass correlation coefficient in the one-way random effects model. *International Statistical Review* 54: 67–82.

Gleason, J. R. 1997. sg65: Computing intraclass correlations and large ANOVAs. *Stata Technical Bulletin* 35: 25–31. Reprinted in *Stata Technical Bulletin Reprints*, vol. 6, pp. 167–176. College Station, TX: Stata Press.

Kish, L. 1965. *Survey Sampling*. New York: Wiley.

Marchenko, Y. V. 2006. Estimating variance components in Stata. *Stata Journal* 6: 1–21.

Winer, B. J., D. R. Brown, and K. M. Michels. 1991. *Statistical Principles in Experimental Design*. 3rd ed. New York: McGraw–Hill.

Also see

[R] **anova** — Analysis of variance and covariance

[R] **oneway** — One-way analysis of variance

Title

> **lowess** — Lowess smoothing

Syntax

lowess *yvar* *xvar* [*if*] [*in*] [, *options*]

options	Description
Main	
<u>mea</u>n	running-mean smooth; default is running-line least squares
<u>nowe</u>ight	suppress weighted regressions; default is tricube weighting function
<u>bw</u>idth(*#*)	use *#* for the bandwidth; default is bwidth(0.8)
<u>logit</u>	transform dependent variable to logits
<u>ad</u>just	adjust smoothed mean to equal mean of dependent variable
<u>nog</u>raph	suppress graph
<u>g</u>enerate(*newvar*)	create *newvar* containing smoothed values of *yvar*
Plot	
marker_options	change look of markers (color, size, etc.)
marker_label_options	add marker labels; change look or position
Smoothed line	
<u>lineopts</u>(*cline_options*)	affect rendition of the smoothed line
Add plots	
<u>addplot</u>(*plot*)	add other plots to generated graph
Y axis, X axis, Titles, Legend, Overall, By	
twoway_options	any of the options documented in [G-3] ***twoway_options***

yvar and *xvar* may contain time-series operators; see [U] **11.4.4 Time-series varlists**.

Menu

Statistics > Nonparametric analysis > Lowess smoothing

Description

lowess carries out a locally weighted regression of *yvar* on *xvar*, displays the graph, and optionally saves the smoothed variable.

Warning: lowess is computationally intensive and may therefore take a long time to run on a slow computer. Lowess calculations on 1,000 observations, for instance, require performing 1,000 regressions.

Options

⌐ Main ⌐

mean specifies running-mean smoothing; the default is running-line least-squares smoothing.

noweight prevents the use of Cleveland's (1979) tricube weighting function; the default is to use the weighting function.

bwidth(#) specifies the bandwidth. Centered subsets of bwidth() × N observations are used for calculating smoothed values for each point in the data except for the end points, where smaller, uncentered subsets are used. The greater the bwidth(), the greater the smoothing. The default is 0.8.

logit transforms the smoothed *yvar* into logits. Predicted values less than 0.0001 or greater than 0.9999 are set to $1/N$ and $1 - 1/N$, respectively, before taking logits.

adjust adjusts the mean of the smoothed *yvar* to equal the mean of *yvar* by multiplying by an appropriate factor. This option is useful when smoothing binary (0/1) data.

nograph suppresses displaying the graph.

generate(*newvar*) creates *newvar* containing the smoothed values of *yvar*.

⌐ Plot ⌐

marker_options affect the rendition of markers drawn at the plotted points, including their shape, size, color, and outline; see [G-3] *marker_options*.

marker_label_options specify if and how the markers are to be labeled; see [G-3] *marker_label_options*.

⌐ Smoothed line ⌐

lineopts(*cline_options*) affects the rendition of the lowess-smoothed line; see [G-3] *cline_options*.

⌐ Add plots ⌐

addplot(*plot*) provides a way to add other plots to the generated graph; see [G-3] *addplot_option*.

⌐ Y axis, X axis, Titles, Legend, Overall, By ⌐

twoway_options are any of the options documented in [G-3] *twoway_options*. These include options for titling the graph (see [G-3] *title_options*), options for saving the graph to disk (see [G-3] *saving_option*), and the by() option (see [G-3] *by_option*).

Remarks

By default, lowess provides locally weighted scatterplot smoothing. The basic idea is to create a new variable (*newvar*) that, for each *yvar* y_i, contains the corresponding smoothed value. The smoothed values are obtained by running a regression of *yvar* on *xvar* by using only the data (x_i, y_i) and a few of the data near this point. In lowess, the regression is weighted so that the central point (x_i, y_i) gets the highest weight and points that are farther away (based on the distance $|x_j - x_i|$) receive less weight. The estimated regression line is then used to predict the smoothed value $\widehat{y}_i$ for y_i only. The procedure is repeated to obtain the remaining smoothed values, which means that a separate weighted regression is performed for every point in the data.

Lowess is a desirable smoother because of its locality—it tends to follow the data. Polynomial smoothing methods, for instance, are global in that what happens on the extreme left of a scatterplot can affect the fitted values on the extreme right.

Example 1

The amount of smoothing is affected by bwidth(#). You are warned to experiment with different values. For instance,

```
. use http://www.stata-press.com/data/r12/lowess1
(example data for lowess)
. lowess h1 depth
```

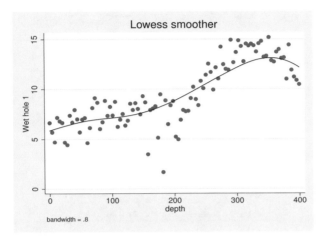

Now compare that with

```
. lowess h1 depth, bwidth(.4)
```

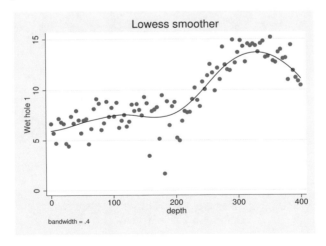

In the first case, the default bandwidth of 0.8 is used, meaning that 80% of the data are used in smoothing each point. In the second case, we explicitly specified a bandwidth of 0.4. Smaller bandwidths follow the original data more closely.   ◁

▷ Example 2

Two lowess options are especially useful with binary (0/1) data: adjust and logit. adjust adjusts the resulting curve (by multiplication) so that the mean of the smoothed values is equal to the mean of the unsmoothed values. logit specifies that the smoothed curve be in terms of the log of the odds ratio:

```
. use http://www.stata-press.com/data/r12/auto
(1978 Automobile Data)
. lowess foreign mpg, ylabel(0 "Domestic" 1 "Foreign") jitter(5) adjust
```

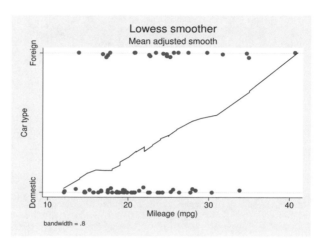

```
. lowess foreign mpg, logit yline(0)
```

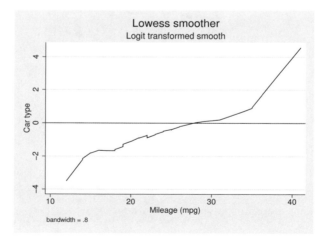

With binary data, if you do not use the logit option, it is a good idea to specify graph's jitter() option; see [G-2] **graph twoway scatter**. Because the underlying data (whether the car was manufactured outside the United States here) take on only two values, raw data points are more likely to be on top of each other, thus making it impossible to tell how many points there are. graph's jitter() option adds some noise to the data to shift the points around. This noise affects only the location of points on the graph, not the lowess curve.

When you specify the `logit` option, the display of the raw data is suppressed.

◁

ꓶ Technical note

`lowess` can be used for more than just lowess smoothing. Lowess can be usefully thought of as a combination of two smoothing concepts: the use of predicted values from regression (rather than means) for imputing a smoothed value and the use of the tricube weighting function (rather than a constant weighting function). `lowess` allows you to combine these concepts freely. You can use line smoothing without weighting (specify `noweight`), mean smoothing with tricube weighting (specify `mean`), or mean smoothing without weighting (specify `mean` and `noweight`).

❏

Methods and formulas

`lowess` is implemented as an ado-file.

Let y_i and x_i be the two variables, and assume that the data are ordered so that $x_i \leq x_{i+1}$ for $i = 1, \ldots, N - 1$. For each y_i, a smoothed value y_i^s is calculated.

The subset used in calculating y_i^s is indices $i_- = \max(1, i - k)$ through $i_+ = \min(i + k, N)$, where $k = \lfloor (N \times \texttt{bwidth} - 0.5)/2 \rfloor$. The weights for each of the observations between $j = i_-, \ldots, i_+$ are either 1 (`noweight`) or the tricube (default),

$$
w_j = \left\{ 1 - \left(\frac{|x_j - x_i|}{\Delta} \right)^3 \right\}^3
$$

where $\Delta = 1.0001 \max(x_{i_+} - x_i, x_i - x_{i_-})$. The smoothed value y_i^s is then the (weighted) mean or the (weighted) regression prediction at x_i.

William Swain Cleveland (1943–) studied mathematics and statistics at Princeton and Yale. He worked for several years at Bell Labs in New Jersey and now teaches statistics and computer science at Purdue. He has made key contributions in many areas of statistics, including graphics and data visualization, time series, environmental applications, and analysis of Internet traffic data.

Acknowledgment

`lowess` is a modified version of a command originally written by Patrick Royston of the MRC Clinical Trials Unit, London.

References

Chambers, J. M., W. S. Cleveland, B. Kleiner, and P. A. Tukey. 1983. *Graphical Methods for Data Analysis*. Belmont, CA: Wadsworth.

Cleveland, W. S. 1979. Robust locally weighted regression and smoothing scatterplots. *Journal of the American Statistical Association* 74: 829–836.

———. 1993. *Visualizing Data*. Summit, NJ: Hobart.

———. 1994. *The Elements of Graphing Data*. Rev. ed. Summit, NJ: Hobart.

Cox, N. J. 2005. Speaking Stata: Smoothing in various directions. *Stata Journal* 5: 574–593.

Goodall, C. 1990. A survey of smoothing techniques. In *Modern Methods of Data Analysis*, ed. J. Fox and J. S. Long, 126–176. Newbury Park, CA: Sage.

Lindsey, C., and S. J. Sheather. 2010. Model fit assessment via marginal model plots. *Stata Journal* 10: 215–225.

Royston, P. 1991. gr6: Lowess smoothing. *Stata Technical Bulletin* 3: 7–9. Reprinted in *Stata Technical Bulletin Reprints*, vol. 1, pp. 41–44. College Station, TX: Stata Press.

Royston, P., and N. J. Cox. 2005. A multivariable scatterplot smoother. *Stata Journal* 5: 405–412.

Salgado-Ugarte, I. H., and M. Shimizu. 1995. snp8: Robust scatterplot smoothing: Enhancements to Stata's ksm. *Stata Technical Bulletin* 25: 23–26. Reprinted in *Stata Technical Bulletin Reprints*, vol. 5, pp. 190–194. College Station, TX: Stata Press.

Sasieni, P. 1994. snp7: Natural cubic splines. *Stata Technical Bulletin* 22: 19–22. Reprinted in *Stata Technical Bulletin Reprints*, vol. 4, pp. 171–174. College Station, TX: Stata Press.

Also see

[D] **ipolate** — Linearly interpolate (extrapolate) values

[R] **smooth** — Robust nonlinear smoother

[R] **lpoly** — Kernel-weighted local polynomial smoothing

Title

> **lpoly** — Kernel-weighted local polynomial smoothing

Syntax

lpoly *yvar xvar* $\left[\,if\,\right]$ $\left[\,in\,\right]$ $\left[\,weight\,\right]$ $\left[\,,\ options\,\right]$

options	Description
Main	
<u>k</u>ernel(*kernel*)	specify kernel function; default is kernel(epanechnikov)
<u>bw</u>idth(# | *varname*)	specify kernel bandwidth
<u>d</u>egree(#)	specify degree of the polynomial smooth; default is degree(0)
<u>g</u>enerate($\left[\,newvar_x\,\right]$ *newvar_s*)	store smoothing grid in *newvar_x* and smoothed points in *newvar_s*
n(#)	obtain the smooth at # points; default is min(N, 50)
at(*varname*)	obtain the smooth at the values specified by *varname*
<u>nog</u>raph	suppress graph
<u>nosc</u>atter	suppress scatterplot only
SE/CI	
ci	plot confidence bands
<u>l</u>evel(#)	set confidence level; default is level(95)
se(*newvar*)	store standard errors in *newvar*
<u>p</u>width(#)	specify pilot bandwidth for standard error calculation
<u>v</u>ar(# | *varname*)	specify estimates of residual variance
Scatterplot	
marker_options	change look of markers (color, size, etc.)
marker_label_options	add marker labels; change look or position
Smoothed line	
lineopts(*cline_options*)	affect rendition of the smoothed line
CI plot	
ciopts(*cline_options*)	affect rendition of the confidence bands
Add plots	
addplot(*plot*)	add other plots to the generated graph
Y axis, X axis, Titles, Legend, Overall	
twoway_options	any options other than by() documented in [G-3] ***twoway_options***

kernel	Description
epanechnikov	Epanechnikov kernel function; the default
epan2	alternative Epanechnikov kernel function
biweight	biweight kernel function
cosine	cosine trace kernel function
gaussian	Gaussian kernel function
parzen	Parzen kernel function
rectangle	rectangle kernel function
triangle	triangle kernel function

fweights and aweights are allowed; see [U] **11.1.6 weight**.

Menu

Statistics > Nonparametric analysis > Local polynomial smoothing

Description

lpoly performs a kernel-weighted local polynomial regression of *yvar* on *xvar* and displays a graph of the smoothed values with (optional) confidence bands.

Options

┌─ Main ┐

kernel(*kernel*) specifies the kernel function for use in calculating the weighted local polynomial estimate. The default is kernel(epanechnikov).

bwidth(*#* | *varname*) specifies the half-width of the kernel—the width of the smoothing window around each point. If bwidth() is not specified, a rule-of-thumb (ROT) bandwidth estimator is calculated and used. A local variable bandwidth may be specified in *varname*, in conjunction with an explicit smoothing grid using the at() option.

degree(*#*) specifies the degree of the polynomial to be used in the smoothing. The default is degree(0), meaning local-mean smoothing.

generate([*newvar_x*] *newvar_s*) stores the smoothing grid in *newvar_x* and the smoothed values in *newvar_s*. If at() is not specified, then both *newvar_x* and *newvar_s* must be specified. Otherwise, only *newvar_s* is to be specified.

n(*#*) specifies the number of points at which the smooth is to be calculated. The default is $\min(N, 50)$, where N is the number of observations.

at(*varname*) specifies a variable that contains the values at which the smooth should be calculated. By default, the smoothing is done on an equally spaced grid, but you can use at() to instead perform the smoothing at the observed *x*'s, for example. This option also allows you to more easily obtain smooths for different variables or different subsamples of a variable and then overlay the estimates for comparison.

nograph suppresses drawing the graph of the estimated smooth. This option is often used with the generate() option.

noscatter suppresses superimposing a scatterplot of the observed data over the smooth. This option is useful when the number of resulting points would be so large as to clutter the graph.

ci plots confidence bands, using the confidence level specified in level().

level(#) specifies the confidence level, as a percentage, for confidence intervals. The default is level(95) or as set by set level; see [U] **20.7 Specifying the width of confidence intervals**.

se(*newvar*) stores the estimates of the standard errors in *newvar*. This option requires specifying generate() or at().

pwidth(#) specifies the pilot bandwidth to be used for standard-error computations. The default is chosen to be 1.5 times the value of the ROT bandwidth selector. If you specify pwidth() without specifying se() or ci, then the ci option is assumed.

var(# | *varname*) specifies an estimate of a constant residual variance or a variable containing estimates of the residual variances at each grid point required for standard-error computation. By default, the residual variance at each smoothing point is estimated by the normalized weighted residual sum of squares obtained from locally fitting a polynomial of order $p + 2$, where p is the degree specified in degree(). var(*varname*) is allowed only if at() is specified. If you specify var() without specifying se() or ci, then the ci option is assumed.

Scatterplot

marker_options affect the rendition of markers drawn at the plotted points, including their shape, size, color, and outline; see [G-3] ***marker_options***.

marker_label_options specify if and how the markers are to be labeled; see [G-3] ***marker_label_options***.

Smoothed line

lineopts(*cline_options*) affects the rendition of the smoothed line; see [G-3] ***cline_options***.

CI plot

ciopts(*cline_options*) affects the rendition of the confidence bands; see [G-3] ***cline_options***.

Add plots

addplot(*plot*) provides a way to add other plots to the generated graph; see [G-3] ***addplot_option***.

Y axis, X axis, Titles, Legend, Overall

twoway_options are any of the options documented in [G-3] ***twoway_options***, excluding by(). These include options for titling the graph (see [G-3] ***title_options***) and for saving the graph to disk (see [G-3] ***saving_option***).

Remarks

Remarks are presented under the following headings:

> *Introduction*
> *Local polynomial smoothing*
> *Choice of a bandwidth*
> *Confidence bands*

Introduction

The last 25 years or so has seen a significant outgrowth in the literature on scatterplot smoothing, otherwise known as univariate nonparametric regression. Of most appeal is the idea of making no assumptions about the functional form for the expected value of a response given a regressor, but instead allowing the data to "speak for themselves". Various methods and estimators fall into the category of nonparametric regression, including local mean smoothing as described independently by Nadaraya (1964) and Watson (1964), the Gasser and Müller (1979) estimator, locally weighted scatterplot smoothing (LOWESS) as described by Cleveland (1979), wavelets (for example, Donoho [1995]), and splines (Eubank 1999), to name a few. Much of the vast literature focuses on automating the amount of smoothing to be performed and dealing with the bias/variance tradeoff inherent to this type of estimation. For example, for Nadaraya–Watson the amount of smoothing is controlled by choosing a *bandwidth*.

Smoothing via local polynomials is by no means a new idea but instead one that has been rediscovered in recent years in articles such as Fan (1992). A natural extension of the local mean smoothing of Nadaraya–Watson, local polynomial regression involves fitting the response to a polynomial form of the regressor via locally weighted least squares. Higher-order polynomials have better bias properties than the zero-degree local polynomials of the Nadaraya–Watson estimator; in general, higher-order polynomials do not require bias adjustment at the boundary of the regression space. For a definitive reference on local polynomial smoothing, see Fan and Gijbels (1996).

Local polynomial smoothing

Consider a set of scatterplot data $\{(x_1, y_1), \ldots, (x_n, y_n)\}$ from the model

$$y_i = m(x_i) + \sigma(x_i)\epsilon_i \tag{1}$$

for some unknown mean and variance functions $m(\cdot)$ and $\sigma^2(\cdot)$, and symmetric errors ϵ_i with $E(\epsilon_i) = 0$ and $\text{Var}(\epsilon_i) = 1$. The goal is to estimate $m(x_0) = E[Y|X = x_0]$, making no assumption about the functional form of $m(\cdot)$.

lpoly estimates $m(x_0)$ as the constant term (intercept) of a regression, weighted by the kernel function specified in kernel(), of *yvar* on the polynomial terms $(xvar - x_0), (xvar - x_0)^2, \ldots, (xvar - x_0)^p$ for each smoothing point x_0. The degree of the polynomial, p, is specified in degree(), the amount of smoothing is controlled by the bandwidth specified in bwidth(), and the chosen kernel function is specified in kernel().

▷ Example 1

Consider the motorcycle data as examined (among other places) in Fan and Gijbels (1996). The data consist of 133 observations and measure the acceleration (accel measured in grams [g]) of a dummy's head during impact over time (time measured in milliseconds). For these data, we use lpoly to fit a local cubic polynomial with the default bandwidth (obtained using the ROT method) and the default Epanechnikov kernel.

```
. use http://www.stata-press.com/data/r12/motorcycle
(Motorcycle data from Fan & Gijbels (1996))

. lpoly accel time, degree(3)
```

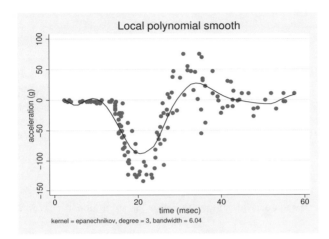

The default bandwidth and kernel settings do not provide a satisfactory fit in this example. To improve the fit, we can either supply a different bandwidth by using the `bwidth()` option or specify a different kernel by using the `kernel()` option. For example, using the alternative Epanechnikov kernel, `kernel(epan2)`, below provides a better fit for these data.

```
. lpoly accel time, degree(3) kernel(epan2)
```

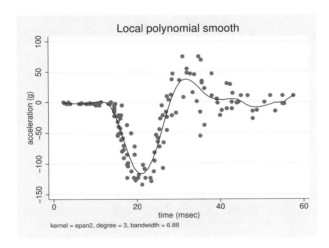

◁

❑ Technical note

`lpoly` allows specifying in `degree()` both odd and even orders of the polynomial to be used for the smoothing. However, the odd-order, $2k + 1$, polynomial approximations are preferable. They have an extra parameter compared with the even-order, $2k$, approximations, which leads to a significant bias reduction and there is no increase of variability associated with adding this extra parameter. Using an odd order when estimating the regression function is therefore usually sufficient. For a more thorough discussion, see Fan and Gijbels (1996).

❑

Choice of a bandwidth

The choice of a bandwidth is crucial for many smoothing techniques, including local polynomial smoothing. In general, using a large bandwidth gives smooths with a large bias, whereas a small bandwidth may result in highly variable smoothed values. Various techniques exist for optimal bandwidth selection. By default, lpoly uses the ROT method to estimate the bandwidth used for the smoothing; see *Methods and formulas* for details.

▷ Example 2

Using the motorcycle data, we demonstrate how a local linear polynomial fit changes using different bandwidths.

```
. lpoly accel time, degree(1) kernel(epan2) bwidth(1) generate(at smooth1)
> nograph
. lpoly accel time, degree(1) kernel(epan2) bwidth(7) at(at) generate(smooth2)
> nograph
. label variable smooth1 "smooth: width = 1"
. label variable smooth2 "smooth: width = 7"
. lpoly accel time, degree(1) kernel(epan2) at(at) addplot(line smooth* at)
> legend(label(2 "smooth: width = 3.42 (ROT)")) note("kernel = epan2, degree = 1")
```

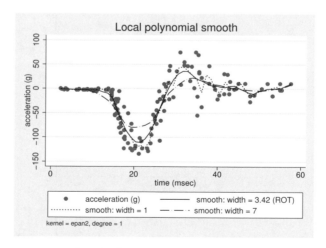

From this graph, we can see that the local linear polynomial fit with larger bandwidth (width = 7) corresponds to a smoother line but fails to fit the curvature of the scatterplot data. The smooth obtained using the width equal to one seems to fit most data points, but the corresponding line has several spikes indicating larger variability. The smooth obtained using the ROT bandwidth estimator seems to have a good tradeoff between the fit and variability in this example.

In the above, we also demonstrated how the `generate()` and `addplot()` options may be used to produce overlaid plots obtained from `lpoly` with different options. The `nograph` option saves time when you need to save only results with `generate()`.

However, to avoid generating variables manually, one can use `twoway lpoly` instead; see [G-2] **graph twoway lpoly** for more details.

```
. twoway scatter accel time ||
>           lpoly accel time, degree(1) kernel(epan2) lpattern(solid) ||
>           lpoly accel time, degree(1) kernel(epan2) bwidth(1)        ||
>           lpoly accel time, degree(1) kernel(epan2) bwidth(7)        ||
>       , legend(label(2 "smooth: width = 3.42 (ROT)") label(3 "smooth: width = 1")
>               label(4 "smooth: width = 7"))
>           title("Local polynomial smooth") note("kernel = epan2, degree = 1")
>           xtitle("time (msec)") ytitle("acceleration (g)")
```

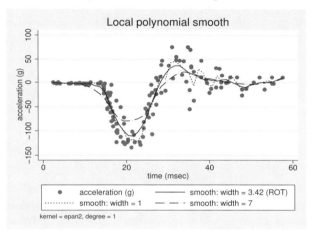

◁

The ROT estimate is commonly used as an initial guess for the amount of smoothing; this approach may be sufficient when the choice of a bandwidth is less important. In other cases, you can pick your own bandwidth.

When the shape of the regression function has a combination of peaked and flat regions, a variable bandwidth may be preferable over the constant bandwidth to allow for different degrees of smoothness in different regions. The `bwidth()` option allows you to specify the values of the local variable bandwidths as those stored in a variable in your data.

Similar issues with bias and variability arise when choosing a pilot bandwidth (the `pwidth()` option) used to compute standard errors of the local polynomial smoother. The default value is chosen to be 1.5 × ROT. For a review of methods for pilot bandwidth selection, see Fan and Gijbels (1996).

Confidence bands

The established asymptotic normality of the local polynomial estimators under certain conditions allows the construction of approximate confidence bands. `lpoly` offers the `ci` option to plot these bands.

▷ Example 3

Let us plot the confidence bands for the local polynomial fit from example 1.

. lpoly accel time, degree(3) kernel(epan2) ci

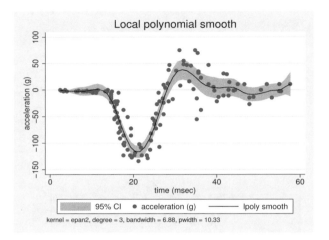

You can obtain graphs with overlaid confidence bands by using `twoway lpolyci`; see [G-2] **graph twoway lpolyci** for examples.

◁

Constructing the confidence intervals involves computing standard errors obtained by taking a square root of the estimate of the conditional variance of the local polynomial estimator at each grid point x_0. Estimating the conditional variance requires fitting a polynomial of a higher order locally by using a different bandwidth, the pilot bandwidth. The value of the pilot bandwidth may be supplied by using `pwidth()`. By default, the value of $1.5 \times$ ROT is used. Also, estimates of the residual variance $\sigma^2(x_0)$ at each grid point, x_0, are required to obtain the estimates of the conditional variances. These estimates may be supplied by using the `var()` option. By default, they are computed using the normalized weighted residual sum of squares from a local polynomial fit of a higher order. See *Methods and formulas* for details. The standard errors may be saved by using `se()`.

Saved results

`lpoly` saves the following in `r()`:

Scalars
r(degree)	smoothing polynomial degree	r(bwidth)	bandwidth of the smooth
r(ngrid)	number of successful regressions	r(pwidth)	pilot bandwidth
r(N)	sample size		

Macros
r(kernel)	name of kernel

Methods and formulas

lpoly is implemented as an ado-file.

Consider model (1), written in matrix notation,

$$\mathbf{y} = m(\mathbf{x}) + \epsilon$$

where $\mathbf{y}$ and $\mathbf{x}$ are the $n \times 1$ vectors of scatterplot values, ϵ is the $n \times 1$ vector of errors with zero mean and covariance matrix $\boldsymbol{\Sigma} = \mathrm{diag}\{\sigma(x_i)\}\mathbf{I}_n$, and $m()$ and $\sigma()$ are some unknown functions. Define $m(x_0) = E[Y|X = x_0]$ and $\sigma^2(x_0) = \mathrm{Var}[Y|X = x_0]$ to be the conditional mean and conditional variance of random variable Y (residual variance), respectively, for some realization x_0 of random variable X.

The method of local polynomial smoothing is based on the approximation of $m(x)$ locally by a pth order polynomial in $(x - x_0)$ for some x in the neighborhood of x_0. For the scatterplot data $\{(x_1, y_1), \ldots, (x_n, y_n)\}$, the pth-order local polynomial smooth $\widehat{m}(x_0)$ is equal to $\widehat{\beta}_0$, an estimate of the intercept of the weighted linear regression,

$$\widehat{\beta} = (\mathbf{X}^T \mathbf{W} \mathbf{X})^{-1} \mathbf{X}^T \mathbf{W} \mathbf{y} \tag{2}$$

where $\widehat{\beta} = (\widehat{\beta}_0, \widehat{\beta}_1, \ldots, \widehat{\beta}_p)^T$ is the vector of estimated regression coefficients (with $\{\widehat{\beta}_j = (j!)^{-1}\widehat{m}^{(j)}(x)|_{x=x_0}, \ j = 0, \ldots, p\}$ also representing estimated coefficients from a corresponding Taylor expansion); $\mathbf{X} = \{(x_i - x_0)^j\}_{i,j=1,0}^{n,p}$ is a design matrix; and $\mathbf{W} = \mathrm{diag}\{K_h(x_i - x_0)\}_{n \times n}$ is a weighting matrix with weights $K_h(\cdot)$ defined as $K_h(x) = h^{-1}K(x/h)$, with $K(\cdot)$ being a kernel function and h defining a bandwidth. The kernels are defined in *Methods and formulas* of [R] **kdensity**.

The default bandwidth is obtained using the ROT method of bandwidth selection. The ROT bandwidth is the plugin estimator of the asymptotically optimal constant bandwidth. This is the bandwidth that minimizes the conditional weighted mean integrated squared error. The ROT plugin bandwidth selector for the smoothing bandwidth h is defined as follows; assuming constant residual variance $\sigma^2(x_0) = \sigma^2$ and odd degree p:

$$\widehat{h} = C_{0,p}(K) \left[\frac{\widehat{\sigma}^2 \int w_0(x)dx}{n \int \{\widehat{m}^{(p+1)}(x)\}^2 w_0(x)f(x)dx} \right]^{1/(2p+3)} \tag{3}$$

where $C_{0,p}(K)$ is a constant, as defined in Fan and Gijbels (1996), that depends on the kernel function $K(\cdot)$, and the degree of a polynomial p and w_0 is chosen to be an indicator function on the interval $[\min_\mathbf{x} + 0.05 \times \mathrm{range}_\mathbf{x}, \max_\mathbf{x} - 0.05 \times \mathrm{range}_\mathbf{x}]$ with $\min_\mathbf{x}$, $\max_\mathbf{x}$, and $\mathrm{range}_\mathbf{x}$ being, respectively, the minimum, maximum, and the range of $\mathbf{x}$. To obtain the estimates of a constant residual variance, $\widehat{\sigma}^2$, and $(p+1)$th order derivative of $m(x)$, denoted as $\widehat{m}^{(p+1)}(x)$, a polynomial in $\mathbf{x}$ of order $(p+3)$ is fit globally to $\mathbf{y}$. $\widehat{\sigma}^2$ is estimated as a standardized residual sum of squares from this fit.

The expression for the asymptotically optimal constant bandwidth used in constructing the ROT bandwidth estimator is derived for the odd-order polynomial approximations. For even-order polynomial fits the expression would depend not only on $m^{(p+1)}(x)$ but also on $m^{(p+2)}(x)$ and the design density and its derivative, $f(x)$ and $f'(x)$. Therefore, the ROT bandwidth selector would require estimation of these additional quantities. Instead, for an even-degree p of the local polynomial, lpoly uses the value of the ROT estimator (3) computed using degree $p + 1$. As such, for even degrees this is not a plugin estimator of the asymptotically optimal constant bandwidth.

The estimates of the conditional variance of local polynomial estimators are obtained using

$$\widehat{\text{Var}}\{\widehat{m}(x_0)|X = x_0\} = \widehat{\sigma}_m^2(x_0) = (\mathbf{X}^T\mathbf{W}\mathbf{X})^{-1}(\mathbf{X}^T\mathbf{W}^2\mathbf{X})(\mathbf{X}^T\mathbf{W}\mathbf{X})^{-1}\widehat{\sigma}^2(x_0) \qquad (4)$$

where $\widehat{\sigma}^2(x_0)$ is estimated by the normalized weighted residual sum of squares from the $(p+2)$th order polynomial fit using pilot bandwidth $h^\star$.

When the bias is negligible the normal-approximation method yields a $(1-\alpha) \times 100\%$ confidence interval for $m(x_0)$,

$$\left\{\widehat{m}(x_0) - z_{(1-\alpha/2)}\widehat{\sigma}_m(x_0), \ \widehat{m}(x_0) + z_{(1-\alpha/2)}\widehat{\sigma}_m(x_0)\right\}$$

where $z_{(1-\alpha/2)}$ is the $(1-\alpha/2)$th quantile of the standard Gaussian distribution, and $\widehat{m}(x_0)$ and $\widehat{\sigma}_m(x_0)$ are as defined in (2) and (4), respectively.

References

Cleveland, W. S. 1979. Robust locally weighted regression and smoothing scatterplots. *Journal of the American Statistical Association* 74: 829–836.

Cox, N. J. 2005. Speaking Stata: Smoothing in various directions. *Stata Journal* 5: 574–593.

Donoho, D. L. 1995. Nonlinear solution of linear inverse problems by wavelet-vaguelette decomposition. *Applied and Computational Harmonic Analysis* 2: 101–126.

Eubank, R. L. 1999. *Nonparametric Regression and Spline Smoothing*. 2nd ed. New York: Dekker.

Fan, J. 1992. Design-adaptive nonparametric regression. *Journal of the American Statistical Association* 87: 998–1004.

Fan, J., and I. Gijbels. 1996. *Local Polynomial Modelling and Its Applications*. London: Chapman & Hall.

Gasser, T., and H. G. Müller. 1979. Kernel estimation of regression functions. In *Smoothing Techniques for Curve Estimation, Lecture Notes in Mathematics*, ed. T. Gasser and M. Rosenblatt, 23–68. New York: Springer.

Gutierrez, R. G., J. M. Linhart, and J. S. Pitblado. 2003. From the help desk: Local polynomial regression and Stata plugins. *Stata Journal* 3: 412–419.

Nadaraya, E. A. 1964. On estimating regression. *Theory of Probability and Its Application* 9: 141–142.

Sheather, S. J., and M. C. Jones. 1991. A reliable data-based bandwidth selection method for kernel density estimation. *Journal of the Royal Statistical Society, Series B* 53: 683–690.

Watson, G. S. 1964. Smooth regression analysis. *Sankhyā Series A* 26: 359–372.

Also see

[R] **kdensity** — Univariate kernel density estimation

[R] **lowess** — Lowess smoothing

[R] **smooth** — Robust nonlinear smoother

[G-2] **graph twoway lpoly** — Local polynomial smooth plots

[G-2] **graph twoway lpolyci** — Local polynomial smooth plots with CIs

> **lrtest** — Likelihood-ratio test after estimation

lrtest *modelspec₁* [*modelspec₂*] [, *options*]

where *modelspec* is

name | . | (*namelist*)

where *name* is the name under which estimation results were saved using estimates store (see [R] **estimates store**), and "." refers to the last estimation results, whether or not these were already stored.

options	Description
stats	display statistical information about the two models
dir	display descriptive information about the two models
df(#)	override the automatic degrees-of-freedom calculation; seldom used
force	force testing even when apparently invalid

Statistics > Postestimation > Tests > Likelihood-ratio test

lrtest performs a likelihood-ratio test of the null hypothesis that the parameter vector of a statistical model satisfies some smooth constraint. To conduct the test, both the unrestricted and the restricted models must be fit using the maximum likelihood method (or some equivalent method), and the results of at least one must be stored using estimates store; see [R] **estimates store**.

modelspec₁ and *modelspec₂* specify the restricted and unrestricted model in any order. *modelspec₁* and *modelspec₂* cannot have names in common; for example, lrtest (A B C) (C D E) is not allowed because both model specifications include C. If *modelspec₂* is not specified, the last estimation result is used; this is equivalent to specifying *modelspec₂* as a period (.).

lrtest supports composite models specified by a parenthesized list of model names. In a composite model, we assume that the log likelihood and dimension (number of free parameters) of the full model are obtained as the sum of the log-likelihood values and dimensions of the constituting models.

lrtest provides an important alternative to test (see [R] **test**) for models fit via maximum likelihood or equivalent methods.

stats displays statistical information about the unrestricted and restricted models, including the information indices of Akaike and Schwarz.

dir displays descriptive information about the unrestricted and restricted models; see `estimates dir` in [R] **estimates store**.

df(#) is seldom specified; it overrides the automatic degrees-of-freedom calculation.

force forces the likelihood-ratio test calculations to take place in situations where `lrtest` would normally refuse to do so and issue an error. Such situations arise when one or more assumptions of the test are violated, for example, if the models were fit with vce(robust), vce(cluster *clustvar*), or pweights; when the dependent variables in the two models differ; when the null log likelihoods differ; when the samples differ; or when the estimation commands differ. If you use the force option, there is no guarantee as to the validity or interpretability of the resulting test.

Remarks

The standard way to use `lrtest` is to do the following:

1. Fit either the restricted model or the unrestricted model by using one of Stata's estimation commands and then store the results using `estimates store` *name*.

2. Fit the alternative model (the unrestricted or restricted model) and then type '`lrtest` *name* .'. `lrtest` determines for itself which of the two models is the restricted model by comparing the degrees of freedom.

Often you may want to store the alternative model with `estimates store` *name₂*, for instance, if you plan additional tests against models yet to be fit. The likelihood-ratio test is then obtained as `lrtest` *name name₂*.

Remarks are presented under the following headings:

> *Nested models*
> *Composite models*

Nested models

`lrtest` may be used with any estimation command that reports a log likelihood, including heckman, logit, poisson, stcox, and streg. You must check that one of the model specifications implies a statistical model that is *nested within* the model implied by the other specification. Usually, this means that both models are fit with the same estimation command (for example, both are fit by logit, with the same dependent variables) and that the set of covariates of one model is a subset of the covariates of the other model. Second, `lrtest` is valid only for models that are fit by maximum likelihood or by some equivalent method, so it does not apply to models that were fit with probability weights or clusters. Specifying the vce(robust) option similarly would indicate that you are worried about the valid specification of the model, so you would not use `lrtest`. Third, `lrtest` assumes that under the null hypothesis, the test statistic is (approximately) distributed as chi-squared. This assumption is not true for likelihood-ratio tests of "boundary conditions", such as tests for the presence of overdispersion or random effects (Gutierrez, Carter, and Drukker 2001).

▷ Example 1

We have data on infants born with low birthweights along with the characteristics of the mother (Hosmer and Lemeshow 2000; see also [R] **logistic**). We fit the following model:

```
. use http://www.stata-press.com/data/r12/lbw
(Hosmer & Lemeshow data)
. logistic low age lwt i.race smoke ptl ht ui
```

Logistic regression

Number of obs	=	189
LR chi2(8)	=	33.22
Prob > chi2	=	0.0001
Pseudo R2	=	0.1416

Log likelihood =   -100.724

low	Odds Ratio	Std. Err.	z	P>|z|	[95% Conf. Interval]	
age	.9732636	.0354759	-0.74	0.457	.9061578	1.045339
lwt	.9849634	.0068217	-2.19	0.029	.9716834	.9984249
race						
2	3.534767	1.860737	2.40	0.016	1.259736	9.918406
3	2.368079	1.039949	1.96	0.050	1.001356	5.600207
smoke	2.517698	1.00916	2.30	0.021	1.147676	5.523162
ptl	1.719161	.5952579	1.56	0.118	.8721455	3.388787
ht	6.249602	4.322408	2.65	0.008	1.611152	24.24199
ui	2.1351	.9808153	1.65	0.099	.8677528	5.2534
_cons	1.586014	1.910496	0.38	0.702	.1496092	16.8134

We now wish to test the constraint that the coefficients on age, lwt, ptl, and ht are all zero or, equivalently here, that the odds ratios are all 1. One solution is to type

```
. test age lwt ptl ht
 ( 1)  [low]age = 0
 ( 2)  [low]lwt = 0
 ( 3)  [low]ptl = 0
 ( 4)  [low]ht = 0
           chi2(  4) =    12.38
         Prob > chi2 =    0.0147
```

This test is based on the inverse of the information matrix and is therefore based on a quadratic approximation to the likelihood function; see [R] **test**. A more precise test would be to refit the model, applying the proposed constraints, and then calculate the likelihood-ratio test.

We first save the current model:

```
. estimates store full
```

We then fit the constrained model, which here is the model omitting age, lwt, ptl, and ht:

```
. logistic low i.race smoke ui
```

Logistic regression

Number of obs	=	189
LR chi2(4)	=	18.80
Prob > chi2	=	0.0009
Pseudo R2	=	0.0801

Log likelihood = -107.93404

low	Odds Ratio	Std. Err.	z	P>|z|	[95% Conf. Interval]	
race						
2	3.052746	1.498087	2.27	0.023	1.166747	7.987382
3	2.922593	1.189229	2.64	0.008	1.316457	6.488285
smoke	2.945742	1.101838	2.89	0.004	1.415167	6.131715
ui	2.419131	1.047359	2.04	0.041	1.035459	5.651788
_cons	.1402209	.0512295	-5.38	0.000	.0685216	.2869447

That done, lrtest compares this model with the model we previously saved:

```
. lrtest full .
Likelihood-ratio test                        LR chi2(4)   =     14.42
(Assumption: . nested in full)               Prob > chi2 =    0.0061
```

Let's compare results. test reported that age, lwt, ptl, and ht were jointly significant at the 1.5% level; lrtest reports that they are significant at the 0.6% level. Given the quadratic approximation made by test, we could argue that lrtest's results are more accurate.

lrtest explicates the assumption that, from a comparison of the degrees of freedom, it has assessed that the last fit model (.) is nested within the model stored as full. In other words, full is the unconstrained model and . is the constrained model.

The names in "(Assumption: . nested in full)" are actually links. Click on a name, and the results for that model are replayed.

◁

Aside: The nestreg command provides a simple syntax for performing likelihood-ratio tests for nested model specifications; see [R] nestreg. In the previous example, we fit a full logistic model, used estimates store to store the full model, fit a constrained logistic model, and used lrtest to report a likelihood-ratio test between two models. To do this with one call to nestreg, use the lrtable option.

❏ Technical note

lrtest determines the degrees of freedom of a model as the rank of the (co)variance matrix e(V). There are two issues here. First, the *numerical* determination of the rank of a matrix is a subtle problem that can, for instance, be affected by the scaling of the variables in the model. The rank of a matrix depends on the number of (independent) linear combinations of coefficients that sum exactly to zero. In the world of numerical mathematics, it is hard to tell whether a very small number is really nonzero or is a real zero that happens to be slightly off because of roundoff error from the finite precision with which computers make floating-point calculations. Whether a small number is being classified as one or the other, typically on the basis of a threshold, affects the determined degrees of freedom. Although Stata generally makes sensible choices, it is bound to make mistakes occasionally. The moral of this story is to make sure that the calculated degrees of freedom are as you expect before interpreting the results.

❏

❏ Technical note

A second issue involves regress and related commands such as anova. Mainly for historical reasons, regress does not treat the residual variance, σ^2, the same way that it treats the regression coefficients. Type estat vce after regress, and you will see the regression coefficients, not $\hat{\sigma}^2$. Most estimation commands for models with ancillary parameters (for example, streg and heckman) treat all parameters as equals. There is nothing technically wrong with regress here; we are usually focused on the regression coefficients, and their estimators are uncorrelated with $\hat{\sigma}^2$. But, formally, σ^2 adds a degree of freedom to the model, which does not matter if you are comparing two regression models by a likelihood-ratio test. This test depends on the difference in the degrees of freedom, and hence being "off by 1" in each does not matter. But, if you are comparing a regression model with a larger model—for example, a heteroskedastic regression model fit by arch—the automatic determination of the degrees of freedom is incorrect, and you must specify the df(#) option.

❏

Example 2

Returning to the low-birthweight data in the example 1, we now wish to test that the coefficient on 2.race is equal to that on 3.race. The base model is still stored under the name full, so we need only fit the constrained model and perform the test. With z as the index of the logit model, the base model is

$$z = \beta_0 + \beta_1 \text{age} + \beta_2 \text{lwt} + \beta_3 2.\text{race} + \beta_4 3.\text{race} + \cdots$$

If $\beta_3 = \beta_4$, this can be written as

$$z = \beta_0 + \beta_1 \text{age} + \beta_2 \text{lwt} + \beta_3 (2.\text{race} + 3.\text{race}) + \cdots$$

We can fit the constrained model as follows:

```
. constraint 1 2.race = 3.race
. logistic low age lwt i.race smoke ptl ht ui, constraints(1)
Logistic regression                              Number of obs   =        189
                                                 Wald chi2(7)    =      25.17
Log likelihood =  -100.9997                      Prob > chi2     =     0.0007
 ( 1)  [low]2.race - [low]3.race = 0
```

low	Odds Ratio	Std. Err.	z	P>|z|	[95% Conf. Interval]	
age	.9716799	.0352638	-0.79	0.429	.9049649	1.043313
lwt	.9864971	.0064627	-2.08	0.038	.9739114	.9992453
race						
2	2.728186	1.080207	2.53	0.011	1.255586	5.927907
3	2.728186	1.080207	2.53	0.011	1.255586	5.927907
smoke	2.664498	1.052379	2.48	0.013	1.228633	5.778414
ptl	1.709129	.5924776	1.55	0.122	.8663666	3.371691
ht	6.116391	4.215585	2.63	0.009	1.58425	23.61385
ui	2.09936	.9699702	1.61	0.108	.8487997	5.192407
_cons	1.309371	1.527398	0.23	0.817	.1330839	12.8825

Comparing this model with our original model, we obtain

```
. lrtest full .
Likelihood-ratio test                          LR chi2(1)  =       0.55
(Assumption: . nested in full)                 Prob > chi2 =     0.4577
```

By comparison, typing test 2.race=3.race after fitting our base model results in a significance level of 0.4572. Alternatively, we can first store the restricted model, here using the name equal. Next lrtest is invoked specifying the names of the restricted and unrestricted models (we do not care about the order). This time, we also add the option stats requesting a table of model statistics, including the model selection indices AIC and BIC.

```
. estimates store equal
. lrtest equal full, stats
Likelihood-ratio test                          LR chi2(1)  =       0.55
(Assumption: equal nested in full)             Prob > chi2 =     0.4577
```

Model	Obs	ll(null)	ll(model)	df	AIC	BIC
equal	189	.	-100.9997	8	217.9994	243.9334
full	189	-117.336	-100.724	9	219.448	248.6237

Note: N=Obs used in calculating BIC; see **[R] BIC note**

Composite models

lrtest supports composite models; that is, models that can be fit by fitting a series of simpler models or by fitting models on subsets of the data. Theoretically, a composite model is one in which the likelihood function, $L(\theta)$, of the parameter vector, θ, can be written as the product

$$L(\theta) = L_1(\theta_1) \times L_2(\theta_2) \times \cdots \times L_k(\theta_k)$$

of likelihood terms with $\theta = (\theta_1, \ldots, \theta_k)$ a partitioning of the full parameter vector. In such a case, the full-model likelihood $L(\theta)$ is maximized by maximizing the likelihood terms $L_j(\theta_j)$ in turn. Obviously, $\log L(\widehat{\theta}) = \sum_{j=1}^{k} \log L_j(\widehat{\theta_j})$. The degrees of freedom for the composite model is obtained as the sum of the degrees of freedom of the constituting models.

▷ Example 3

As an example of the application of composite models, we consider a test of the hypothesis that the coefficients of a statistical model do not differ between different portions ("regimes") of the covariate space. Economists call a test for such a hypothesis a *Chow test*.

We continue the analysis of the data on children of low birthweight by using logistic regression modeling and study whether the regression coefficients are the same among the three races: white, black, and other. A likelihood-ratio Chow test can be obtained by fitting the logistic regression model for each of the races and then comparing the combined results with those of the model previously stored as full. Because the full model included dummies for the three races, this version of the Chow test allows the intercept of the logistic regression model to vary between the regimes (races).

```
. logistic low age lwt smoke ptl ht ui if 1.race, nolog
Logistic regression                             Number of obs   =         96
                                                LR chi2(6)      =      13.86
                                                Prob > chi2     =     0.0312
Log likelihood = -45.927061                     Pseudo R2       =     0.1311
```

low	Odds Ratio	Std. Err.	z	P>|z|	[95% Conf. Interval]	
age	.9869674	.0527757	-0.25	0.806	.8887649	1.096021
lwt	.9900874	.0106101	-0.93	0.353	.9695089	1.011103
smoke	4.208697	2.680133	2.26	0.024	1.20808	14.66222
ptl	1.592145	.7474264	0.99	0.322	.6344379	3.995544
ht	2.900166	3.193537	0.97	0.334	.3350554	25.1032
ui	1.229523	.9474768	0.27	0.789	.2715165	5.567715
_cons	.4891008	.993785	-0.35	0.725	.0091175	26.23746

```
. estimates store white
```

```
. logistic low age lwt smoke ptl ht ui if 2.race, nolog
```

Logistic regression

```
Number of obs   =         26
LR chi2(6)      =      10.12
Prob > chi2     =     0.1198
```
Log likelihood = -12.654157

```
Pseudo R2       =     0.2856
```

low	Odds Ratio	Std. Err.	z	P>|z|	[95% Conf. Interval]	
age	.8735313	.1377846	-0.86	0.391	.6412332	1.189983
lwt	.9747736	.016689	-1.49	0.136	.9426065	1.008038
smoke	16.50373	24.37044	1.90	0.058	.9133647	298.2083
ptl	4.866916	9.33151	0.83	0.409	.1135573	208.5895
ht	85.05605	214.6382	1.76	0.078	.6049308	11959.27
ui	67.61338	133.3313	2.14	0.033	1.417399	3225.322
_cons	48.7249	169.9216	1.11	0.265	.0523961	45310.94

```
. estimates store black

. logistic low age lwt smoke ptl ht ui if 3.race, nolog
```

Logistic regression

```
Number of obs   =         67
LR chi2(6)      =      14.06
Prob > chi2     =     0.0289
```
Log likelihood = -37.228444

```
Pseudo R2       =     0.1589
```

low	Odds Ratio	Std. Err.	z	P>|z|	[95% Conf. Interval]	
age	.9263905	.0665386	-1.06	0.287	.8047407	1.06643
lwt	.9724499	.015762	-1.72	0.085	.9420424	1.003839
smoke	.7979034	.6340585	-0.28	0.776	.1680885	3.787586
ptl	2.845675	1.777944	1.67	0.094	.8363053	9.682908
ht	7.767503	10.00537	1.59	0.112	.6220764	96.98826
ui	2.925006	2.046473	1.53	0.125	.7423107	11.52571
_cons	49.09444	113.9165	1.68	0.093	.5199275	4635.769

```
. estimates store other
```

We are now ready to perform the likelihood-ratio Chow test:

```
. lrtest (full) (white black other), stats
```

Likelihood-ratio test

```
LR chi2(12) =       9.83
Prob > chi2 =     0.6310
```

Assumption: (full) nested in (white, black, other)

Model	Obs	ll(null)	ll(model)	df	AIC	BIC
full	189	-117.336	-100.724	9	219.448	248.6237
white	96	-52.85752	-45.92706	7	105.8541	123.8046
black	26	-17.71291	-12.65416	7	39.30831	48.11499
other	67	-44.26039	-37.22844	7	88.45689	103.8897

Note: N=Obs used in calculating BIC; see **[R] BIC note**

We cannot reject the hypothesis that the logistic regression model applies to each of the races at any reasonable significance level. By specifying the stats option, we can verify the degrees of freedom of the test: $12 = 7 + 7 + 7 - 9$. We can obtain the same test by fitting an expanded model with interactions between all covariates and race.

```
. logistic low race##c.(age lwt smoke ptl ht ui)
```

Logistic regression

Number of obs	=	189
LR chi2(20)	=	43.05
Prob > chi2	=	0.0020

Log likelihood = -95.809661

Pseudo R2	=	0.1835

low	Odds Ratio	Std. Err.	z	P>|z|	[95% Conf. Interval]	
race						
2	99.62137	402.0829	1.14	0.254	.0365434	271578.9
3	100.3769	309.586	1.49	0.135	.2378638	42358.38
age	.9869674	.0527757	-0.25	0.806	.8887649	1.096021
lwt	.9900874	.0106101	-0.93	0.353	.9695089	1.011103
smoke	4.208697	2.680133	2.26	0.024	1.20808	14.66222
ptl	1.592145	.7474264	0.99	0.322	.6344379	3.995544
ht	2.900166	3.193537	0.97	0.334	.3350554	25.1032
ui	1.229523	.9474768	0.27	0.789	.2715165	5.567715
race#c.age						
2	.885066	.1474079	-0.73	0.464	.638569	1.226714
3	.9386232	.0840486	-0.71	0.479	.7875366	1.118695
race#c.lwt						
2	.9845329	.0198857	-0.77	0.440	.9463191	1.02429
3	.9821859	.0190847	-0.93	0.355	.9454839	1.020313
race#c.smoke						
2	3.921338	6.305992	0.85	0.395	.167725	91.67917
3	.1895844	.1930601	-1.63	0.102	.025763	1.395113
race#c.ptl						
2	3.05683	6.034089	0.57	0.571	.0638301	146.3918
3	1.787322	1.396789	0.74	0.457	.3863582	8.268285
race#c.ht						
2	29.328	80.7482	1.23	0.220	.1329492	6469.623
3	2.678295	4.538712	0.58	0.561	.0966916	74.18702
race#c.ui						
2	54.99155	116.4274	1.89	0.058	.8672471	3486.977
3	2.378976	2.476124	0.83	0.405	.309335	18.29579
_cons	.4891008	.993785	-0.35	0.725	.0091175	26.23746

```
. lrtest full .
```

Likelihood-ratio test	LR chi2(12) =	9.83
(Assumption: full nested in .)	Prob > chi2 =	0.6310

Applying `lrtest` for the full model against the model with all interactions yields the same test statistic and p-value as for the full model against the composite model for the three regimes. Here the specification of the model with interactions was convenient, and `logistic` had no problem computing the estimates for the expanded model. In models with more complicated likelihoods, such as Heckman's selection model (see [R] **heckman**) or complicated survival-time models (see [ST] **streg**), fitting the models with all interactions may be numerically demanding and may be much more time consuming than fitting a series of models separately for each regime.

Given the model with all interactions, we could also test the hypothesis of no differences among the regions (races) by a Wald version of the Chow test by using the `testparm` command; see [R] **test**.

```
. testparm race#c.(age lwt smoke ptl ht ui)
 ( 1)   [low]2.race#c.age = 0
 ( 2)   [low]3.race#c.age = 0
 ( 3)   [low]2.race#c.lwt = 0
 ( 4)   [low]3.race#c.lwt = 0
 ( 5)   [low]2.race#c.smoke = 0
 ( 6)   [low]3.race#c.smoke = 0
 ( 7)   [low]2.race#c.ptl = 0
 ( 8)   [low]3.race#c.ptl = 0
 ( 9)   [low]2.race#c.ht = 0
 (10)   [low]3.race#c.ht = 0
 (11)   [low]2.race#c.ui = 0
 (12)   [low]3.race#c.ui = 0
            chi2( 12) =     8.24
          Prob > chi2 =    0.7663
```

We conclude that, here, the Wald version of the Chow test is similar to the likelihood-ratio version of the Chow test.

◁

Saved results

lrtest saves the following in r():

Scalars
r(p)	level of significance
r(df)	degrees of freedom
r(chi2)	LR test statistic

Programmers wishing their estimation commands to be compatible with lrtest should note that lrtest requires that the following results be returned:

e(cmd)	name of estimation command
e(ll)	log likelihood
e(V)	variance–covariance matrix of the estimators
e(N)	number of observations

lrtest also verifies that e(N), e(ll_0), and e(depvar) are consistent between two noncomposite models.

Methods and formulas

lrtest is implemented as an ado-file.

Let L_0 and L_1 be the log-likelihood values associated with the full and constrained models, respectively. The test statistic of the likelihood-ratio test is LR $= -2(L_1 - L_0)$. If the constrained model is true, LR is approximately χ^2 distributed with $d_0 - d_1$ degrees of freedom, where d_0 and d_1 are the model degrees of freedom associated with the full and constrained models, respectively (Greene 2012, 526–527).

lrtest determines the degrees of freedom of a model as the rank of e(V), computed as the number of nonzero diagonal elements of invsym(e(V)).

References

Greene, W. H. 2012. *Econometric Analysis*. 7th ed. Upper Saddle River, NJ: Prentice Hall.

Gutierrez, R. G., S. Carter, and D. M. Drukker. 2001. sg160: On boundary-value likelihood-ratio tests. *Stata Technical Bulletin* 60: 15–18. Reprinted in *Stata Technical Bulletin Reprints*, vol. 10, pp. 269–273. College Station, TX: Stata Press.

Hosmer, D. W., Jr., and S. Lemeshow. 2000. *Applied Logistic Regression*. 2nd ed. New York: Wiley.

Kleinbaum, D. G., and M. Klein. 2010. *Logistic Regression: A Self-Learning Text*. 3rd ed. New York: Springer.

Pérez-Hoyos, S., and A. Tobías. 1999. sg111: A modified likelihood-ratio test command. *Stata Technical Bulletin* 49: 24–25. Reprinted in *Stata Technical Bulletin Reprints*, vol. 9, pp. 171–173. College Station, TX: Stata Press.

Wang, Z. 2000. sg133: Sequential and drop one term likelihood-ratio tests. *Stata Technical Bulletin* 54: 46–47. Reprinted in *Stata Technical Bulletin Reprints*, vol. 9, pp. 332–334. College Station, TX: Stata Press.

Also see

[R] **test** — Test linear hypotheses after estimation

[R] **testnl** — Test nonlinear hypotheses after estimation

[R] **nestreg** — Nested model statistics

lv — Letter-value displays

lv [*varlist*] [*if*] [*in*] [, generate tail(*#*)]

by is allowed; see [D] **by**.

Statistics > Summaries, tables, and tests > Distributional plots and tests > Letter-value display

lv shows a letter-value display (Tukey 1977, 44–49; Hoaglin 1983) for each variable in *varlist*. If no variables are specified, letter-value displays are shown for each numeric variable in the data.

⌐ Main └

generate adds four new variables to the data: _mid, containing the midsummaries; _spread, containing the spreads; _psigma, containing the pseudosigmas; and _z2, containing the squared values from a standard normal distribution corresponding to the particular letter value. If the variables _mid, _spread, _psigma, and _z2 already exist, their contents are replaced. At most, only the first 11 observations of each variable are used; the remaining observations contain missing. If *varlist* specifies more than one variable, the newly created variables contain results for the last variable specified. The generate option may not be used with the by prefix.

tail(*#*) indicates the inverse of the tail density through which letter values are to be displayed: 2 corresponds to the median (meaning half in each tail), 4 to the fourths (roughly the 25th and 75th percentiles), 8 to the eighths, and so on. *#* may be specified as 4, 8, 16, 32, 64, 128, 256, 512, or 1,024 and defaults to a value of *#* that has corresponding depth just greater than 1. The default is taken as 1,024 if the calculation results in a number larger than 1,024. Given the intelligent default, this option is rarely specified.

Letter-value displays are a collection of observations drawn systematically from the data, focusing especially on the tails rather than the middle of the distribution. The displays are called letter-value displays because letters have been (almost arbitrarily) assigned to tail densities:

Letter	Tail area	Letter	Tail area
M	1/2	B	1/64
F	1/4	A	1/128
E	1/8	Z	1/256
D	1/16	Y	1/512
C	1/32	X	1/1024

▷ Example 1

We have data on the mileage ratings of 74 automobiles. To obtain a letter-value display, we type

```
. use http://www.stata-press.com/data/r12/auto
(1978 Automobile Data)
. lv mpg
```

#	74	Mileage (mpg)			spread	pseudosigma
M	37.5		20			
F	19	18	21.5	25	7	5.216359
E	10	15	21.5	28	13	5.771728
D	5.5	14	22.25	30.5	16.5	5.576303
C	3	14	24.5	35	21	5.831039
B	2	12	23.5	35	23	5.732448
A	1.5	12	25	38	26	6.040635
	1	12	26.5	41	29	6.16562

				# below	# above
inner fence		7.5	35.5	0	1
outer fence		-3	46	0	0

The decimal points can be made to line up and thus the output made more readable by specifying a display format for the variable; see [U] **12.5 Formats: Controlling how data are displayed**.

```
. format mpg %9.2f
. lv mpg
```

#	74	Mileage (mpg)			spread	pseudosigma
M	37.5		20.00			
F	19	18.00	21.50	25.00	7.00	5.22
E	10	15.00	21.50	28.00	13.00	5.77
D	5.5	14.00	22.25	30.50	16.50	5.58
C	3	14.00	24.50	35.00	21.00	5.83
B	2	12.00	23.50	35.00	23.00	5.73
A	1.5	12.00	25.00	38.00	26.00	6.04
	1	12.00	26.50	41.00	29.00	6.17

				# below	# above
inner fence		7.50	35.50	0	1
outer fence		-3.00	46.00	0	0

At the top, the number of observations is indicated as 74. The first line shows the statistics associated with M, the letter value that puts half the density in each tail, or the median. The median has *depth* 37.5 (that is, in the ordered data, M is 37.5 observations in from the extremes) and has value 20. The next line shows the statistics associated with F or the fourths. The fourths have depth 19 (that is, in the ordered data, the lower fourth is observation 19, and the upper fourth is observation $74 - 19 + 1$), and the values of the lower and upper fourths are 18 and 25. The number in the middle is the point halfway between the fourths—called a midsummary. If the distribution were perfectly symmetric, the midsummary would equal the median. The spread is the difference between the lower and upper summaries ($25 - 18 = 7$). For fourths, half the data lie within a 7-mpg band. The pseudosigma is a calculation of the standard deviation using only the lower and upper summaries and assuming that the variable is normally distributed. If the data really were normally distributed, all the pseudosigmas would be roughly equal.

After the letter values, the line labeled with depth 1 reports the minimum and maximum values. Here the halfway point between the extremes is 26.5, which is greater than the median, indicating that 41 is more extreme than 12, at least relative to the median. And with each letter value, the midsummaries are increasing — our data are skewed. The pseudosigmas are also increasing, indicating that the data are spreading out relative to a normal distribution, although, given the evident skewness, this elongation may be an artifact of the skewness.

At the end is an attempt to identify outliers, although the points so identified are merely outside some predetermined cutoff. Points outside the inner fence are called *outside values* or *mild outliers*. Points outside the outer fence are called *severe outliers*. The inner fence is defined as $(3/2)\text{IQR}$ and the outer fence as 3IQR above and below the F summaries, where the interquartile range (IQR) is the spread of the fourths.

◁

Technical note

The form of the letter-value display has varied slightly with different authors. `lv` displays appear as described by Hoaglin (1983) but as modified by Emerson and Stoto (1983), where they included the midpoint of each of the spreads. This format was later adopted by Hoaglin (1985). If the distribution is symmetric, the midpoints will all be roughly equal. On the other hand, if the midpoints vary systematically, the distribution is skewed.

The pseudosigmas are obtained from the lower and upper summaries for each letter value. For each letter value, they are the standard deviation a normal distribution would have if its spread for the given letter value were to equal the observed spread. If the pseudosigmas are all roughly equal, the data are said to have *neutral elongation*. If the pseudosigmas increase systematically, the data are said to be more elongated than a normal, that is, have thicker tails. If the pseudosigmas decrease systematically, the data are said to be less elongated than a normal, that is, have thinner tails.

Interpretation of the number of mild and severe outliers is more problematic. The following discussion is drawn from Hamilton (1991):

Obviously, the presence of any such outliers does not rule out that the data have been drawn from a normal distribution; in large datasets, there will most certainly be observations outside $(3/2)\text{IQR}$ and 3IQR. Severe outliers, however, make up about two per million (0.0002%) of a normal population. In samples, they lie far enough out to have substantial effects on means, standard deviations, and other classical statistics. The 0.0002%, however, should be interpreted carefully; outliers appear more often in small samples than one might expect from population proportions because of sampling variation in estimated quartiles. Monte Carlo simulation by Hoaglin, Iglewicz, and Tukey (1986) obtained these results on the percentages and numbers of outliers in random samples from a normal population:

n	percentage		number	
	any outliers	severe	any outliers	severe
10	2.83	.362	.283	.0362
20	1.66	.074	.332	.0148
50	1.15	.011	.575	.0055
100	.95	.002	.95	.002
200	.79	.001	1.58	.002
300	.75	.001	2.25	.003
∞	.70	.0002	∞	∞

Thus the presence of any severe outliers in samples of less than 300 is sufficient to reject normality. Hoaglin, Iglewicz, and Tukey (1981) suggested the approximation $0.00698 + 0.4/n$ for the fraction of mild outliers in a sample of size n or, equivalently, $0.00698n + 0.4$ for the number of outliers.

❑

▷ Example 2

The generate option adds the _mid, _spread, _psigma, and _z2 variables to our data, making possible many of the diagnostic graphs suggested by Hoaglin (1985).

```
. lv mpg, generate
(output omitted)
. list _mid _spread _psigma _z2 in 1/12
```

	_mid	_spread	_psigma	_z2
1.	20	.	.	.
2.	21.5	7	5.216359	.4501955
3.	21.5	13	5.771728	1.26828
4.	22.25	16.5	5.576303	2.188846
5.	24.5	21	5.831039	3.24255
6.	23.5	23	5.732448	4.024532
7.	25	26	6.040635	4.631499
8.	.	.	.	.
9.	.	.	.	.
10.	.	.	.	.
11.	26.5	29	6.16562	5.53073
12.	.	.	.	.

Observations 12 through the end are missing for these new variables. The definition of the observations is always the same. The first observation contains the M summary; the second, the F; the third, the E; and so on. Observation 11 always contains the summary for depth 1. Observations 8–10—corresponding to letter values Z, Y, and X—contain missing because these statistics were not calculated. We have only 74 observations, and their depth would be 1.

Hoaglin (1985) suggests graphing the midsummary against z^2. If the distribution is not skewed, the points in the resulting graph will be along a horizontal line:

```
. scatter _mid _z2
```

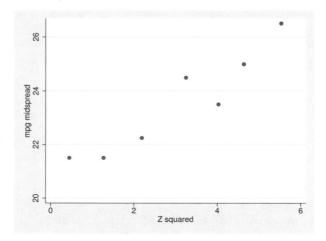

The graph clearly indicates the skewness of the distribution. We might also graph _psigma against _z2 to examine elongation.

◁

aved results

lv saves the following in r():

Scalars

r(N)	number of observations	r(u_C)	upper 32nd
r(min)	minimum	r(1_B)	lower 64th
r(max)	maximum	r(u_B)	upper 64th
r(median)	median	r(1_A)	lower 128th
r(1_F)	lower 4th	r(u_A)	upper 128th
r(u_F)	upper 4th	r(1_Z)	lower 256th
r(1_E)	lower 8th	r(u_Z)	upper 256th
r(u_E)	upper 8th	r(1_Y)	lower 512th
r(1_D)	lower 16th	r(u_Y)	upper 512th
r(u_D)	upper 16th	r(1_X)	lower 1024th
r(1_C)	lower 32nd	r(u_X)	upper 1024th

The lower/upper 8ths, 16ths, ..., 1024ths will be defined only if there are sufficient data.

Methods and formulas

lv is implemented as an ado-file.

Let N be the number of (nonmissing) observations on x, and let $x_{(i)}$ refer to the ordered data when i is an integer. Define $x_{(i+0.5)} = (x_{(i)} + x_{(i+1)})/2$; the median is defined as $x_{\{(N+1)/2\}}$.

Define $x_{[d]}$ as the pair of numbers $x_{(d)}$ and $x_{(N+1-d)}$, where d is called the *depth*. Thus $x_{[1]}$ refers to the minimum and maximum of the data. Define $m = (N+1)/2$ as the depth of the median, $f = (\lfloor m \rfloor + 1)/2$ as the depth of the fourths, $e = (\lfloor f \rfloor + 1)/2$ as the depth of the eighths, and so on. Depths are reported on the far left of the letter-value display. The corresponding fourths of the data are $x_{[f]}$, the eighths are $x_{[e]}$, and so on. These values are reported inside the display. The middle value is defined as the corresponding midpoint of $x_{[.]}$. The spreads are defined as the difference in $x_{[.]}$.

The corresponding point z_i on a standard normal distribution is obtained as (Hoaglin 1985, 456–457)

$$z_i = \begin{cases} F^{-1}\{(d_i - 1/3)/(N + 1/3)\} & \text{if } d_i > 1 \\ F^{-1}\{0.695/(N + 0.390)\} & \text{otherwise} \end{cases}$$

where d_i is the depth of the letter value. The corresponding pseudosigma is obtained as the ratio of the spread to $-2z_i$ (Hoaglin 1985, 431).

Define $(F_l, F_u) = x_{[f]}$. The inner fence has cutoffs $F_l - \frac{3}{2}(F_u - F_l)$ and $F_u + \frac{3}{2}(F_u - F_l)$. The outer fence has cutoffs $F_l - 3(F_u - F_l)$ and $F_u + 3(F_u - F_l)$.

The inner-fence values reported by lv are almost equal to those used by graph, box to identify outside points. The only difference is that graph uses a slightly different definition of fourths, namely, the 25th and 75th percentiles as defined by summarize; see [R] **summarize**.

References

Emerson, J. D., and M. A. Stoto. 1983. Transforming data. In *Understanding Robust and Exploratory Data Analysis*, ed. D. C. Hoaglin, F. Mosteller, and J. W. Tukey, 97–128. New York: Wiley.

Fox, J. 1990. Describing univariate distributions. In *Modern Methods of Data Analysis*, ed. J. Fox and J. S. Long, 58–125. Newbury Park, CA: Sage.

Hamilton, L. C. 1991. sed4: Resistant normality check and outlier identification. *Stata Technical Bulletin* 3: 15–18. Reprinted in *Stata Technical Bulletin Reprints*, vol. 1, pp. 86–90. College Station, TX: Stata Press.

Hoaglin, D. C. 1983. Letter values: A set of selected order statistics. In *Understanding Robust and Exploratory Data Analysis*, ed. D. C. Hoaglin, F. Mosteller, and J. W. Tukey, 33–57. New York: Wiley.

——. 1985. Using quantiles to study shape. In *Exploring Data Tables, Trends, and Shapes*, ed. D. C. Hoaglin, F. Mosteller, and J. W. Tukey, 417–460. New York: Wiley.

Hoaglin, D. C., B. Iglewicz, and J. W. Tukey. 1981. Small-sample performance of a resistant rule for outlier detection. In *1980 Proceedings of the Statistical Computing Section*. Washington, DC: American Statistical Association.

——. 1986. Performance of some resistant rules for outlier labeling. *Journal of the American Statistical Association* 81: 991–999.

Tukey, J. W. 1977. *Exploratory Data Analysis*. Reading, MA: Addison–Wesley.

Also see

margins — Marginal means, predictive margins, and marginal effects

margins [*marginlist*] [*if*] [*in*] [*weight*] [, *response_options options*]

where *marginlist* is a list of factor variables or interactions that appear in the current estimation results. The variables may be typed with or without the i. prefix, and you may use any factor-variable syntax:

```
. margins i.sex i.group i.sex#i.group
. margins sex group sex#i.group
. margins sex##group
```

response_options	Description
Main	
<u>p</u>redict(*pred_opt*)	estimate margins for predict, *pred_opt*
<u>e</u>xpression(*pnl_exp*)	estimate margins for *pnl_exp*
dydx(*varlist*)	estimate marginal effect of variables in *varlist*
eyex(*varlist*)	estimate elasticities of variables in *varlist*
dyex(*varlist*)	estimate semielasticity—$d(y)/d(\ln x)$
eydx(*varlist*)	estimate semielasticity—$d(\ln y)/d(x)$
<u>cont</u>inuous	treat factor-level indicators as continuous

options	Description
Main	
grand	add the overall margin; default if no *marginlist*
At	
at(*atspec*)	estimate margins at specified values of covariates
atmeans	estimate margins at the means of covariates
<u>asbal</u>anced	treat all factor variables as balanced
if/in/over	
over(*varlist*)	estimate margins at unique values of *varlist*
subpop(*subspec*)	estimate margins for subpopulation
Within	
within(*varlist*)	estimate margins at unique values of the nesting factors in *varlist*
SE	
vce(delta)	estimate SEs using delta method; the default
vce(unconditional)	estimate SEs allowing for sampling of covariates
nose	do not estimate SEs

1027

Advanced

noweights	ignore weights specified in estimation
noesample	do not restrict margins to the estimation sample
emptycells(*empspec*)	treatment of empty cells for balanced factors
estimtolerance(*tol*)	specify numerical tolerance used to determine estimable functions; default is estimtolerance(1e-5)
noestimcheck	suppress estimability checks
force	estimate margins despite potential problems
chainrule	use the chain rule when computing derivatives
nochainrule	do not use the chain rule

Reporting

level(#)	confidence level; default is level(95)
mcompare(*method*)	multiple comparisons; default is mcompare(noadjust)
noatlegend	suppress legend of fixed covariate values
post	post margins and their VCE as estimation results
display_options	control column formats, row spacing, and line width

method	Description
noadjust	do not adjust for multiple comparisons; the default
bonferroni [adjustall]	Bonferroni's method; adjust across all terms
sidak [adjustall]	Šidák's method; adjust across all terms
scheffe	Scheffé's method

Time-series operators are allowed if they were used in the estimation.

See at() under *Options* for a description of *atspec*.

fweights, aweights, iweights, and pweights are allowed; see [U] **11.1.6 weight**.

Menu

Statistics > Postestimation > Marginal means and predictive margins

Statistics > Postestimation > Marginal effects

Description

Margins are statistics calculated from predictions of a previously fit model at fixed values of some covariates and averaging or otherwise integrating over the remaining covariates.

The margins command estimates margins of responses for specified values of covariates and presents the results as a table.

Capabilities include estimated marginal means, least-squares means, average and conditional marginal and partial effects (which may be reported as derivatives or as elasticities), average and conditional adjusted predictions, and predictive margins.

▶ptions

Warning: The option descriptions are brief and use jargon. Skip to Remarks if you are reading about margins *for the first time.*

Main

predict(*pred_opt*) and expression(*pnl_exp*) are mutually exclusive; they specify the response. If neither is specified, the response will be the default prediction that would be produced by predict after the underlying estimation command.

predict(*pred_opt*) specifies the option(s) to be specified with the predict command to produce the variable that will be used as the response. After estimation by logistic, you could specify predict(xb) to obtain linear predictions rather than the predict command's default, the probabilities.

expression(*pnl_exp*) specifies the response as an expression. See [R] **predictnl** for a full description of *pnl_exp*. After estimation by logistic, you might specify expression(exp(predict(xb))) to use relative odds rather than probabilities as the response. For examples, see *Example 12: Margins of a specified expression*.

dydx(*varlist*), eyex(*varlist*), dyex(*varlist*), and eydx(*varlist*) request that margins report derivatives of the response with respect to *varlist* rather than on the response itself. eyex(), dyex(), and eydx() report derivatives as elasticities; see *Expressing derivatives as elasticities*.

continuous is relevant only when one of dydx() or eyex() is also specified. It specifies that the levels of factor variables be treated as continuous; see *Derivatives versus discrete differences*. This option is implied if there is a single-level factor variable specified in dydx() or eyex().

grand specifies that the overall margin be reported. grand is assumed when *marginlist* is empty.

At

at(*atspec*) specifies values for covariates to be treated as fixed.

at(age=20) fixes covariate age to the value specified. at() may be used to fix continuous or factor covariates.

at(age=20 sex=1) simultaneously fixes covariates age and sex at the values specified.

at(age=(20 30 40 50)) fixes age first at 20, then at 30, ... margins produces separate results for each specified value.

at(age=(20(10)50)) does the same as at(age=(20 30 40 50)); that is, you may specify a numlist.

at((mean) age (median) distance) fixes the covariates at the summary statistics specified. at((p25) _all) fixes all covariates at their 25th percentile values. See *Syntax of at()* for the full list of summary-statistic modifiers.

at((mean) _all (median) x x2=1.2 z=(1 2 3)) is read from left to right, with latter specifiers overriding former ones. Thus all covariates are fixed at their means except for x (fixed at its median), x2 (fixed at 1.2), and z (fixed first at 1, then at 2, and finally at 3).

at((means) _all (asobserved) x2) is a convenient way to set all covariates except x2 to the mean.

Multiple at() options can be specified, and each will produce a different set of margins.

See *Syntax of at()* for more information.

atmeans specifies that covariates be fixed at their means and is shorthand for at((mean) _all). atmeans differs from at((mean) _all) in that atmeans will affect subsequent at() options. For instance,

. margins ..., atmeans at((p25) x) at((p75) x)

produces two sets of margins with both sets evaluated at the means of all covariates except x.

asbalanced is shorthand for at((asbalanced) _factor) and specifies that factor covariates be evaluated as though there were an equal number of observations in each level; see *Obtaining margins as though the data were balanced.* asbalanced differs from at((asbalanced) _factor) in that asbalanced will affect subsequent at() options in the same way as atmeans does.

⌐ if/in/over ⌐

over(*varlist*) specifies that separate sets of margins be estimated for the groups defined by *varlist*. The variables in *varlist* must contain nonnegative integer values. The variables need not be covariates in your model. When over() is combined with the vce(unconditional) option, each group is treated as a subpopulation; see [SVY] **subpopulation estimation**.

subpop([*varname*] [*if*]) is intended for use with the vce(unconditional) option. It specifies that margins be estimated for the single subpopulation identified by the indicator variable or by the if expression or by both. Zero indicates that the observation be excluded; nonzero, that it be included; and missing value, that it be treated as outside of the population (and so ignored). See [SVY] **subpopulation estimation** for why subpop() is preferred to if expressions and in ranges when also using vce(unconditional). If subpop() is used without vce(unconditional), it is treated merely as an additional if qualifier.

⌐ Within ⌐

within(*varlist*) allows for nested designs. *varlist* contains the nesting variable(s) over which margins are to be estimated. See *Obtaining margins with nested designs.* As with over(*varlist*), when within(*varlist*) is combined with vce(unconditional), each level of the variables in *varlist* is treated as a subpopulation.

⌐ SE ⌐

vce(delta) and vce(unconditional) specify how the VCE and, correspondingly, standard errors are calculated.

vce(delta) is the default. The delta method is applied to the formula for the response and the VCE of the estimation command. This method assumes that values of the covariates used to calculate the response are given or, if all covariates are not fixed using at(), that the data are given.

vce(unconditional) specifies that the covariates that are not fixed be treated in a way that accounts for their having been sampled. The VCE is estimated using the linearization method. This method allows for heteroskedasticity or other violations of distributional assumptions and allows for correlation among the observations in the same manner as vce(robust) and vce(cluster ...), which may have been specified with the estimation command. This method also accounts for complex survey designs if the data are svyset. See *Obtaining margins with survey data and representative samples.*

nose suppresses calculation of the VCE and standard errors. See *Requirements for model specification* for an example of the use of this option.

Advanced

noweights specifies that any weights specified on the previous estimation command be ignored by margins. By default, margins uses the weights specified on the estimator to average responses and to compute summary statistics. If weights are specified on the margins command, they override previously specified weights, making it unnecessary to specify noweights. The noweights option is not allowed after svy: estimation when the vce(unconditional) option is specified.

noesample specifies that margins not restrict its computations to the estimation sample used by the previous estimation command. See *Example 15: Margins evaluated out of sample.*

With the default delta-method VCE, noesample margins may be estimated on samples other than the estimation sample; such results are valid under the assumption that the data used are treated as being given.

You can specify noesample and vce(unconditional) together, but if you do, you should be sure that the data in memory correspond to the original e(sample). To show that you understand that, you must also specify the force option. Be aware that making the vce(unconditional) calculation on a sample different from the estimation sample would be equivalent to estimating the coefficients on one set of data and computing the scores used by the linearization on another set; see [P] _robust.

emptycells(strict) and emptycells(reweight) are relevant only when the asbalanced option is also specified. emptycells() specifies how empty cells are handled in interactions involving factor variables that are being treated as balanced; see *Obtaining margins as though the data were balanced.*

    emptycells(strict) is the default; it specifies that margins involving empty cells be treated as not estimable.

    emptycells(reweight) specifies that the effects of the observed cells be increased to accommodate any missing cells. This makes the margin estimable but changes its interpretation. emptycells(reweight) is implied when the within() option is specified.

estimtolerance(*tol*) specifies the numerical tolerance used to determine estimable functions. The default is estimtolerance(1e-5).

    A linear combination of the model coefficients z is found to be not estimable if

$$\text{mreldif}(z,\ z \times H) > tol$$

    where H is defined in *Methods and formulas.*

noestimcheck specifies that margins not check for estimability. By default, the requested margins are checked and those found not estimable are reported as such. Nonestimability is usually caused by empty cells. If noestimcheck is specified, estimates are computed in the usual way and reported even though the resulting estimates are manipulable, which is to say they can differ across equivalent models having different parameterizations. See *Estimability of margins.*

force instructs margins to proceed in some situations where it would otherwise issue an error message because of apparent violations of assumptions. Do not be casual about specifying force. You need to understand and fully evaluate the statistical issues. For an example of the use of force, see *Using margins after the estimates use command.*

chainrule and nochainrule specify whether margins uses the chain rule when numerically computing derivatives. You need not specify these options when using margins after any official Stata estimator; margins will choose the appropriate method automatically.

    Specify nochainrule after estimation by a user-written command. We recommend using nochainrule, even though chainrule is usually safe and is always faster. nochainrule is safer because it makes no assumptions about how the parameters and covariates join to form the response.

`nochainrule` is implied when the `expression()` option is specified.

┌─ Reporting ┐

`level(#)` specifies the confidence level, as a percentage, for confidence intervals. The default is `level(95)` or as set by `set level`; see [U] **20.7 Specifying the width of confidence intervals**.

`mcompare(`*method*`)` specifies the method for computing p-values and confidence intervals that account for multiple comparisons within a factor-variable term.

Most methods adjust the comparisonwise error rate, α_c, to achieve a prespecified experimentwise error rate, α_e.

`mcompare(noadjust)` is the default; it specifies no adjustment.

$$\alpha_c = \alpha_e$$

`mcompare(bonferroni)` adjusts the comparisonwise error rate based on the upper limit of the Bonferroni inequality

$$\alpha_e \leq m\alpha_c$$

where m is the number of comparisons within the term.

The adjusted comparisonwise error rate is

$$\alpha_c = \alpha_e/m$$

`mcompare(sidak)` adjusts the comparisonwise error rate based on the upper limit of the probability inequality

$$\alpha_e \leq 1 - (1 - \alpha_c)^m$$

where m is the number of comparisons within the term.

The adjusted comparisonwise error rate is

$$\alpha_c = 1 - (1 - \alpha_e)^{1/m}$$

This adjustment is exact when the m comparisons are independent.

`mcompare(scheffe)` controls the experimentwise error rate using the F or χ^2 distribution with degrees of freedom equal to the rank of the term.

`mcompare(`*method* `adjustall)` specifies that the multiple-comparison adjustments count all comparisons across all terms rather than performing multiple comparisons term by term. This leads to more conservative adjustments when multiple variables or terms are specified in *marginslist*. This option is compatible only with the `bonferroni` and `sidak` methods.

`noatlegend` specifies that the legend showing the fixed values of covariates be suppressed.

`post` causes `margins` to behave like a Stata estimation (e-class) command. `margins` posts the vector of estimated margins along with the estimated variance–covariance matrix to `e()`, so you can treat the estimated margins just as you would results from any other estimation command. For example, you could use `test` to perform simultaneous tests of hypotheses on the margins, or you could use `lincom` to create linear combinations. See *Example 10: Testing margins—contrasts of margins*.

display_options: `vsquish`, `cformat(`*%fmt*`)`, `pformat(`*%fmt*`)`, `sformat(`*%fmt*`)`, and `nolstretch`.

`vsquish` specifies that the blank space separating factor-variable terms or time-series–operated variables from other variables in the model be suppressed.

`cformat(`*%fmt*`)` specifies how to format margins, standard errors, and confidence limits in the table of estimated margins.

pformat(*% fmt*) specifies how to format p-values in the table of estimated margins.

sformat(*% fmt*) specifies how to format test statistics in the table of estimated margins.

nolstretch specifies that the width of the table of estimated margins not be automatically widened to accommodate longer variable names. The default, lstretch, is to automatically widen the table of estimated margins up to the width of the Results window. To change the default, use set lstretch off. nolstretch is not shown in the dialog box.

emarks

Remarks are presented under the following headings:

Introduction

margins is a postestimation command, a command for use after you have fit a model using an estimation command such as regress or logistic, or using almost any other estimation command.

margins estimates and reports margins of responses and margins of derivatives of responses, also known as marginal effects. A margin is a statistic based on a fitted model in which some of or all the covariates are fixed. Marginal effects are changes in the response for change in a covariate, which can be reported as a derivative, elasticity, or semielasticity.

Obtaining margins of responses

What we call margins of responses are also known as predictive margins, adjusted predictions, and recycled predictions. When applied to balanced data, margins of responses are also called estimated marginal means and least-squares means.

A margin is a statistic based on a fitted model calculated over a dataset in which some of or all the covariates are fixed at values different from what they really are. For instance, after a linear regression fit on males and females, the marginal mean (margin of mean) for males is the predicted mean of the dependent variable, where every observation is treated as if it represents a male; thus those observations that in fact do represent males are included, as well as those observations that represent females. The marginal mean for female would be similarly obtained by treating all observations as if they represented females.

In making the calculation, sex is treated as male or female everywhere it appears in the model. The model might be

. regress y age bp i.sex sex#c.age sex#c.bp

and then, in making the marginal calculation of the mean for males and females, margins not only accounts for the direct effect of i.sex but also for the indirect effects of sex#c.age and sex#c.bp.

The response being margined can be any statistic produced by [R] **predict**, or any expression of those statistics.

Standard errors are obtained by the delta method, at least by default. The delta method assumes that the values at which the covariates are evaluated to obtain the marginal responses are fixed. When your sample represents a population, whether you are using svy or not (see [SVY] **svy**), you can specify margins' vce(unconditional) option and margins will produce standard errors that account for the sampling variability of the covariates. Some researchers reserve the term predictive margins to describe this.

The best way to understand margins is to see some examples. You can run the following examples yourself if you type

. use http://www.stata-press.com/data/r12/margex
(Artificial data for margins)

ample 1: A simple case after regress

```
. regress y i.sex i.group
 (output omitted )

. margins sex
Predictive margins                        Number of obs    =        3000
Model VCE      : OLS
Expression     : Linear prediction, predict()
```

	Margin	Delta-method Std. Err.	z	P>|z|	[95% Conf. Interval]	
sex						
0	60.56034	.5781782	104.74	0.000	59.42713	61.69355
1	78.88236	.5772578	136.65	0.000	77.75096	80.01377

The numbers reported in the "Margin" column are average values of y. Based on a linear regression of y on sex and group, 60.6 would be the average value of y if everyone in the data were treated as if they were male, and 78.9 would be the average value if everyone were treated as if they were female.

xample 2: A simple case after logistic

margins may be used after almost any estimation command.

```
. logistic outcome i.sex i.group
 (output omitted )
. margins sex
Predictive margins                        Number of obs    =        3000
Model VCE      : OIM
Expression     : Pr(outcome), predict()
```

	Margin	Delta-method Std. Err.	z	P>|z|	[95% Conf. Interval]	
sex						
0	.1286796	.0111424	11.55	0.000	.106841	.1505182
1	.1905087	.0089719	21.23	0.000	.1729241	.2080933

The numbers reported in the "Margin" column are average predicted probabilities. Based on a logistic regression of outcome on sex and group, 0.13 would be the average probability of outcome if everyone in the data were treated as if they were male, and 0.19 would be the average probability if everyone were treated as if they were female.

margins reports average values after regress and average probabilities after logistic. By default, margins makes tables of whatever it is that predict (see [R] **predict**) predicts by default. Alternatively, margins can make tables of anything that predict can produce if you use margins' predict() option; see *Example 11: Margins of a specified prediction.*

Example 3: Average response versus response at average

In example 2, `margins` reported average probabilities of `outcome` for `sex = 0` and `sex = 1`. If we instead wanted the predicted probabilities evaluated at the mean of the covariates, we would specify `margins`' `atmeans` option. We previously typed

```
. logistic outcome i.sex i.group
(output omitted )
. margins sex
(output omitted )
```

and now we type

```
. margins sex, atmeans
```

| Adjusted predictions | Number of obs | = | 3000 |
| Model VCE   : OIM | | | |

```
Expression  : Pr(outcome), predict()
at          : 0.sex           =     .4993333 (mean)
              1.sex           =     .5006667 (mean)
              1.group         =     .3996667 (mean)
              2.group         =     .3726667 (mean)
              3.group         =     .2276667 (mean)
```

	Margin	Delta-method Std. Err.	z	P>|z|	[95% Conf. Interval]
sex					
0	.0966105	.0089561	10.79	0.000	.0790569 .1141641
1	.1508362	.0118064	12.78	0.000	.127696 .1739764

The prediction at the average of the covariates is different from the average of the predictions. The first is the expected probability of a person with average characteristics, a person who, in another problem, might be 3/4 married and have 1.2 children. The second is the average of the probability among actual persons in the data.

When you specify `atmeans` or any other at option, `margins` reports the values used for the covariates in the legend above the table. `margins` lists the values for all the covariates, including values it may not use, in the results that follow. In this example, `margins` reported means for `sex` even though those means were not used. They were not used because we asked for the margins of `sex`, so `sex` was fixed first at 0 and then at 1.

If you wish to suppress this legend, specify the `nolegend` option.

Example 4: Multiple margins from one command

More than one margin can be reported by just one `margins` command. You can type

```
. margins sex group
```

and doing that is equivalent in terms of the output to typing

```
. margins sex
. margins group
```

When multiple margins are requested on the same command, each is estimated separately. There is, however, a difference when you also specify `margins`' `post` option. Then the variance–covariance matrix for all margins requested is posted, and that is what allows you to test equality of margins, etc. Testing equality of margins is covered in *Example 10: Testing margins—contrasts of margins*.

In any case, below we request margins for sex and for group.

```
. margins sex group
Predictive margins                              Number of obs   =       3000
Model VCE      : OIM
Expression    : Pr(outcome), predict()
```

	Margin	Delta-method Std. Err.	z	P>|z|	[95% Conf. Interval]	
sex						
0	.1286796	.0111424	11.55	0.000	.106841	.1505182
1	.1905087	.0089719	21.23	0.000	.1729241	.2080933
group						
1	.2826207	.0146234	19.33	0.000	.2539593	.311282
2	.1074814	.0094901	11.33	0.000	.0888812	.1260817
3	.0291065	.0073417	3.96	0.000	.0147169	.043496

Example 5: Margins with interaction terms

The estimation command on which margins bases its calculations may contain interaction terms, such as an interaction of sex and group:

```
. logistic outcome i.sex i.group sex#group
  (output omitted )
. margins sex group
Predictive margins                              Number of obs   =       3000
Model VCE      : OIM
Expression    : Pr(outcome), predict()
```

	Margin	Delta-method Std. Err.	z	P>|z|	[95% Conf. Interval]	
sex						
0	.1561738	.0132774	11.76	0.000	.1301506	.182197
1	.1983749	.0101546	19.54	0.000	.1784723	.2182776
group						
1	.3211001	.0176403	18.20	0.000	.2865257	.3556744
2	.1152127	.0099854	11.54	0.000	.0956417	.1347838
3	.0265018	.0109802	2.41	0.016	.0049811	.0480226

We fit the model by typing logistic outcome i.sex i.group sex#group, but the meaning would have been the same had we typed logistic outcome sex##group.

As mentioned in example 4, the results for sex and the results for group are calculated independently, and we would have obtained the same results had we typed margins sex followed by margins group.

The margin for male (sex = 0) is 0.16. The probability 0.16 is the average probability if everyone in the data were treated as if sex = 0, including sex = 0 in the main effect and sex = 0 in the interaction of sex with group.

Had we specified margins sex, atmeans, we would have obtained not average probabilities but the probabilities evaluated at the average. Rather than obtaining 0.16, we would have obtained 0.10

for `sex = 0`. The 0.10 is calculated by taking the fitted model, plugging in `sex = 0` everywhere, and plugging in the average value of the group indicator variables everywhere they are used. That is, rather than treating the group indicators as being $(1, 0, 0)$, $(0, 1, 0)$, or $(0, 0, 1)$ depending on observation, the group indicators are treated as being $(0.40, 0.37, 0.23)$, which are the average values of `group = 1`, `group = 2`, and `group = 3`.

Example 6: Margins with continuous variables

To the above example, we will add the continuous covariate `age` to the model and then rerun `margins sex group`.

```
. logistic outcome i.sex i.group sex#group age
  (output omitted )
. margins sex group
```

Predictive margins   Number of obs   =   3000
Model VCE   : OIM

Expression   : Pr(outcome), predict()

	Margin	Delta-method Std. Err.	z	P>|z|	[95% Conf. Interval]	
sex						
0	.1600644	.0125653	12.74	0.000	.1354368	.184692
1	.1966902	.0100043	19.66	0.000	.1770821	.2162983
group						
1	.2251302	.0123233	18.27	0.000	.200977	.2492834
2	.150603	.0116505	12.93	0.000	.1277685	.1734376
3	.0736157	.0337256	2.18	0.029	.0075147	.1397167

Compared with the results presented in example 5, results for sex change little, but results for groups 1 and 3 change markedly. The tables differ because now we are adjusting for the continuous covariate `age`, as well as for `sex` and `group`.

We will continue examining interactions in example 8. Because we have added a continuous variable, let's take a detour to explain how to obtain margins for continuous variables and to explain their interpretation.

Example 7: Margins of continuous variables

Continuing with our example of

```
. logistic outcome i.sex i.group sex#group age
```

let's examine the continuous covariate `age`.

You are not allowed to type `margins age`; doing that will produce an error:

```
. margins age
factor age not found in e(b)
r(111);
```

The message "factor age not found in e(b)" is `margins`' way of saying, "Yes, age might be in the model, but if it is, it is not included as a factor variable; it is in as a continuous variable." Sometimes, Stata is overly terse. `margins` might also say that because age is continuous there are an infinite number of values at which it could evaluate the margins. At what value(s) should age be fixed? `margins` requires more guidance with continuous covariates. We can provide that guidance by using the `at()` option and typing

```
. margins, at(age=40)
```

To understand why that yields the desired result, let us tell you that if you were to type

```
. margins
```

margins would report the overall margin—the margin that holds nothing constant. Because our model is logistic, the average value of the predicted probabilities would be reported. The at() option fixes one or more covariates to the value(s) specified and can be used with both factor and continuous variables. Thus, if you typed margins, at(age=40), then margins would average over the data the responses for everybody, setting age=40. Here is what happens when you type that:

```
. margins, at(age=40)
```

Predictive margins				Number of obs	=	3000
Model VCE	: OIM					
Expression	: Pr(outcome), predict()					
at	: age	=	40			

	Margin	Delta-method Std. Err.	z	P>|z|	[95% Conf. Interval]	
_cons	.1133603	.0070731	16.03	0.000	.0994972	.1272234

Reported is the margin for age $= 40$, adjusted for the other covariates in our model.

If we wanted to obtain the margins for age 30, 35, 40, 45, and 50, we could type

```
. margins, at(age=(30 35 40 45 50))
```

or, equivalently,

```
. margins, at(age=(30(5)50))
```

ample 8: Margins of interactions

Our model is

```
. logistic outcome i.sex i.group sex#group age
```

We can obtain the margins of all possible combinations of the levels of sex and the levels of group by typing

```
. margins sex#group
```

Predictive margins				Number of obs	=	3000
Model VCE	: OIM					
Expression	: Pr(outcome), predict()					

	Margin	Delta-method Std. Err.	z	P>|z|	[95% Conf. Interval]	
sex#group						
0 1	.2379605	.0237178	10.03	0.000	.1914745	.2844465
0 2	.0658294	.0105278	6.25	0.000	.0451953	.0864636
0 3	.0538001	.0136561	3.94	0.000	.0270347	.0805656
1 1	.2158632	.0112968	19.11	0.000	.1937218	.2380045
1 2	.2054406	.0183486	11.20	0.000	.1694781	.2414032
1 3	.085448	.0533914	1.60	0.110	-.0191973	.1900932

The first line in the table reports the marginal probability for sex = 0 and group = 1. That is, it reports the estimated probability if everyone in the data were treated as if they were sex = 0 and group = 1.

Also reported are all the other combinations of sex and group.

By the way, we could have typed margins sex#group even if our fitted model did not include sex#group. Estimation is one thing, and asking questions about the nature of the estimates is another. margins does, however, require that i.sex and i.group appear somewhere in the model, because fixing a value outside the model would just produce the grand margin, and you can separately ask for that if you want it by typing margins without arguments.

Example 9: Decomposing margins

We have the model

        . logistic outcome i.sex i.group sex#group age

In example 6, we typed margins sex and obtained 0.160 for males and 0.197 for females. We are going to decompose each of those numbers. Let us explain:

1. The margin for males, 0.160, treats everyone as if they were male, and that amounts to simultaneously

        1a. treating males as males and

        1b. treating females as males.

2. The margin for females, 0.197, treats everyone as if they were female, and that amounts to simultaneously

        2a. treating males as females and

        2b. treating females as females.

The margins 1a and 1b are the decomposition of 1, and the margins 2a and 2b are the decomposition of 2.

We could obtain 1a and 2a by typing

        . margins if sex==0, at(sex=(0 1))

because the qualifier if sex==0 would restrict margins to running on only the males. Similarly, we could obtain 1b and 2b by typing

        . margins if sex==1, at(sex=(0 1))

We run these examples below:

```
. margins if sex==0, at(sex=(0 1))
Predictive margins                            Number of obs    =       1498
Model VCE     : OIM
Expression    : Pr(outcome), predict()
1._at         : sex               =          0
2._at         : sex               =          1
```

	Margin	Delta-method Std. Err.	z	P>|z|	[95% Conf. Interval]
_at					
1	.0794393	.0062147	12.78	0.000	.0672586	.0916199
2	.1335584	.0127351	10.49	0.000	.1085981	.1585187

```
. margins if sex==1, at(sex=(0 1))
Predictive margins                            Number of obs    =       1502
Model VCE     : OIM
Expression    : Pr(outcome), predict()
1._at         : sex               =          0
2._at         : sex               =          1
```

	Margin	Delta-method Std. Err.	z	P>|z|	[95% Conf. Interval]
_at					
1	.2404749	.0199709	12.04	0.000	.2013326	.2796171
2	.2596538	.0104756	24.79	0.000	.2391219	.2801857

Putting together the results from example 6 and the results above, we have

Margin treating everybody as themself	0.170
Margin treating everybody as male	0.160
Margin treating male as male	0.079
Margin treating female as male	0.240
Margin treating everybody as female	0.197
Margin treating male as female	0.134
Margin treating female as female	0.260

Example 10: Testing margins—contrasts of margins

Continuing with the previous example, it would be interesting to test the equality of 2b and 1b, to test whether the average probability of a positive outcome for females treated as females is equal to that for females treated as males. That test would be different from testing the overall significance of sex in our model. The test performed on our model would be a test of whether the probability of a positive outcome differs between males and females when they have equal values of the other covariates. The test of equality of margins is a test of whether the *average* probabilities differ given the different pattern of values of the other covariates that the two sexes have in our data.

We can also perform such tests by treating the results from margins as estimation results. There are three steps required to perform tests on margins. First, you must arrange it so that all the margins of interest are reported by just one margins command. Second, you must specify margins' post option. Third, you perform the test with the test command.

Such tests and comparisons can be readily performed by contrasting margins; see [R] **margins, contrast**. Also see *Contrasts of margins—effects (discrete marginal effects)* in [R] **marginsplot**.

In the previous example, we used two commands to obtain our results, namely,

```
. margins if sex==0, at(sex=(0 1))
. margins if sex==1, at(sex=(0 1))
```

We could, however, have obtained the same results by typing just one command:

```
. margins, over(sex) at(sex=(0 1))
```

Performing `margins, over(sex)` first restricts the sample to `sex==0` and then restricts it to `sex==1`, and that is equivalent to the two different `if` conditions that we specified before.

To test whether females treated as females is equal to females treated as males, we will need to type

```
. margins, over(sex) at(sex=(0 1)) post
. test _b[2._at#1.sex] = _b[1._at#1.sex]
```

We admit that the second command may seem to have come out of nowhere. When we specify `post` on the `margins` command, `margins` behaves as if it were an estimation command, which means that 1) it posts its estimates and full VCE to `e()`, 2) it gains the ability to replay results just as any estimation command can, and 3) it gains access to the standard postestimation commands. Item 3 explains why we could use `test`. We learned that we wanted to test `_b[2._at#1.sex]` and `_b[1._at#1.sex]` by replaying the estimation results, but this time with the standard estimation command `coeflegend` option. So what we typed was

```
. margins, over(sex) at(sex=(0 1)) post
. margins, coeflegend
. test _b[2._at#1.sex] = _b[1._at#1.sex]
```

We will let you try `margins, coeflegend` for yourself. The results of running the other two commands are

```
. margins, over(sex) at(sex=(0 1)) post
Predictive margins                              Number of obs   =        3000
Model VCE     : OIM

Expression    : Pr(outcome), predict()
over          : sex
1._at         : 0.sex
                    sex         =          0
              1.sex
                    sex         =          0
2._at         : 0.sex
                    sex         =          1
              1.sex
                    sex         =          1
```

	Margin	Delta-method Std. Err.	z	P>|z|	[95% Conf. Interval]	
_at#sex						
1 0	.0794393	.0062147	12.78	0.000	.0672586	.0916199
1 1	.2404749	.0199709	12.04	0.000	.2013326	.2796171
2 0	.1335584	.0127351	10.49	0.000	.1085981	.1585187
2 1	.2596538	.0104756	24.79	0.000	.2391219	.2801857

```
. test _b[2._at#1.sex] = _b[1._at#1.sex]

 ( 1)  - 1bn._at#1.sex + 2._at#1.sex = 0

          chi2(  1) =      0.72
        Prob > chi2 =    0.3951
```

We can perform the same test in one command using contrasts of margins:

```
. logistic outcome i.sex i.group sex#group age
 (output omitted )
. margins, over(sex) at(sex=(0 1)) contrast(atcontrast(r._at) wald)

Contrasts of predictive margins
Model VCE    : OIM

Expression   : Pr(outcome), predict()
over         : sex

1._at        : 0.sex
                    sex               =          0
               1.sex
                    sex               =          0

2._at        : 0.sex
                    sex               =          1
               1.sex
                    sex               =          1
```

	df	chi2	P>chi2
_at@sex			
(2 vs 1) 0	1	14.59	0.0001
(2 vs 1) 1	1	0.72	0.3951
Joint	2	16.13	0.0003

	Contrast	Delta-method Std. Err.	[95% Conf. Interval]	
_at@sex				
(2 vs 1) 0	.0541192	.0141706	.0263453	.081893
(2 vs 1) 1	.0191789	.0225516	-.0250215	.0633793

We refitted our logistic model because its estimation results were replaced when we posted our margins. The syntax to perform the contrast we want is admittedly not obvious. Contrasting (testing) across at() groups is more difficult than contrasting across the margins themselves or across over() groups, because we have no natural place for the contrast operators (r., in our case). We also explicitly requested Wald tests of the contrasts, which are not provided by default. Nevertheless, the chi-squared statistic and its p-value for (2 vs 1) for sex = 1 matches the results of our test command. We also obtain the test of whether the response of males treated as males is equal to the response of males treated as females.

For a gentler introduction to contrasts of margins, see [R] **margins, contrast**.

Example 11: Margins of a specified prediction

We will fit the model

```
. use http://www.stata-press.com/data/r12/margex
. tobit ycn i.sex i.group sex#group age, ul(90)
```

and we will tell the following story about the variables: We run a peach orchard where we allow people to pick their own peaches. A person receives one empty basket in exchange for $20, along with the right to enter the orchard. There is no official limit on how many peaches a person can pick, but only 90 peaches will fit into a basket. The dependent variable in the above tobit model, ycn, is the number of peaches picked. We use tobit, a special case of censored-normal regression, because ycn is censored at 90.

After fitting this model, if we typed

. margins sex

we would obtain the margins for males and for females of the uncensored number of peaches picked. We would obtain that because predict after tobit produces the uncensored number by default. To obtain the censored prediction, we would have to specify predict's ystar(.,90) option. If we want the margins based on that response, we type

. margins sex, predict(ystar(.,90))

The results of typing that are

```
. tobit ycn i.sex i.group sex#group age, ul(90)
  (output omitted )
. margins sex, predict(ystar(.,90))
Predictive margins                            Number of obs   =      3000
Model VCE    : OIM

Expression   : E(ycn*|ycn<90), predict(ystar(.,90))
```

	Margin	Delta-method Std. Err.	z	P>|z|	[95% Conf. Interval]
sex						
0	62.21804	.5996928	103.75	0.000	61.04266 63.39342
1	78.34272	.455526	171.98	0.000	77.4499 79.23553

In our previous examples, sex = 1 has designated females, so evidently the females visiting our orchard are better at filling baskets than the men.

Example 12: Margins of a specified expression

Continuing with our peach orchard example and the previously fit model

```
. use http://www.stata-press.com/data/r12/margex
. tobit ycn i.sex i.group sex#group age, ul(90)
```

let's examine how well our baskets are working for us. What is the proportion of the number of peaches actually picked to the number that would have been picked were the baskets larger? As mentioned in example 11, predict, ystar(.,90) produces the expected number picked given the limit of basket size. predict, xb would predict the expected number without a limit. We want the ratio of those two predictions. That ratio will measure as a proportion how well the baskets work. Thus we could type

. margins sex, expression(predict(ystar(.,90))/predict(xb))

That would give us the proportion for everyone treated as male and everyone treated as female, but what we want to know is how well baskets work for true males and true females, so we will type

. margins, over(sex) expression(predict(ystar(.,90))/predict(xb))

```
. margins, over(sex) expression(predict(ystar(0,90))/predict(xb))
Predictive margins                              Number of obs    =      3000
Model VCE     : OIM

Expression    : predict(ystar(0,90))/predict(xb)
over          : sex
```

	Margin	Delta-method Std. Err.	z	P>|z|	[95% Conf. Interval]	
sex						
0	.9811785	.0013037	752.60	0.000	.9786233	.9837337
1	.9419962	.0026175	359.88	0.000	.936866	.9471265

By the way, we could count the number of peaches saved by the limited basket size during the period of data collection by typing

```
. count
  3000
. margins, expression(3000*(predict(xb)-predict(ystar(.,90))))
  (output omitted )
```

The number of peaches saved turns outs to be 9,183.

Example 13: Margins with multiple outcomes (responses)

Estimation commands such as mlogit and mprobit (see [R] **mlogit** and [R] **mprobit**) calculate multiple responses, and those multiple responses are reflected in the options available with predict after estimation. Obtaining margins for such estimators is thus the same as obtaining margins of a specified prediction, which was demonstrated in example 11. The solution is to include the *predict_opt* that selects the desired response in margins' predict(*predict_opt*) option.

If we fit the multinomial logistic model

```
. mlogit group i.sex age
```

then to obtain the margins for the probability that group = 1, we would type

```
. margins sex, predict(outcome(1))
```

and to obtain the margins for the probability that group = 3, we would type

```
. margins sex, predict(outcome(3))
```

We learned about the outcome(1) and outcome(3) options by looking in [R] **mlogit postestimation**. For an example using margins with a multiple-outcome estimator, see example 4 in [R] **mlogit postestimation**.

Example 14: Margins with multiple equations

Estimation commands such as mvreg, manova, sureg, and reg3 (see [R] **mvreg**, [MV] **manova**, [R] **sureg**, and [R] **reg3**) fit multiple equations. Obtaining margins for such estimators is the same as obtaining margins with multiple outcomes (see example 13), which in turn is the same as obtaining margins of a specified prediction (see example 11). You place the relevant option from the estimator's predict command into margins' predict(*predict_opt*) option.

If we fit the seemingly unrelated regression model

```
. sureg (y = i.sex age) (distance = i.sex i.group)
```

we can obtain the marginal means of y for males and females by typing

> . margins sex, predict(equation(y))

and we can obtain the marginal means of `distance` by typing

> . margins sex, predict(equation(distance))

We could obtain the difference between the margins of y and `distance` by typing

> . margins sex, expression(predict(equation(y)) -
> > predict(equation(distance)))

More examples can be found in [MV] **manova** and [MV] **manova postestimation**.

Example 15: Margins evaluated out of sample

You can fit your model on one dataset and use `margins` on another if you specify margins'
`noesample` option. Remember that `margins` reports estimated average responses, and, unless you
lock all the covariates at fixed values by using the `at()` option, the remaining variables are allowed
to vary as they are observed to vary in the data. That is indeed the point of using `margins`. The
fitted model provides the basis for adjusting for the remaining variables, and the data provide their
values. The predictions produced by `margins` are of interest assuming the data used by `margins`
are in some sense interesting or representative. In some cases, you might need to fit your model on
one set of data and perform `margins` on another.

In example 11, we fit the model

> . tobit ycn i.sex i.group sex#group age, ul(90)

and we told a story about our peach orchard in which we charged people $20 to collect a basket of
peaches, where baskets could hold at most 90 peaches. Let us now tell you that we believe the data on
which we estimated those margins were unrepresentative, or at least, we have a more representative
sample stored in another .dta file. That dataset includes the demographics of our customers but does
not include counts of peaches picked. It is a lot of work counting those peaches.

Thus we will fit our model just as we did previously using the detailed data, but we will bring the other,
more representative dataset into memory before issuing the `margins sex, predict(ystar(.,90))`
command, and we will add `noesample` to it.

```
. use http://www.stata-press.com/data/r12/margex
(Artificial data for margins)
. tobit ycn i.sex i.group sex#group age, ul(90)
  (output omitted )
. use http://www.stata-press.com/data/r12/peach
. margins sex, predict(ystar(.,90)) noesample
```

Predictive margins   Number of obs   =   2727
Model VCE   : OIM

Expression   : E(ycn*|ycn<90), predict(ystar(.,90))

	Margin	Delta-method Std. Err.	z	P>|z|	[95% Conf. Interval]	
sex						
0	56.79774	1.003727	56.59	0.000	54.83047	58.76501
1	75.02146	.643742	116.54	0.000	73.75975	76.28317

In example 12, we produced an estimate of the number of peaches saved by the limited-size baskets. We can update that estimate using the new demographic data by typing

```
. count
2727
. margins, exp(2727*(predict(xb)-predict(ystar(.,90)))) noesample
(output omitted)
```

By running the above, we find that the updated number of peaches saved is 6,408.

Obtaining margins of derivatives of responses (a.k.a. marginal effects)

Derivatives of responses are themselves responses, so everything said above in *Obtaining margins of responses* is equally true of derivatives of responses, and every example above could be repeated here substituting the derivative of the response for the response.

Derivatives are of interest because they are an informative way of summarizing fitted results. The change in a response for a change in the covariate is easy to understand and to explain. In simple models, one hardly needs `margins` to assist in obtaining such margins. Consider the simple linear regression

$$y = \beta_0 + \beta_1 \times \text{sex} + \beta_2 \times \text{age} + \epsilon$$

The derivatives of the responses are

$$dy/d(\text{sex}) = \beta_1$$
$$dy/d(\text{age}) = \beta_2$$

The derivatives are the fitted coefficients. How does y change between males and females? It changes by β_1. How does y change with age? It changes by β_2 per year.

If you make the model a little more complicated, however, the need for margins arises. Consider the model

$$y = \beta_0 + \beta_1 \times \text{sex} + \beta_2 \times \text{age} + \beta_3 \times \text{age}^2 + \epsilon$$

Now the derivative with respect to age is

$$dy/d(\text{age}) = \beta_2 + 2 \times \beta_3 \times \text{age}$$

The change in y for a change in age itself changes with age, and so to better understand the fitted results, you might want to make a table of the change in y for a change in age for age $= 30$, age $= 40$, and age $= 50$. `margins` can do that.

Consider an even more complicated model, such as

$$y = \beta_0 + \beta_1 \times \text{sex} + \beta_2 \times \text{age} + \beta_3 \times \text{age}^2 + \beta_4 \times \text{bp} + \beta_5 \times \text{sex} \times \text{bp} + \beta_6 \times \text{tmt} \qquad (1)$$
$$+ \beta_7 \times \text{tmt} \times \text{age} + \beta_8 \times \text{tmt} \times \text{age}^2 + \epsilon$$

The derivatives are

$$dy/d(\text{sex}) = \beta_1 + \beta_5 \times \text{bp}$$

$$dy/d(\text{age}) = \beta_2 + 2 \times \beta_3 \times \text{age} + \beta_7 \times \text{tmt} + 2 \times \beta_8 \times \text{tmt} \times \text{age}$$
$$dy/d(\text{bp}) = \beta_4 + \beta_5 \times \text{sex}$$

$$dy/d(\text{tmt}) = \beta_6 + \beta_7 \times \text{age} + \beta_8 \times \text{age}^2$$

At this point, `margins` becomes indispensable.

Do not specify marginlist when you mean over()

margins has the same syntax when used with derivatives of responses as when used with responses. To obtain derivatives, one specifies the dydx() option. If we wanted to examine the response variable dy/d(tmt), we would specify margins' dydx(tmt) option. The rest of the margins command has the same syntax as ordinarily, although one tends to specify different syntactical elements. For instance, one usually does not specify a *marginlist*. If we typed

. margins sex, dydx(tmt)

we would obtain dy/d(tmt) calculated first as if everyone were male and then as if everyone were female. At the least, we would probably want to specify

. margins sex, dydx(tmt) grand

so as also to obtain dy/d(tmt), the overall margin, the margin with everyone having their own value of sex. Usually, however, all we want is the overall margin, and because grand is the default when the *marginlist* is not specified, we would just type

. margins, dydx(tmt)

Alternatively, if we were interested in the decomposition by sex, then rather than type margins sex, dydx(tmt), we probably want to type

. margins, over(sex) dydx(tmt)

This command gives us the average effect of tmt for males and again for females rather than the average effect with everyone treated as male and then again with everyone treated as female.

Use at() freely, especially with continuous variables

Another option one tends to use more often with derivatives of responses than one does with responses is at(). Such use is often to better understand or to communicate how the response varies, or, in technical jargon, to explore the nature of the response surface.

For instance, the effect dy/d(tmt) in (1) is equal to $\beta_6 + \beta_7 \times \text{age} + \beta_8 \times \text{age}^2$, and so simply to understand how treatment varies with age, we may want to fix age at various values. We might type

. margins, dydx(tmt) over(sex) at(age=(30 40 50))

Expressing derivatives as elasticities

You specify the dydx(*varname*) option on the margins command to use dy/d(*varname*) as the response variable. If you want that derivative expressed as an elasticity, you can specify eyex(*varname*), eydx(*varname*), or dyex(*varname*). You substitute e for d where you want an elasticity. The formulas are

$$\text{dydx()} = dy/dx$$
$$\text{eyex()} = dy/dx \times (x/y)$$
$$\text{eydx()} = dy/dx \times (1/y)$$
$$\text{dyex()} = dy/dx \times (x)$$

and the interpretations are

		change in y for a		change in x
dydx():		change in y for a		change in x
eyex():	proportional	change in y for a	proportional	change in x
eydx():	proportional	change in y for a		change in x
dyex():		change in y for a	proportional	change in x

As margins always does with response functions, calculations are made at the observational level and are then averaged. Let's assume that in observation 5, $dy/dx = 0.5$, $y = 15$, and $x = 30$; then

$$\text{dydx}() = 0.5$$

$$\text{eyex}() = 1.0$$

$$\text{eydx}() = 0.03$$

$$\text{dyex}() = 15.0$$

Many social scientists would informally explain the meaning of eyex() = 1 as "y increases 100% when x increases 100%" or as "y doubles when x doubles", although neither statement is literally true. eyex(), eydx(), and dyex() are rates evaluated at a point, just as dydx() is a rate, and all such interpretations are valid only for small (infinitesimal) changes in x. It is true that eyex() = 1 means y increases with x at a rate such that, if the rate were constant, y would double if x doubled. This issue of casual interpretation is no different from casually interpreting dydx() as if it represents the response to a unit change. It is not necessarily true that dydx() = 0.5 means that "y increases by 0.5 if x increases by 1". It is true that "y increases with x at a rate such that, if the rate were constant, y would increase by 0.5 if x increased by 1".

dydx(), eyex(), eydx(), and dyex() may be used with continuous x variables. dydx() and eydx() may also be used with factor variables.

Derivatives versus discrete differences

In (1),

$$y = \beta_0 + \beta_1 \times \text{sex} + \beta_2 \times \text{age} + \beta_3 \times \text{age}^2 + \beta_4 \times \text{bp} + \beta_5 \times \text{sex} \times \text{bp} + \beta_6 \times \text{tmt}$$
$$+ \beta_7 \times \text{tmt} \times \text{age} + \beta_8 \times \text{tmt} \times \text{age}^2 + \epsilon$$

Let us call your attention to the derivatives of y with respect to age and sex:

$$dy/d(\text{age}) = \beta_2 + 2 \times \beta_3 \times \text{age} + \beta_7 \times \text{tmt} + 2 \times \beta_8 \times \text{tmt} \times \text{age} \tag{2}$$

$$dy/d(\text{sex}) = \beta_1 + \beta_5 \times \text{bp} \tag{3}$$

age is presumably a continuous variable and (2) is precisely how margins calculates its derivatives when you type margins, dydx(age). sex, however, is presumably a factor variable, and margins does not necessarily make the calculation using (3) were you to type margins, dydx(sex). We will explain, but let us first clarify what we mean by a continuous and a factor variable. Say that you fit (1) by typing

```
. regress y i.sex age c.age#c.age i.bp bp#sex
> i.tmt tmt#c.age tmt#c.age#c.age
```

It is important that `sex` entered the model as a factor variable. It would not do to type `regress y sex` ... because then `sex` would be a continuous variable, or at least it would be a continuous variable from Stata's point of view. The model estimates would be the same, but `margins`' understanding of those estimates would be a little different. With the model estimated using `i.sex`, `margins` understands that either `sex` is 0 or `sex` is 1. With the model estimated using `sex`, `margins` thinks `sex` is continuous and, for instance, `sex` = 1.5 is a possibility.

`margins` calculates `dydx()` differently for continuous and for factor variables. For continuous variables, `margins` calculates dy/dx. For factor variables, `margins` calculates the discrete first-difference from the base category. To obtain that for `sex`, write down the model and then subtract from it the model evaluated at the base category for `sex`, which is `sex` = 0. If you do that, you will get the same formula as we obtained for the derivative, namely,

$$\text{discrete difference}\{(\text{sex} = 1) - (\text{sex} = 0)\} = \beta_1 + \beta_5 \times \text{bp}$$

We obtain the same formula because our model is linear regression. Outside of linear regression, and outside of linear response functions generally, the discrete difference is not equal to the derivative. The discrete difference is not equal to the derivative for logistic regression, probit, etc. The discrete difference calculation is generally viewed as better for factor variables than the derivative calculation because the discrete difference is what would actually be observed.

If you want the derivative calculation for your factor variables, specify the `continuous` option on the `margins` command.

Example 16: Average marginal effect (partial effects)

Concerning the title of this example, the way we use the term marginal effect, the effects of factor variables are calculated using discrete first-differences. If you wanted the continuous calculation, you would specify `margins`' `continuous` option in what follows.

```
. use http://www.stata-press.com/data/r12/margex
(Artificial data for margins)
. logistic outcome treatment##group age c.age#c.age treatment#c.age
  (output omitted)
. margins, dydx(treatment)
Average marginal effects                       Number of obs    =       3000
Model VCE      : OIM

Expression     : Pr(outcome), predict()
dy/dx w.r.t.   : 1.treatment
```

		Delta-method				
	dy/dx	Std. Err.	z	P>|z|	[95% Conf. Interval]	
1.treatment	.0385625	.0162848	2.37	0.018	.0066449	.0704801

Note: dy/dx for factor levels is the discrete change from the base level.

The average marginal effect of treatment on the probability of a positive outcome is 0.039.

Example 17: Average marginal effect of all covariates

We will continue with the model

```
. logistic outcome treatment##group age c.age#c.age treatment#c.age
```

if we wanted the average marginal effects for all covariates, we would type margins, dydx(*) or margins, dydx(_all); they mean the same thing. This is probably the most common way margins, dydx() is used.

```
. margins, dydx(*)
Average marginal effects                    Number of obs   =      3000
Model VCE    : OIM

Expression   : Pr(outcome), predict()
dy/dx w.r.t. : 1.treatment 2.group 3.group age
```

	dy/dx	Delta-method Std. Err.	z	P>|z|	[95% Conf. Interval]	
1.treatment	.0385625	.0162848	2.37	0.018	.0066449	.0704801
group						
2	-.0776906	.0181584	-4.28	0.000	-.1132805	-.0421007
3	-.1505652	.0400882	-3.76	0.000	-.2291366	-.0719937
age	.0095868	.0007796	12.30	0.000	.0080589	.0111148

Note: dy/dx for factor levels is the discrete change from the base level.

Example 18: Evaluating marginal effects over the response surface

Continuing with the model

```
. logistic outcome treatment##group age c.age#c.age treatment#c.age
```

What follows maps out the entire response surface of our fitted model. We report the marginal effect of treatment evaluated at age $= 20, 30, \ldots, 60$, by each level of group.

```
. margins group, dydx(treatment) at(age=(20(10)60))
```

| Conditional marginal effects | | Number of obs | = | 3000 |
Model VCE   : OIM

Expression : Pr(outcome), predict()
dy/dx w.r.t. : 1.treatment

1._at	: age	=	20
2._at	: age	=	30
3._at	: age	=	40
4._at	: age	=	50
5._at	: age	=	60

	dy/dx	Delta-method Std. Err.	z	P>|z|	[95% Conf. Interval]	
1.treatment						
_at#group						
1 1	-.0208409	.0152862	-1.36	0.173	-.0508013	.0091196
1 2	.009324	.0059896	1.56	0.120	-.0024155	.0210635
1 3	.0006558	.0048682	0.13	0.893	-.0088856	.0101972
2 1	-.0436964	.0279271	-1.56	0.118	-.0984325	.0110397
2 2	.0382959	.0120405	3.18	0.001	.014697	.0618949
2 3	.0064564	.0166581	0.39	0.698	-.0261929	.0391057
3 1	-.055676	.0363191	-1.53	0.125	-.1268601	.015508
3 2	.1152235	.0209858	5.49	0.000	.074092	.156355
3 3	.0284808	.0471293	0.60	0.546	-.0638908	.1208524
4 1	-.027101	.0395501	-0.69	0.493	-.1046177	.0504158
4 2	.2447682	.0362623	6.75	0.000	.1736954	.315841
4 3	.0824401	.1025028	0.80	0.421	-.1184616	.2833418
5 1	.0292732	.0587751	0.50	0.618	-.0859239	.1444703
5 2	.3757777	.0578106	6.50	0.000	.2624709	.4890844
5 3	.1688268	.1642191	1.03	0.304	-.1530368	.4906904

Note: dy/dx for factor levels is the discrete change from the base level.

Obtaining margins with survey data and representative samples

The standard errors and confidence intervals produced by margins are based by default on the delta method applied to the VCE of the current estimates. Delta-method standard errors treat the covariates at which the response is evaluated as given or fixed. Such standard errors are appropriate if you specify at() to fix the covariates, and they are appropriate when you are making inferences about groups exactly like your sample whether you specify at() or not.

On the other hand, if you have a representative sample of the population or if you have complex survey data and if you want to make inferences about the underlying population, you need to account for the variation in the covariates that would arise in repeated sampling. You do that using vce(unconditional), which invokes a different standard-error calculation based on Korn and Graubard (1999). Syntactically, there are three cases. They all involve specifying the vce(unconditional) option on the margins command:

1. *You have a representative random sample, and you have not* svyset *your data.*
   When you fit the model, you need to specify the vce(robust) or vce(cluster *clustvar*) option. When you issue the margins command, you need to specify the vce(unconditional) option.

2. *You have a weighted sample, and you have not* svyset *your data.*
   You need to specify [pw=weight] when you fit the model and, of course, specify the vce(unconditional) option on the margins command. You do not need to specify the weights on the margins command because margins will obtain them from the estimation results.

3. *You have* svyset *your data, whether it be a simple random sample or something more complex including weights, strata, sampling units, or poststratification.*
   You need to use the svy prefix when you fit the model. You need to specify vce(unconditional) when you issue the margins command. You do not need to respecify the weights.

   Even though the data are svyset, and even though the estimation was svy estimation, margins does not default to vce(unconditional). It does not default to vce(unconditional) because there are valid reasons to want the data-specific, vce(delta) standard-error estimates. Whether you specify vce(unconditional) or not, margins uses the weights, so you do not need to respecify them even if you are using vce(unconditional).

   vce(unconditional) is allowed only after estimation with vce(robust), vce(cluster ...), or the svy prefix. If the VCE of the current estimates was specified as clustered, so will be the VCE estimates of margins. If the estimates were from a survey estimation, the survey settings in the dataset will be used by margins.

   When you use vce(unconditional), never specify if *exp* or in *range* on the margins command; instead, specify the subpop(if *exp*) option. You do that for the usual reasons; see [SVY] **subpopulation estimation**. If you specify over(*varlist*) to examine subgroups, the subgroups will automatically be treated as subpopulations.

Example 19: Inferences for populations, margins of response

In example 6, we fit the model

```
. logistic outcome i.sex i.group sex#group age
```

and we obtained margins by sex and margins by group,

```
. margins sex group
```

If our data were randomly drawn from the population of interest and we wanted to account for this, we would have typed

```
. logistic outcome i.sex i.group sex#group age, vce(robust)
. margins sex group, vce(unconditional)
```

We do that below:

```
. logistic outcome i.sex i.group sex#group age, vce(robust)
(output omitted)
. margins sex group, vce(unconditional)
```
Predictive margins   Number of obs   =   3000
Expression   : Pr(outcome), predict()

	Margin	Unconditional Std. Err.	z	P>|z|	[95% Conf. Interval]	
sex						
0	.1600644	.0131685	12.16	0.000	.1342546	.1858743
1	.1966902	.0104563	18.81	0.000	.1761963	.2171841
group						
1	.2251302	.0127069	17.72	0.000	.200225	.2500354
2	.150603	.0118399	12.72	0.000	.1273972	.1738088
3	.0736157	.0343188	2.15	0.032	.0063522	.1408793

The estimated margins are the same as they were in example 6, but the standard errors and confidence intervals differ, although not by much. Given that we have 3,000 observations in our randomly drawn sample, we should expect this.

Example 20: Inferences for populations, marginal effects

In example 17, we fit a logistic model and then obtained the average marginal effects for all covariates by typing

```
. logistic outcome treatment##group age c.age#c.age treatment#c.age
. margins, dydx(*)
```

To repeat that and also obtain standard errors for our population, we would type

```
. logistic outcome treatment##group age c.age#c.age treatment#c.age,
> vce(robust)
. margins, dydx(*) vce(unconditional)
```

The results are

```
. logistic outcome treatment##group age c.age#c.age treatment#c.age, vce(robust)
(output omitted)
. margins, dydx(*) vce(unconditional)
```
Average marginal effects   Number of obs   =   3000
Expression   : Pr(outcome), predict()
dy/dx w.r.t. : 1.treatment 2.group 3.group age

	dy/dx	Unconditional Std. Err.	z	P>|z|	[95% Conf. Interval]	
1.treatment	.0385625	.0163872	2.35	0.019	.0064442	.0706808
group						
2	-.0776906	.0179573	-4.33	0.000	-.1128863	-.0424949
3	-.1505652	.0411842	-3.66	0.000	-.2312848	-.0698456
age	.0095868	.0007814	12.27	0.000	.0080553	.0111183

Note: dy/dx for factor levels is the discrete change from the base level.

Example 21: Inferences for populations with svyset data

See example 3 in [SVY] **svy postestimation**.

Standardizing margins

A standardized margin is the margin calculated on data different from the data used to fit the model. Typically, the word standardized is reserved for situations in which the alternate population is a reference population, which may be real or artificial, and which is treated as fixed.

Say that you work for a hospital and have fit a model of mortality on the demographic characteristics of the hospital's patients. At this stage, were you to type

```
. margins
```

you would obtain the mortality rate for your hospital. You have another dataset, `hstandard.dta`, that contains demographic characteristics of patients across all hospitals along with the population of each hospital recorded in the `pop` variable. You could obtain the expected mortality rate at your hospital if your patients matched the characteristics of the standard population by typing

```
. use http://www.stata-press.com/data/r12/hstandard, clear
. margins [fw=pop], noesample
```

You specified `noesample` because the margin is being calculated on data other than the data used to estimate the model. You specified `[fw=pop]` because the reference dataset you are using included population counts, as many reference datasets do.

Obtaining margins as though the data were balanced

Here we discuss what are commonly called estimated marginal means or least-squares means. These are margins assuming that all levels of factor variables are equally likely or, equivalently, that the design is balanced. The seminal reference on these margins is Searle, Speed, and Milliken (1980).

In designed experiments, observations are often allocated in a balanced way so that the variances can be easily compared and decomposed. At the Acme Portable Widget Company, they are experimenting with a new machine. The machine has three temperature settings and two pressure settings; a combination of settings will be optimal on any particular day, determined by the weather. At start-up, one runs a quick test and chooses the optimal setting for the day. Across different days, each setting will be used about equally, says the manufacturer.

In experiments with the machine, 10 widgets were collected for stress testing at each of the settings over a six-week period. We wish to know the average stress-test value that can be expected from these machines over a long period.

Balancing using asbalanced

The data were intended to be balanced, but unfortunately, the stress test sometimes destroys samples before the stress can be measured. Thus even though the experiment was designed to be balanced, the data are not balanced. You specify the `asbalanced` option to estimate the margins as if the data were balanced. We will type

```
. use http://www.stata-press.com/data/r12/acmemanuf
. regress y pressure##temp
. margins, asbalanced
```

So that you can compare the asbalanced results with the observed results, we will also include margins without the asbalanced option in what follows:

```
. use http://www.stata-press.com/data/r12/acmemanuf
. regress y pressure##temp
  (output omitted )
. margins
```

| Predictive margins | | | | Number of obs | = | 49 |
| Model VCE   : OLS | | | | | | |

Expression   : Linear prediction, predict()

	Margin	Delta-method Std. Err.	z	P>|z|	[95% Conf. Interval]	
_cons	109.9214	1.422629	77.27	0.000	107.1331	112.7097

```
. margins, asbalanced
```

| Adjusted predictions | | | | Number of obs | = | 49 |
| Model VCE   : OLS | | | | | | |

Expression   : Linear prediction, predict()
at   : pressure   (asbalanced)
               temp   (asbalanced)

	Margin	Delta-method Std. Err.	z	P>|z|	[95% Conf. Interval]	
_cons	115.3758	1.530199	75.40	0.000	112.3767	118.375

❑ Technical note

Concerning how asbalanced calculations are performed, if a factor variable has l levels, then each level's coefficient contributes to the response weighted by $1/l$. If two factors, a and b, interact, then each coefficient associated with their interaction is weighted by $1/(l_a \times l_b)$.

If a balanced factor interacts with a continuous variable, then each coefficient in the interaction is applied to the value of the continuous variable, and the results are weighted equally. So, if the factor being interacted has l_a levels, the effect of each coefficient on the value of the continuous covariate is weighted by $1/l_a$.

❑

Balancing by standardization

To better understand the balanced results, we can perform the balancing ourselves by using the standardizing method shown in *Standardizing margins*. To do that, we will input a balanced dataset and then type margins, noesample.

```
. use http://www.stata-press.com/data/r12/acmemanuf
. regress y pressure##temp
  (output omitted)
. drop _all
. input pressure temp

       pressure        temp
  1. 1 1
  2. 1 2
  3. 1 3
  4. 2 1
  5. 2 2
  6. 2 3
  7. end
. margins, noesample
Predictive margins                              Number of obs   =        6
Model VCE    : OLS
Expression   : Linear prediction, predict()
```

		Delta-method				
	Margin	Std. Err.	z	P>|z|	[95% Conf.	Interval]
_cons	115.3758	1.530199	75.40	0.000	112.3767	118.375

We obtain the same results as previously.

Balancing nonlinear responses

If our testing had produced a binary outcome, say, acceptable/unacceptable, rather than a continuous variable, we would type

```
. use http://www.stata-press.com/data/r12/acmemanuf, clear
. logistic acceptable pressure##temp
. margins, asbalanced
```

The result of doing that would be 0.680. If we omitted the asbalanced option, the result would have been 0.667. The two results are so similar because acmemanuf.dta is nearly balanced.

Even though the asbalanced option can be used on both linear and nonlinear responses, such as probabilities, there is an issue of which you should be aware. The most widely used formulas for balancing responses apply the balancing to the linear prediction, average that as if it were balanced, and then apply the nonlinear transform. That is the calculation that produced 0.680.

An alternative would be to apply the standardization method. That amounts to making the linear predictions observation by observation, applying the nonlinear transform to each, and then averaging the nonlinear result as if it were balanced. You could do that by typing

```
. use http://www.stata-press.com/data/r12/acmemanuf, clear
. logistic acceptable pressure##temp
. clear
. input pressure temp
  (see above for entered data)
. margins, noesample
```

The result from the standardization procedure would be 0.672. These two ways of averaging nonlinear responses are discussed in detail in Lane and Nelder (1982) within the context of general linear models.

Concerning the method used by the asbalanced option, if your data start balanced and you have a nonlinear response, you will get different results with and without the asbalanced option!

Treating a subset of covariates as balanced

So far, we have treated all the covariates as if they were balanced. margins will allow you to treat a subset of the covariates as balanced, too. For instance, you might be performing an experiment in which you are randomly allocating patients to a treatment arm and so want to balance on arm, but you do not want to balance the other characteristics because you want mean effects for the experiment's population.

In this example, we will imagine that the outcome of the experiment is continuous. We type

```
. use http://www.stata-press.com/data/r12/margex, clear
. regress y arm##sex sex##agegroup
. margins, at((asbalanced) arm)
```

If we wanted results balanced on agegroup as well, we could type

```
. margins, at((asbalanced) arm agegroup)
```

If we wanted results balanced on all three covariates, we could type

```
. margins, at((asbalanced) arm agegroup sex)
```

or we could type

```
. margins, at((asbalanced) _factor)
```

or we could type

```
. margins, asbalanced
```

Using fvset design

As a convenience feature, equivalent to

```
. regress y arm##sex sex##agegroup
. margins, at((asbalanced) arm sex)
```

is

```
. fvset design asbalanced arm sex
. regress y arm##sex sex##agegroup
. margins
```

The advantage of the latter is that you have to set the variables as balanced only once. This is useful when balancing is a design characteristic of certain variables and you wish to avoid accidentally treating them as unbalanced.

If you save your data after fvsetting, the settings will be remembered in future sessions. If you want to clear the setting(s), type

```
. fvset clear varlist
```

See [R] **fvset**.

alancing in the presence of empty cells

The issue of empty cells is not exclusively an issue of balancing, but there are special considerations when balancing. Empty cells are discussed generally in *Estimability of margins*.

An empty cell is an interaction of levels of two or more factor variables for which you have no data. Usually, margins involving empty cells cannot be estimated. When balancing, there is an alternate definition of the margin that allows the margin to be estimated. margins makes the alternate calculation when you specify the emptycells(reweight) option. By default, margins uses the emptycells(strict) option.

If you have empty cells in your data and you request margins involving the empty cells, those margins will be marked as not estimable even if you specify the asbalanced option.

```
. use http://www.stata-press.com/data/r12/estimability, clear
(margins estimability)

. regress y sex##group
 (output omitted )

. margins sex, asbalanced
Adjusted predictions                          Number of obs   =         69
Model VCE    : OLS
Expression   : Linear prediction, predict()
Empty cells  : reweight
at           : sex                (asbalanced)
               group              (asbalanced)
```

	Margin	Delta-method Std. Err.	z	P>|z|	[95% Conf. Interval]
sex					
0	21.91389	1.119295	19.58	0.000	19.72011 24.10767
1	.	(not estimable)			

This example is discussed in *Estimability of margins*, although without the asbalanced option. What is said there is equally relevant to the asbalanced case. For reasons explained there, the margin for sex $= 1$ cannot be estimated.

The margin for sex $= 1$ can be estimated in the asbalanced case if you are willing to make an assumption. Remember that margins makes the balanced calculation by summing the responses associated with the levels and then dividing by the number of levels. If you specify emptycells(reweight), margins sums what is available and divides by the number available. Thus you are assuming that, whatever the responses in the empty cells, those responses are such that they would not change the overall mean of what is observed.

The results of specifying emptycells(reweight) are

```
. margins sex, asbalanced emptycells(reweight)
Adjusted predictions                           Number of obs   =        69
Model VCE    : OLS

Expression   : Linear prediction, predict()
Empty cells  : reweight
at           : sex                (asbalanced)
               group              (asbalanced)
```

	Margin	Delta-method Std. Err.	z	P>|z|	[95% Conf. Interval]
sex					
0	21.91389	1.119295	19.58	0.000	19.72011 24.10767
1	24.85185	1.232304	20.17	0.000	22.43658 27.26712

Obtaining margins with nested designs

Introduction

Factors whose meaning depends on other factors are called nested factors, and the factors on which their meaning depends are called the nesting factors. For instance, assume that we have a sample of patients and each patient is assigned to one doctor. Then patient is nested within doctor. Let the identifiers of the first 5 observations of our data be

Doctor	Patient	Name
1	1	Fred
1	2	Mary
1	3	Bob
2	1	Karen
2	2	Hank

The first patient on one doctor's list has nothing whatsoever to do with the first patient on another doctor's list. The meaning of patient = 1 is defined only when the value of doctor is supplied.

Nested factors enter into models as interactions of nesting and nested; the nested factor does not appear by itself. We might estimate a model such as

```
. regress y ... i.doctor doctor#patient ...
```

You do not include i.patient because the coding for patient has no meaning except within doctor. Patient 1 is Fred for doctor 1 and Karen for doctor 2, etc.

margins provides an option to help account for the structure of nested models. The within(*varlist*) option specifies that margins estimate and report a set of margins for the value combinations of *varlist*. We might type

```
. margins, within(doctor)
```

Margin calculations are performed first for doctor = 1, then for doctor = 2, and so on.

Sometimes you need to specify within(), and other times you do not. Let's consider the particular model

```
. regress y i.doctor doctor#patient i.sex sex#doctor#patient
```

The guidelines are the following:

1. You may compute overall margins by typing

   `margins`.

2. You may compute overall margins within levels of a nesting factor by typing

   `margins, within(doctor)`.

3. You may compute margins of a nested factor within levels of its nesting factor by typing

   `margins patient, within(doctor)`.

4. You may compute margins of factors in your model, as long as the factor does not nest other factors and is not nested within other factors, by typing

   `margins sex`.

5. You may not compute margins of a nesting factor, such as `margins doctor`, because they are not estimable.

For examples using `within()`, see [R] **anova**.

argins with nested designs as though the data were balanced

To obtain margins with nested designs as though the data were balanced, the guidelines are the same as above except that 1) you add the `asbalanced` option and 2) whenever you do not specify `within()`, you specify `emptycells(reweight)`. The updated guidelines are

1. You may compute overall margins by typing

   `margins, asbalanced emptycells(reweight)`.

2. You may compute overall margins within levels of a nesting factor by typing

   `margins, asbalanced within(doctor)`.

3. You may compute margins of a nested factor within levels of its nesting factor by typing

   `margins patient, asbalanced within(doctor)`.

4. You may compute margins of factors in your model, as long as the factor does not nest other factors and is not nested within other factors, by typing

   `margins sex, asbalanced emptycells(reweight)`.

5. You may not compute margins of a nesting factor, such as `margins doctor`, because they are not estimable.

Just as explained in *Using fvset design*, rather than specifying the `asbalanced` option, you may set the balancing characteristic on the factor variables once and for all by using the command `fvset design asbalanced` *varlist*.

❏ Technical note

Specifying either `emptycells(reweight)` or `within(`*varlist*`)` causes `margins` to rebalance over all empty cells in your model. If you have interactions in your model that are not involved in the nesting, `margins` will lose its ability to detect estimability.

❏

❏ Technical note

Careful readers will note that the description of `within(`*varlist*`)` matches closely the description of `over(`*varlist*`)`. The concept of nesting is similar to the concept of subpopulations. `within()` differs from `over()` in that it gracefully handles the missing cells when margins are computed as balanced.

❏

Coding of nested designs

In the *Introduction* to this section, we showed a coding of the nested variable `patient`, where the coding started over with each `doctor`:

Doctor	Patient	Name
1	1	Fred
1	2	Mary
1	3	Bob
2	1	Karen
2	2	Hank

That coding style is not required. The data could just as well have been coded

Doctor	Patient	Name
1	1	Fred
1	2	Mary
1	3	Bob
2	4	Karen
2	5	Hank

or even

Doctor	Patient	Name
1	1037239	Fred
1	2223942	Mary
1	0611393	Bob
2	4433329	Karen
2	6110271	Hank

Actually, either of the above two alternatives are better than the first one because `margins` will be better able to give you feedback about estimability should you make a mistake following the guidelines. On the other hand, both of these two alternatives require more memory at the estimation step. If you run short of memory, you will need to recode your patient ID to the first coding style, which you could do by typing

    . sort doctor patient
    . by doctor: gen newpatient = _n

Alternatively, you can `set emptycells drop` and continue to use your patient ID variable just as it is coded. If you do this, we recommend that you remember to type `set emptycells keep` when you are finished; `margins` is better able to determine estimability that way. If you regularly work with large nested models, you can `set emptycells keep, permanently` so that the setting persists across sessions. See [R] **set emptycells**.

ecial topics

quirements for model specification

The results that `margins` reports are based on the most recently fit model or, in Stata jargon, the most recently issued estimation command. Here we discuss 1) mechanical requirements for how you specify that estimation command, 2) work-arounds to use when those restrictions prove impossible, and 3) requirements for `margins`' `predict(`*pred_opt*`)` option to work.

Concerning 1, when you specify the estimation command, covariates that are logically factor variables must be Stata factor variables, and that includes indicator variables, binary variables, and dummies. It will not do to type

```
. regress y ... female ...
```

even if `female` is a 0/1 variable. You must type

```
. regress y ... i.female ...
```

If you violate this rule, you will not get incorrect results, but you will discover that you will be unable to obtain margins on `female`:

```
. margins female
factor female not found in e(b)
r(111);
```

It is also important that if the same continuous variable appears in your model more than once, differently transformed, those transforms be performed via Stata's factor-variable notation. It will not do to type

```
. generate age2 = age^2
. regress y ... age age2 ...
```

You must type

```
. regress y ... age c.age#c.age ...
```

You must do that because `margins` needs to know everywhere that variable appears in the model if it is to be able to set covariates to fixed values.

Concerning 2, sometimes the transformations you desire may not be achievable using the factor-variable notation; in those situations, there is a work-around. Let's assume you wish to estimate

```
. generate age1_5 = age^1.5
. regress y ... age age1_5 ...
```

There is no factor-variable notation for including `age` and $age^{1.5}$ in a model, so obviously you are going to obtain the estimates by typing just what we have shown. In what follows, it would be okay if there are interactions of `age` and `age1_5` with other variables specified by the factor-variable notation, so the model could just as well be

```
. regress y ... age age1_5 sex#c.age sex#c.age1_5 ...
```

Let's assume you have fit one of these two models. On any subsequent `margins` command where you leave age free to vary, there will be no issue. You can type

```
. margins female
```

and results will be correct. Issues arise when you attempt to fix age at predetermined values. The following would produce incorrect results:

        . margins female, at(age=20)

The results would be incorrect because they leave age1_5 free to vary, and, logically, fixing age implies that age1_5 should also be fixed. Because we were unable to state the relationship between age and age1_5 using the factor-variable notation, margins does not know to fix age1_5 at $20^{1.5}$ when it fixes age at 20. To get the correct results, you must fix the value of age1_5 yourself:

        . margins female, at(age=20 age1_5=89.442719)

That command produces correct results. In the command, 89.442719 is $20^{1.5}$.

In summary, when there is a functional relationship between covariates of your model and that functional relationship is not communicated to margins via the factor-variable notation, then it becomes your responsibility to ensure that all variables that are functionally related are set to the appropriate fixed values when any one of them is set to a fixed value.

Concerning 3, we wish to amend our claim that you can calculate margins for anything that predict will produce. We need to add a qualifier. Let us show you an example where the statement is not true. After regress, predict will predict something it calls pr(a,b), which is the probability $a \le y \le b$. Yet if we attempted to use pr() with margins after estimation by regress, we would obtain

        . margins sex, predict(pr(10,20))
        prediction is a function of possibly stochastic quantities other than e(b)
        r(498);

What we should have stated was that you can calculate margins for anything that predict will produce for which all the estimated quantities used in its calculation appear in e(V), the estimated VCE. pr() is a function of β, the estimated coefficients, and of s^2, the estimated variance of the residual. regress does not post the variance of the residual variance (sic) in e(V), or even estimate it, and therefore, predict(pr(10,20)) cannot be specified with margins after estimation by regress.

It is unlikely that you will ever encounter these kinds of problems because there are so few predictions where the components are not posted to e(V). If you do encounter the problem, the solution may be to specify nose to suppress the standard-error calculation. If the problem is not with computing the margin, but with computing its standard error, margins will report the result:

        . margins sex, predict(pr(10,20)) nose
          (output appears with SEs, tests, and CIs left blank)

❑ Technical note

Programmers: If you run into this after running an estimation command that you have written, be aware that as of Stata 11, you are supposed to set in e(marginsok) the list of options allowed with predict that are okay to use with margins. When that list is not set, margins looks for violations of its assumptions and, if it finds any, refuses to proceed.

                                                                                              ❑

stimability of margins

Sometimes `margins` will report that a margin cannot be estimated:

```
. use http://www.stata-press.com/data/r12/estimability, clear
(margins estimability)
. regress y sex##group
(output omitted)
. margins sex
```

Predictive margins   Number of obs   =   69
Model VCE   : OLS

Expression   : Linear prediction, predict()

	Margin	Delta-method Std. Err.	z	P>|z|	[95% Conf. Interval]
sex					
0	21	.8500245	24.71	0.000	19.33398 22.66602
1	.	(not estimable)			

In the above output, the margin for sex = 0 is estimated, but the margin for sex = 1 is not estimable. This occurs because of empty cells. An empty cell is an interaction of levels of two or more factor variables for which you have no data. In the example, the lack of estimability arises because we have two empty cells:

```
. table sex group
```

sex	1	2	group 3	4	5
0	2	9	27	8	2
1	9	9	3		

To calculate the marginal mean response for sex = 1, we have no responses to average over for group = 4 and group = 5. We obviously could calculate that mean for the observations that really are sex = 1, but remember, the marginal calculation for sex = 1 treats everyone as if female, and we will thus have 8 and 2 observations for which we have no basis for estimating the response.

There is no solution for this problem unless you are willing to treat the data as if it were balanced and adjust your definition of a margin; see *Balancing in the presence of empty cells.*

Manipulability of tests

Manipulability is a problem that arises with some tests, and in particular, arises with Wald tests. Tests of margins are based on Wald tests, hence our interest. This is a generic issue and not specific to the `margins` command.

Let's understand the problem. Consider performing a test of whether some statistic ϕ is 0. Whatever the outcome of that test, it would be desirable if the outcome were the same were we to test whether the sqrt(ϕ) were 0, or whether ϕ^2 were 0, or whether any other monotonic transform of ϕ were 0 (for ϕ^2, we were considering only the positive half of the number line). If a test does not have that property, it is manipulable.

Wald tests are manipulable, and that means the tests produced by `margins` are manipulable. You can see this for yourself by typing

```
. use http://www.stata-press.com/data/r12/margex, clear
. replace y = y - 65
. regress y sex##group
. margins
. margins, expression(predict(xb)^2)
```

We would prefer if the test against zero produced by margins was equal to the test produced by margins, expression(predict(xb)^2). But alas, they produce different results. The first produces $z = 12.93$, and the second produces $z = 12.57$.

The difference is not much in our example, but behind the scenes, we worked to make it small. We subtracted 65 from y so that the experiment would be for a case where it might be reasonable that you would be testing against 0. One does not typically test whether the mean income in the United States is zero or whether the mean blood pressure of live patients is zero. Had we left y as it was originally, we would have obtained $z = 377$ and $z = 128$. We did not want to show that comparison to you first because the mean of y is so far from 0 that you probably would never be testing it. The corresponding difference in ϕ is tiny.

Regardless of the example, it is important that you base your tests in the metric where the likelihood surface is most quadratic. For further discussion on manipulability, see *Manipulability* in [R] **predictnl**.

This manipulability is not limited to Wald tests after estimation; you can also see the manipulability of results produced by linear regression just by applying nonlinear transforms to a covariate (Phillips and Park 1988; Gould 1996).

Using margins after the estimates use command

Assume you fit and used estimates save (see [R] **estimates save**) to save the estimation results:

```
. regress y sex##group age c.age*c.age if site==1
. ...
. estimates save mymodel
(file mymodel.ster saved)
```

Later, perhaps in a different Stata session, you reload the estimation results by typing

```
. estimates use mymodel
```

You plan to use margins with the reloaded results. You must remember that margins bases its results not only on the current estimation results but also on the current data in memory. Before you can use margins, you must reload the dataset on which you fit the model or, if you wish to produce standardized margins, some other dataset.

```
. use mydata, clear
(data for fitting models)
```

If the dataset you loaded contained the data for standardization, you can stop reading; you know that to produce standardized margins, you need to specify the noesample option.

We reloaded the original data and want to produce margins for the estimation sample. In addition to the data, margins requires that e(sample) be set, as margins will remind us:

```
. margins sex
e(sample) does not identify the estimation sample
r(322);
```

The best solution is to use `estimates esample` to rebuild `e(sample)`:

```
. estimates esample: y sex group age if site==1
```

If we knew we had no missing values in y and the covariates, we could type

```
. estimates esample: if site==1
```

Either way, `margins` would now work:

```
. margins sex
(usual output appears)
```

There is an alternative. We do not recommend it, but we admit that we have used it. Rather than rebuilding `e(sample)`, you can use `margins`' `noesample` option to tell `margins` to skip using `e(sample)`. You could then specify the appropriate `if` statement (if necessary) to identify the estimation sample:

```
. estimates use mymodel
. use mydata, clear
(data for fitting models)
. margins sex if !missing(y, sex, group age) & site==1, noesample
(usual output appears)
```

In the above, we are not really running on a sample different from the estimation sample; we are merely using `noesample` to fool `margins`, and then we are specifying on the `margins` command the conditions equivalent to re-create `e(sample)`.

If we wish to obtain `vce(unconditional)` results, however, `noesample` will be insufficient. We must also specify the `force` option,

```
. margins sex if !missing(y, sex, group age) & site==1,
> vce(unconditional) noesample force
(usual output appears)
```

Regardless of the approach you choose—resetting `e(sample)` or specifying `noesample` and possibly `force`—make sure you are right. In the `vce(delta)` case, you want to be right to ensure that you obtain the results you want. In the `vce(unconditional)` case, you need to be right because otherwise results will be statistically invalid.

Syntax of at()

In `at`(*atspec*), *atspec* may contain one or more of the following specifications:

> *varlist*
>
> (*stat*) *varlist*
>
> *varname* = #
>
> *varname* = (*numlist*)

where

1. *varname*s must be covariates in the previously fit model (estimation command).

2. Variable names (whether in *varname* or *varlist*) may be continuous variables, factor variables, or specific level variables, such as age, group, or 3.group.

3. *varlist* may also be one of three standard lists:

    a. _all (all covariates),

    b. _factor (all factor-variable covariates), or

    c. _continuous (all continuous covariates).

4. Specifications are processed from left to right with latter specifications overriding previous ones.

5. *stat* can be any of the following:

stat	Description	Variables allowed
asobserved	at observed values in the sample (default)	all
mean	means (default for *varlist*)	all
median	medians	continuous
p1	1st percentile	continuous
p2	2nd percentile	continuous
. . .	3rd–49th percentiles	continuous
p50	50th percentile (same as median)	continuous
. . .	51st–97th percentiles	continuous
p98	98th percentile	continuous
p99	99th percentile	continuous
min	minimums	continuous
max	maximums	continuous
zero	fixed at zero	continuous
base	base level	factors
asbalanced	all levels equally probable and sum to 1	factors

Any *stat* except zero, base, and asbalanced may be prefixed with an o to get the overall statistic—the sample over all over() groups. For example, omean, omedian, and op25. Overall statistics differ from their correspondingly named statistics only when the over() or within() option is specified. When no *stat* is specified, mean is assumed.

Estimation commands that may be used with margins

margins may be used after most estimation commands.

margins cannot be used after estimation commands that do not produce full variance matrices, such as exlogistic and expoisson (see [R] **exlogistic** and [R] **expoisson**).

margins is all about covariates and cannot be used after estimation commands that do not post the covariates, which eliminates gmm (see [R] **gmm**).

margins cannot be used after estimation commands that have an odd data organization, and that excludes asclogit, asmprobit, asroprobit, and nlogit (see [R] **asclogit**, [R] **asmprobit**, [R] **asroprobit**, and [R] **nlogit**).

lossary

adjusted mean. A *margin* when the response is the linear predictor from linear regression, ANOVA, etc. For some authors, adjusting also implies adjusting for unbalanced data. See *Obtaining margins of responses* and see *Obtaining margins as though the data were balanced*.

average marginal effect. See *marginal effect and average marginal effect*.

average partial effect. See *partial effect and average partial effect*.

conditional margin. A *margin* when the response is evaluated at fixed values of all the covariates. If any covariates are left to vary, the margin is called a predictive margin.

effect. The effect of x is the derivative of the *response* with respect to covariate x, or it is the difference in responses caused by a discrete change in x. Also see *marginal effect*.

The effect of x measures the change in the response for a change in x. Derivatives or differences might be reported as elasticities. If x is continuous, the effect is measured continuously. If x is a factor, the effect is measured with respect to each level of the factor and may be calculated as a discrete difference or as a continuous change, as measured by the derivative. `margins` calculates the discrete difference by default and calculates the derivative if the `continuous` option is specified.

elasticity and **semielasticity**. The elasticity of y with respect to x is $d(\ln y)/d(\ln x) = (x/y) \times (dy/dx)$, which is approximately equal to the proportional change in y for a proportional change in x.

The semielasticity of y with respect to x is either 1) $dy/d(\ln x) = x \times (dy/dx)$ or 2) $d(\ln y)/dx = (1/y) \times (dy/dx)$, which is approximately 1) the change in y for a proportional change in x or 2) the proportional change in y for a change in x.

empty cell. An interaction of levels of two or more factor variables for which you have no data. For instance, you have sex interacted with group in your model, and in your data there are no females in group 1. Empty cells affect which margins can be estimated; see *Estimability of margins*.

estimability. Estimability concerns whether a margin can be uniquely estimated (identified); see *Estimability of margins*.

estimated marginal mean. This is one of the few terms that has the same meaning across authors. An estimated marginal mean is a margin assuming the levels of each factor covariate are equally likely (balanced), including interaction terms. This is obtained using `margins'` `asbalanced` option. In addition, there is an alternate definition of estimated marginal mean in which margins involving empty cells are redefined so that they become estimable. This is invoked by `margins'` `emptycells(reweight)` option. See *Balancing in the presence of empty cells*.

least-squares mean. Synonym for *estimated marginal mean*.

margin. A statistic calculated from predictions or other statistics of a previously fit model at fixed values of some covariates and averaging or otherwise integrating over the remaining covariates. The prediction or other statistic on which the margin is based is called the response.

If all the covariates are fixed, then the margin is called a conditional margin. If any covariates are left to vary, the margin is called a predictive margin.

In this documentation, we divide margins on the basis of whether the statistic is a response or a derivative of a response; see *Obtaining margins of responses* and *Obtaining margins of derivatives of responses*.

marginal effect and **average marginal effect**. The marginal effect of x is the *margin* of the *effect* of x. The term is popular with social scientists, and because of that, you might think the word marginal in marginal effect means derivative because of terms like marginal cost and marginal revenue. Marginal used in that way, however, refers to the derivative of revenue and the derivative of cost; it refers to the numerator, whereas marginal effect refers to the denominator. Moreover, *effect* is already a derivative or difference.

Some researchers interpret marginal in marginal effect to mean instantaneous, and thus a marginal effect is the instantaneous derivative rather than the discrete first-difference, corresponding to margins' continuous option. Researchers who use marginal in this way refer to the discrete difference calculation of an effect as a partial effect.

Other researchers define marginal effect to be the margin when all covariates are held fixed and the average marginal effect when some covariates are not fixed.

out-of-sample prediction. Predictions made in one dataset using the results from a model fit on another. Sample here refers to the sample on which the model was fit, and out-of-sample refers to the dataset on which the predictions are made.

partial effect and **average partial effect**. Some authors restrict the term *marginal effect* to mean derivatives and use the term partial effect to denote discrete differences; see *marginal effect and average marginal effect*.

population marginal mean. The theoretical (true) value that is estimated by *estimated marginal mean*. We avoid this term because it can be confused with the concept of a population in survey statistics, with which the population marginal mean has no connection.

posting results, **posting margins**. A Stata concept having to do with saving the results from the margins command in e() so that those results can be used as if they were estimation results, thus allowing the subsequent use of postestimation commands, such as test, testnl, lincom, and nlcom (see [R] **test**, [R] **testnl**, [R] **lincom**, and [R] **nlcom**). This is achieved by specifying margins' post option. See *Example 10: Testing margins—contrasts of margins*.

predictive margin. A *margin* in which all the covariates are not fixed. When all covariates are fixed, it is called a *conditional margin*.

recycled prediction. A synonym for *predictive margin*.

response. A prediction or other statistic derived from combining the parameter estimates of a fitted model with data or specified values on covariates. Derivatives of responses are themselves responses. Responses are what we take *margins* of.

standardized margin. The margin calculated on data different from the data used to fit the model. The term standardized is usually reserved for situations in which the alternate population is a reference population, which may be real or artificial, and which is treated as fixed.

subpopulation. A subset of your sample that represents a subset of the population, such as the males in a sample of people. In survey contexts when it is desired to account for sampling of the covariates, standard errors for marginal statistics and effects need to account for both the population and the subpopulation. This is accomplished by specifying the vce(unconditional) option and one of the subpop() or over() options. In fact, the above is allowed even when your data are not svyset because vce(unconditional) implies that the sample represents a population.

aved results

margins saves the following in r():

Scalars

r(N)	number of observations
r(N_sub)	subpopulation observations
r(N_clust)	number of clusters
r(N_psu)	number of samples PSUs, survey data only
r(N_strata)	number of strata, survey data only
r(df_r)	variance degrees of freedom, survey data only
r(N_poststrata)	number of post strata, survey data only
r(k_margins)	number of terms in *marginlist*
r(k_by)	number of subpopulations
r(k_at)	number of at() options
r(level)	confidence level of confidence intervals

Macros

r(cmd)	margins
r(cmdline)	command as typed
r(est_cmd)	e(cmd) from original estimation results
r(est_cmdline)	e(cmdline) from original estimation results
r(title)	title in output
r(subpop)	*subspec* from subpop()
r(model_vce)	*vcetype* from estimation command
r(model_vcetype)	Std. Err. title from estimation command
r(vce)	*vcetype* specified in vce()
r(vcetype)	title used to label Std. Err.
r(clustvar)	name of cluster variable
r(margins)	*marginlist*
r(predict_label)	label from predict()
r(expression)	response expression
r(xvars)	*varlist* from dydx(), dyex(), eydx(), or eyex()
r(derivatives)	"", "dy/dx", "dy/ex", "ey/dx", "ey/ex"
r(over)	*varlist* from over()
r(within)	*varlist* from within()
r(by)	union of r(over) and r(within) lists
r(by#)	interaction notation identifying the #th subpopulation
r(atstats#)	the #th at() specification
r(emptycells)	*empspec* from emptycells()
r(mcmethod)	*method* from mcompare()
r(mcadjustall)	adjustall or empty

Matrices

r(b)	estimates
r(V)	variance–covariance matrix of the estimates
r(Jacobian)	Jacobian matrix
r(_N)	sample size corresponding to each margin estimate
r(at)	matrix of values from the at() options
r(chainrule)	chainrule information from the fitted model
r(error)	margin estimability codes;
	0 means estimable,
	8 means not estimable
r(table)	matrix containing the margins with their standard errors, test statistics, *p*-values, and confidence intervals

margins with the post option also saves the following in e():

Scalars

e(N)	number of observations
e(N_sub)	subpopulation observations
e(N_clust)	number of clusters
e(N_psu)	number of samples PSUs, survey data only
e(N_strata)	number of strata, survey data only
e(df_r)	variance degrees of freedom, survey data only
e(N_poststrata)	number of post strata, survey data only
e(k_margins)	number of terms in *marginlist*
e(k_by)	number of subpopulations
e(k_at)	number of at() options

Macros

e(cmd)	margins
e(cmdline)	command as typed
e(est_cmd)	e(cmd) from original estimation results
e(est_cmdline)	e(cmdline) from original estimation results
e(title)	title in estimation output
e(subpop)	*subspec* from subpop()
e(model_vce)	*vcetype* from estimation command
e(model_vcetype)	Std. Err. title from estimation command
e(vce)	*vcetype* specified in vce()
e(vcetype)	title used to label Std. Err.
e(clustvar)	name of cluster variable
e(margins)	*marginlist*
e(predict_label)	label from predict()
e(expression)	prediction expression
e(xvars)	*varlist* from dydx(), dyex(), eydx(), or eyex()
e(derivatives)	" ", "dy/dx", "dy/ex", "ey/dx", "ey/ex"
e(over)	*varlist* from over()
e(within)	*varlist* from within()
e(by)	union of r(over) and r(within) lists
e(by#)	interaction notation identifying the #th subpopulation
e(atstats#)	the #th at() specification
e(emptycells)	*empspec* from emptycells()
e(mcmethod)	*method* from mcompare()
e(mcadjustall)	adjustall or empty

Matrices

e(b)	estimates
e(V)	variance–covariance matrix of the estimates
e(Jacobian)	Jacobian matrix
e(_N)	sample size corresponding to each margin estimate
e(at)	matrix of values from the at() options
e(chainrule)	chainrule information from the fitted model

Methods and formulas

margins is implemented as an ado-file.

Margins are statistics calculated from predictions of a previously fit model at fixed values of some covariates and averaging or otherwise integrating over the remaining covariates. There are many names for the different statistics that margins can compute: estimates marginal means (see Searle, Speed, and Milliken [1980]), predictive margins (see Graubard and Korn [2004]), marginal effects (see Greene [2012]), and average marginal/partial effects (see (Wooldridge 2010) and Bartus [2005]).

Methods and formulas are presented under the following headings:

Notation
Marginal effects
Fixing covariates and balancing factors
Estimable functions
Standard errors conditional on the covariates
Unconditional standard errors

Notation

Let θ be the vector of parameters in the current model fit, let $\mathbf{z}$ be a vector of covariate values, and let $f(\mathbf{z}, \theta)$ be a scalar-valued function returning the value of the predictions of interest. The following table illustrates the parameters and default prediction for several of Stata's estimation commands.

Command	θ	$\mathbf{z}$	$f(\mathbf{z}, \theta)$
regress	β	$\mathbf{x}$	$\mathbf{x}\beta$
cloglog	β	$\mathbf{x}$	$1 - e^{-e^{x\beta}}$
logit	β	$\mathbf{x}$	$1/(1 + e^{-x\beta})$
poisson	β	$\mathbf{x}$	$e^{x\beta}$
probit	β	$\mathbf{x}$	$\Phi(\mathbf{x}\beta)$
biprobit	β_1, β_2, ρ	$\mathbf{x}_1, \mathbf{x}_2$	$\Phi_2(\mathbf{x}_1\beta_1, \mathbf{x}_2\beta_2, \rho)$
mlogit	$\beta_1, \beta_2, \ldots, \beta_k$	$\mathbf{x}$	$e^{-x\beta_1}/(\sum_i e^{-x\beta_i})$
nbreg	$\beta, \ln\alpha$	$\mathbf{x}$	$e^{x\beta}$

$\Phi()$ and $\Phi_2()$ are cumulative distribution functions: $\Phi()$ for the standard normal distribution and $\Phi_2()$ for the standard bivariate normal distribution.

margins computes estimates of

$$p(\theta) = \frac{1}{M_{S_\mathrm{p}}} \sum_{j=1}^{M} \delta_j(S_\mathrm{p}) f(\mathbf{z}_j, \theta)$$

where $\delta_j(S_\mathrm{p})$ identifies elements within the subpopulation S_p (for the prediction of interest),

$$\delta_j(S_\mathrm{p}) = \begin{cases} 1, & j \in S_\mathrm{p} \\ 0, & j \notin S_\mathrm{p} \end{cases}$$

M_{S_p} is the subpopulation size,

$$M_{S_\mathrm{p}} = \sum_{j=1}^{M} \delta_j(S_\mathrm{p})$$

and M is the population size.

Let $\widehat{\theta}$ be the vector of parameter estimates. Then margins estimates $p(\theta)$ via

$$\widehat{p} = \frac{1}{w.} \sum_{j=1}^{N} \delta_j(S_\mathrm{p}) w_j f(\mathbf{z}_j, \widehat{\theta})$$

where

$$w. = \sum_{j=1}^{N} \delta_j(S_\mathrm{p})w_j$$

$\delta_j(S_\mathrm{p})$ indicates whether observation j is in subpopulation S_p, w_j is the weight for the jth observation, and N is the sample size.

Marginal effects

margins also computes marginal/partial effects. For the marginal effect of continuous covariate x, margins computes

$$\widehat{p} = \frac{1}{w.} \sum_{j=1}^{N} \delta_j(S_\mathrm{p})w_j h(\mathbf{z}_j, \widehat{\boldsymbol{\theta}})$$

where

$$h(\mathbf{z}, \boldsymbol{\theta}) = \frac{\partial f(\mathbf{z}, \boldsymbol{\theta})}{\partial x}$$

The marginal effect for level k of factor variable A is the simple contrast (a.k.a. difference) comparing its margin with the margin at the base level.

$$h(\mathbf{z}, \boldsymbol{\theta}) = f(\mathbf{z}, \boldsymbol{\theta}|A = k) - f(\mathbf{z}, \boldsymbol{\theta}|A = \mathrm{base})$$

Fixing covariates and balancing factors

margins controls the values in each $\mathbf{z}$ vector through the *marginlist*, the at() option, the atmeans option, and the asbalanced and emptycells() options. Suppose $\mathbf{z}$ is composed of the elements from the equation specification

A##B x

where A is a factor variable with a levels, B is a factor variable with b levels, and x is a continuous covariate. To simplify the notation for this discussion, assume the levels of A and B start with 1 and are contiguous. Then

$$\mathbf{z} = (A_1, \ldots, A_a, B_1, \ldots, B_b, A_1 B_1, A_1 B_2, \ldots, A_a B_b, x, 1)$$

where A_i, B_j, and $A_i B_j$ represent the indicator values for the factor variables A and B and the interaction A#B.

When factor A is in the *marginlist*, margins replaces A with i and then computes the mean of the subsequent prediction, for $i = 1, \ldots, a$. When the interaction term A#B is in the *marginlist*, margins replaces A with i and B with j, and then computes the mean of the subsequent prediction, for all combinations of $i = 1, \ldots, a$ and $j = 1, \ldots, b$.

The at() option sets model covariates to fixed values. For example, at(x=15) causes margins to temporarily set x to 15 for each observation in the dataset before computing any predictions. Similarly, at((median) x) causes margins to temporarily set x to the median of x using the current dataset.

When factor variable A is specified as asbalanced, margins sets each A_i to $1/a$. Thus each z vector will look like

$$\mathbf{z} = (1/a, \ldots, 1/a, B_1, \ldots, B_b, B_1/a, B_2/a, \ldots, B_b/a, x, 1)$$

If B is also specified as asbalanced, then each B_j is set to $1/b$, and each z vector will look like

$$\mathbf{z} = (1/a, \ldots, 1/a, 1/b, \ldots, 1/b, 1/ab, 1/ab, \ldots, 1/ab, x, 1)$$

If emptycells(reweight) is also specified, then margins uses a different balancing weight for each element of z, depending on how many empty cells the element is associated with. Let δ_{ij} indicate that the ijth cell of A#B was observed in the estimation sample.

$$\delta_{ij} = \begin{cases} 0, & \text{A} = i \text{ and B} = j \text{ was an empty cell} \\ 1, & \text{otherwise} \end{cases}$$

For the grand margin, the affected elements of z and their corresponding balancing weights are

$$A_i = \frac{\sum_j \delta_{ij}}{\sum_k \sum_j \delta_{kj}}$$

$$B_j = \frac{\sum_i \delta_{ij}}{\sum_i \sum_k \delta_{ik}}$$

$$A_i B_j = \frac{\delta_{ij}}{\sum_k \sum_l \delta_{kl}}$$

For the jth margin of B, the affected elements of z and their corresponding balancing weights are

$$A_i = \frac{\delta_{ij}}{\sum_k \delta_{kj}}$$

$$B_l = \begin{cases} 1, & \text{if } l = j \text{ and not all } \delta_{ij} \text{ are zero} \\ 0, & \text{otherwise} \end{cases}$$

$$A_i B_l = \frac{\delta_{il}}{\sum_k \delta_{kl}} B_l$$

Estimable functions

The fundamental idea behind estimable functions is clearly defined in the statistical literature for linear models; see Searle (1971). Assume that we are working with the following linear model:

$$\mathbf{y} = \mathbf{Xb} + \mathbf{e}$$

where $\mathbf{y}$ is an $N \times 1$ vector of responses, $\mathbf{X}$ is an $N \times p$ matrix of covariate values, $\mathbf{b}$ is a $p \times 1$ vector of coefficients, and $\mathbf{e}$ is a vector of random errors. Assuming a constant variance for the random errors, the normal equations for the least-squares estimator, $\widehat{\mathbf{b}}$, are

$$\mathbf{X}'\mathbf{X}\widehat{\mathbf{b}} = \mathbf{X}'\mathbf{y}$$

When $\mathbf{X}$ is not of full column rank, we will need a generalized inverse (g-inverse) of $\mathbf{X}'\mathbf{X}$ to solve for $\widehat{\mathbf{b}}$. Let $\mathbf{G}$ be a g-inverse of $\mathbf{X}'\mathbf{X}$.

Searle (1971) defines a linear function of the parameters as *estimable* if it is identically equal to some linear function of the expected values of the $\mathbf{y}$ vector. Let $\mathbf{H} = \mathbf{GX}'\mathbf{X}$. Then this definition simplifies to the following rule:

$$\mathbf{zb} \text{ is estimable if } \mathbf{z} = \mathbf{zH}$$

margins generalizes this to nonlinear functions by assuming the prediction function $f(\mathbf{z}, \boldsymbol{\theta})$ is a function of one or more of the linear predictions from the equations in the model that $\boldsymbol{\theta}$ represents.

$$f(\mathbf{z}, \boldsymbol{\theta}) = h(\mathbf{z}_1\boldsymbol{\beta}_1, \mathbf{z}_2\boldsymbol{\beta}_2, \ldots, \mathbf{z}_k\boldsymbol{\beta}_k)$$

$\mathbf{z}_i\boldsymbol{\beta}_i$ is considered estimable if $\mathbf{z}_i = \mathbf{z}_i\mathbf{H}_i$, where $\mathbf{H}_i = \mathbf{G}_i\mathbf{X}_i'\mathbf{X}_i$, $\mathbf{G}_i$ is a g-inverse for $\mathbf{X}_i'\mathbf{X}_i$, and $\mathbf{X}_i$ is the matrix of covariates from the ith equation of the fitted model. margins considers $p(\boldsymbol{\theta})$ to be estimable if every $\mathbf{z}_i\boldsymbol{\beta}_i$ is estimable.

Standard errors conditional on the covariates

By default, margins uses the delta method to estimate the variance of $\widehat{p}$.

$$\widehat{\mathrm{Var}}(\widehat{p} \mid \mathbf{z}) = \mathbf{v}'\mathbf{V}\mathbf{v}$$

where $\mathbf{V}$ is a variance estimate for $\widehat{\boldsymbol{\theta}}$ and

$$\mathbf{v} = \left.\frac{\partial \widehat{p}}{\partial \boldsymbol{\theta}}\right|_{\boldsymbol{\theta} = \widehat{\boldsymbol{\theta}}}$$

This variance estimate is conditional on the $\mathbf{z}$ vectors used to compute the marginalized predictions.

nconditional standard errors

`margins` with the `vce(unconditional)` option uses linearization to estimate the unconditional variance of $\widehat{\boldsymbol{\theta}}$. Linearization uses the variance estimator for the total of a score variable for $\widehat{p}$ as an approximate estimator for $\text{Var}(\widehat{p})$; see [SVY] **variance estimation**. `margins` requires that the model was fit using some form of linearized variance estimator and that `predict, scores` computes the appropriate score values for the linearized variance estimator.

The score for $\widehat{p}$ from the jth observation is given by

$$s_j = \frac{\partial \widehat{p}}{\partial w_j} = -\frac{\delta_j(S_{\mathrm{p}})}{w.}\widehat{p} + \frac{\delta_j(S_{\mathrm{p}})}{w.}f(\mathbf{z}_j, \widehat{\boldsymbol{\theta}}) + \frac{1}{w.}\sum_{i=1}^{N}\delta_i(S_{\mathrm{p}})w_i\frac{\partial f(\mathbf{z}_i, \widehat{\boldsymbol{\theta}})}{\partial w_j}$$

The remaining partial derivative can be decomposed using the chain rule.

$$\frac{\partial f(\mathbf{z}_i, \widehat{\boldsymbol{\theta}})}{\partial w_j} = \left(\left.\frac{\partial f(\mathbf{z}_i, \boldsymbol{\theta})}{\partial \boldsymbol{\theta}}\right|_{\theta=\widehat{\theta}}\right)\left(\frac{\partial \widehat{\boldsymbol{\theta}}}{\partial w_j}\right)'$$

This is the inner product of two vectors, the second of which is not a function of the i index. Thus the score is

$$s_j = -\frac{\delta_j(S_{\mathrm{p}})}{w.}\widehat{p} + \frac{\delta_j(S_{\mathrm{p}})}{w.}f(\mathbf{z}_j, \widehat{\boldsymbol{\theta}}) + \left(\left.\frac{\partial \widehat{p}}{\partial \boldsymbol{\theta}}\right|_{\theta=\widehat{\theta}}\right)\left(\frac{\partial \widehat{\boldsymbol{\theta}}}{\partial w_j}\right)'$$

If $\widehat{\boldsymbol{\theta}}$ was derived from a system of equations (such as in linear regression or maximum likelihood estimation), then $\widehat{\boldsymbol{\theta}}$ is the solution to

$$\mathbf{G}(\boldsymbol{\theta}) = \sum_{j=1}^{N}\delta_j(S_{\mathrm{m}})w_j\mathbf{g}(\boldsymbol{\theta}, \mathbf{y}_j, \mathbf{x}_j) = \mathbf{0}$$

where S_{m} identifies the subpopulation used to fit the model, $\mathbf{g}()$ is the model's gradient function, and $\mathbf{y}_j$ and $\mathbf{x}_j$ are the values of the dependent and independent variables for the jth observation. We can use linearization to derive a first-order approximation for $\partial\widehat{\boldsymbol{\theta}}/\partial w_j$.

$$\mathbf{G}(\widehat{\boldsymbol{\theta}}) \approx \mathbf{G}(\boldsymbol{\theta}_0) + \left.\frac{\partial \mathbf{G}(\boldsymbol{\theta})}{\partial \boldsymbol{\theta}}\right|_{\theta=\theta_0}(\widehat{\boldsymbol{\theta}} - \boldsymbol{\theta}_0)$$

Let $\mathbf{H}$ be the Hessian matrix

$$\mathbf{H} = \left.\frac{\partial \mathbf{G}(\boldsymbol{\theta})}{\partial \boldsymbol{\theta}}\right|_{\theta=\theta_0}$$

Then

$$\widehat{\boldsymbol{\theta}} \approx \boldsymbol{\theta}_0 + (-\mathbf{H})^{-1}\mathbf{G}(\boldsymbol{\theta}_0)$$

and

$$\frac{\partial \widehat{\boldsymbol{\theta}}}{\partial w_j} \approx (-\mathbf{H})^{-1} \left. \frac{\partial \mathbf{G}(\boldsymbol{\theta})}{\partial w_j} \right|_{\theta=\widehat{\theta}} = (-\mathbf{H})^{-1} \delta_j(S_\mathrm{m}) \mathbf{g}(\widehat{\boldsymbol{\theta}}, \mathbf{y}_j, \mathbf{x}_j)$$

The computed value of the score for $\widehat{p}$ for the jth observation is

$$s_j = \mathbf{v}' \mathbf{u}_j$$

where

$$\mathbf{v} = \begin{bmatrix} -\frac{\widehat{p}}{w.} \\ \frac{1}{w.} \\ \frac{\partial \widehat{p}}{\partial \boldsymbol{\theta}}(-\mathbf{H})^{-1} \end{bmatrix}$$

and

$$\mathbf{u}_j = \begin{bmatrix} \delta_j(S_\mathrm{p}) \\ \delta_j(S_\mathrm{p}) f(\mathbf{z}_j, \widehat{\boldsymbol{\theta}}) \\ \delta_j(S_\mathrm{m}) \mathbf{g}(\widehat{\boldsymbol{\theta}}, \mathbf{y}_j, \mathbf{x}_j) \end{bmatrix}$$

Thus the variance estimate for $\widehat{p}$ is

$$\widehat{\mathrm{Var}}(\widehat{p}) = \mathbf{v}' \widehat{\mathrm{Var}}(\widehat{\mathbf{U}}) \mathbf{v}$$

where

$$\widehat{\mathbf{U}} = \sum_{j=1}^{N} w_j \mathbf{u}_j$$

margins uses the model-based variance estimates for $(-\mathbf{H})^{-1}$ and the scores from predict for $\mathbf{g}(\widehat{\boldsymbol{\theta}}, \mathbf{y}_j, \mathbf{x}_j)$.

References

Bartus, T. 2005. Estimation of marginal effects using margeff. *Stata Journal* 5: 309–329.

Baum, C. F. 2010. Stata tip 88: Efficiently evaluating elasticities with the margins command. *Stata Journal* 10: 309–312.

Buis, M. L. 2010. Stata tip 87: Interpretation of interactions in nonlinear models. *Stata Journal* 10: 305–308.

Chang, I. M., R. Gelman, and M. Pagano. 1982. Corrected group prognostic curves and summary statistics. *Journal of Chronic Diseases* 35: 669–674.

Gould, W. W. 1996. crc43: Wald test of nonlinear hypotheses after model estimation. *Stata Technical Bulletin* 29: 2–4. Reprinted in *Stata Technical Bulletin Reprints*, vol. 5, pp. 15–18. College Station, TX: Stata Press.

Graubard, B. I., and E. L. Korn. 2004. Predictive margins with survey data. *Biometrics* 55: 652–659.

Greene, W. H. 2012. *Econometric Analysis*. 7th ed. Upper Saddle River, NJ: Prentice Hall.

Korn, E. L., and B. I. Graubard. 1999. *Analysis of Health Surveys.* New York: Wiley.

Lane, P. W., and J. A. Nelder. 1982. Analysis of covariance and standardization as instances of prediction. *Biometrics* 38: 613–621.

Phillips, P. C. B., and J. Y. Park. 1988. On the formulation of Wald tests of nonlinear restrictions. *Econometrica* 56: 1065–1083.

Searle, S. R. 1971. *Linear Models.* New York: Wiley.

——. 1997. *Linear Models for Unbalanced Data.* New York: Wiley.

Searle, S. R., F. M. Speed, and G. A. Milliken. 1980. Population marginal means in the linear model: An alternative to least squares means. *American Statistician* 34: 216–221.

Wooldridge, J. M. 2010. *Econometric Analysis of Cross Section and Panel Data.* 2nd ed. Cambridge, MA: MIT Press.

lso see

[R] **contrast** — Contrasts and linear hypothesis tests after estimation

[R] **margins, contrast** — Contrasts of margins

[R] **margins, pwcompare** — Pairwise comparisons of margins

[R] **margins postestimation** — Postestimation tools for margins

[R] **marginsplot** — Graph results from margins (profile plots, etc.)

[R] **lincom** — Linear combinations of estimators

[R] **nlcom** — Nonlinear combinations of estimators

[R] **predict** — Obtain predictions, residuals, etc., after estimation

[R] **predictnl** — Obtain nonlinear predictions, standard errors, etc., after estimation

[U] **20 Estimation and postestimation commands**

Title

margins postestimation — Postestimation tools for margins

Description

The following standard postestimation command is available after `margins`:

Command	Description
marginsplot	graph the results from margins—profile plots, interaction plots, etc.

For information on `marginsplot`, see [R] **marginsplot**.

The following standard postestimation commands are available after `margins, post`:

Command	Description
contrast	contrasts and ANOVA-style joint tests of estimates
estat	estimation sample summary; `estat summarize` only
estimates	cataloging estimation results
lincom	point estimates, standard errors, testing, and inference for linear combinations of coefficients
nlcom	point estimates, standard errors, testing, and inference for nonlinear combinations of coefficients
pwcompare	pairwise comparisons of estimates
test	Wald tests of simple and composite linear hypotheses
testnl	Wald tests of nonlinear hypotheses

Remarks

Continuing with the example from *Example 8: Margins of interactions* in [R] **margins**, we use the dataset and reestimate the logistic model of `outcome`:

```
. use http://www.stata-press.com/data/r12/margex
(Artificial data for margins)
. logistic outcome sex##group age
  (output omitted )
```

We then estimate the margins for males and females and post the margins as estimation results with a full VCE.

```
. margins sex, post
Predictive margins                              Number of obs   =      3000
Model VCE   : OIM
Expression  : Pr(outcome), predict()
```

	Margin	Delta-method Std. Err.	z	P>|z|	[95% Conf. Interval]
sex					
0	.1600644	.0125653	12.74	0.000	.1354368 .184692
1	.1966902	.0100043	19.66	0.000	.1770821 .2162983

We can now use nlcom (see [R] **nlcom**) to estimate a risk ratio of females to males using the average probabilities for females and males posted by margins:

```
. nlcom (risk_ratio: _b[1.sex] / _b[0.sex])
   risk_ratio:  _b[1.sex] / _b[0.sex]
```

	Coef.	Std. Err.	z	P>|z|	[95% Conf. Interval]
risk_ratio	1.228819	.1149538	10.69	0.000	1.003514 1.454124

We could similarly estimate the average risk difference between females and males:

```
. nlcom (risk_diff: _b[1.sex] - _b[0.sex])
   risk_diff:  _b[1.sex] - _b[0.sex]
```

	Coef.	Std. Err.	z	P>|z|	[95% Conf. Interval]
risk_diff	.0366258	.0160632	2.28	0.023	.0051425 .068109

Also see

[R] **margins** — Marginal means, predictive margins, and marginal effects

[U] **20 Estimation and postestimation commands**

Title

> **margins, contrast** — Contrasts of margins

Syntax

> margins [*marginlist*] [*if*] [*in*] [*weight*] [, contrast *margins_options*]
>
> margins [*marginlist*] [*if*] [*in*] [*weight*] [, contrast(*suboptions*) *margins_options*]

where *marginlist* is a list of factor variables or interactions that appear in the current estimation results. The variables may be typed with or without contrast operators, and you may use any factor-variable syntax:

> . margins sex##group, contrast
>
> . margins sex##g.group, contrast
>
> . margins sex@group, contrast

See the operators (*op.*) table in [R] **contrast** for the list of contrast operators. Contrast operators may also be specified on the variables in margins' over() and within() options to perform contrasts across the levels of those variables.

See [R] **margins** for the available *margins_options*.

suboptions	Description
Contrast	
overall	add a joint hypothesis test for all specified contrasts
lincom	treat user-defined contrasts as linear combinations
atcontrast(*op*[._at])	apply the *op.* contrast operator to the groups defined by at()
atjoint	test jointly across all groups defined by at()
overjoint	test jointly across all levels of the unoperated over() variables
withinjoint	test jointly across all levels of the unoperated within() variables
marginswithin	perform contrasts within the levels of the unoperated terms in *marginlist*
cieffects	show effects table with confidence intervals
pveffects	show effects table with *p*-values
effects	show effects table with confidence intervals and *p*-values
nowald	suppress table of Wald tests
noatlevels	report only the overall Wald test for terms that use the within @ or nested	operator
nosvyadjust	compute unadjusted Wald tests for survey results

fweights, aweights, iweights, and pweights are allowed; see [U] **11.1.6 weight**.

Menu

Statistics > Postestimation > Contrasts of margins

Description

margins with the contrast option or with contrast operators performs contrasts of margins. This extends the capabilities of contrast to any of the nonlinear responses, predictive margins, or other margins that can be estimated by margins.

Suboptions

Contrast

overall specifies that a joint hypothesis test over all terms be performed.

lincom specifies that user-defined contrasts be treated as linear combinations. The default is to require that all user-defined contrasts sum to zero. (Summing to zero is part of the definition of a contrast.)

atcontrast($op[\ ._at\ ]$) specifies that the $op.$ contrast operator be applied to the groups defined by the at() option(s). The default behavior, by comparison, is to perform tests and contrasts within the groups defined by the at() option(s).

See example 6 in *Remarks*.

atjoint specifies that joint tests be performed across all groups defined by the at() option. The default behavior, by comparison, is to perform contrasts and tests within each group.

See example 5 in *Remarks*.

overjoint specifies how unoperated variables in the over() option are treated.

Each variable in the over() option may be specified either with or without a contrast operator. For contrast-operated variables, the specified contrast comparisons are always performed.

overjoint specifies that joint tests be performed across all levels of the unoperated variables. The default behavior, by comparison, is to perform contrasts and tests within each combination of levels of the unoperated variables.

See example 3 in *Remarks*.

withinjoint specifies how unoperated variables in the within() option are treated.

Each variable in the within() option may be specified either with or without a contrast operator. For contrast-operated variables, the specified contrast comparisons are always performed.

withinjoint specifies that joint tests be performed across all levels of the unoperated variables. The default behavior, by comparison, is to perform contrasts and tests within each combination of levels of the unoperated variables.

marginswithin specifies how unoperated variables in *marginlist* are treated.

Each variable in *marginlist* may be specified either with or without a contrast operator. For contrast-operated variables, the specified contrast comparisons are always performed.

marginswithin specifies that contrasts and tests be performed within each combination of levels of the unoperated variables. The default behavior, by comparison, is to perform joint tests across all levels of the unoperated variables.

See example 4 in *Remarks*.

cieffects specifies that a table containing a confidence interval for each individual contrast be reported.

pveffects specifies that a table containing a p-value for each individual contrast be reported.

effects specifies that a single table containing a confidence interval and p-value for each individual contrast be reported.

nowald suppresses the table of Wald tests.

noatlevels indicates that only the overall Wald test be reported for each term containing within or nested (@ or |) operators.

nosvyadjust is for use with svy estimation commands. It specifies that the Wald test be carried out without the default adjustment for the design degrees of freedom. That is to say the test is carried out as $W/k \sim F(k, d)$ rather than as $(d - k + 1)W/(kd) \sim F(k, d - k + 1)$, where k is the dimension of the test and d is the total number of sampled PSUs minus the total number of strata.

Remarks

Remarks are presented under the following headings:

> *Contrasts of margins*
> *Contrasts and the over() option*
> *The overjoint suboption*
> *The marginswithin suboption*
> *Contrasts and the at() option*
> *Conclusion*

Contrasts of margins

▷ Example 1

Estimating contrasts of margins is as easy as adding a contrast operator to the variable name. Let's review *Example 2: A simple case after logistic* of [R] **margins**. Variable sex is coded 0 for males and 1 for females.

```
. use http://www.stata-press.com/data/r12/margex
. logistic outcome i.sex i.group
  (output omitted)
. margins sex
```

Predictive margins   Number of obs   =   3000
Model VCE   : OIM

Expression   : Pr(outcome), predict()

		Delta-method			
	Margin	Std. Err.	z	P>|z|	[95% Conf. Interval]
sex					
0	.1286796	.0111424	11.55	0.000	.106841 .1505182
1	.1905087	.0089719	21.23	0.000	.1729241 .2080933

The first margin, 0.13, is the average probability of a positive outcome, treating everyone as if they were male. The second margin, 0.19, is the average probability of a positive outcome, treating everyone as if they were female. We can compare females with males by rerunning margins and adding a contrast operator:

```
. margins r.sex
```
Contrasts of predictive margins
Model VCE   : OIM
Expression   : Pr(outcome), predict()

	df	chi2	P>chi2
sex	1	16.61	0.0000

	Contrast	Delta-method Std. Err.	[95% Conf. Interval]	
sex (1 vs 0)	.0618291	.0151719	.0320927	.0915656

The `r.` prefix for `sex` is the reference-category contrast operator—see [R] **contrast**. (The default reference category is zero, the lowest value of `sex`.) Contrast operators in a *marginlist* work just as they do in the *termlist* of a `contrast` command.

The contrast estimate of 0.06 says that unconditional on group, females on average are about 6% more likely than males to have a positive outcome. The chi-squared statistic of 16.61 shows that the contrast is significantly different from zero.

You may be surprised that we did not need to include the `contrast` option to estimate our contrast. If we had included the option, our output would not have changed:

```
. margins r.sex, contrast
```
Contrasts of predictive margins
Model VCE   : OIM
Expression   : Pr(outcome), predict()

	df	chi2	P>chi2
sex	1	16.61	0.0000

	Contrast	Delta-method Std. Err.	[95% Conf. Interval]	
sex (1 vs 0)	.0618291	.0151719	.0320927	.0915656

The `contrast` option is useful mostly for its suboptions, which control the output and how contrasts are estimated in more complicated situations. But `contrast` may be specified on its own (without contrast operators or suboptions) if we do not need estimates or confidence intervals:

```
. margins sex group, contrast
```
Contrasts of predictive margins
Model VCE   : OIM
Expression   : Pr(outcome), predict()

	df	chi2	P>chi2
sex	1	16.61	0.0000
group	2	225.76	0.0000

Each chi-squared statistic is a joint test of constituent contrasts. The test for group has two degrees of freedom because group has three levels.

◁

Contrasts and the over() option

▷ Example 2

It is common to estimate margins at combinations of factor levels, and margins, contrast includes several suboptions for contrasting such margins. Let's fit a model with two categorical predictors and their interaction:

```
. logistic outcome group##agegroup
```

Logistic regression

Log likelihood = -1105.7504

Number of obs	=	3000
LR chi2(8)	=	520.64
Prob > chi2	=	0.0000
Pseudo R2	=	0.1906

outcome	Odds Ratio	Std. Err.	z	P>|z|	[95% Conf. Interval]	
group						
2	.834507	.5663738	-0.27	0.790	.2206611	3.15598
3	.2146729	.1772897	-1.86	0.062	.0425407	1.083303
agegroup						
2	3.54191	2.226951	2.01	0.044	1.032882	12.14576
3	16.23351	9.61188	4.71	0.000	5.086452	51.80955
group# agegroup						
2 2	.4426927	.3358505	-1.07	0.283	.1000772	1.958257
2 3	.440672	.3049393	-1.18	0.236	.1135259	1.71055
3 2	1.160885	1.103527	0.16	0.875	.1801543	7.480553
3 3	.4407912	.4034688	-0.89	0.371	.0733	2.650709
_cons	.0379747	.0223371	-5.56	0.000	.0119897	.1202762

Each of group and agegroup has three levels. To compare each age group with the reference category on the probability scale, we can again use margins with the r. contrast operator.

```
. margins r.agegroup
```

Contrasts of predictive margins
Model VCE   : OIM

Expression   : Pr(outcome), predict()

	df	chi2	P>chi2
agegroup			
(2 vs 1)	1	10.04	0.0015
(3 vs 1)	1	224.44	0.0000
Joint	2	238.21	0.0000

	Contrast	Delta-method Std. Err.	[95% Conf. Interval]	
agegroup				
(2 vs 1)	.044498	.0140448	.0169706	.0720253
(3 vs 1)	.2059281	.0137455	.1789874	.2328688

Our model includes an interaction, though, so it would be nice to estimate the contrasts separately for each value of group. We need the over() option:

```
. margins r.agegroup, over(group)
```

Contrasts of predictive margins
Model VCE   : OIM

Expression   : Pr(outcome), predict()
over   : group

	df	chi2	P>chi2
agegroup@group			
(2 vs 1) 1	1	6.94	0.0084
(2 vs 1) 2	1	1.18	0.2783
(2 vs 1) 3	1	3.10	0.0783
(3 vs 1) 1	1	173.42	0.0000
(3 vs 1) 2	1	57.77	0.0000
(3 vs 1) 3	1	5.12	0.0236
Joint	6	266.84	0.0000

	Contrast	Delta-method Std. Err.	[95% Conf. Interval]	
agegroup@group				
(2 vs 1) 1	.0819713	.0311208	.0209757	.142967
(2 vs 1) 2	.0166206	.0153309	-.0134275	.0466686
(2 vs 1) 3	.0243462	.0138291	-.0027583	.0514508
(3 vs 1) 1	.3447797	.0261811	.2934658	.3960937
(3 vs 1) 2	.1540882	.0202722	.1143554	.193821
(3 vs 1) 3	.0470319	.0207774	.006309	.0877548

The effect of agegroup appears to be greatest for the first level of group.

Including a variable in the over() option is not equivalent to including the variable in the main *marginlist*. The variables in the *marginlist* are manipulated in the analysis, so that we can measure, for example, the effect of being in age group 3 and not age group 1. (The manipulation could be mimicked by running replace and then predict, but the manipulations actually performed by margins do not change the data in memory.) The variables in the over() option are not so manipulated—the values of the over() variables are left as they were observed, and the *marginlist* variables are manipulated separately for each observed over() group. For more information, see *Do not specify marginlist when you mean over()* in [R] **margins**.

◁

The overjoint suboption

▷ Example 3

Each variable in an over() option may be specified with or without contrast operators. Our option over(group) did not include a contrast operator, so margins estimated the contrasts separately for each level of group. If we had instead specified over(r.group), we would have received differences of the contrasts:

```
. margins r.agegroup, over(r.group)
Contrasts of predictive margins
Model VCE     : OIM
Expression    : Pr(outcome), predict()
over          : group
```

	df	chi2	P>chi2
group#agegroup			
(2 vs 1) (2 vs 1)	1	3.55	0.0596
(2 vs 1) (3 vs 1)	1	33.17	0.0000
(3 vs 1) (2 vs 1)	1	2.86	0.0906
(3 vs 1) (3 vs 1)	1	79.36	0.0000
Joint	4	83.88	0.0000

	Contrast	Delta-method Std. Err.	[95% Conf. Interval]	
group#agegroup				
(2 vs 1) (2 vs 1)	-.0653508	.0346921	-.133346	.0026445
(2 vs 1) (3 vs 1)	-.1906915	.0331121	-.25559	-.1257931
(3 vs 1) (2 vs 1)	-.0576251	.0340551	-.1243719	.0091216
(3 vs 1) (3 vs 1)	-.2977479	.0334237	-.3632572	-.2322385

The contrasts are double differences: the estimate of -0.19, for example, says that the difference in the probability of success between age group 3 and age group 1 is smaller in group 2 than in group 1. We can jointly test pairs of the double differences with the overjoint suboption:

```
. margins r.agegroup, over(group) contrast(overjoint)
Contrasts of predictive margins
Model VCE     : OIM
Expression    : Pr(outcome), predict()
over          : group
```

	df	chi2	P>chi2
group#agegroup			
(joint) (2 vs 1)	2	3.62	0.1641
(joint) (3 vs 1)	2	79.45	0.0000
Joint	4	83.88	0.0000

The contrast(overjoint) option overrides the default behavior of over() and requests joint tests over the levels of the unoperated variable group. The chi-squared statistic of 3.62 tests that the first and third contrasts from the previous table are jointly zero. The chi-squared statistic of 79.45 jointly tests the other pair of contrasts.

◁

The marginswithin suboption

Example 4

Another suboption that may usefully be combined with over() is marginswithin. margins-within requests that contrasts be performed within the levels of unoperated variables in the main *marginlist*, instead of performing them jointly across the levels. marginswithin affects only unop-erated variables because contrast operators take precedence over suboptions.

Let's first look at the default behavior, which occurs when marginswithin is not specified:

```
. margins agegroup, over(r.group) contrast(effects)
```

Contrasts of predictive margins
Model VCE   : OIM

Expression   : Pr(outcome), predict()
over   : group

	df	chi2	P>chi2
group#agegroup			
(2 vs 1) (joint)	2	33.94	0.0000
(3 vs 1) (joint)	2	83.38	0.0000
Joint	4	83.88	0.0000

	Contrast	Delta-method Std. Err.	z	P>|z|	[95% Conf. Interval]
group# agegroup (2 vs 1) (2 vs base)	-.0653508	.0346921	-1.88	0.060	-.133346	.0026445
(2 vs 1) (3 vs base)	-.1906915	.0331121	-5.76	0.000	-.25559	-.1257931
(3 vs 1) (2 vs base)	-.0576251	.0340551	-1.69	0.091	-.1243719	.0091216
(3 vs 1) (3 vs base)	-.2977479	.0334237	-8.91	0.000	-.3632572	-.2322385

Here agegroup in the main *marginlist* is an unoperated variable, so margins by default performs joint tests across the levels of agegroup: the chi-squared statistic of 33.94, for example, jointly tests whether the first two contrast estimates in the lower table differ significantly from zero.

When we specify `marginswithin`, the contrasts will instead be performed within the levels of agegroup:

```
. margins agegroup, over(r.group) contrast(marginswithin effects)
Contrasts of predictive margins
Model VCE    : OIM

Expression   : Pr(outcome), predict()
over         : group
```

	df	chi2	P>chi2
group@agegroup			
(2 vs 1) 1	1	0.06	0.7991
(2 vs 1) 2	1	7.55	0.0060
(2 vs 1) 3	1	68.39	0.0000
(3 vs 1) 1	1	1.80	0.1798
(3 vs 1) 2	1	10.47	0.0012
(3 vs 1) 3	1	159.89	0.0000
Joint	6	186.87	0.0000

	Contrast	Delta-method Std. Err.	z	P>|z|	[95% Conf. Interval]	
group@ agegroup						
(2 vs 1) 1	-.0058686	.0230533	-0.25	0.799	-.0510523	.039315
(2 vs 1) 2	-.0712194	.0259246	-2.75	0.006	-.1220308	-.0204081
(2 vs 1) 3	-.1965602	.0237688	-8.27	0.000	-.2431461	-.1499742
(3 vs 1) 1	-.0284991	.0212476	-1.34	0.180	-.0701436	.0131453
(3 vs 1) 2	-.0861243	.0266137	-3.24	0.001	-.1382862	-.0339624
(3 vs 1) 3	-.326247	.0258009	-12.64	0.000	-.3768159	-.2756781

The joint tests in the top table have been replaced by one-degree-of-freedom tests, one for each combination of the two reference comparisons and three levels of agegroup. The reference-category contrasts for group have been performed within levels of agegroup.

◁

Contrasts and the at() option

▷ Example 5

The at() option of margins is used to set predictors to particular values. When at() is used, contrasts are by default performed within each at() level:

```
. margins r.agegroup, at(group=(1/3))
```

Contrasts of adjusted predictions
Model VCE   : OIM

Expression   : Pr(outcome), predict()

1._at	: group	=	1
2._at	: group	=	2
3._at	: group	=	3

	df	chi2	P>chi2
agegroup@_at			
(2 vs 1) 1	1	6.94	0.0084
(2 vs 1) 2	1	1.18	0.2783
(2 vs 1) 3	1	3.10	0.0783
(3 vs 1) 1	1	173.42	0.0000
(3 vs 1) 2	1	57.77	0.0000
(3 vs 1) 3	1	5.12	0.0236
Joint	6	266.84	0.0000

	Contrast	Delta-method Std. Err.	[95% Conf. Interval]	
agegroup@_at				
(2 vs 1) 1	.0819713	.0311208	.0209757	.142967
(2 vs 1) 2	.0166206	.0153309	-.0134275	.0466686
(2 vs 1) 3	.0243462	.0138291	-.0027583	.0514508
(3 vs 1) 1	.3447797	.0261811	.2934658	.3960937
(3 vs 1) 2	.1540882	.0202722	.1143554	.193821
(3 vs 1) 3	.0470319	.0207774	.006309	.0877548

Our option at(group=(1/3)) manipulates the values of group and is therefore not equivalent to over(group). We see that the reference-category contrasts for agegroup have been performed within each at() level. For a similar example that uses the ._at operator instead of the at() option, see *Contrasts of at() groups—discrete effects* in [R] **marginsplot**.

The default within behavior of at() may be changed to joint behavior with the atjoint suboption:

```
. margins r.agegroup, at(group=(1/3)) contrast(atjoint)
```

Contrasts of adjusted predictions
Model VCE   : OIM

Expression   : Pr(outcome), predict()

1._at	: group	=	1
2._at	: group	=	2
3._at	: group	=	3

	df	chi2	P>chi2
_at#agegroup			
(joint) (2 vs 1)	2	3.62	0.1641
(joint) (3 vs 1)	2	79.45	0.0000
Joint	4	83.88	0.0000

Now the tests are performed jointly over the levels of group, the at() variable. The atjoint suboption is the analogue for at() of the overjoint suboption from example 3.

◁

▷ Example 6

What if we would like to apply a contrast operator, like r., to the at() levels? It is not possible to specify the operator inside the at() option. Instead, we need a new suboption, atcontrast():

```
. margins r.agegroup, at(group=(1/3)) contrast(atcontrast(r))
Contrasts of adjusted predictions
Model VCE     : OIM
Expression    : Pr(outcome), predict()
1._at         : group         =         1
2._at         : group         =         2
3._at         : group         =         3
```

	df	chi2	P>chi2
_at#agegroup			
(2 vs 1) (2 vs 1)	1	3.55	0.0596
(2 vs 1) (3 vs 1)	1	33.17	0.0000
(3 vs 1) (2 vs 1)	1	2.86	0.0906
(3 vs 1) (3 vs 1)	1	79.36	0.0000
Joint	4	83.88	0.0000

	Contrast	Delta-method Std. Err.	[95% Conf. Interval]	
_at#agegroup				
(2 vs 1) (2 vs 1)	-.0653508	.0346921	-.133346	.0026445
(2 vs 1) (3 vs 1)	-.1906915	.0331121	-.25559	-.1257931
(3 vs 1) (2 vs 1)	-.0576251	.0340551	-.1243719	.0091216
(3 vs 1) (3 vs 1)	-.2977479	.0334237	-.3632572	-.2322385

When we specify contrast(atcontrast(r)), margins will apply the r. reference-category operator to the levels of group, the variable specified inside at(). The default reference category is 1, the lowest level of group.

Conclusion

margins, contrast is a powerful command, and its abundance of suboptions may seem daunting. The suboptions are in the service of only three goals, however. There are three things that margins, contrast can do with a factor variable or a set of at() definitions:

1. Perform contrasts across the levels of the factor or set (as in example 1).

2. Perform a joint test across the levels of the factor or set (as in example 5).

3. Perform other tests and contrasts within each level of the factor or set (as in example 4).

The default behavior for variables specified inside at(), over(), and within() is to perform contrasts within groups; the default behavior for variables in the *marginlist* is to perform joint tests across groups.

◁

Saved results

margins, contrast saves the following additional items in r():

Scalars
    r(k_terms)   number of terms participating in contrasts

Macros
    r(cmd)   contrast
    r(cmd2)   margins
    r(overall)   overall or empty

Matrices
    r(L)   matrix of contrasts applied to the margins
    r(chi2)   vector of χ^2 statistics
    r(p)   vector of p-values corresponding to r(chi2)
    r(df)   vector of degrees of freedom corresponding to r(p)

margins, contrast with the post option also saves the following additional items in e():

Scalars
    e(k_terms)   number of terms participating in contrasts

Macros
    e(cmd)   contrast
    e(cmd2)   margins
    e(overall)   overall or empty

Matrices
    e(L)   matrix of contrasts applied to the margins
    e(chi2)   vector of χ^2 statistics
    e(p)   vector of p-values corresponding to e(chi2)
    e(df)   vector of degrees of freedom corresponding to e(p)

Methods and formulas

See *Methods and formulas* in [R] **margins** and *Methods and formulas* in [R] **contrast**.

Also see

[R] **contrast** — Contrasts and linear hypothesis tests after estimation

[R] **margins** — Marginal means, predictive margins, and marginal effects

[R] **margins postestimation** — Postestimation tools for margins

[R] **margins, pwcompare** — Pairwise comparisons of margins

[R] **nlcom** — Nonlinear combinations of estimators

[R] **predict** — Obtain predictions, residuals, etc., after estimation

[R] **predictnl** — Obtain nonlinear predictions, standard errors, etc., after estimation

[R] **pwcompare** — Pairwise comparisons

Title

> **margins, pwcompare** — Pairwise comparisons of margins

Syntax

margins [*marginlist*] [*if*] [*in*] [*weight*] [, <u>pwcomp</u>are *margins_options*]

margins [*marginlist*] [*if*] [*in*] [*weight*] [, <u>pwcomp</u>are(*suboptions*) *margins_options*]

where *marginlist* is a list of factor variables or interactions that appear in the current estimation results. The variables may be typed with or without the i. prefix, and you may use any factor-variable syntax:

. margins i.sex i.group i.sex#i.group, pwcompare

. margins sex group sex#i.group, pwcompare

. margins sex##group, pwcompare

See [R] **margins** for the available *margins_options*.

suboptions	Description
Pairwise comparisons	
<u>ci</u>effects	show effects table with confidence intervals; the default
<u>pv</u>effects	show effects table with *p*-values
<u>eff</u>ects	show effects table with confidence intervals and *p*-values
<u>cim</u>argins	show table of margins and confidence intervals
<u>group</u>s	show table of margins and group codes
sort	sort the margins or contrasts in each term

fweights, aweights, iweights, and pweights are allowed; see [U] **11.1.6 weight**.

Menu

Statistics > Postestimation > Pairwise comparisons of margins

Description

margins with the pwcompare option performs pairwise comparisons of margins. margins, pwcompare extends the capabilities of pwcompare to any of the nonlinear responses, predictive margins, or other margins that can be estimated by margins.

Suboptions

⌐Pairwise comparisons⌐

cieffects specifies that a table of the pairwise comparisons with their standard errors and confidence intervals be reported. This is the default.

pveffects specifies that a table of the pairwise comparisons with their standard errors, test statistics, and *p*-values be reported.

effects specifies that a table of the pairwise comparisons with their standard errors, test statistics, *p*-values, and confidence intervals be reported.

cimargins specifies that a table of the margins with their standard errors and confidence intervals be reported.

groups specifies that a table of the margins with their standard errors and group codes be reported. Margins with the same letter in the group code are not significantly different at the specified significance level.

sort specifies that the reported tables be sorted on the margins or contrasts in each term.

Remarks

You should be familiar with the concepts and syntax of both margins and pwcompare before using the pwcompare option of margins. These remarks build on those in [R] **margins** and [R] **pwcompare**.

margins can perform pairwise comparisons of any of the margins that it estimates.

In the *Continuous covariates* example in [R] **marginsplot**, we fit a logistic regression model using the NHANES II dataset, ignoring the complex survey nature of the data. Our dependent variable is highbp, an indicator for whether a person has high blood pressure. We fit a fully interacted model including two factor variables representing gender and age group as well as the continuous covariate, bmi.

```
. use http://www.stata-press.com/data/r12/nhanes2
. logistic highbp sex##agegrp##c.bmi
  (output omitted )
```

By default, margins will compute the predictive margins of the probability of a positive outcome for each of the terms in *marginlist* after logistic regression. We will margin on agegrp so that margins will estimate the average predicted probabilities of having high blood pressure conditional on being in each of the six age groups and unconditional on sex and BMI. We can specify the pwcompare option to obtain all possible pairwise comparisons of these predictive margins:

```
. margins agegrp, pwcompare
```

Pairwise comparisons of predictive margins
Model VCE   : OIM
Expression   : Pr(highbp), predict()

	Contrast	Delta-method Std. Err.	Unadjusted [95% Conf. Interval]	
agegrp				
2 vs 1	.0182344	.0069751	.0045635	.0319054
3 vs 1	.08395	.0097271	.0648852	.1030148
4 vs 1	.1443977	.0111944	.122457	.1663383
5 vs 1	.1517272	.0082323	.1355922	.1678622
6 vs 1	.1443064	.0126661	.1194813	.1691314
3 vs 2	.0657156	.010205	.0457141	.0857171
4 vs 2	.1261632	.0116121	.1034039	.1489225
5 vs 2	.1334928	.0087919	.116261	.1507245
6 vs 2	.1260719	.0130367	.1005205	.1516234
4 vs 3	.0604477	.0134464	.0340932	.0868021
5 vs 3	.0677772	.0111023	.046017	.0895374
6 vs 3	.0603564	.0146942	.0315562	.0891565
5 vs 4	.0073296	.012408	-.0169898	.0316489
6 vs 4	-.0000913	.0157041	-.0308707	.0306882
6 vs 5	-.0074208	.0137504	-.0343712	.0195295

This table gives each of the pairwise differences with confidence intervals. We can see that the confidence interval in the row labeled (2 vs 1) does not include zero. At the 5% level, the predictive margins for the first and second age groups are significantly different. The same is true of many of the other comparisons. With many pairwise comparisons, output in this format can be difficult to sort through. We can put some structure on this by adding the group suboption:

```
. margins agegrp, pwcompare(group)
```

Pairwise comparisons of predictive margins
Model VCE   : OIM
Expression   : Pr(highbp), predict()

	Margin	Delta-method Std. Err.	Unadjusted Groups
agegrp			
1	.0375314	.004423	
2	.0557658	.0053934	
3	.1214814	.0086634	
4	.181929	.0102836	A
5	.1892586	.0069432	A
6	.1818377	.0118687	A

Note: Margins sharing a letter in the group label
      are not significantly different at the 5%
      level.

The group output includes the predictive margins for each age group and letters denoting margins that are not significantly different from each other. In this case, there is not a letter associated with the first age group in the "Unadjusted Groups" column. This missingness indicates that the average predicted probability for this age group is significantly different from the average predicted probability for each of the other age groups at the 5% significance level. The absence of a letter next to the second and third age groups is interpreted in a similar manner. The fourth, fifth, and sixth age groups each have an A in the "Unadjusted Groups" column, which indicates that the average predicted probabilities for these groups are not significantly different at our 5% level.

We can also include the mcompare(bonferroni) option to perform tests using Bonferroni's method to account for making multiple comparisons.

```
. margins agegrp, pwcompare(group) mcompare(bonferroni)
Pairwise comparisons of predictive margins
Model VCE    : OIM
Expression   : Pr(highbp), predict()
```

	Number of Comparisons
agegrp	15

	Margin	Delta-method Std. Err.	Bonferroni Groups
agegrp			
1	.0375314	.004423	B
2	.0557658	.0053934	B
3	.1214814	.0086634	
4	.181929	.0102836	A
5	.1892586	.0069432	A
6	.1818377	.0118687	A

Note: Margins sharing a letter in the group label are not significantly different at the 5% level.

We now see the letter B on the rows corresponding to the first and second age groups. At the 5% level and using Bonferroni's adjustment, the predictive margins for the probability in the first and second age groups are not significantly different.

Saved results

margins, pwcompare saves the following additional items in r():

Scalars
    r(k_terms)   number of terms participating in pairwise comparisons

Macros
    r(cmd)   pwcompare
    r(cmd2)   margins
    r(group#)   group code for the #th margin in r(b)
    r(mcmethod_vs)   *method* from mcompare()
    r(mctitle_vs)   title for *method* from mcompare()
    r(mcadjustall_vs)   adjustall or empty

Matrices
    r(b)   margin estimates
    r(V)   variance–covariance matrix of the margin estimates
    r(b_vs)   margin difference estimates
    r(V_vs)   variance–covariance margin difference of the margin estimates
    r(error_vs)   margin difference estimability codes;
                         0 means estimable,
                         8 means not estimable
    r(table_vs)   matrix containing the margin differences with their standard errors, test statistics,
                         p-values, and confidence intervals
    r(L)   matrix that produces the margin differences

margins, pwcompare with the post option also saves the following additional items in e():

Scalars
    e(k_terms)   number of terms participating in pairwise comparisons

Macros
    e(cmd)   pwcompare
    e(cmd2)   margins

Matrices
    e(b)   margin estimates
    e(V)   variance–covariance matrix of the margin estimates
    e(b_vs)   margin difference estimates
    e(V_vs)   variance–covariance margin difference of the margin estimates
    e(error_vs)   margin difference estimability codes;
                                0 means estimable,
                                8 means not estimable
    e(L)   matrix that produces the margin differences

Methods and formulas

See *Methods and formulas* in [R] **margins** and *Methods and formulas* in [R] **pwcompare**.

Also see

[R] **contrast** — Contrasts and linear hypothesis tests after estimation

[R] **margins** — Marginal means, predictive margins, and marginal effects

[R] **margins postestimation** — Postestimation tools for margins

[R] **nlcom** — Nonlinear combinations of estimators

[R] **predict** — Obtain predictions, residuals, etc., after estimation

[R] **predictnl** — Obtain nonlinear predictions, standard errors, etc., after estimation

[R] **pwcompare** — Pairwise comparisons

Title

| marginsplot — Graph results from margins (profile plots, etc.) |

Syntax

marginsplot [, *options*]

options	Description

Main

<u>x</u>dimension(*dimlist* [, *dimopts*])	use *dimlist* to define x axis
<u>plotd</u>imension(*dimlist* [, *dimopts*])	create plots for groups in *dimlist*
<u>byd</u>imension(*dimlist* [, *dimopts*])	create subgraphs for groups in *dimlist*
<u>graphd</u>imension(*dimlist* [, *dimopts*])	create graphs for groups in *dimlist*
<u>horiz</u>ontal	swap x and y axes
noci	do not plot confidence intervals
name(*name* | *stub* [, replace])	name of graph, or stub if multiple graphs

Labels

<u>allxl</u>abels	place ticks and labels on the x axis for each value
<u>nol</u>abels	label groups with their values, not their labels
<u>allsi</u>mplelabels	forgo variable name and equal signs in all labels
<u>nosi</u>mplelabels	include variable name and equal signs in all labels
<u>sep</u>arator(*string*)	separator for labels when multiple variables are specified in a dimension
<u>nosep</u>arator	do not use a separator

Plot

<u>ploto</u>pts(*plot_options*)	affect rendition of all margin plots
<u>plot#o</u>pts(*plot_options*)	affect rendition of #th margin plot
recast(*plottype*)	plot margins using *plottype*

CI plot

<u>cio</u>pts(*rcap_options*)	affect rendition of all confidence-interval plots
<u>ci#o</u>pts(*rcap_options*)	affect rendition of #th confidence-interval plot
recastci(*plottype*)	plot confidence intervals using *plottype*

Pairwise

| <u>uni</u>que | plot only unique pairwise comparisons |
| csort | sort comparison categories first |

Add plots

| addplot(*plot*) | add other plots to the graph |

Y axis, X axis, Titles, Legend, Overall, By

| *twoway_options* | any options documented in [G-3] ***twoway_options*** |
| <u>byo</u>pts(*byopts*) | how subgraphs are combined, labeled, etc. |

1099

where *dimlist* may be any of the dimensions across which margins were computed in the immediately preceding
margins command; see [R] **margins**. That is to say, *dimlist* may be any variable used in the margins command,
including variables specified in the at(), over(), and within() options. More advanced specifications of *dimlist*
are covered in *Addendum: Advanced uses of dimlist*.

dimopts	Description
<u>labe</u>ls(*lablist*)	list of quoted strings to label each level of the dimension
<u>elab</u>els(*elablist*)	list of enumerated labels
<u>nol</u>abels	label groups with their values, not their labels
<u>alls</u>implelabels	forgo variable name and equal signs in all labels
<u>nosi</u>mplelabels	include variable name and equal signs in all labels
<u>sep</u>arator(*string*)	separator for labels when multiple variables are specified in the dimension
<u>nose</u>parator	do not use a separator

where *lablist* is defined as

"*label*" ["*label*" [...]]

elablist is defined as

"*label*" [# "*label*" [...]]

and the #s are the indices of the levels of the dimension—1 is the first level, 2 is the second level,
and so on.

plot_options	Description
marker_options	change look of markers (color, size, etc.)
marker_label_options	add marker labels; change look or position
cline_options	change look of the line

Menu

Statistics > Postestimation > Margins plots and profile plots

Description

marginsplot graphs the results of the immediately preceding margins command; see [R] **margins**.
Common names for some of the graphs that marginsplot can produce are profile plots and interaction
plots.

Options

<u>┌ Main ┐</u>

xdimension(), plotdimension(), bydimension(), and graphdimension() specify the variables
from the preceding margins command whose group levels will be used for the graph's *x* axis,
plots, by() subgraphs, and graphs.

marginsplot chooses default dimensions based on the margins command. In most cases, the first variable appearing in an at() option and evaluated over more than one value is used for the x axis. If no at() variable meets this condition, the first variable in the *marginlist* is usually used for the x axis and the remaining variables determine the plotted lines or markers. Pairwise comparisons and graphs of marginal effects (derivatives) have different defaults. In all cases, you may override the defaults and explicitly control which variables are used on each dimension of the graph by using these dimension options.

Each of these options supports suboptions that control the labeling of the dimension—axis labels for xdimension(), plot labels for plotdimension(), subgraph titles for bydimension(), and graph titles for graphdimension() titles.

For examples using the dimension options, see *Controlling the graph's dimensions*.

> xdimension(*dimlist* [, *dimopts*]) specifies the variables for the x axis in *dimlist* and controls the content of those labels with *dimopts*.

> plotdimension(*dimlist* [, *dimopts*]) specifies in *dimlist* the variables whose group levels determine the plots and optionally specifies in *dimopts* the content of the plots' labels.

> bydimension(*dimlist* [, *dimopts*]) specifies in *dimlist* the variables whose group levels determine the by() subgraphs and optionally specifies in *dimopts* the content of the subgraphs' titles. For an example using by(), see *Three-way interactions*.

> graphdimension(*dimlist* [, *dimopts*]) specifies in *dimlist* the variables whose group levels determine the graphs and optionally specifies in *dimopts* the content of the graphs' titles.

horizontal reverses the default x and y axes. By default, the y axis represents the estimates of the margins and the x axis represents one or more factors or continuous covariates. Specifying horizontal swaps the axes so that the x axis represents the estimates of the margins. This option can be useful if the labels on the factor or continuous covariates are long.

The horizontal option is discussed in *Horizontal is sometimes better*.

noci removes plots of the pointwise confidence intervals. The default is to plot the confidence intervals.

name(*name* | *stub* [, replace]) specifies the name of the graph or graphs. If the graphdimension() option is specified, or if the default action is to produce multiple graphs, then the argument of name() is taken to be *stub* and graphs named *stub*1, *stub*2, ... are created.

The replace suboption causes existing graphs with the specified name or names to be replaced.

If name() is not specified, default names are used and the graphs may be replaced by subsequent marginsplot or other graphing commands.

⌐ Labels ⌐

With the exception of allxlabels, all these options may be specified either directly as options or as *dimopts* within options xdimension(), plotdimension(), bydimension(), and graphdimension(). When specified in one of the dimension options, only the labels for that dimension are affected. When specified outside the dimension options, all labels on all dimensions are affected. Specifications within the dimension options take precedence.

allxlabels specifies that tick marks and labels be placed on the x axis for each value of the x-dimension variables. By default, if there are more than 25 ticks, default graph axis labeling rules are applied. Labeling may also be specified using the standard graph twoway x-axis label rules and options—xlabel(); see [G-3] *axis_label_options*.

`nolabels` specifies that value labels not be used to construct graph labels and titles for the group levels in the dimension. By default, if a variable in a dimension has value labels, those labels are used to construct labels and titles for axis ticks, plots, subgraphs, and graphs.

Graphs of contrasts and pairwise comparisons are an exception to this rule and are always labeled with values rather than value labels.

`allsimplelabels` and `nosimplelabels` control whether graphs' labels and titles include just the values of the variables or include variable names and equal signs. The default is to use just the value label for variables that have value labels and to use variable names and equal signs for variables that do not have value labels. An example of the former is "Female" and the latter is "country=2".

Sometimes value labels are universally descriptive, and sometimes they have meaning only when considered in relation to their variable. For example, "Male" and "Female" are typically universal, regardless of the variable from which they are taken. "High" and "Low" may not have meaning unless you know they are in relation to a specific measure, say, blood-pressure level. The `allsimplelabels` and `nosimplelabels` options let you override the default labeling.

> `allsimplelabels` specifies that all titles and labels use just the value or value label of the variable.

> `nosimplelabels` specifies that all titles and labels include *varname*= before the value or value label of the variable.

`separator(`*string*`)` and `noseparator` control the separator between label sections when more than one variable is used to specify a dimension. The default separator is a comma followed by a space, but no separator may be requested with `noseparator` or the default may be changed to any string with `separator()`.

For example, if `plotdimension(a b)` is specified, the plot labels in our graph legend might be "a=1, b=1", "a=1, b=2", ... Specifying `separator(:)` would create labels "a=1:b=1", "a=1:b=2", ...

Plot

`plotopts(`*plot_options*`)` affects the rendition of all margin plots. The *plot_options* can affect the size and color of markers, whether and how the markers are labeled, and whether and how the points are connected; see [G-3] *marker_options*, [G-3] *marker_label_options*, and [G-3] *cline_options*.

These settings may be overridden for specific plots by using the `plot#opts()` option.

`plot#opts(`*plot_options*`)` affects the rendition of the #th margin plot. The *plot_options* can affect the size and color of markers, whether and how the markers are labeled, and whether and how the points are connected; see [G-3] *marker_options*, [G-3] *marker_label_options*, and [G-3] *cline_options*.

`recast(`*plottype*`)` specifies that margins be plotted using *plottype*. *plottype* may be `scatter`, `line`, `connected`, `bar`, `area`, `spike`, `dropline`, or `dot`; see [G-2] **graph twoway**. When `recast()` is specified, the plot-rendition options appropriate to the specified *plottype* may be used in lieu of *plot_options*. For details on those options, follow the appropriate link from [G-2] **graph twoway**.

For an example using `recast()`, see *Continuous covariates*.

You may specify `recast()` within a `plotopts()` or `plot#opts()` option. It is better, however, to specify it as documented here, outside those options. When specified outside those options, you have greater access to the plot-specific rendition options of your specified *plottype*.

◻ CI plot ◻

ciopts(*rcap_options*) affects the rendition of all confidence-interval plots; see [G-3] ***rcap_options***.

These settings may be overridden for specific confidence-interval plots with the ci#opts() option.

ci#opts(*rcap_options*) affects the rendition of the #th confidence interval; see [G-3] ***rcap_options***.

recastci(*plottype*) specifies that confidence intervals be plotted using *plottype*. *plottype* may be rarea, rbar, rspike, rcap, rcapsym, rline, rconnected, or rscatter; see [G-2] **graph twoway**. When recastci() is specified, the plot-rendition options appropriate to the specified *plottype* may be used in lieu of *rcap_options*. For details on those options, follow the appropriate link from [G-2] **graph twoway**.

For an example using recastci(), see *Continuous covariates*.

You may specify recastci() within a ciopts() or ci#opts() option. It is better, however, to specify it as documented here, outside those options. When specified outside those options, you have greater access to the plot-specific rendition options of your specified *plottype*.

◻ Pairwise ◻

These options have an effect only when the pwcompare option was specified on the preceding margins command.

unique specifies that only unique pairwise comparisons be plotted. The default is to plot all pairwise comparisons, including those that are mirror images of each other—"male" versus "female" and "female" versus "male". margins reports only the unique pairwise comparisons. unique also changes the default xdimension() for graphs of pairwise comparisons from the reference categories (_pw0) to the comparisons of each pairwise category (_pw).

Unique comparisons are often preferred with horizontal graphs that put all pairwise comparisons on the *x* axis, whereas including the full matrix of comparisons is preferred for charts showing the reference groups on an axis and the comparison groups as plots; see *Pairwise comparisons* and *Horizontal is sometimes better*.

csort specifies that comparison categories are sorted first, and then reference categories are sorted within comparison category. The default is to sort reference categories first, and then sort comparison categories within reference categories. This option has an observable effect only when _pw is also specified in one of the dimension options. It then determines the order of the labeling in the dimension where _pw is specified.

◻ Add plots ◻

addplot(*plot*) provides a way to add other plots to the generated graph; see [G-3] ***addplot_option***.

For an example using addplot(), see *Adding scatterplots of the data*.

If multiple graphs are drawn by a single marginsplot command or if *plot* specifies plots with multiple *y* variables, for example, scatter y1 y2 x, then the graph's legend will not clearly identify all the plots and will require customization using the legend() option; see [G-3] ***legend_options***.

◻ Y axis, X axis, Titles, Legend, Overall, By ◻

twoway_options are any of the options documented in [G-3] ***twoway_options***. These include options for titling the graph (see [G-3] ***title_options***); for saving the graph to disk (see [G-3] ***saving_option***); for controlling the labeling and look of the axes (see [G-3] ***axis_options***); for controlling the look, contents, position, and organization of the legend (see [G-3] ***legend_options***); for adding lines (see [G-3] ***added_line_options***) and text (see [G-3] ***added_text_options***); and for controlling other aspects of the graph's appearance (see [G-3] ***twoway_options***).

The `label()` suboption of the `legend()` option has no effect on `marginsplot`. Use the `order()` suboption instead.

`byopts(`*byopts*`)` affects the appearance of the combined graph when `bydimension()` is specified or when the default graph has subgraphs, including the overall graph title, the position of the legend, and the organization of subgraphs. See [G-3] ***by_option***.

Remarks

Remarks are presented under the following headings:

> *Introduction*
> *Dataset*
> *Profile plots*
> *Interaction plots*
> *Contrasts of margins—effects (discrete marginal effects)*
> *Three-way interactions*
> *Continuous covariates*
> *Plots at every value of a continuous covariate*
> *Contrasts of at() groups—discrete effects*
> *Controlling the graph's dimensions*
> *Pairwise comparisons*
> *Horizontal is sometimes better*
> *Marginal effects*
> *Plotting a subset of the results from margins*
> *Advanced usage*
> > *Plots with multiple terms*
> > *Plots with multiple at() options*
> > *Adding scatterplots of the data*

Introduction

`marginsplot` is a post-`margins` command. It graphs the results of the `margins` command, whether those results are marginal means, predictive margins, marginal effects, contrasts, pairwise comparisons, or other statistics; see [R] **margins**.

By default, the margins are plotted on the y axis, and all continuous and factor covariates specified in the `margins` command will usually be placed on the x axis or used to identify plots. Exceptions are discussed in the following sections and in *Addendum: Advanced uses of dimlist* below.

`marginsplot` produces classic plots, such as profile plots and interaction plots. Beyond that, anything that `margins` can compute, `marginsplot` can graph.

We will be using some relatively complicated `margins` commands with little explanation of the syntax. We will also avoid lengthy interpretations of the results of margins. See [R] **margins** for the complete syntax of `margins` and discussions of its results.

All graphs in this entry were drawn using the `s2gmanual` scheme; see [G-4] **scheme s2**.

Dataset

For continuity, we will use one dataset for most examples—the Second National Health and Nutrition Examination Survey (NHANES II) (McDowell et al. 1981). NHANES II is part of a study to assess the health and nutritional status of adults and children in the United States. It is designed to be a nationally representative sample of the U.S. population. This particular sample is from 1976 to 1980.

The survey nature of the dataset—weights, strata, and sampling units—will be ignored in our analyses. We are discussing graphing, not survey statistics. If you would like to see the results with the appropriate adjustments for the survey design, just add `svy:` before each estimation command, and if you wish, add `vce(unconditional)` as an option to each `margins` command. See [R] **margins**, particularly the discussion and examples under *Obtaining margins with survey data and representative samples*, for reasons why you probably would want to add `vce(unconditional)` when analyzing survey data. For the most part, adjusting for survey design produces moderately larger confidence intervals and relatively small changes in point estimates.

rofile plots

What does my estimation say about how my response varies as one (or more) of my covariates changes? That is the question that is answered by profile plots. Profile plots are also referred to as plots of estimated (or expected, or least-squares) means, though that is unnecessarily restrictive when considering models of binary, count, and ordered outcomes. In the latter cases, we might prefer to say they plot conditional expectations of responses, where a response might be a probability.

What we do with the other covariates depends on the questions we wish to answer. Sometimes we wish to hold other covariates at fixed values, and sometimes we wish to average the response over their values. `margins` can do either, so you can graph either.

We can fit a fully factorial two-way ANOVA of systolic blood pressure on age group and sex using the NHANES II data.

```
. use http://www.stata-press.com/data/r12/nhanes2
. anova bpsystol agegrp##sex
```

| | Number of obs = | | 10351 | R-squared | = 0.2497 |
| | Root MSE | | = 20.2209 | Adj R-squared = | 0.2489 |
Source	Partial SS	df	MS	F	Prob > F
Model	1407229.28	11	127929.935	312.88	0.0000
agegrp	1243037.82	5	248607.565	608.02	0.0000
sex	27728.3794	1	27728.3794	67.81	0.0000
agegrp#sex	88675.043	5	17735.0086	43.37	0.0000
Residual	4227440.75	10339	408.882943		
Total	5634670.03	10350	544.412563		

If you are more comfortable with regression than ANOVA, then type

```
. regress bpsystol agegrp##sex
```

The `anova` and `regress` commands fit identical models. The output from `anova` displays all the terms in the model and thus tends to be more conducive to exploration with `margins` and `marginsplot`.

We estimate the predictive margins of systolic blood pressure for each age group using `margins`.

. margins agegrp

Predictive margins   Number of obs   =   10351

Expression   : Linear prediction, predict()

	Margin	Delta-method Std. Err.	z	P>|z|	[95% Conf. Interval]	
agegrp						
1	117.2684	.419845	279.31	0.000	116.4455	118.0913
2	120.2383	.5020813	239.48	0.000	119.2542	121.2224
3	126.9255	.56699	223.86	0.000	125.8142	128.0368
4	135.682	.5628593	241.06	0.000	134.5788	136.7852
5	141.5285	.3781197	374.30	0.000	140.7874	142.2696
6	148.1096	.6445073	229.80	0.000	146.8464	149.3728

The six predictive margins are just the averages of the predictions over the estimation sample, holding `agegrp` to each of its six levels. If this were a designed experiment rather than survey data, we might wish to assume the cells are balanced—that they have the same number of observations—and thus estimate what are often called expected means or least-squares means. To do that, we would simply add the `asbalanced` option to the `margins` command. The NHANES II data are decidedly unbalanced over `sex#agegrp` cells. So much so that it is unreasonable to assume the cells are balanced.

We graph the results:

. marginsplot

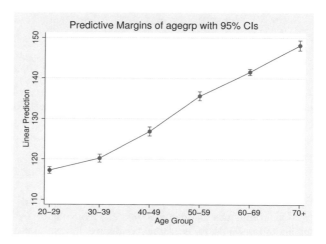

Profile plots are often drawn without confidence intervals (CIs). The CIs may be removed by adding the `noci` option. We prefer to see the CIs.

Disciplines vary widely in their use of the term profile plot. Some disciplines consider any connected plot of a response over values of other variables to be a profile plot. By that definition, most graphs in this entry are profile plots.

teraction plots

Interaction plots are often used to explore the form of an interaction. The interaction term in our ANOVA results is highly significant. Are the interaction effects also large enough to matter? What form do they take? We can answer these questions by fixing agegrp and sex to each possible combination of the two covariates and estimating the margins for those cells.

```
. margins agegrp#sex
```

Then we can graph the results:

```
. marginsplot
```

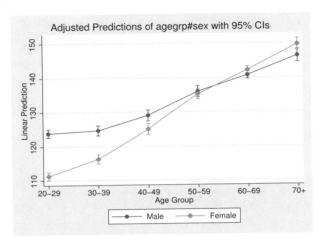

It is clear that the effect of age differs by sex—there is an interaction. If there were no interaction, then the two lines would be parallel.

While males start out with higher systolic blood pressure, females catch up to the males as age increases and may even surpass males in the upper age groups. We say "may" because we cannot tell if the differences are statistically significant. The CIs overlap for the top three age groups. It is tempting to conclude from this overlap that the differences are not statistically significant. Do not fall into this trap. Likewise, do not fall into the trap that the first three age groups are different because their CIs do not overlap. The CIs are for the point estimates, not the differences. There is a covariance between the differences that we must consider if we are to make statements about those differences.

Contrasts of margins—effects (discrete marginal effects)

To assess the differences, all we need do is ask margins to contrast the sets of effects that we just estimated; see [R] **margins, contrast**. With only two groups in sex, it does not matter much which contrast operator we choose. We will use the reference contrast. It will compare the difference between males and females, with males (the first category) as the reference category.

```
. margins r.sex@agegrp
Contrasts of adjusted predictions
Expression    : Linear prediction, predict()
```

	df	chi2	P>chi2
sex@agegrp			
(2 vs 1) 1	1	224.92	0.0000
(2 vs 1) 2	1	70.82	0.0000
(2 vs 1) 3	1	12.15	0.0005
(2 vs 1) 4	1	0.47	0.4949
(2 vs 1) 5	1	3.88	0.0488
(2 vs 1) 6	1	6.37	0.0116
Joint	6	318.62	0.0000

	Contrast	Delta-method Std. Err.	[95% Conf. Interval]	
sex@agegrp				
(2 vs 1) 1	-12.60132	.8402299	-14.24814	-10.9545
(2 vs 1) 2	-8.461161	1.005448	-10.4318	-6.490518
(2 vs 1) 3	-3.956451	1.134878	-6.180771	-1.732131
(2 vs 1) 4	-.7699782	1.128119	-2.98105	1.441094
(2 vs 1) 5	1.491684	.756906	.0081759	2.975193
(2 vs 1) 6	3.264762	1.293325	.7298908	5.799633

Because we are looking for effects that are different from 0, we will add a reference line at 0 to our graph.

```
. marginsplot, yline(0)
```

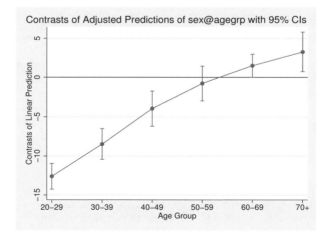

We can now say that females' systolic blood pressure is substantially and significantly lower than males' in the first three age groups but is significantly higher in the last two age groups. Despite the overlapping CIs for the last two age groups in the interaction graph, the effect of sex is significant in these age groups.

The terminology for what we just estimated and graphed varies widely across disciplines. Those versed in design of experiments refer to these values as contrasts or effects. Economists and some other social scientists call them marginal or partial effects. The latter groups might be more comfortable if we avoided the whole concept of contrasts and instead estimated the effects by typing

```
. margins agegrp, dydx(sex)
```

This will produce estimates that are identical to those shown above, and we can graph them by typing `marginsplot`.

The advantage of using the contrast notation and thinking in contrasts is most evident when we take marginal effects with respect to a categorical covariate with more than two levels. Marginal effects for each level of the covariate will be taken with respect to a specified base level. Contrasts are much more flexible. Using the `r.` operator, we can reproduce the marginal-effects results by taking derivatives with respect to a reference level (as we saw above.) We can also estimate the marginal effect of first moving from level 1 to level 2, then from level 2 to level 3, then from level 3 to level 4, ... using the `ar.` or "reverse adjacent" operator. Adjacent effects (marginal effects) can be valuable when evaluating an ordinal covariate, such as `agegrp` in our current model. For a discussion of contrasts, see [R] **contrast** and [R] **margins, contrast**.

Three-way interactions

`marginsplot` can handle any number of covariates in your `margins` command. Consider the three-way ANOVA model that results from adding an indicator for whether an individual has been diagnosed with diabetes. We will fully interact the new covariate with the others in the model.

```
. anova bpsystol agegrp##sex##diabetes
```

	Number of obs =	10349	R-squared	=	0.2572
	Root MSE =	20.131	Adj R-squared =		0.2556

Source	Partial SS	df	MS	F	Prob > F
Model	1448983.17	23	62999.2681	155.45	0.0000
agegrp	107963.582	5	21592.7164	53.28	0.0000
sex	1232.79267	1	1232.79267	3.04	0.0812
agegrp#sex	11679.5925	5	2335.91849	5.76	0.0000
diabetes	7324.98924	1	7324.98924	18.07	0.0000
agegrp#diabetes	5484.54623	5	1096.90925	2.71	0.0189
sex#diabetes	102.988239	1	102.988239	0.25	0.6142
agegrp#sex#diabetes	4863.14971	5	972.629943	2.40	0.0349
Residual	4184296.88	10325	405.258778		
Total	5633280.05	10348	544.38346		

The three-way interaction is significant, as is the main effect of `diabetes` and its interaction with `agegrp`.

Again, if you are more comfortable with regression than ANOVA, you may type

```
. regress bpsystol agegrp##sex##diabetes
```

The `margins` and `marginsplot` results will be the same.

We estimate the expected cell means for each combination of `agegrp`, `sex`, and `diabetes`, and then graph the results by typing

. `margins agegrp#sex#diabetes`
(*output omitted*)

. `marginsplot`

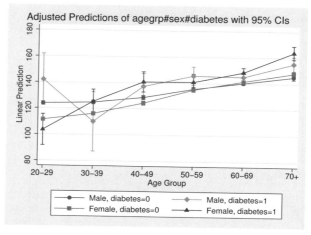

The graph is busy and difficult to interpret.

We can make it better by putting those with diabetes on one subgraph and those without on another:

. `marginsplot, by(diabetes)`

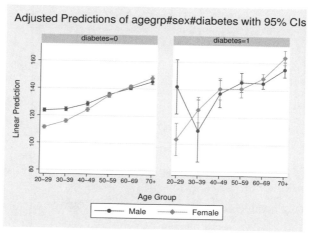

We notice much larger CIs for diabetics. That is not surprising because our sample contains only 499 diabetics compared with 9,850 nondiabetics.

A more interesting way to arrange the plots is by grouping the subgraphs on `sex`:

```
. marginsplot, by(sex)
```

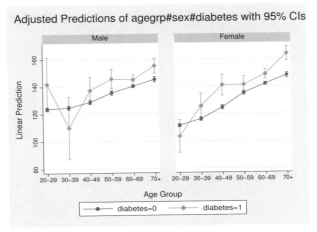

Aside from increased systolic blood pressure in the upper-age groups, which we saw earlier, it appears that those with diabetes are at greater risk of higher systolic blood pressure for many upper-age groups. We can check that by having `margins` estimate the differences between diabetics and nondiabetics, and graphing the results.

```
. margins r.diabetes@agegrp#sex
(output omitted )
. marginsplot, by(sex) yline(0)
```

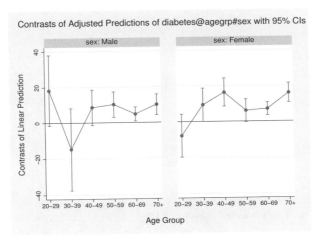

With CIs above 0 for six of eight age groups over 40, this graph provides evidence that diabetes is related to higher blood pressure in those over 40.

Continuous covariates

margins and marginsplot are just as useful with continuous covariates as they are with factor variables. As a variation on our ANOVA/regression models, let's move to a logistic regression, using as our dependent variable an indicator for whether a person has high blood pressure. We introduce a continuous covariate—body mass index (BMI), a measure of weight relative to height. High BMI is often associated with high blood pressure. We will allow the effect of BMI to vary across sexes, age groups, and sex/age combinations by fully interacting the covariates.

```
. logistic highbp sex##agegrp##c.bmi
```

If we wished, we could perform all the analyses above on this model. Instead of estimating margins, contrasts, and marginal effects on the level of systolic blood pressure, we would be estimating margins, contrasts, and marginal effects on the probability of having high blood pressure. You can see those results by repeating any of the prior commands that involve sex and agegrp. In this section, we will focus on the continuous covariate bmi.

With continuous covariates, rather than specify them in the *marginlist* of margins, we specify the specific values at which we want the covariate evaluated in an at() option. at() options are very flexible, and there are many ways to specify values; see *Syntax of at()* in [R] **margins**.

BMI in our sample ranges from 12.4 to 61.2. Let's estimate the predictive margins for males and females at levels of BMI from 10 through 65 at intervals of 5 and graph the results:

```
. margins sex, at(bmi=(10(5)65))
  (output omitted)
. marginsplot, xlabel(10(10)60)
```

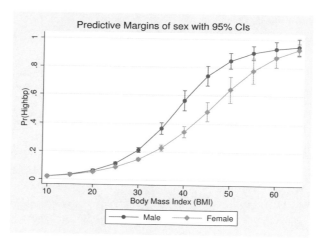

We added the xlabel(10(10)60) option to improve the labeling of the x axis. You may add any *twoway_options* (see [G-3] ***twoway_options***) to the marginsplot command.

For a given BMI, males are generally more susceptible to high blood pressure, though the effect is attenuated by the logistic response when the probabilities approach 0 or 1.

Because bmi is continuous, we might prefer to see the response graphed using a line. We might also prefer that the CIs be plotted as areas. We change the plottype of the response by using the recast() option and the plottype of the CI by using the recastci() option:

```
. marginsplot, xlabel(10(10)60) recast(line) recastci(rarea)
```

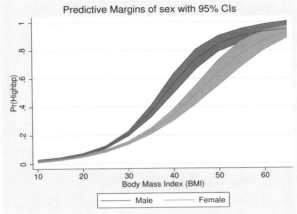

The CIs are a little dark for our tastes. You can dim them a bit by reducing the intensity of their color. Adding `ciopts(color(*.8))` to our `marginsplot` command will do that. Any plot option accepted by `twoway rarea` (see [G-2] **graph twoway rarea**) may be specified in a `ciopts()` option.

Given their confidence regions, the male and female profiles appear to be statistically different over most of the range of BMI. As with the profiles of categorical covariates, we can check that assertion by contrasting the two profiles on `sex` and graphing the results. Let's improve the smoothness of the response by specifying intervals of 1 instead of 5.

```
. margins r.sex, at(bmi=(10(1)65))
  (output omitted )
. marginsplot, xlabel(10(10)60) recast(line) recastci(rarea)
```

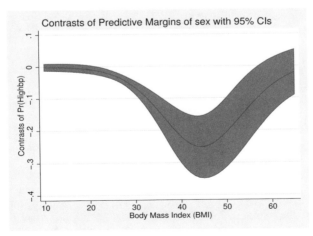

We see that the difference between the sexes is largest at a BMI of about 45 and that the sexes respond more similarly with very high and very low BMI. This shape is largely determined by the response of the logistic function, which is attenuated near probabilities 0 and 1, combined with the fact that the lowest measured BMIs are associated with extremely low probabilities of high blood pressure and the highest measured BMIs are associated with high probabilities of high blood pressure.

As when we contrasted profiles of categorical variables, different disciplines will think of this graph differently. Those familiar with designed experiments will be comfortable with the terms used above—this is a contrast of profiles, or a profile of effects, or a profile of a contrast. Many social scientists will prefer to think of this as a graph of marginal or partial effects. For them, this is a plot of the discrete marginal effect of being female for various levels of BMI. They can obtain an identical graph, with labeling more appropriate for the marginal effect's interpretation, by typing

```
. margins, at(bmi=(10(1)65)) dydx(sex)
. marginsplot, xlabel(10(10)60) recast(line) recastci(rarea)
```

We can also plot profiles of the response of BMI by levels of another continuous covariate (rather than by the categorical variable sex). To do so, we will need another continuous variable in our model. We have been using age groups as a covariate to emphasize the treatment of categorical variables and to allow the effect of age to be flexible. Our dataset also has age recorded in integer years. We replace agegrp with continuous age in our logistic regression.

```
. logistic highbp sex##c.age##c.bmi
```

We can now obtain profiles of BMI for different ages by specifying ranges for both bmi and age in a single at() option on the margins command:

```
. margins sex, at(bmi=(10(5)60) age=(20(10)80))
```

With seven ages specified, we have many profiles, so we will dispense with the CIs by adding the noci option and also tidy up the graph by asking for four columns in the legend:

```
. marginsplot, noci by(sex) legend(cols(4))
```

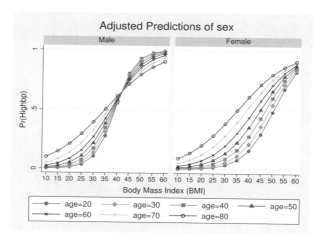

Our model seems to indicate that males have a much sharper reaction to body mass indices than do females. Likewise, younger subjects display a sharper response, while older subjects have a more gradual response with earlier onset. That interpretation might be a result of our parametric treatment of age. As it turns out, the interpretation holds if we allow age to take more flexible forms or return to our use of age groups, which allows each of seven age groups to have unique BMI profiles. Here are the commands to perform that analysis:

```
. logistic highbp sex##agegrp##c.bmi
  (output omitted)
. margins sex#agegrp, at(bmi=(10(5)60))
  (output omitted)
. marginsplot, noci by(sex) legend(cols(4))
```

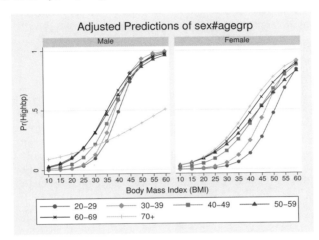

Plots at every value of a continuous covariate

In some cases, the specific values of a continuous covariate are important, and we want to plot the response at those specific values. Return to our logistic example with age treated as a continuous covariate.

```
. logistic highbp sex##c.age##c.bmi
```

We can use a programming trick to extract all the values of age and then supply them in an at() option, just as we would any list of values.

```
. levelsof age
. margins sex, at(age=('r(levels)'))
```

See [P] **levelsof** for a discussion of the levelsof command. levelsof returns in r(levels) the sorted list of unique values of the specified *varlist*, in our case, age.

We can then plot the results using marginsplot.

This is not a very interesting trick when using our age variable, which is recorded as integers from 20 to 74, but the approach will work with almost any continuous variable. In our model, bmi might seem more interesting, but there are 9,941 unique values of bmi in our dataset. A graph cannot resolve so many different values. For that reason, we usually recommend against plotting at every value of a covariate. Instead, graph at reasonable values over the range of the covariate by using the at() option, as we did earlier. This trick is best reserved for variables with a few, or at most a few dozen, unique values.

Contrasts of at() groups—discrete effects

We have previously contrasted across the values of factor variables in our model. Put another way, we have estimated the discrete marginal effects of factor variables. We can do the same for the levels of variables in at() specifications and across separate at() specifications.

Returning to one of our logistic models and its margins, we earlier estimated the predictive margins of BMI at 5-unit intervals for both sexes. These are the commands we typed:

```
. logistic highbp sex##agegrp##c.bmi
. margins sex, at(bmi=(10(5)65))
. marginsplot, xlabel(10(10)60)
```

We can estimate the discrete effects by sex of bmi moving from 10 to 15, then from 15 to 20, . . . , and then from 60 to 65 by contrasting the levels of the at() groups using the reverse-adjacent contrast operator (ar.). We specify the operator within the atcontrast() suboption of the contrast() option. We need to specify one other option. By default, margins, contrast will apply a contrast to all variables in its *marginlist* when a contrast has been requested. In this case, we do not want to contrast across sexes but rather to contrast across the levels of BMI within each sex. To prevent margins from contrasting across the sexes, we specify the marginswithin option. Our margins command is

```
. margins sex, at(bmi=(10(5)65)) contrast(atcontrast(ar._at) marginswithin)
```

And we graph the results using marginsplot:

```
. marginsplot
```

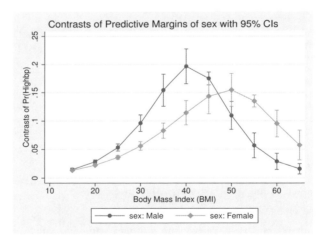

The graph shows the contrasts (or if you prefer, discrete changes) in the probability of high blood pressure by sex as one increases BMI in 5-unit increments.

We can even estimate contrasts (discrete effects) across at() options. To start, let's compare the age-group profiles of the probability of high blood pressure for those in the 25th and 75th percentile of BMI.

```
. margins agegrp, at((p25) bmi) at((p75) bmi)
  (output omitted)

. marginsplot
```

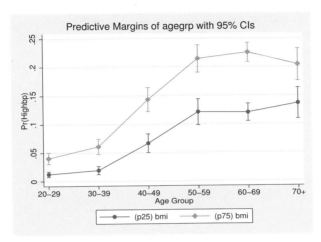

For each age group, people whose BMI is at the 75th percentile have a much higher probability of high blood pressure than those at the 25th percentile. What is that difference in probability and its CI? To contrast across the percentiles of BMI within age groups, we again specify a contrast operator on the at() groups using atcontrast(), and we also tell margins to perform that contrast within the levels of the *marginlist* by using the marginswithin option.

```
. margins agegrp, at((p25) bmi) at((p75) bmi)
> contrast(atcontrast(r._at) marginswithin)
  (output omitted)

. marginsplot
```

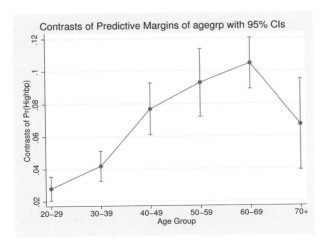

The differences in probability between 25th and 75th BMI percentiles are clearly significantly greater than 0 and appear to be larger for those in higher age groups. The point estimate for those over 70 drops but has a large CI.

Controlling the graph's dimensions

Thus far, `marginsplot` has miraculously done almost exactly what we want in most cases. The things we want on the x axis have been there, the choice of plots has made sense, etc. Some of that luck sprang from the relatively simple analyses we were performing, and some was from careful specification of our `margins` command. Sometimes, we will not be so lucky.

Consider the following `regress`, `margins`, and `marginsplot` commands:

```
. regress bpsystol agegrp##sex##c.bmi
```
(*output omitted*)
```
. margins agegrp, over(sex) at(bmi=(10(10)60))
```
(*output omitted*)
```
. marginsplot
```

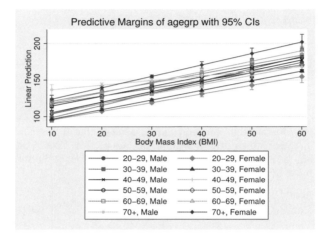

By default, `marginsplot` places the levels of the first multilevel `at()` specification on the x axis, and then usually plots the levels of all remaining variables as connected lines. That is what we see in the graph above—bmi, the `at()` variable, is on the x axis, and each combination of `agegrp` and `sex` is plotted as a separate connected line. If there is no multilevel `at()` specification, then the first variable in *marginlist* becomes the x axis. There are many more rules, but it is usually best to simply type `marginsplot` and see what happens. If you do not like `marginsplot`'s choices, change them.

What if we wanted `agegrp` on the x axis instead of BMI? We tell `marginsplot` to make that change by specifying `agegrp` in the `xdimension()` option:

```
. marginsplot, xdimension(agegrp)
```

Variables that uniquely identify margins: bmi agegrp sex

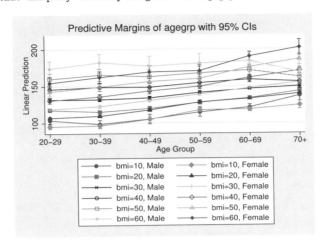

We have been suppressing the Results window output for marginsplot, but that output is helpful if we want to change how things are plotted. You may specify any variable used in your margins command in any of the dimension options—xdimension(), plotdimension(), bydimension(), and graphdimension(). (In fact, there are some pseudovariables that you may also specify in some cases; see *Addendum: Advanced uses of dimlist* for details.) marginsplot tries to help you narrow your choices by listing a set of variables that uniquely identify all your margins. You are not restricted to this list.

We have a different x axis and a different set of plots, but our graph is still busy and difficult to read. We can make it better by creating separate graph panels for each sex. We do that by adding a bydimension() option with sex as the argument.

```
. marginsplot, xdimension(agegrp) bydimension(sex)
```

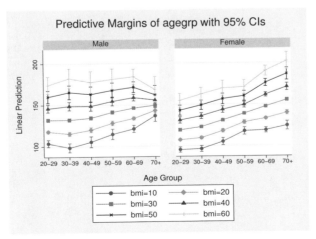

The patterns and the differences between males and females are now easier to see.

If our interest is in comparing males and females, we might even choose to create a separate panel for each level of BMI:

```
. marginsplot, xdimension(agegrp) bydimension(bmi) xlabel(, angle(45))
```

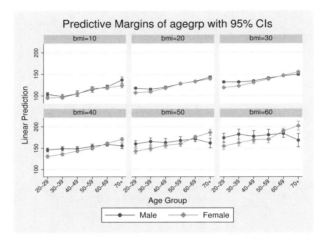

The x-axis labels did not fit, so we angled them.

We leave you to explore the use of the graphdimension() option. It is much like bydimension() but creates separate graphs rather than separate panels. Operationally, the plotdimension() option is rarely used. All variables not in the x dimension and not specified elsewhere become the plotted connected lines.

You will likely use the dimension options frequently. This is one of the rare cases where we recommend using the minimal abbreviations of the options—x() for xdimension(), plot() for plotdimension(), by() for bydimension(), and graph() for graphdimension(). The abbreviations are easy to read and just as meaningful as the full option names. The full names exist to reinforce the relationship between the dimension options.

Pairwise comparisons

marginsplot can graph the results of margins, pwcompare; see [R] **margins, pwcompare**. We return to one of our ANOVA examples. Here we request pairwise comparisons with the pwcompare option of margins, and we request Bonferroni-adjusted CIs with the mcompare() option:

```
. anova bpsystol agegrp##sex
(output omitted )
. margins agegrp, pwcompare mcompare(bonferroni)
(output omitted )
. marginsplot
```

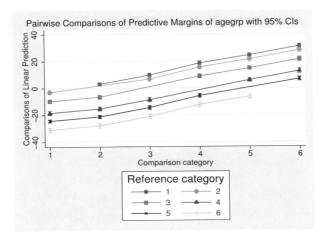

Each connected line plot in the graph represents a reference age-group category for the pairwise comparison. The ticks on the x axis represent comparison age-group categories. So, each plot is a profile for a reference category showing its comparison to each other category.

Horizontal is sometimes better

Another interesting way to graph pairwise comparisons is to simply plot each comparison and label the two categories being compared. This type of graph works better if it is oriented horizontally rather than vertically.

Continuing with the example above, we will switch the graph to horizontal. We will also make several changes to display the graph better. We specify that only unique comparisons be plotted. The graph above plotted both 1 versus 2 and 2 versus 1, which are the same comparison with opposite signs. We add a reference line at 0 because we are interested in comparisons that differ from 0. This graph looks better without the connecting lines, so we add the option recast(scatter). We also reverse the y scale so that the smallest levels of age group appear at the top of the axis.

. `marginsplot, horizontal unique xline(0) recast(scatter) yscale(reverse)`

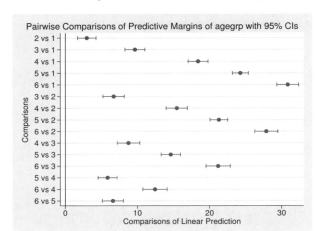

All the comparisons differ from 0, so all our age groups are statistically different from each other.

The `horizontal` option can be useful outside of pairwise comparisons. Profile plots are usually oriented vertically. However, when your covariates have long labels or there are many levels at which the margins are being evaluated, the graph may be easier to read when rendered horizontally.

Marginal effects

We have seen how to graph discrete effects for factor variables and continuous variables by using contrasts, and optionally by using the `dydx()` option of `margins`: *Contrasts of margins—effects (discrete marginal effects)* and *Continuous covariates*. Let's now consider graphing instantaneous marginal effects for continuous covariates. Begin by refitting our logistic model of high blood pressure as a function of sex, age, and BMI:

. `logistic highbp sex##agegrp##c.bmi`

We estimate the average marginal effect of BMI on the probability of high blood pressure for each age group and then graph the results by typing

```
. margins agegrp, dydx(bmi)
(output omitted )
. marginsplot
```

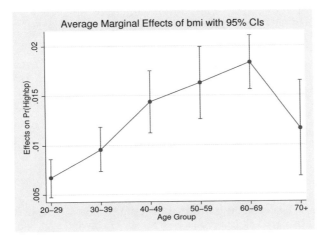

These are the conditional expectations of the marginal effects treating everyone in the sample as though they were in each age group. We can estimate fully conditional marginal effects that do not depend on averaging over the sample by also margining on our one remaining covariate—sex.

```
. margins agegrp#sex, dydx(bmi)
(output omitted )
. marginsplot
```

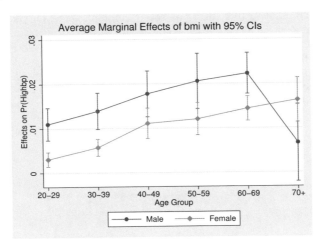

The effect of BMI on the probability of high blood pressure looks to increase with age and is also higher for males than for females.

You may want to confirm that assertion by contrasting across sexes within agegrp:

```
. margins r.sex@agegrp, dydx(bmi)
```

Plotting a subset of the results from margins

marginsplot plots all the margins produced by the preceding margins command. If you want a graph that does not include all the margins, then enter a margins command that produces a reduced set of margins. Obvious ways to reduce the number of margins include not specifying some factors or interactions in the *marginlist* of margins, not specifying some at() or over() options, or reducing the values specified in an at() option. A less obvious technique uses selection lists in factor operators to select specific sets of levels from factor variables specified in the *marginlist*.

Instead of typing

. margins agegrp

which will give you margins for all six age groups in our sample, type

. margins i(2/4).agegrp

which will give you only three margins—those for groups 2, 3, and 4. See [U] **11.4.3.4 Selecting levels**.

Advanced usage

margins is incredibly flexible in the statistics it can estimate and in the grouping of those estimates. Many of the estimates that margins can produce do not make convincing graphs. marginsplot plots the results of any margins command, regardless of whether the resulting graph is easily interpreted. Here we demonstrate some options that can make complicated margins into graphs that are somewhat more useful than those produced by marginsplot's defaults. Others may find truly useful applications for these approaches.

Plots with multiple terms

Margins plots are rarely interesting when you specify multiple terms on your margins command, for example, margins a b. Such plots often compare things that are not comparable. The defaults for marginsplot rarely produce useful plots with multiple terms. Perhaps the most interesting graph in such cases puts all the levels of all the terms together on the vertical axis and plots their margins on the horizontal axis. We do that by including the *marginlist* from margins in an xdimension() option on marginsplot. The long labels on such graphs look better with a horizontal orientation, and there is no need to connect the margin estimates, so we specify the recast(scatter) option.

Using one of our ANOVA examples from earlier,

```
. anova bpsystol agegrp##sex
(output omitted )
. margins agegrp sex
(output omitted )
. marginsplot, xdimension(agegrp sex) horizontal recast(scatter)
```

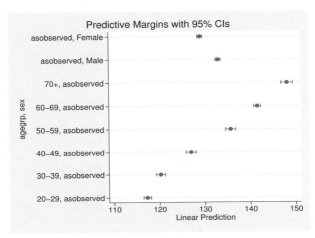

The "asobserved" notations in the y-axis labels are informing us that, for example, when the margin for females is evaluated, the values of age group are taken as they are observed in the dataset. The margin is computed as an average over those values.

Plots with multiple at() options

Some disciplines like to compute margins at the means of other covariates in their model and others like to compute the response for each observation and then take the means of the response. These correspond to the `margins` options `at((mean) _all)` and `at((asobserved) _all)`. For responses that are linear functions of the coefficients, such as `predict` after `regress`, the two computations yield identical results. For responses that are nonlinear functions of the coefficients, the two computations estimate different things.

Using one of our logistic models of high blood pressure,

```
. logistic highbp sex##agegrp##c.bmi
```

and computing both sets of margins for each age group,

```
. margins agegrp, at((mean) _all) at((asobserved) _all)
```

we can use `marginsplot` to compare the approaches:

. `marginsplot`

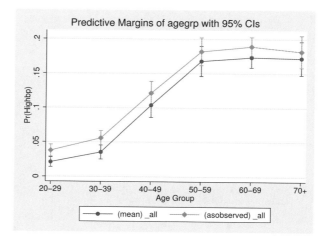

In this case, the probabilities of high blood pressure are lower for each age group at the means of `sex` and `bpi` than are the mean probabilities of high blood pressure averaged over the observed values of `sex` and `bpi`.

Such comparisons come up even more frequently when evaluating marginal effects. We can estimate the marginal effects of `sex` at each age group and graph the results by adding `dydx(sex)` to our `margins` command:

. `margins agegrp, at((mean) _all) at((asobserved) _all) dydx(sex)`
  (*output omitted*)

. `marginsplot`

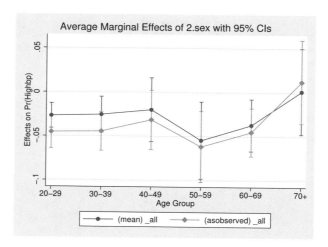

The average marginal effect is smaller for most age groups, but the CIs for both sets of estimates are wide. Can we tell the difference between the estimates? To answer that, we use the now-familiar tactic of taking the contrast of our estimated marginal-effects profiles. That means adding `contrast(atjoint`

marginswithin) to our margins command. We will also add mcompare(bonferroni) to account
for the fact that we will be comparing six contrasts.

```
. margins agegrp, at((mean) _all) at((asobserved) _all) dydx(sex)
> contrast(atjoint marginswithin) mcompare(bonferroni)
```

We will also add the familiar reference line at 0 to our graph of the contrasts.

```
. marginsplot, yline(0)
```

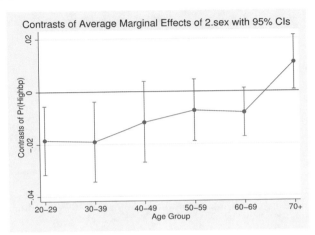

Contrasts of Average Marginal Effects of 2.sex with 95% CIs

While the difference in the estimates of marginal effects is not large, we can distinguish the estimates
for three of the six age groups.

The at() option of margins provides far more flexibility than demonstrated above. It can be
used to evaluate a response or marginal effect at almost any point of interest or combinations of such
points. See *Syntax of at()* in [R] **margins**.

Adding scatterplots of the data

We can add scatterplots of the observed data to our plots of the margins. The NHANES II dataset
is too large for this to be interesting, so for this example, we will use auto.dta. We fit mileage
on whether the care is foreign and on a quadratic in the weight of the car. We convert the weight
into tons (U.S. definition) to improve the scaling, and we format the new tons variable to improve
its labels on the graph. For our graph, we create separate variables for mileage of domestic and of
foreign cars. We fit a fully interacted model so that the effect of weight on mileage can be different
for foreign and for domestic cars.

```
. use http://www.stata-press.com/data/r12/auto
. generate tons = weight/2000
. format tons %6.2f
. separate mpg, by(foreign)
. regress mpg foreign##c.tons##c.tons
```

We then estimate the margins over the range of tons, using the option over(foreign) to obtain
separate estimates for foreign and domestic cars.

```
. margins, at(tons=(.8(.05)2.4)) over(foreign)
```

Adding scatterplots of mileage for domestic and foreign cars is easy. We insert into an `addplot()` option of `marginsplot` the same scatterplot syntax for `twoway` that we would type to produce a scatterplot of the data:

```
. marginsplot, addplot(scatter mpg0 tons || scatter mpg1 tons) recast(line) noci
```

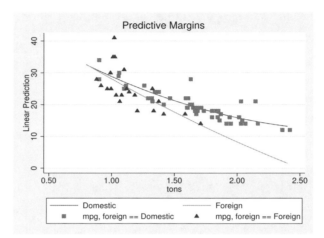

Many will be surprised that the mileage profile is higher in 1978 for domestic (U.S. built) cars. Is the difference significant?

```
. margins, at(tons=(.8(.05)2.4)) over(r.for)
  (output omitted)
. marginsplot, yline(0)
```

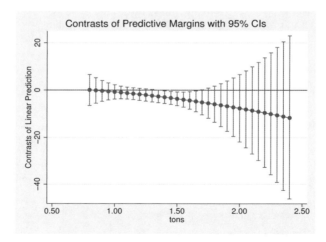

As we did earlier, we contrast the two profiles. We can discern some difference between the two profiles for midweight vehicles, but otherwise there is insufficient information to believe mileage differs across domestic and foreign cars.

ddendum: Advanced uses of dimlist

dimlist specifies the dimensions from the immediately preceding margins command that are to be used for the marginsplot's x axis, plots, subgraphs, and graphs. *dimlist* may contain:

dim	Description
varname	Any variable referenced in the preceding margins command.
at(*varname*)	If a variable is specified in both the *marginlist* or the over() option and in the at() option of margins, then the two uses can be distinguished in marginsplot by typing the at() variables as at(*varname*) in *dimlist*.
_deriv	If the preceding margins command included a dydx(), eyex(), dyex(), or eydx() option, *dimlist* may also contain _deriv to specify all the variables over which derivatives were taken.
_term	If the preceding margins command included multiple terms (for example, margins a b), then *dimlist* may contain _term to enumerate those terms.
_atopt	If the preceding margins command included multiple at() options, then *dimlist* may contain _atopt to enumerate those at() options.

When the pairwise option is specified on margins, you may specify dimensions that enumerate the pairwise comparisons.

_pw	enumerates all the pairwise comparisons
_pw0	enumerates the reference categories of the comparisons
_pw1	enumerates the comparison categories of the comparisons

ethods and formulas

marginsplot is implemented as an ado-file.

cknowledgments

StataCorp thanks Phil Ender for consultations and for his programs that demonstrated what could be done in this area. We also thank Michael Mitchell for his generous advice and comprehensive insight into the application of margins and their plots.

eference

McDowell, A., A. Engel, J. T. Massey, and K. Maurer. 1981. Plan and operation of the Second National Health and Nutrition Examination Survey, 1976–1980. *Vital and Health Statistics* 1(15): 1–144.

lso see

[R] **margins** — Marginal means, predictive margins, and marginal effects

[R] **margins, contrast** — Contrasts of margins

[R] **margins, pwcompare** — Pairwise comparisons of margins

[R] **margins postestimation** — Postestimation tools for margins

Title

> **matsize** — Set the maximum number of variables in a model

Syntax

<u>set</u> <u>mat</u>size # [, <u>perm</u>anently]

where $10 \leq \# \leq 11000$ for Stata/MP and Stata/SE and where $10 \leq \# \leq 800$ for Stata/IC.

Description

set matsize sets the maximum number of variables that can be included in any of Stata's estimation commands.

For Stata/MP and Stata/SE, the default value is 400, but it may be changed upward or downward. The upper limit is 11,000.

For Stata/IC, the initial value is 400, but it may be changed upward or downward. The upper limit is 800.

This command may not be used with Small Stata; matsize is permanently frozen at 100.

Changing matsize has no effect on Mata.

Option

permanently specifies that, in addition to making the change right now, the matsize setting be remembered and become the default setting when you invoke Stata.

Remarks

set matsize controls the internal size of matrices that Stata uses. The default of 400 for Stata/IC, for instance, means that linear regression models are limited to 198 independent variables—198 because the constant uses one position and the dependent variable another, making a total of 200.

You may change matsize with data in memory, but increasing matsize increases the amount of memory consumed by Stata, increasing the probability of page faults and thus of making Stata run more slowly.

▷ Example 1

We wish to fit a model of y on the variables x1 through x400. Without thinking, we type

```
. regress y x1-x400
matsize too small
    You have attempted to create a matrix with more than 400 rows or columns
    or to fit a model with more than 400 variables plus ancillary parameters.
    You need to increase matsize by using the set matsize command; see help
    matsize.
r(908);
```

We realize that we need to increase matsize, so we type

```
. set matsize 450
. regress y x1-x400
  (output omitted )
```
◁

Programmers should note that the current setting of matsize is stored as the c-class value c(matsize); see [P] **creturn**.

lso see

[R] **query** — Display system parameters

[D] **memory** — Memory management

[U] **6 Managing memory**

Title

> **maximize** — Details of iterative maximization

Syntax

Maximum likelihood optimization

> mle_cmd ... [, *options*]

Set default maximum iterations

> set maxiter # [, <u>permanently</u>]

options	Description
<u>diff</u>icult	use a different stepping algorithm in nonconcave regions
<u>tech</u>nique(*algorithm_spec*)	maximization technique
<u>iter</u>ate(#)	perform maximum of # iterations; default is iterate(16000)
[<u>no</u>]<u>log</u>	display an iteration log of the log likelihood; typically, the default
<u>tr</u>ace	display current parameter vector in iteration log
<u>grad</u>ient	display current gradient vector in iteration log
<u>shows</u>tep	report steps within an iteration in iteration log
<u>hess</u>ian	display current negative Hessian matrix in iteration log
<u>showtol</u>erance	report the calculated result that is compared to the effective convergence criterion
<u>tol</u>erance(#)	tolerance for the coefficient vector; see *Options* for the defaults
<u>ltol</u>erance(#)	tolerance for the log likelihood; see *Options* for the defaults
nrtolerance(#)	tolerance for the scaled gradient; see *Options* for the defaults
<u>qtol</u>erance(#)	when specified with algorithms bhhh, dfp, or bfgs, the $q - H$ matrix is used as the final check for convergence rather than nrtolerance() and the H matrix; seldom used
<u>nonrtol</u>erance	ignore the nrtolerance() option
from(*init_specs*)	initial values for the coefficients

where *algorithm_spec* is

> *algorithm* [# [*algorithm* [#]]...]

algorithm is { nr | bhhh | dfp | bfgs }

and *init_specs* is one of

> *matname* [, skip copy]

> { [*eqname*:]*name* = # | /*eqname* = # } [...]

> # [# ...], copy

escription

All Stata commands maximize likelihood functions using `moptimize()` and `optimize()`; see *Methods and formulas* below. Commands use the Newton–Raphson method with step halving and special fixups when they encounter nonconcave regions of the likelihood. For details, see [M-5] **moptimize()** and [M-5] **optimize()**. For more information about programming maximum likelihood estimators in ado-files and Mata, see [R] **ml** and the fourth edition of *Maximum Likelihood Estimation with Stata* (Gould, Pitblado, and Poi 2010).

`set maxiter` specifies the default maximum number of iterations for estimation commands that iterate. The initial value is 16000, and # can be 0 to 16000. To change the maximum number of iterations performed by a particular estimation command, you need not reset `maxiter`; you can specify the `iterate(#)` option. When `iterate(#)` is not specified, the `maxiter` value is used.

aximization options

`difficult` specifies that the likelihood function is likely to be difficult to maximize because of nonconcave regions. When the message "not concave" appears repeatedly, `ml`'s standard stepping algorithm may not be working well. `difficult` specifies that a different stepping algorithm be used in nonconcave regions. There is no guarantee that `difficult` will work better than the default; sometimes it is better and sometimes it is worse. You should use the `difficult` option only when the default stepper declares convergence and the last iteration is "not concave" or when the default stepper is repeatedly issuing "not concave" messages and producing only tiny improvements in the log likelihood.

`technique(`*algorithm_spec*`)` specifies how the likelihood function is to be maximized. The following algorithms are allowed. For details, see Gould, Pitblado, and Poi (2010).

`technique(nr)` specifies Stata's modified Newton–Raphson (NR) algorithm.

`technique(bhhh)` specifies the Berndt–Hall–Hall–Hausman (BHHH) algorithm.

`technique(dfp)` specifies the Davidon–Fletcher–Powell (DFP) algorithm.

`technique(bfgs)` specifies the Broyden–Fletcher–Goldfarb–Shanno (BFGS) algorithm.

The default is `technique(nr)`.

You can switch between algorithms by specifying more than one in the `technique()` option. By default, an algorithm is used for five iterations before switching to the next algorithm. To specify a different number of iterations, include the number after the technique in the option. For example, specifying `technique(bhhh 10 nr 1000)` requests that `ml` perform 10 iterations with the BHHH algorithm followed by 1000 iterations with the NR algorithm, and then switch back to BHHH for 10 iterations, and so on. The process continues until convergence or until the maximum number of iterations is reached.

`iterate(#)` specifies the maximum number of iterations. When the number of iterations equals `iterate()`, the optimizer stops and presents the current results. If convergence is declared before this threshold is reached, it will stop when convergence is declared. Specifying `iterate(0)` is useful for viewing results evaluated at the initial value of the coefficient vector. Specifying `iterate(0)` and `from()` together allows you to view results evaluated at a specified coefficient vector; however, not all commands allow the `from()` option. The default value of `iterate(#)` for both estimators programmed internally and estimators programmed with `ml` is the current value of `set maxiter`, which is `iterate(16000)` by default.

`log` and `nolog` specify whether an iteration log showing the progress of the log likelihood is to be displayed. For most commands, the log is displayed by default, and `nolog` suppresses it. For a few commands (such as the svy maximum likelihood estimators), you must specify log to see the log.

trace adds to the iteration log a display of the current parameter vector.

gradient adds to the iteration log a display of the current gradient vector.

showstep adds to the iteration log a report on the steps within an iteration. This option was added so that developers at StataCorp could view the stepping when they were improving the ml optimizer code. At this point, it mainly provides entertainment.

hessian adds to the iteration log a display of the current negative Hessian matrix.

showtolerance adds to the iteration log the calculated value that is compared with the effective convergence criterion at the end of each iteration. Until convergence is achieved, the smallest calculated value is reported.

shownrtolerance is a synonym of showtolerance.

Below we describe the three convergence tolerances. Convergence is declared when the nrtolerance() criterion is met and either the tolerance() or the ltolerance() criterion is also met.

tolerance(#) specifies the tolerance for the coefficient vector. When the relative change in the coefficient vector from one iteration to the next is less than or equal to tolerance(), the tolerance() convergence criterion is satisfied.

tolerance(1e-4) is the default for estimators programmed with ml.

tolerance(1e-6) is the default.

ltolerance(#) specifies the tolerance for the log likelihood. When the relative change in the log likelihood from one iteration to the next is less than or equal to ltolerance(), the ltolerance() convergence is satisfied.

ltolerance(0) is the default for estimators programmed with ml.

ltolerance(1e-7) is the default.

nrtolerance(#) specifies the tolerance for the scaled gradient. Convergence is declared when $\mathbf{g}\mathbf{H}^{-1}\mathbf{g}' < $ nrtolerance(). The default is nrtolerance(1e-5).

qtolerance(#) when specified with algorithms bhhh, dfp, or bfgs uses the $\mathbf{q} - \mathbf{H}$ matrix as the final check for convergence rather than nrtolerance() and the $\mathbf{H}$ matrix.

Beginning with Stata 12, by default, Stata now computes the $\mathbf{H}$ matrix when the $\mathbf{q} - \mathbf{H}$ matrix passes the convergence tolerance, and Stata requires that $\mathbf{H}$ be concave and pass the nrtolerance() criterion before concluding convergence has occurred.

qtolerance() provides a way for the user to obtain Stata's earlier behavior.

nonrtolerance specifies that the default nrtolerance() criterion be turned off.

---

from() specifies initial values for the coefficients. Not all estimators in Stata support this option. You can specify the initial values in one of three ways: by specifying the name of a vector containing the initial values (for example, from(b0), where b0 is a properly labeled vector); by specifying coefficient names with the values (for example, from(age=2.1 /sigma=7.4)); or by specifying a list of values (for example, from(2.1 7.4, copy)). from() is intended for use when doing bootstraps (see [R] **bootstrap**) and in other special situations (for example, with iterate(0)).

Even when the values specified in from() are close to the values that maximize the likelihood, only a few iterations may be saved. Poor values in from() may lead to convergence problems.

skip specifies that any parameters found in the specified initialization vector that are not also found in the model be ignored. The default action is to issue an error message.

copy specifies that the list of values or the initialization vector be copied into the initial-value vector by position rather than by name.

ption for set maxiter

permanently specifies that, in addition to making the change right now, the maxiter setting be remembered and become the default setting when you invoke Stata.

emarks

Only in rare circumstances would you ever need to specify any of these options, except nolog. The nolog option is useful for reducing the amount of output appearing in log files.

The following is an example of an iteration log:

```
Iteration 0:   log likelihood = -3791.0251
Iteration 1:   log likelihood =  -3761.738
Iteration 2:   log likelihood = -3758.0632   (not concave)
Iteration 3:   log likelihood = -3758.0447
Iteration 4:   log likelihood = -3757.5861
Iteration 5:   log likelihood =  -3757.474
Iteration 6:   log likelihood = -3757.4613
Iteration 7:   log likelihood = -3757.4606
Iteration 8:   log likelihood = -3757.4606
       (table of results omitted )
```

At iteration 8, the model converged. The message "not concave" at the second iteration is notable. This example was produced using the heckman command; its likelihood is not globally concave, so it is not surprising that this message sometimes appears. The other message that is occasionally seen is "backed up". Neither of these messages should be of any concern unless they appear at the final iteration.

If a "not concave" message appears at the last step, there are two possibilities. One is that the result is valid, but there is collinearity in the model that the command did not otherwise catch. Stata checks for obvious collinearity among the independent variables before performing the maximization, but strange collinearities or near collinearities can sometimes arise between coefficients and ancillary parameters. The second, more likely cause for a "not concave" message at the final step is that the optimizer entered a flat region of the likelihood and prematurely declared convergence.

If a "backed up" message appears at the last step, there are also two possibilities. One is that Stata found a perfect maximum and could not step to a better point; if this is the case, all is fine, but this is a highly unlikely occurrence. The second is that the optimizer worked itself into a bad concave spot where the computed gradient and Hessian gave a bad direction for stepping.

If either of these messages appears at the last step, perform the maximization again with the gradient option. If the gradient goes to zero, the optimizer has found a maximum that may not be unique but is a maximum. From the standpoint of maximum likelihood estimation, this is a valid result. If the gradient is not zero, it is not a valid result, and you should try tightening up the convergence criterion, or try ltol(0) tol(1e-7) to see if the optimizer can work its way out of the bad region.

If you get repeated "not concave" steps with little progress being made at each step, try specifying the difficult option. Sometimes difficult works wonderfully, reducing the number of iterations and producing convergence at a good (that is, concave) point. Other times, difficult works poorly, taking much longer to converge than the default stepper.

Saved results

Maximum likelihood estimators save the following in e():

Scalars

e(N)	number of observations	always saved
e(k)	number of parameters	always saved
e(k_eq)	number of equations in e(b)	usually saved
e(k_eq_model)	number of equations in overall model test	usually saved
e(k_dv)	number of dependent variables	usually saved
e(df_m)	model degrees of freedom	always saved
e(r2_p)	pseudo-R-squared	sometimes saved
e(ll)	log likelihood	always saved
e(ll_0)	log likelihood, constant-only model	saved when constant-only model is fit
e(N_clust)	number of clusters	saved when vce(cluster *clustvar*) is specified; see [U] **20.20 Obtaining robust variance estimates**
e(chi2)	χ^2	usually saved
e(p)	significance of model of test	usually saved
e(rank)	rank of e(V)	always saved
e(rank0)	rank of e(V) for constant-only model	saved when constant-only model is fit
e(ic)	number of iterations	usually saved
e(rc)	return code	usually saved
e(converged)	1 if converged, 0 otherwise	usually saved

Macros

e(cmd)	name of command	always saved
e(cmdline)	command as typed	always saved
e(depvar)	names of dependent variables	always saved
e(wtype)	weight type	saved when weights are specified or implied
e(wexp)	weight expression	saved when weights are specified or implied
e(title)	title in estimation output	usually saved by commands using ml
e(clustvar)	name of cluster variable	saved when vce(cluster *clustvar*) is specified; see [U] **20.20 Obtaining robust variance estimates**
e(chi2type)	Wald or LR; type of model χ^2 test	usually saved
e(vce)	*vcetype* specified in vce()	saved when command allows (vce())
e(vcetype)	title used to label Std. Err.	sometimes saved
e(opt)	type of optimization	always saved
e(which)	max or min; whether optimizer is to perform maximization or minimization	always saved
e(ml_method)	type of ml method	always saved by commands using ml
e(user)	name of likelihood-evaluator program	always saved
e(technique)	from technique() option	sometimes saved
e(singularHmethod)	m-marquardt or hybrid; method used when Hessian is singular	sometimes saved[1]
e(crittype)	optimization criterion	always saved[1]
e(properties)	estimator properties	always saved
e(predict)	program used to implement predict	usually saved

Matrices

e(b)	coefficient vector	always saved
e(Cns)	constraints matrix	sometimes saved
e(ilog)	iteration log (up to 20 iterations)	usually saved
e(gradient)	gradient vector	usually saved
e(V)	variance–covariance matrix of the estimators	always saved
e(V_modelbased)	model-based variance	only saved when e(V) is neither the OIM nor OPG variance

Functions

e(sample)	marks estimation sample	always saved

1. Type ereturn list, all to view these results; see [P] **return**.

See *Saved results* in the manual entry for any maximum likelihood estimator for a list of returned results.

Methods and formulas

Optimization is currently performed by moptimize() and optimize(), with the former implemented in terms of the latter; see [M-5] **moptimize()** and [M-5] **optimize()**. Some estimators use moptimize() and optimize() directly, and others use the ml ado-file interface to moptimize().

Prior to Stata 11, Stata had three separate optimization engines: an internal one used by estimation commands implemented in C code; ml implemented in ado-code separately from moptimize() and used by most estimators; and moptimize() and optimize() used by a few recently written estimators. These days, the internal optimizer and the old version of `ml` are used only under version control. In addition, `arch` and `arima` (see [TS] **arch** and [TS] **arima**) are currently implemented using the old `ml`.

Let L_1 be the log likelihood of the full model (that is, the log-likelihood value shown on the output), and let L_0 be the log likelihood of the "constant-only" model. The likelihood-ratio χ^2 model test is defined as $2(L_1 - L_0)$. The pseudo-R^2 (McFadden 1974) is defined as $1 - L_1/L_0$. This is simply the log likelihood on a scale where 0 corresponds to the "constant-only" model and 1 corresponds to perfect prediction for a discrete model (in which case the overall log likelihood is 0).

Some maximum likelihood routines can report coefficients in an exponentiated form, for example, odds ratios in `logistic`. Let b be the unexponentiated coefficient, s its standard error, and b_0 and b_1 the reported confidence interval for b. In exponentiated form, the point estimate is e^b, the standard error $e^b s$, and the confidence interval e^{b_0} and e^{b_1}. The displayed Z (or t) statistics and p-values are the same as those for the unexponentiated results. This is justified because $e^b = 1$ and $b = 0$ are equivalent hypotheses, and normality is more likely to hold in the b metric.

References

Gould, W. W., J. S. Pitblado, and B. P. Poi. 2010. *Maximum Likelihood Estimation with Stata*. 4th ed. College Station, TX: Stata Press.

McFadden, D. L. 1974. Conditional logit analysis of qualitative choice behavior. In *Frontiers in Econometrics*, ed. P. Zarembka, 105–142. New York: Academic Press.

Also see

[R] **ml** — Maximum likelihood estimation

[SVY] **ml for svy** — Maximum pseudolikelihood estimation for survey data

[M-5] **moptimize( )** — Model optimization

[M-5] **optimize( )** — Function optimization

```
mean — Estimate means
```

mean *varlist* [*if*] [*in*] [*weight*] [, *options*]

options	Description
Model	
<u>stdize</u>(*varname*)	variable identifying strata for standardization
<u>stdweight</u>(*varname*)	weight variable for standardization
<u>nostdr</u>escale	do not rescale the standard weight variable
if/in/over	
over(*varlist*[, <u>nolabel</u>])	group over subpopulations defined by *varlist*; optionally, suppress group labels
SE/Cluster	
vce(*vcetype*)	*vcetype* may be analytic, <u>clus</u>ter *clustvar*, <u>boot</u>strap, or jackknife
Reporting	
<u>level</u>(#)	set confidence level; default is level(95)
<u>nohe</u>ader	suppress table header
<u>nol</u>egend	suppress table legend
display_options	control column formats and line width
<u>coefl</u>egend	display legend instead of statistics

bootstrap, jackknife, mi estimate, rolling, statsby, and svy are allowed; see [U] **11.1.10 Prefix commands**.
vce(bootstrap) and vce(jackknife) are not allowed with the mi estimate prefix; see [MI] **mi estimate**.
Weights are not allowed with the bootstrap prefix; see [R] **bootstrap**.
aweights are not allowed with the jackknife prefix; see [R] **jackknife**.
vce() and weights are not allowed with the svy prefix; see [SVY] **svy**.
fweights, aweights, iweights, and pweights are allowed; see [U] **11.1.6 weight**.
coeflegend does not appear in the dialog box.
See [U] **20 Estimation and postestimation commands** for more capabilities of estimation commands.

Menu

Statistics > Summaries, tables, and tests > Summary and descriptive statistics > Means

Description

mean produces estimates of means, along with standard errors.

Options

    ⌐ Model ⌐

stdize(*varname*) specifies that the point estimates be adjusted by direct standardization across the strata identified by *varname*. This option requires the stdweight() option.

stdweight(*varname*) specifies the weight variable associated with the standard strata identified in the stdize() option. The standardization weights must be constant within the standard strata.

nostdrescale prevents the standardization weights from being rescaled within the over() groups. This option requires stdize() but is ignored if the over() option is not specified.

    ⌐ if/in/over ⌐

over(*varlist* [, nolabel]) specifies that estimates be computed for multiple subpopulations, which are identified by the different values of the variables in *varlist*.

When this option is supplied with one variable name, such as over(*varname*), the value labels of *varname* are used to identify the subpopulations. If *varname* does not have labeled values (or there are unlabeled values), the values themselves are used, provided that they are nonnegative integers. Noninteger values, negative values, and labels that are not valid Stata names are substituted with a default identifier.

When over() is supplied with multiple variable names, each subpopulation is assigned a unique default identifier.

nolabel requests that value labels attached to the variables identifying the subpopulations be ignored.

    ⌐ SE/Cluster ⌐

vce(*vcetype*) specifies the type of standard error reported, which includes types that are derived from asymptotic theory, that allow for intragroup correlation, and that use bootstrap or jackknife methods; see [R] *vce_option*.

vce(analytic), the default, uses the analytically derived variance estimator associated with the sample mean.

    ⌐ Reporting ⌐

level(#); see [R] **estimation options**.

noheader prevents the table header from being displayed. This option implies nolegend.

nolegend prevents the table legend identifying the subpopulations from being displayed.

display_options: cformat(%*fmt*) and nolstretch; see [R] **estimation options**.

The following option is available with mean but is not shown in the dialog box:

coeflegend; see [R] **estimation options**.

Remarks

Example 1

Using the fuel data from example 2 of [R] **ttest**, we estimate the average mileage of the cars without the fuel treatment (mpg1) and those with the fuel treatment (mpg2).

```
. use http://www.stata-press.com/data/r12/fuel
. mean mpg1 mpg2
Mean estimation                     Number of obs    =     12
```

	Mean	Std. Err.	[95% Conf.	Interval]
mpg1	21	.7881701	19.26525	22.73475
mpg2	22.75	.9384465	20.68449	24.81551

Using these results, we can test the equality of the mileage between the two groups of cars.

```
. test mpg1 = mpg2
 ( 1)   mpg1 - mpg2 = 0
       F(  1,    11) =    5.04
            Prob > F =    0.0463
```

◁

Example 2

In example 1, the joint observations of mpg1 and mpg2 were used to estimate a covariance between their means.

```
. matrix list e(V)
symmetric e(V)[2,2]
           mpg1        mpg2
mpg1   .62121212
mpg2    .4469697   .88068182
```

If the data were organized this way out of convenience but the two variables represent independent samples of cars (coincidentally of the same sample size), we should reshape the data and use the over() option to ensure that the covariance between the means is zero.

```
. use http://www.stata-press.com/data/r12/fuel
. stack mpg1 mpg2, into(mpg) clear
. mean mpg, over(_stack)
Mean estimation                     Number of obs    =     24
           1: _stack = 1
           2: _stack = 2
```

	Over	Mean	Std. Err.	[95% Conf.	Interval]
mpg					
	1	21	.7881701	19.36955	22.63045
	2	22.75	.9384465	20.80868	24.69132

```
. matrix list e(V)
symmetric e(V)[2,2]
              mpg:      mpg:
                1          2
mpg:1    .62121212
mpg:2            0   .88068182
```

Now we can test the equality of the mileage between the two independent groups of cars.

```
. test [mpg]1 = [mpg]2
 ( 1)  [mpg]1 - [mpg]2 = 0
        F(  1,    23) =     2.04
              Prob > F =   0.1667
```

◁

▷ Example 3: standardized means

Suppose that we collected the blood pressure data from example 2 of [R] **dstdize**, and we wish to obtain standardized high blood pressure rates for each city in 1990 and 1992, using, as the standard, the age, sex, and race distribution of the four cities and two years combined. Our rate is really the mean of a variable that indicates whether a sampled individual has high blood pressure. First, we generate the strata and weight variables from our standard distribution, and then use mean to compute the rates.

```
. use http://www.stata-press.com/data/r12/hbp, clear
. egen strata = group(age race sex) if inlist(year, 1990, 1992)
(675 missing values generated)
. by strata, sort: gen stdw = _N
. mean hbp, over(city year) stdize(strata) stdweight(stdw)
Mean estimation
N. of std strata =       24          Number of obs    =      455
            Over: city year
      _subpop_1: 1 1990
      _subpop_2: 1 1992
      _subpop_3: 2 1990
      _subpop_4: 2 1992
      _subpop_5: 3 1990
      _subpop_6: 3 1992
      _subpop_7: 5 1990
      _subpop_8: 5 1992
```

Over	Mean	Std. Err.	[95% Conf. Interval]	
hbp				
_subpop_1	.058642	.0296273	.0004182	.1168657
_subpop_2	.0117647	.0113187	-.0104789	.0340083
_subpop_3	.0488722	.0238958	.0019121	.0958322
_subpop_4	.014574	.007342	.0001455	.0290025
_subpop_5	.1011211	.0268566	.0483425	.1538998
_subpop_6	.0810577	.0227021	.0364435	.1256719
_subpop_7	.0277778	.0155121	-.0027066	.0582622
_subpop_8	.0548926	.	.	.

The standard error of the high blood pressure rate estimate is missing for city 5 in 1992 because there was only one individual with high blood pressure; that individual was the only person observed in the stratum of white males 30–35 years old.

By default, mean rescales the standard weights within the over() groups. In the following, we use the nostdrescale option to prevent this, thus reproducing the results in [R] **dstdize**.

```
. mean hbp, over(city year) nolegend stdize(strata) stdweight(stdw)
> nostdrescale

Mean estimation

N. of std strata =        24        Number of obs    =       455
```

Over	Mean	Std. Err.	[95% Conf. Interval]	
hbp				
_subpop_1	.0073302	.0037034	.0000523	.0146082
_subpop_2	.0015432	.0014847	-.0013745	.004461
_subpop_3	.0078814	.0038536	.0003084	.0154544
_subpop_4	.0025077	.0012633	.000025	.0049904
_subpop_5	.0155271	.0041238	.007423	.0236312
_subpop_6	.0081308	.0022772	.0036556	.012606
_subpop_7	.0039223	.0021904	-.0003822	.0082268
_subpop_8	.0088735	0	.	.

◁

Saved results

mean saves the following in e():

Scalars
e(N)	number of observations
e(N_over)	number of subpopulations
e(N_stdize)	number of standard strata
e(N_clust)	number of clusters
e(k_eq)	number of equations in e(b)
e(df_r)	sample degrees of freedom
e(rank)	rank of e(V)

Macros
e(cmd)	mean
e(cmdline)	command as typed
e(varlist)	*varlist*
e(stdize)	*varname* from stdize()
e(stdweight)	*varname* from stdweight()
e(wtype)	weight type
e(wexp)	weight expression
e(title)	title in estimation output
e(cluster)	name of cluster variable
e(over)	*varlist* from over()
e(over_labels)	labels from over() variables
e(over_namelist)	names from e(over_labels)
e(vce)	*vcetype* specified in vce()
e(vcetype)	title used to label Std. Err.
e(properties)	b V
e(estat_cmd)	program used to implement estat
e(marginsnotok)	predictions disallowed by margins

Matrices
e(b)	vector of mean estimates
e(V)	(co)variance estimates
e(_N)	vector of numbers of nonmissing observations
e(_N_stdsum)	number of nonmissing observations within the standard strata
e(_p_stdize)	standardizing proportions
e(error)	error code corresponding to e(b)

Functions
e(sample)	marks estimation sample

Methods and formulas

mean is implemented as an ado-file.

Methods and formulas are presented under the following headings:

> *The mean estimator*
> *Survey data*
> *The survey mean estimator*
> *The standardized mean estimator*
> *The poststratified mean estimator*
> *The standardized poststratified mean estimator*
> *Subpopulation estimation*

The mean estimator

Let y be the variable on which we want to calculate the mean and y_j an individual observation on y, where $j = 1, \ldots, n$ and n is the sample size. Let w_j be the weight, and if no weight is specified, define $w_j = 1$ for all j. For aweights, the w_j are normalized to sum to n. See *The survey mean estimator* for pweighted data.

Let W be the sum of the weights

$$W = \sum_{j=1}^{n} w_j$$

The mean is defined as

$$\bar{y} = \frac{1}{W} \sum_{j=1}^{n} w_j y_j$$

The default variance estimator for the mean is

$$\widehat{V}(\bar{y}) = \frac{1}{W(W-1)} \sum_{j=1}^{n} w_j (y_j - \bar{y})^2$$

The standard error of the mean is the square root of the variance.

If x, x_j, and $\bar{x}$ are similarly defined for another variable (observed jointly with y), the covariance estimator between $\bar{x}$ and $\bar{y}$ is

$$\widehat{\text{Cov}}(\bar{x}, \bar{y}) = \frac{1}{W(W-1)} \sum_{j=1}^{n} w_j (x_j - \bar{x})(y_j - \bar{y})$$

Survey data

See [SVY] **variance estimation**, [SVY] **direct standardization**, and [SVY] **poststratification** for discussions that provide background information for the following formulas. The following formulas are derived from the fact that the mean is a special case of the ratio estimator where the denominator variable is one, $x_j = 1$; see [R] **ratio**.

The survey mean estimator

Let Y_j be a survey item for the jth individual in the population, where $j = 1, \dots, M$ and M is the size of the population. The associated population mean for the item of interest is $\overline{Y} = Y/M$ where

$$Y = \sum_{j=1}^{M} Y_j$$

Let y_j be the survey item for the jth sampled individual from the population, where $j = 1, \dots, m$ and m is the number of observations in the sample.

The estimator for the mean is $\overline{y} = \widehat{Y}/\widehat{M}$, where

$$\widehat{Y} = \sum_{j=1}^{m} w_j y_j \quad \text{and} \quad \widehat{M} = \sum_{j=1}^{m} w_j$$

and w_j is a sampling weight. The score variable for the mean estimator is

$$z_j(\overline{y}) = \frac{y_j - \overline{y}}{\widehat{M}} = \frac{\widehat{M} y_j - \widehat{Y}}{\widehat{M}^2}$$

The standardized mean estimator

Let D_g denote the set of sampled observations that belong to the gth standard stratum and define $I_{D_g}(j)$ to indicate if the jth observation is a member of the gth standard stratum; where $g = 1, \dots, L_D$ and L_D is the number of standard strata. Also, let π_g denote the fraction of the population that belongs to the gth standard stratum, thus $\pi_1 + \cdots + \pi_{L_D} = 1$. π_g is derived from the `stdweight()` option.

The estimator for the standardized mean is

$$\overline{y}^D = \sum_{g=1}^{L_D} \pi_g \frac{\widehat{Y}_g}{\widehat{M}_g}$$

where

$$\widehat{Y}_g = \sum_{j=1}^{m} I_{D_g}(j) \, w_j y_j \quad \text{and} \quad \widehat{M}_g = \sum_{j=1}^{m} I_{D_g}(j) \, w_j$$

The score variable for the standardized mean is

$$z_j(\overline{y}^D) = \sum_{g=1}^{L_D} \pi_g I_{D_g}(j) \frac{\widehat{M}_g y_j - \widehat{Y}_g}{\widehat{M}_g^2}$$

The poststratified mean estimator

Let P_k denote the set of sampled observations that belong to poststratum k and define $I_{P_k}(j)$ to indicate if the jth observation is a member of poststratum k; where $k = 1, \ldots, L_P$ and L_P is the number of poststrata. Also let M_k denote the population size for poststratum k. P_k and M_k are identified by specifying the poststrata() and postweight() options on svyset; see [SVY] **svyset**.

The estimator for the poststratified mean is

$$\overline{y}^P = \frac{\widehat{Y}^P}{\widehat{M}^P} = \frac{\widehat{Y}^P}{M}$$

where

$$\widehat{Y}^P = \sum_{k=1}^{L_P} \frac{M_k}{\widehat{M}_k} \widehat{Y}_k = \sum_{k=1}^{L_P} \frac{M_k}{\widehat{M}_k} \sum_{j=1}^{m} I_{P_k}(j)\, w_j y_j$$

and

$$\widehat{M}^P = \sum_{k=1}^{L_P} \frac{M_k}{\widehat{M}_k} \widehat{M}_k = \sum_{k=1}^{L_P} M_k = M$$

The score variable for the poststratified mean is

$$z_j(\overline{y}^P) = \frac{z_j(\widehat{Y}^P)}{M} = \frac{1}{M} \sum_{k=1}^{L_P} I_{P_k}(j) \frac{M_k}{\widehat{M}_k} \left(y_j - \frac{\widehat{Y}_k}{\widehat{M}_k} \right)$$

The standardized poststratified mean estimator

The estimator for the standardized poststratified mean is

$$\overline{y}^{DP} = \sum_{g=1}^{L_D} \pi_g \frac{\widehat{Y}_g^P}{\widehat{M}_g^P}$$

where

$$\widehat{Y}_g^P = \sum_{k=1}^{L_p} \frac{M_k}{\widehat{M}_k} \widehat{Y}_{g,k} = \sum_{k=1}^{L_p} \frac{M_k}{\widehat{M}_k} \sum_{j=1}^{m} I_{D_g}(j) I_{P_k}(j)\, w_j y_j$$

and

$$\widehat{M}_g^P = \sum_{k=1}^{L_p} \frac{M_k}{\widehat{M}_k} \widehat{M}_{g,k} = \sum_{k=1}^{L_p} \frac{M_k}{\widehat{M}_k} \sum_{j=1}^{m} I_{D_g}(j) I_{P_k}(j)\, w_j$$

The score variable for the standardized poststratified mean is

$$z_j(\overline{y}^{DP}) = \sum_{g=1}^{L_D} \pi_g \frac{\widehat{M}_g^P z_j(\widehat{Y}_g^P) - \widehat{Y}_g^P z_j(\widehat{M}_g^P)}{(\widehat{M}_g^P)^2}$$

where

$$z_j(\widehat{Y}_g^P) = \sum_{k=1}^{L_P} I_{P_k}(j) \frac{M_k}{\widehat{M}_k} \left\{ I_{D_g}(j) y_j - \frac{\widehat{Y}_{g,k}}{\widehat{M}_k} \right\}$$

and

$$z_j(\widehat{M}_g^P) = \sum_{k=1}^{L_P} I_{P_k}(j) \frac{M_k}{\widehat{M}_k} \left\{ I_{D_g}(j) - \frac{\widehat{M}_{g,k}}{\widehat{M}_k} \right\}$$

Subpopulation estimation

Let S denote the set of sampled observations that belong to the subpopulation of interest, and define $I_S(j)$ to indicate if the jth observation falls within the subpopulation.

The estimator for the subpopulation mean is $\overline{y}^S = \widehat{Y}^S / \widehat{M}^S$, where

$$\widehat{Y}^S = \sum_{j=1}^{m} I_S(j) \, w_j y_j \quad \text{and} \quad \widehat{M}^S = \sum_{j=1}^{m} I_S(j) \, w_j$$

Its score variable is

$$z_j(\overline{y}^S) = I_S(j) \frac{y_j - \overline{y}^S}{\widehat{M}^S} = I_S(j) \frac{\widehat{M}^S y_j - \widehat{Y}^S}{(\widehat{M}^S)^2}$$

The estimator for the standardized subpopulation mean is

$$\overline{y}^{DS} = \sum_{g=1}^{L_D} \pi_g \frac{\widehat{Y}_g^S}{\widehat{M}_g^S}$$

where

$$\widehat{Y}_g^S = \sum_{j=1}^{m} I_{D_g}(j) I_S(j) \, w_j y_j \quad \text{and} \quad \widehat{M}_g^S = \sum_{j=1}^{m} I_{D_g}(j) I_S(j) \, w_j$$

Its score variable is

$$z_j(\overline{y}^{DS}) = \sum_{g=1}^{L_D} \pi_g I_{D_g}(j) I_S(j) \frac{\widehat{M}_g^S y_j - \widehat{Y}_g^S}{(\widehat{M}_g^S)^2}$$

The estimator for the poststratified subpopulation mean is

$$\overline{y}^{PS} = \frac{\widehat{Y}^{PS}}{\widehat{M}^{PS}}$$

where

$$\widehat{Y}^{PS} = \sum_{k=1}^{L_P} \frac{M_k}{\widehat{M}_k} \widehat{Y}_k^S = \sum_{k=1}^{L_P} \frac{M_k}{\widehat{M}_k} \sum_{j=1}^{m} I_{P_k}(j) I_S(j) \, w_j y_j$$

and

$$\widehat{M}^{PS} = \sum_{k=1}^{L_P} \frac{M_k}{\widehat{M}_k} \widehat{M}_k^S = \sum_{k=1}^{L_P} \frac{M_k}{\widehat{M}_k} \sum_{j=1}^{m} I_{P_k}(j) I_S(j) \, w_j$$

Its score variable is

$$z_j(\overline{y}^{PS}) = \frac{\widehat{M}^{PS} z_j(\widehat{Y}^{PS}) - \widehat{Y}^{PS} z_j(\widehat{M}^{PS})}{(\widehat{M}^{PS})^2}$$

where

$$z_j(\widehat{Y}^{PS}) = \sum_{k=1}^{L_P} I_{P_k}(j) \frac{M_k}{\widehat{M}_k} \left\{ I_S(j) \, y_j - \frac{\widehat{Y}_k^S}{\widehat{M}_k} \right\}$$

and

$$z_j(\widehat{M}^{PS}) = \sum_{k=1}^{L_P} I_{P_k}(j) \frac{M_k}{\widehat{M}_k} \left\{ I_S(j) - \frac{\widehat{M}_k^S}{\widehat{M}_k} \right\}$$

The estimator for the standardized poststratified subpopulation mean is

$$\bar{y}^{DPS} = \sum_{g=1}^{L_D} \pi_g \frac{\widehat{Y}_g^{PS}}{\widehat{M}_g^{PS}}$$

where

$$\widehat{Y}_g^{PS} = \sum_{k=1}^{L_p} \frac{M_k}{\widehat{M}_k} \widehat{Y}_{g,k}^S = \sum_{k=1}^{L_p} \frac{M_k}{\widehat{M}_k} \sum_{j=1}^m I_{D_g}(j) I_{P_k}(j) I_S(j)\, w_j y_j$$

and

$$\widehat{M}_g^{PS} = \sum_{k=1}^{L_p} \frac{M_k}{\widehat{M}_k} \widehat{M}_{g,k}^S = \sum_{k=1}^{L_p} \frac{M_k}{\widehat{M}_k} \sum_{j=1}^m I_{D_g}(j) I_{P_k}(j) I_S(j)\, w_j$$

Its score variable is

$$z_j(\bar{y}^{DPS}) = \sum_{g=1}^{L_D} \pi_g \frac{\widehat{M}_g^{PS} z_j(\widehat{Y}_g^{PS}) - \widehat{Y}_g^{PS} z_j(\widehat{M}_g^{PS})}{(\widehat{M}_g^{PS})^2}$$

where

$$z_j(\widehat{Y}_g^{PS}) = \sum_{k=1}^{L_P} I_{P_k}(j) \frac{M_k}{\widehat{M}_k} \left\{ I_{D_g}(j) I_S(j)\, y_j - \frac{\widehat{Y}_{g,k}^S}{\widehat{M}_k} \right\}$$

and

$$z_j(\widehat{M}_g^{PS}) = \sum_{k=1}^{L_P} I_{P_k}(j) \frac{M_k}{\widehat{M}_k} \left\{ I_{D_g}(j) I_S(j) - \frac{\widehat{M}_{g,k}^S}{\widehat{M}_k} \right\}$$

References

Bakker, A. 2003. The early history of average values and implications for education. *Journal of Statistics Education* 11(1). http://www.amstat.org/publications/jse/v11n1/bakker.html.

Cochran, W. G. 1977. *Sampling Techniques*. 3rd ed. New York: Wiley.

Stuart, A., and J. K. Ord. 1994. *Kendall's Advanced Theory of Statistics: Distribution Theory, Vol I*. 6th ed. London: Arnold.

Also see

[R] **mean postestimation** — Postestimation tools for mean

[R] **ameans** — Arithmetic, geometric, and harmonic means

[R] **proportion** — Estimate proportions

[R] **ratio** — Estimate ratios

[R] **summarize** — Summary statistics

[R] **total** — Estimate totals

[MI] **estimation** — Estimation commands for use with mi estimate

[SVY] **direct standardization** — Direct standardization of means, proportions, and ratios

[SVY] **poststratification** — Poststratification for survey data

[SVY] **subpopulation estimation** — Subpopulation estimation for survey data

[SVY] **svy estimation** — Estimation commands for survey data

[SVY] **variance estimation** — Variance estimation for survey data

[U] **20 Estimation and postestimation commands**

Title

> **mean postestimation** — Postestimation tools for mean

Description

The following postestimation commands are available after mean:

Command	Description
estat	VCE
estat (svy)	postestimation statistics for survey data
estimates	cataloging estimation results
lincom	point estimates, standard errors, testing, and inference for linear combinations of coefficients
nlcom	point estimates, standard errors, testing, and inference for nonlinear combinations of coefficients
test	Wald tests of simple and composite linear hypotheses
testnl	Wald tests of nonlinear hypotheses

See the corresponding entries in the *Base Reference Manual* for details, but see [SVY] **estat** for details about estat (svy).

Remarks

▷ Example 1

We have a dataset with monthly rates of returns on the Dow and NASDAQ stock indices. We can use mean to compute the average quarterly rates of return for the two indices separately;

```
. use http://www.stata-press.com/data/r12/rates
. mean dow nasdaq
Mean estimation                      Number of obs    =      357
```

	Mean	Std. Err.	[95% Conf. Interval]	
dow	.2489137	6.524386	-12.58227	13.0801
nasdaq	10.78477	4.160821	2.601887	18.96765

If you chose just one of the indices for your portfolio, you either did rather well or rather poorly, depending on which one you picked. However, as we now show with the postestimation command lincom, if you diversified your portfolio, you would have earned a respectable 5.5% rate of return without having to guess which index would be the better performer.

```
. lincom .5*dow + .5*nasdaq
 ( 1)   .5 dow + .5 nasdaq = 0
```

	Coef.	Std. Err.	t	P>|t|	[95% Conf. Interval]	
(1)	5.51684	4.262673	1.29	0.196	-2.866347	13.90003

◁

Methods and formulas

All postestimation commands listed above are implemented as ado-files.

Also see

[R] **mean** — Estimate means

[SVY] **svy postestimation** — Postestimation tools for svy

Title

meta — Meta-analysis

Remarks

Stata does not have a meta-analysis command. Stata users, however, have developed an excellent suite of commands for performing meta-analysis, including commands for performing standard and cumulative meta-analysis, commands for producing forest plots and contour-enhanced funnel plots, and commands for nonparametric analysis of publication bias.

Many articles describing these commands have been published in the *Stata Technical Bulletin* and the *Stata Journal*. These articles were updated and published in a cohesive collection: *Meta-Analysis in Stata: An Updated Collection from the Stata Journal*.

In this collection, editor Jonathan Sterne discusses how these articles relate to each other and how they fit in the overall literature of meta-analysis. Sterne has organized the collection into four areas: classic meta-analysis; meta-regression; graphical and analytic tools for detecting bias; and recent advances such as meta-analysis for dose–response curves, diagnostic accuracy, multivariate analysis, and studies containing missing values.

All meta-analysis commands discussed in this collection may be downloaded by visiting http://www.stata-press.com/books/mais.html.

We highly recommend that Stata users interested in meta-analysis read this book. Since the publication of the meta-analysis collection, Kontopantelis and Reeves (2010) published an article in the *Stata Journal* describing a new command `metaan` that performs fixed- or random-effects meta-analysis.

Please also see the following FAQ on the Stata website:

What meta-analysis features are available in Stata?
http://www.stata.com/support/faqs/stat/meta.html

References

Borenstein, M., L. V. Hedges, J. P. T. Higgins, and H. R. Rothstein. 2009. *Introduction to Meta-Analysis*. Chichester, UK: Wiley.

Egger, M., G. Davey Smith, and D. G. Altman, ed. 2001. *Systematic Reviews in Health Care: Meta-analysis in Context*. 2nd ed. London: BMJ Books.

Kontopantelis, E., and D. Reeves. 2010. metaan: Random-effects meta-analysis. *Stata Journal* 10: 395–407.

Sterne, J. A. C., ed. 2009. *Meta-Analysis in Stata: An Updated Collection from the Stata Journal*. College Station, TX: Stata Press.

Sutton, A. J., K. R. Abrams, D. R. Jones, T. A. Sheldon, and F. Song. 2000. *Methods for Meta-Analysis in Medical Research*. New York: Wiley.

mfp — Multivariable fractional polynomial models

mfp [, *options*] : *regression_cmd* [*yvar*$_1$ [*yvar*$_2$]] *xvarlist* [*if*] [*in*] [*weight*]

[, *regression_cmd_options*]

options	Description
Model 2	
sequential	use the Royston and Altman model-selection algorithm; default uses closed-test procedure
cycles(#)	maximum number of iteration cycles; default is cycles(5)
dfdefault(#)	default maximum degrees of freedom; default is dfdefault(4)
center(cent_list)	specification of centering for the independent variables
alpha(alpha_list)	p-values for testing between FP models; default is alpha(0.05)
df(df_list)	degrees of freedom for each predictor
powers(numlist)	list of FP powers to use; default is powers(-2 -1(.5)1 2 3)
Adv. model	
xorder(+ | - | n)	order of entry into model-selection algorithm; default is xorder(+)
select(select_list)	nominal p-values for selection on each predictor
xpowers(xp_list)	FP powers for each predictor
zero(varlist)	treat nonpositive values of specified predictors as zero when FP transformed
catzero(varlist)	add indicator variable for specified predictors
all	include out-of-sample observations in generated variables
Reporting	
level(#)	set confidence level; default is level(95)
display_options	control column formats and line width

regression_cmd_options	Description
Adv. model	
regression_cmd_options	options appropriate to the regression command in use

All weight types supported by *regression_cmd* are allowed; see [U] **11.1.6 weight**.

See [U] **20 Estimation and postestimation commands** for more capabilities of estimation commands.

fracgen may be used to create new variables containing fractional polynomial powers. See [R] **fracpoly**.

where

> *regression_cmd* may be clogit, glm, intreg, logistic, logit, mlogit, nbreg, ologit, oprobit, poisson, probit, qreg, regress, rreg, stcox, stcrreg, streg, or xtgee.
>
> *yvar₁* is not allowed for streg, stcrreg, and stcox. For these commands, you must first stset your data.
>
> *yvar₁* and *yvar₂* must both be specified when *regression_cmd* is intreg.
>
> *xvarlist* has elements of type *varlist* and/or (*varlist*), for example, x1 x2 (x3 x4 x5)
>
> Elements enclosed in parentheses are tested jointly for inclusion in the model and are not eligible for fractional polynomial transformation.

Menu

Statistics > Linear models and related > Fractional polynomials > Multivariable fractional polynomial models

Description

mfp selects the multivariable fractional polynomial (MFP) model that best predicts the outcome variable from the right-hand-side variables in *xvarlist*.

Options

┌─── Model 2 ──

sequential chooses the sequential fractional polynomial (FP) selection algorithm (see *Methods of FP model selection*).

cycles(*#*) sets the maximum number of iteration cycles permitted. cycles(5) is the default.

dfdefault(*#*) determines the default maximum degrees of freedom (df) for a predictor. The default is dfdefault(4) (second-degree FP).

center(*cent_list*) defines the centering of the covariates *xvar₁*, *xvar₂*, ... of *xvarlist*. The default is center(mean), except for binary covariates, where it is center(*#*), with *#* being the lower of the two distinct values of the covariate. A typical item in *cent_list* is *varlist*:{mean|*#*|no}. Items are separated by commas. The first item is special in that *varlist* is optional, and if it is omitted, the default is reset to the specified value (mean, *#*, or no). For example, center(no, age:mean) sets the default to no (that is, no centering) and the centering of age to mean.

alpha(*alpha_list*) sets the significance levels for testing between FP models of different degrees. The rules for *alpha_list* are the same as those for *df_list* in the df() option (see below). The default nominal *p*-value (significance level, selection level) is 0.05 for all variables.

Example: alpha(0.01) specifies that all variables have an FP selection level of 1%.

Example: alpha(0.05, weight:0.1) specifies that all variables except weight have an FP selection level of 5%; weight has a level of 10%.

df(*df_list*) sets the df for each predictor. The df (not counting the regression constant, _cons) is twice the degree of the FP, so, for example, an *xvar* fit as a second-degree FP (FP2) has 4 df. The first item in *df_list* may be either *#* or *varlist*:*#*. Subsequent items must be *varlist*:*#*. Items are separated by commas, and *varlist* is specified in the usual way for variables. With the first type of item, the df for all predictors is taken to be *#*. With the second type of item, all members of *varlist* (which must be a subset of *xvarlist*) have *#* df.

The default number of degrees of freedom for a predictor of type *varlist* specified in *xvarlist* but not in *df_list* is assigned according to the number of distinct (unique) values of the predictor, as follows:

# of distinct values	Default df
1	(invalid predictor)
2–3	1
4–5	min(2, dfdefault())
≥ 6	dfdefault()

Example: df(4)
All variables have 4 df.

Example: df(2, weight displ:4)
weight and displ have 4 df; all other variables have 2 df.

Example: df(weight displ:4, mpg:2)
weight and displ have 4 df, mpg has 2 df; all other variables have default df.

powers(*numlist*) is the set of FP powers to be used. The default set is $-2, -1, -0.5, 0, 0.5, 1, 2, 3$ (0 means log).

⌐ Adv. model ⌐

xorder(+ | - | n) determines the order of entry of the covariates into the model-selection algorithm. The default is xorder(+), which enters them in decreasing order of significance in a multiple linear regression (most significant first). xorder(-) places them in reverse significance order, whereas xorder(n) respects the original order in *xvarlist*.

select(*select_list*) sets the nominal p-values (significance levels) for variable selection by backward elimination. A variable is dropped if its removal causes a nonsignificant increase in deviance. The rules for *select_list* are the same as those for *df_list* in the df() option (see above). Using the default selection level of 1 for all variables forces them all into the model. Setting the nominal p-value to be 1 for a given variable forces it into the model, leaving others to be selected or not. The nominal p-value for elements of *xvarlist* bound by parentheses is specified by including (*varlist*) in *select_list*.

Example: select(0.05)
All variables have a nominal p-value of 5%.

Example: select(0.05, weight:1)
All variables except weight have a nominal p-value of 5%; weight is forced into the model.

Example: select(a (b c):0.05)
All variables except a, b, and c are forced into the model. b and c are tested jointly with 2 df at the 5% level, and a is tested singly at the 5% level.

xpowers(*xp_list*) sets the permitted FP powers for covariates individually. The rules for *xp_list* are the same as for *df_list* in the df() option. The default selection is the same as that for the powers() option.

Example: xpowers(-1 0 1)
All variables have powers $-1, 0, 1$.

Example: xpowers(x5:-1 0 1)
All variables except x5 have default powers; x5 has powers $-1, 0, 1$.

zero(*varlist*) treats negative and zero values of members of *varlist* as zero when FP transformations are applied. By default, such variables are subjected to a preliminary linear transformation to avoid negative and zero values (see [R] **fracpoly**). *varlist* must be part of *xvarlist*.

catzero(*varlist*) is a variation on zero(); see *Zeros and zero categories* below. *varlist* must be part of *xvarlist*.

regression_cmd_options may be any of the options appropriate to *regression_cmd*.

all includes out-of-sample observations when generating the FP variables. By default, the generated FP variables contain missing values outside the estimation sample.

___

⌐ Reporting ⌐

level(*#*) specifies the confidence level, as a percentage, for confidence intervals. The default is level(95) or as set by set level; see [U] **20.7 Specifying the width of confidence intervals**.

display_options: cformat(%*fmt*), pformat(%*fmt*), sformat(%*fmt*), and nolstretch; see [R] **estimation options**.

Remarks

Remarks are presented under the following headings:

> *Iteration report*
> *Estimation algorithm*
> *Methods of FP model selection*
> *Zeros and zero categories*

For elements in *xvarlist* not enclosed in parentheses, mfp leaves variables in the data named I*xvar*__1, I*xvar*__2, . . . , where *xvar* represents the first four letters of the name of *xvar*$_1$, and so on, for *xvar*$_2$, *xvar*$_3$, etc. The new variables contain the best-fitting FP powers of *xvar*$_1$, *xvar*$_2$, ...

Iteration report

By default, for each continuous predictor, x, mfp compares null, linear, and FP1 models for x with an FP2 model. The deviance for each of these nested submodels is given in the column labeled "Deviance". The line labeled "Final" gives the deviance for the selected model and its powers. All the other predictors currently selected are included, with their transformations (if any). For models specified as having 1 df, the only choice is whether the variable enters the model.

Estimation algorithm

The estimation algorithm in mfp processes the *xvars* in turn. Initially, mfp silently arranges *xvarlist* in order of increasing p-value (that is, of decreasing statistical significance) for omitting each predictor from the model comprising *xvarlist*, with each term linear. The aim is to model relatively important variables before unimportant ones. This approach may help to reduce potential model-fitting difficulties caused by collinearity or, more generally, "concurvity" among the predictors. See the xorder() option above for details on how to change the ordering.

At the initial cycle, the best-fitting FP function for *xvar*$_1$ (the first of *xvarlist*) is determined, with all the other variables assumed to be linear. Either the default or the alternative procedure is used (see *Methods of FP model selection* below). The functional form (but not the estimated regression coefficients) for *xvar*$_1$ is kept, and the process is repeated for *xvar*$_2$, *xvar*$_3$, etc. The first iteration concludes when all the variables have been processed in this way. The next cycle is similar, except that the functional forms from the initial cycle are retained for all variables except the one currently being processed.

A variable whose functional form is prespecified to be linear (that is, to have 1 df) is tested for exclusion within the above procedure when its nominal p-value (selection level) according to select() is less than 1; otherwise, it is included.

Updating of FP functions and candidate variables continues until the functions and variables included in the overall model do not change (convergence). Convergence is usually achieved within 1–4 cycles.

ethods of FP model selection

mfp includes two algorithms for FP model selection, both of which combine backward elimination with the selection of an FP function. For each continuous variable in turn, they start from a most-complex permitted FP model and attempt to simplify the model by reducing the degree. The default algorithm resembles a closed-test procedure, a sequence of tests maintaining the overall type I error rate at a prespecified nominal level, such as 5%. All significance tests are approximate; therefore, the algorithm is not precisely a closed-test procedure (Royston and Sauerbrei 2008, chap. 6).

The closed-test algorithm for choosing an FP model with maximum permitted degree $m = 2$ (that is, an FP2 model with 4 df) for one continuous predictor, x, is as follows:

1. Inclusion: Test FP2 against the null model for x on 4 df at the significance level determined by select(). If x is significant, continue; otherwise, drop x from the model.

2. Nonlinearity: Test FP2 against a straight line in x on 3 df at the significance level determined by alpha(). If significant, continue; otherwise, stop, with the chosen model for x being a straight line.

3. Simplification: Test FP2 against FP1 on 2 df at the significance level determined by alpha(). If significant, the final model is FP2; otherwise, it is FP1.

The first step is omitted if x is to be retained in the model, that is, if its nominal p-value, according to the select() option, is 1.

An alternative algorithm is available with the sequential option, as originally suggested by Royston and Altman (1994):

1. Test FP2 against FP1 on 2 df at the alpha() significance level. If significant, the final model is FP2; otherwise, continue.

2. Test FP1 against a straight line on 1 df at the alpha() level. If significant, the final model is FP1; otherwise, continue.

3. Test a straight line against omitting x on 1 df at the select() level. If significant, the final model is a straight line; otherwise, drop x.

The final step is omitted if x is to be retained in the model, that is, if its nominal p-value, according to the select() option, is 1.

If x is uninfluential, the overall type I error rate of this procedure is about double that of the closed-test procedure, for which the rate is close to the nominal value. This inflated type I error rate confers increased apparent power to detect nonlinear relationships.

eros and zero categories

The zero() option permits fitting an FP model to the positive values of a covariate, taking nonpositive values as zero. An application is the assessment of the effect of cigarette smoking as a risk factor in an epidemiological study. Nonsmokers may be qualitatively different from smokers, so the effect of smoking (regarded as a continuous variable) may not be continuous between one and zero cigarettes. To allow for this, the risk may be modeled as constant for the nonsmokers and as an FP function of the number of cigarettes for the smokers:

```
. generate byte nonsmokr = cond(n_cigs==0, 1, 0) if n_cigs != .
. mfp, zero(n_cigs) df(4, nonsmokr:1): logit case n_cigs nonsmokr age
```

Omission of `zero(n_cigs)` would cause `n_cigs` to be transformed before analysis by the addition of a suitable constant, probably 1.

A closely related approach involves the `catzero()` option. The command

```
. mfp, catzero(n_cigs): logit case n_cigs age
```

would achieve a similar result to the previous command but with important differences. First, `mfp` would create the equivalent of the binary variable `nonsmokr` automatically and include it in the model. Second, the two smoking variables would be treated as one predictor in the model. With the `select()` option active, the two variables would be tested jointly for inclusion in the model. A modified version is described in Royston and Sauerbrei (2008, sec. 4.15).

▷ Example 1

We illustrate two of the analyses performed by Sauerbrei and Royston (1999). We use `brcancer.dta`, which contains prognostic factors data from the German Breast Cancer Study Group of patients with node-positive breast cancer. The response variable is recurrence-free survival time (`rectime`), and the censoring variable is `censrec`. There are 686 patients with 299 events. We use Cox regression to predict the log hazard of recurrence from prognostic factors, of which five are continuous (x1, x3, x5, x6, x7) and three are binary (x2, x4a, x4b). Hormonal therapy (`hormon`) is known to reduce recurrence rates and is forced into the model. We use `mfp` to build a model from the initial set of eight predictors by using the backfitting model-selection algorithm. We set the nominal p-value for variable and FP selection to 0.05 for all variables except `hormon`, for which it is set to 1:

```
. use http://www.stata-press.com/data/r12/brcancer
(German breast cancer data)
. stset rectime, fail(censrec)
```
(*output omitted*)
```
. mfp, alpha(.05) select(.05, hormon:1): stcox x1 x2 x3 x4a x4b x5 x6 x7 hormon,
> nohr
```

Deviance for model with all terms untransformed = 3471.637, 686 observations

Variable	Model	(vs.)	Deviance	Dev diff.	P	Powers	(vs.)
x5	null	FP2	3503.610	61.366	0.000*	.	.5 3
	lin.		3471.637	29.393	0.000+	1	
	FP1		3449.203	6.959	0.031+	0	
	Final		3442.244			.5 3	
x6	null	FP2	3464.113	29.917	0.000*	.	-2 .5
	lin.		3442.244	8.048	0.045+	1	
	FP1		3435.550	1.354	0.508	.5	
	Final		3435.550			.5	
[hormon included with 1 df in model]							
x4a	null	lin.	3440.749	5.199	0.023*	.	1
	Final		3435.550			1	
x3	null	FP2	3436.832	3.560	0.469	.	-2 3
	Final		3436.832			.	
x2	null	lin.	3437.589	0.756	0.384	.	1
	Final		3437.589			.	
x4b	null	lin.	3437.848	0.259	0.611	.	1

```
            Final          3437.848                    .
x1          null    FP2    3437.893    18.085  0.001*   .          -2 -.5
            lin.           3437.848    18.040  0.000+   1
            FP1            3433.628    13.820  0.001+  -2
            Final          3419.808                    -2 -.5
x7          null    FP2    3420.805     3.715  0.446    .          -.5 3
            Final          3420.805                     .
```

```
End of Cycle 1: deviance =      3420.805
```

```
x5          null    FP2    3494.867    74.143  0.000*   .          -2 -1
            lin.           3451.795    31.071  0.000+   1
            FP1            3428.023     7.299  0.026+   0
            Final          3420.724                    -2 -1
x6          null    FP2    3452.093    32.704  0.000*   .           0 0
            lin.           3427.703     8.313  0.040+   1
            FP1            3420.724     1.334  0.513    .5
            Final          3420.724                    .5
[hormon included with 1 df in model]
x4a         null    lin.   3425.310     4.586  0.032*   .           1
            Final          3420.724                     1
x3          null    FP2    3420.724     5.305  0.257    .          -.5 0
            Final          3420.724                     .
x2          null    lin.   3420.724     0.214  0.644    .           1
            Final          3420.724                     .
x4b         null    lin.   3420.724     0.145  0.703    .           1
            Final          3420.724                     .
x1          null    FP2    3440.057    19.333  0.001*   .          -2 -.5
            lin.           3440.038    19.314  0.000+   1
            FP1            3436.949    16.225  0.000+  -2
            Final          3420.724                    -2 -.5
x7          null    FP2    3420.724     2.152  0.708    .          -1 3
            Final          3420.724                     .
```

Fractional polynomial fitting algorithm converged after 2 cycles.

Transformations of covariates:

```
-> gen double Ix1__1 = X^-2-.0355294635 if e(sample)
-> gen double Ix1__2 = X^-.5-.4341573547 if e(sample)
   (where: X = x1/10)
-> gen double Ix5__1 = X^-2-3.983723313 if e(sample)
-> gen double Ix5__2 = X^-1-1.99592668 if e(sample)
   (where: X = x5/10)
-> gen double Ix6__1 = X^.5-.3331600619 if e(sample)
   (where: X = (x6+1)/1000)
```

Final multivariable fractional polynomial model for _t

Variable	Initial			Final		
	df	Select	Alpha	Status	df	Powers
x1	4	0.0500	0.0500	in	4	-2 -.5
x2	1	0.0500	0.0500	out	0	
x3	4	0.0500	0.0500	out	0	
x4a	1	0.0500	0.0500	in	1	1
x4b	1	0.0500	0.0500	out	0	
x5	4	0.0500	0.0500	in	4	-2 -1
x6	4	0.0500	0.0500	in	2	.5
x7	4	0.0500	0.0500	out	0	
hormon	1	1.0000	0.0500	in	1	1

```
Cox regression -- Breslow method for ties
Entry time _t0                                 Number of obs    =         686
                                               LR chi2(7)       =      155.62
                                               Prob > chi2      =      0.0000
Log likelihood = -1710.3619                    Pseudo R2        =      0.0435
```

_t	Coef.	Std. Err.	z	P>|z|	[95% Conf. Interval]	
Ix1__1	44.73377	8.256682	5.42	0.000	28.55097	60.91657
Ix1__2	-17.92302	3.909611	-4.58	0.000	-25.58571	-10.26032
x4a	.5006982	.2496324	2.01	0.045	.0114276	.9899687
Ix5__1	.0387904	.0076972	5.04	0.000	.0237041	.0538767
Ix5__2	-.5490645	.0864255	-6.35	0.000	-.7184554	-.3796736
Ix6__1	-1.806966	.3506314	-5.15	0.000	-2.494191	-1.119741
hormon	-.4024169	.1280843	-3.14	0.002	-.6534575	-.1513763

```
Deviance: 3420.724.
```

Some explanation of the output from the model-selection algorithm is desirable. Consider the first few lines of output in the iteration log:

```
1. Deviance for model with all terms untransformed = 3471.637, 686 observations
      Variable     Model (vs.)    Deviance  Dev diff.    P      Powers   (vs.)

2. x5             null    FP2     3503.610   61.366   0.000*     .         .5 3
3.                lin.            3471.637   29.393   0.000+     1
4.                FP1             3449.203    6.959   0.031+     0
5.                Final           3442.244                      .5 3
```

Line 1 gives the deviance ($-2 \times \log$ partial likelihood) for the Cox model with all terms linear, the place where the algorithm starts. The model is modified variable by variable in subsequent steps. The most significant linear term turns out to be x5, which is therefore processed first. Line 2 compares the best-fitting FP2 for x5 with a model omitting x5. The FP has powers (0.5, 3), and the test for inclusion of x5 is highly significant. The reported deviance of 3,503.610 is of the null model, not for the FP2 model. The deviance for the FP2 model may be calculated by subtracting the deviance difference (Dev diff.) from the reported deviance, giving $3,503.610 - 61.366 = 3,442.244$. Line 3 shows that the FP2 model is also a significantly better fit than a straight line (lin.) and line 4 that FP2 is also somewhat better than FP1 ($p = 0.031$). Thus at this stage in the model-selection procedure, the final model for x5 (line 5) is FP2 with powers (0.5, 3). The overall model with an FP2 for x5 and all other terms linear has a deviance of 3,442.244.

After all the variables have been processed (cycle 1) and reprocessed (cycle 2) in this way, convergence is achieved because the functional forms (FP powers and variables included) after cycle 2 are the same as they were after cycle 1. The model finally chosen is Model II as given in tables 3 and 4 of Sauerbrei and Royston (1999). Because of scaling of variables, the regression coefficients reported there are different, but the model and its deviance are identical. The model includes x1 with powers $(-2, -0.5)$, x4a, x5 with powers $(-2, -1)$, and x6 with power 0.5. There is strong evidence of nonlinearity for x1 and for x5, the deviance differences for comparison with a straight-line model (FP2 vs lin.) being, respectively, 19.3 and 31.1 at convergence (cycle 2). Predictors x2, x3, x4b, and x7 are dropped, as may be seen from their status out in the table Final multivariable fractional polynomial model for _t (the assumed *depvar* when using stcox).

All predictors except x4a and hormon, which are binary, have been centered on the mean of the original variable. For example, the mean of x1 (age) is 53.05 years. The first FP-transformed variable for x1 is x1^-2 and is created by the expression gen double Ix1__1 = X^-2-.0355 if e(sample). The value 0.0355 is obtained from $(53.05/10)^{-2}$. The division by 10 is applied automatically to improve the scaling of the regression coefficient for Ix1__1.

According to Sauerbrei and Royston (1999), medical knowledge dictates that the estimated risk function for x5 (number of positive nodes), which was based on the above FP with powers $(-2, -1)$, should be monotonic, but it was not. They improved Model II by estimating a preliminary exponential transformation, $x5e = \exp(-0.12 \cdot x5)$, for x5 and fitting a degree 1 FP for x5e, thus obtaining a monotonic risk function. The value of -0.12 was estimated univariately using nonlinear Cox regression with the ado-file boxtid (Royston and Ambler 1999b, 1999d). To ensure a negative exponent, Sauerbrei and Royston (1999) restricted the powers for x5e to be positive. Their Model III may be fit by using the following command:

```
. mfp, alpha(.05) select(.05, hormon:1) df(x5e:2) xpowers(x5e:0.5 1 2 3):
> stcox x1 x2 x3 x4a x4b x5e x6 x7 hormon
```

Other than the customization for x5e, the command is the same as it was before. The resulting model is as reported in table 4 of Sauerbrei and Royston (1999):

```
. use http://www.stata-press.com/data/r12/brcancer, clear
(German breast cancer data)
. stset rectime, fail(censrec)
  (output omitted )
. mfp, alpha(.05) select(.05, hormon:1) df(x5e:2) xpowers(x5e:0.5 1 2 3):
> stcox x1 x2 x3 x4a x4b x5e x6 x7 hormon, nohr
  (output omitted )
```

Final multivariable fractional polynomial model for _t

Variable	Initial			Final		
	df	Select	Alpha	Status	df	Powers
x1	4	0.0500	0.0500	in	4	-2 -.5
x2	1	0.0500	0.0500	out	0	
x3	4	0.0500	0.0500	out	0	
x4a	1	0.0500	0.0500	in	1	1
x4b	1	0.0500	0.0500	out	0	
x5e	2	0.0500	0.0500	in	1	1
x6	4	0.0500	0.0500	in	2	.5
x7	4	0.0500	0.0500	out	0	
hormon	1	1.0000	0.0500	in	1	1

```
Cox regression -- Breslow method for ties
Entry time _t0                              Number of obs   =        686
                                            LR chi2(6)      =     153.11
                                            Prob > chi2     =     0.0000
Log likelihood = -1711.6186                 Pseudo R2       =     0.0428
```

_t	Coef.	Std. Err.	z	P>|z|	[95% Conf. Interval]	
Ix1__1	43.55382	8.253433	5.28	0.000	27.37738	59.73025
Ix1__2	-17.48136	3.911882	-4.47	0.000	-25.14851	-9.814212
x4a	.5174351	.2493739	2.07	0.038	.0286713	1.006199
Ix5e__1	-1.981213	.2268903	-8.73	0.000	-2.425909	-1.536516
Ix6__1	-1.84008	.3508432	-5.24	0.000	-2.52772	-1.15244
hormon	-.3944998	.128097	-3.08	0.002	-.6455654	-.1434342

Deviance: 3423.237.

◁

Saved results

In addition to what *regression_cmd* saves, `mfp` saves the following in `e()`:

Scalars
e(fp_nx)	number of predictors in *xvarlist*
e(fp_dev)	deviance of final model fit
e(Fp_id#)	initial degrees of freedom for the #th element of *xvarlist*
e(Fp_fd#)	final degrees of freedom for the #th element of *xvarlist*
e(Fp_al#)	FP selection level for the #th element of *xvarlist*
e(Fp_se#)	backward elimination selection level for the #th element of *xvarlist*

Macros
e(fp_cmd)	fracpoly
e(fp_cmd2)	mfp
e(cmdline)	command as typed
e(fracpoly)	command used to fit the selected model using fracpoly
e(fp_fvl)	variables in final model
e(fp_depv)	*yvar₁* (*yvar₂*)
e(fp_opts)	estimation command options
e(fp_x1)	first variable in *xvarlist*
e(fp_x2)	second variable in *xvarlist*
...	
e(fp_x*N*)	last variable in *xvarlist*, $N=$e(fp_nx)
e(fp_k1)	power for first variable in *xvarlist* (*)
e(fp_k2)	power for second variable in *xvarlist* (*)
...	
e(fp_k*N*)	power for last var. in *xvarlist* (*), $N=$e(fp_nx)

Note: (*) contains '.' if the variable is not selected in the final model.

Methods and formulas

`mfp` is implemented as an ado-file.

Acknowledgments

`mfp` is an update of `mfracpol` by Royston and Ambler (1998).

References

Ambler, G., and P. Royston. 2001. Fractional polynomial model selection procedures: Investigation of Type I error rate. *Journal of Statistical Computation and Simulation* 69: 89–108.

Royston, P., and D. G. Altman. 1994. Regression using fractional polynomials of continuous covariates: Parsimonious parametric modelling. *Applied Statistics* 43: 429–467.

Royston, P., and G. Ambler. 1998. sg81: Multivariable fractional polynomials. *Stata Technical Bulletin* 43: 24–32. Reprinted in *Stata Technical Bulletin Reprints*, vol. 8, pp. 123–132. College Station, TX: Stata Press.

———. 1999a. sg112: Nonlinear regression models involving power or exponential functions of covariates. *Stata Technical Bulletin* 49: 25–30. Reprinted in *Stata Technical Bulletin Reprints*, vol. 9, pp. 173–179. College Station, TX: Stata Press.

———. 1999b. sg81.1: Multivariable fractional polynomials: Update. *Stata Technical Bulletin* 49: 17–23. Reprinted in *Stata Technical Bulletin Reprints*, vol. 9, pp. 161–168. College Station, TX: Stata Press.

———. 1999c. sg112.1: Nonlinear regression models involving power or exponential functions of covariates: Update. *Stata Technical Bulletin* 50: 26. Reprinted in *Stata Technical Bulletin Reprints*, vol. 9, p. 180. College Station, TX: Stata Press.

———. 1999d. sg81.2: Multivariable fractional polynomials: Update. *Stata Technical Bulletin* 50: 25. Reprinted in *Stata Technical Bulletin Reprints*, vol. 9, p. 168. College Station, TX: Stata Press.

Royston, P., and W. Sauerbrei. 2007. Multivariable modeling with cubic regression splines: A principled approach. *Stata Journal* 7: 45–70.

———. 2008. *Multivariable Model-building: A Pragmatic Approach to Regression Analysis Based on Fractional Polynomials for Modelling Continuous Variables*. Chichester, UK: Wiley.

———. 2009a. Two techniques for investigating interactions between treatment and continuous covariates in clinical trials. *Stata Journal* 9: 230–251.

———. 2009b. Bootstrap assessment of the stability of multivariable models. *Stata Journal* 9: 547–570.

Sauerbrei, W., and P. Royston. 1999. Building multivariable prognostic and diagnostic models: Transformation of the predictors by using fractional polynomials. *Journal of the Royal Statistical Society, Series A* 162: 71–94.

———. 2002. Corrigendum: Building multivariable prognostic and diagnostic models: Transformation of the predictors by using fractional polynomials. *Journal of the Royal Statistical Society, Series A* 165: 399–400.

Also see

[R] **mfp postestimation** — Postestimation tools for mfp

[R] **fracpoly** — Fractional polynomial regression

[U] **20 Estimation and postestimation commands**

Title

> **mfp postestimation** — Postestimation tools for mfp

Description

The following postestimation commands are of special interest after `mfp`:

Command	Description
fracplot	plot data and fit from most recently fit fractional polynomial model
fracpred	create variable containing prediction, deviance residuals, or SEs of fitted values

For `fracplot` and `fracpred`, see [R] **fracpoly postestimation**.

The following standard postestimation commands are also available if available after *regression_cmd*:

Command	Description
estat	AIC, BIC, VCE, and estimation sample summary
estimates	cataloging estimation results
lincom	point estimates, standard errors, testing, and inference for linear combinations of coefficients
linktest	link test for model specification
lrtest	likelihood-ratio test
nlcom	point estimates, standard errors, testing, and inference for nonlinear combinations of coefficients
test	Wald tests of simple and composite linear hypotheses
testnl	Wald tests of nonlinear hypotheses

See the corresponding entries in the *Base Reference Manual* for details.

Methods and formulas

All postestimation commands listed above are implemented as ado-files.

Also see

[R] **mfp** — Multivariable fractional polynomial models

[R] **fracpoly postestimation** — Postestimation tools for fracpoly

[U] **20 Estimation and postestimation commands**

> **misstable** — Tabulate missing values

Report counts of missing values

> misstable <u>sum</u>marize [*varlist*] [*if*] [*in*] [, *summarize_options*]

Report pattern of missing values

> misstable <u>pat</u>terns [*varlist*] [*if*] [*in*] [, *patterns_options*]

Present a tree view of the pattern of missing values

> misstable tree [*varlist*] [*if*] [*in*] [, *tree_options*]

List the nesting rules that describe the missing-value pattern

> misstable <u>nest</u>ed [*varlist*] [*if*] [*in*] [, *nested_options*]

summarize_options	Description
all	show all variables
<u>show</u>zeros	show zeros in table
<u>g</u>enerate(*stub* [, exok])	generate missing-value indicators

patterns_options	Description
asis	use variables in order given
<u>freq</u>uency	report frequencies instead of percentages
exok	treat .a, .b, ..., .z as nonmissing
replace	replace data in memory with dataset of patterns
clear	okay to replace even if original unsaved
<u>by</u>patterns	list by patterns rather than by frequency

tree_options	Description
asis	use variables in order given
<u>freq</u>uency	report frequencies instead of percentages
exok	treat .a, .b, ..., .z as nonmissing

nested_options	Description
exok	treat .a, .b, ..., .z as nonmissing

In addition, programmer's option `nopreserve` is allowed with all syntaxes; see [P] **nopreserve option**.

Menu

Statistics > Summaries, tables, and tests > Tables > Tabulate missing values

Description

`misstable` makes tables that help you understand the pattern of missing values in your data.

Options for misstable summarize

`all` specifies that the table should include all the variables specified or all the variables in the dataset. The default is to include only numeric variables that contain missing values.

`showzeros` specifies that zeros in the table should display as 0 rather than being omitted.

`generate(`*stub* [, `exok`]`)` requests that a missing-value indicator *newvar*, a new binary variable containing 0 for complete observations and 1 for incomplete observations, be generated for every numeric variable in *varlist* containing missing values. If the `all` option is specified, missing-value indicators are created for all the numeric variables specified or for all the numeric variables in the dataset. If `exok` is specified within `generate()`, the extended missing values .a, .b, ..., .z are treated as if they do not designate missing.

For each variable in *varlist*, *newvar* is the corresponding variable name *varname* prefixed with *stub*. If the total length of *stub* and *varname* exceeds 32 characters, *newvar* is abbreviated so that its name does not exceed 32 characters.

Options for misstable patterns

`asis`, `frequency`, and `exok` – see *Common options* below.

`replace` specifies that the data in memory be replaced with a dataset corresponding to the table just displayed; see *misstable patterns* under *Remarks* below.

`clear` is for use with `replace`; it specifies that it is okay to change the data in memory even if they have not been saved to disk.

`bypatterns` specifies the table be ordered by pattern rather than by frequency. That is, `bypatterns` specifies that patterns containing one incomplete variable be listed first, followed by those for two incomplete variables, and so on. The default is to list the most frequent pattern first, followed by the next most frequent pattern, etc.

Options for misstable tree

`asis`, `frequency`, and `exok` – see *Common options* below.

Option for misstable nested

exok – see *Common options* below.

Common options

asis specifies that the order of the variables in the table be the same as the order in which they are specified on the misstable command. The default is to order the variables by the number of missing values, and within that, by the amount of overlap of missing values.

frequency specifies that the table should report frequencies instead of percentages.

exok specifies that the extended missing values .a, .b, ..., .z should be treated as if they do not designate missing. Some users use extended missing values to designate values that are missing for a known and valid reason.

nopreserve is a programmer's option allowed with all misstable commands; see [P] **nopreserve option**.

Remarks

Remarks are presented under the following headings:

> *misstable summarize*
> *misstable patterns*
> *misstable tree*
> *misstable nested*
> *Execution time of misstable nested*

In what follows, we will use data from a 125-observation, fictional, student-satisfaction survey:

```
. use http://www.stata-press.com/data/r12/studentsurvey
(Student Survey)
. summarize
```

Variable	Obs	Mean	Std. Dev.	Min	Max
m1	125	2.456	.8376619	1	4
m2	125	2.472	.8089818	1	4
age	122	18.97541	.8763477	17	21
female	122	.5245902	.5014543	0	1
dept	116	2.491379	1.226488	1	4
offcampus	125	.36	.4819316	0	1
comment	0				

The m1 and m2 variables record the student's satisfaction with teaching and with academics. comment is a string variable recording any comments the student might have had.

misstable summarize

▷ Example 1

misstable summarize reports counts of missing values:

. misstable summarize

				Obs<.		
Variable	Obs=.	Obs>.	Obs<.	Unique values	Min	Max
age	3		122	5	17	21
female	3		122	2	0	1
dept	9		116	4	1	4

Stata provides 27 different missing values, namely, ., .a, .b, . . . , .z. The first of those, ., is often called system missing. The remaining missing values are called extended missings. The nonmissing and missing values are ordered *nonmissing* $< . < .a < .b < \cdots < .z$. Thus reported in the column "Obs=." are counts of system missing values; in the column "Obs>.", extended missing values; and in the column "Obs<.", nonmissing values.

The rightmost portion of the table is included to remind you how the variables are encoded.

Our data contain seven variables and yet misstable reported only three of them. The omitted variables contain no missing values or are string variables. Even if we specified the varlist explicitly, those variables would not appear in the table unless we specified the all option.

We can also create missing-value indicators for each of the variables above using the generate() option:

. quietly misstable summarize, generate(miss_)

. describe miss_*

variable name	storage type	display format	value label	variable label
miss_age	byte	%8.0g		(age>=.)
miss_female	byte	%8.0g		(female>=.)
miss_dept	byte	%8.0g		(dept>=.)

For each variable containing missing values, the generate() option creates a new binary variable containing 0 for complete observations and 1 for incomplete observations. In our example, three new missing-value indicators are generated, one for each of the incomplete variables age, female, and dept. The naming convention of generate() is to prefix the corresponding variable names with the specified *stub*, which is miss_ in this example.

Missing-value indicators are useful, for example, for checking whether data are missing completely at random. They are also often used within the multiple-imputation context to identify the observed and imputed data; see [MI] **intro substantive** for a general introduction to multiple imputation. Within Stata's multiple-imputation commands, an incomplete value is identified by the system missing value, a dot. By default, misstable summarize, generate() marks the extended missing values as incomplete values, as well. You can use exok within generate() to treat extended missing values as complete when creating missing-value identifiers.

◁

misstable patterns

Example 2

misstable patterns reports the pattern of missing values:

```
. misstable patterns
   Missing-value patterns
                      Pattern
     Percent     1   2   3

         93%     1   1   1

           5     1   1   0
           2     0   0   0

        100%
   Variables are  (1) age  (2) female  (3) dept
```

There are three patterns in these data: (1,1,1), (1,1,0), and (0,0,0). By default, the rows of the table are ordered by frequency. In larger tables that have more patterns, it is sometimes useful to order the rows by pattern. We could have obtained that by typing mi misstable patterns, bypatterns.

In a pattern, 1 indicates that all values of the variable are nonmissing and 0 indicates that all values are missing. Thus pattern (1,1,1) means no missing values, and 93% of our data have that pattern. There are two patterns in which variables are missing, (1,1,0) and (0,0,0). Pattern (1,1,0) means that age is nonmissing, female is nonmissing, and dept is missing. The order of the variables in the patterns appears in the key at the bottom of the table. Five percent of the observations have pattern (1,1,0). The remaining 2% have pattern (0,0,0), meaning that all three variables contain missing.

As with misstable summarize, only numeric variables that contain missing are listed, so had we typed misstable patterns comments age female offcampus dept, we still would have obtained the same table. Variables that are automatically omitted contain no missing values or are string variables.

The variables in the table are ordered from lowest to highest frequency of missing values, although you cannot see that from the information presented in the table. The variables are ordered this way even if you explicitly specify the varlist with a different ordering. Typing misstable patterns dept female age would produce the same table as above. Specify the asis option if you want the variables in the order in which you specify them.

You can obtain a dataset of the patterns by specifying the replace option:

```
. misstable patterns, replace clear
   Missing-value patterns
                      Pattern
     Percent     1   2   3

         93%     1   1   1

           5     1   1   0
           2     0   0   0

        100%
   Variables are  (1) age  (2) female  (3) dept
(summary data now in memory)
```

```
. list
```

	_freq	age	female	dept
1.	3	0	0	0
2.	6	1	1	0
3.	116	1	1	1

The differences between the dataset and the printed table are that 1) the dataset always records frequency and 2) the rows are reversed.

◁

misstable tree

▷ Example 3

misstable tree presents a tree view of the pattern of missing values:

```
. use http://www.stata-press.com/data/r12/studentsurvey, clear
(Student Survey)
. misstable tree, frequency
```

Nested pattern of missing values		
dept	age	female
------	-----	--------
9	3	3
		0
	6	0
		6
116	0	0
		0
	116	0
		116

(number missing listed first)

In this example, we specified the frequency option to see the table in frequency rather than percentage terms. In the table, each column sums to the total number of observations in the data, 125. Variables are ordered from those with the most missing values to those with the least. Start with the first column. The dept variable is missing in 9 observations and, farther down, the table reports that it is not missing in 116 observations.

Go back to the first row and read across, but only to the second column. The dept variable is missing in 9 observations. Within those 9, age is missing in 3 of them and is not missing in the remaining 6. Reading down the second column, within the 116 observations that dept is not missing, age is missing in 0 and not missing in 116.

Reading straight across the first row again, dept is missing in 9 observations, and within the 9, age is missing in 3, and within the 3, female is also missing in 3. Skipping down just a little, within the 6 observations for which dept is missing and age is not missing, female is not missing, too.

◁

misstable nested

> Example 4

misstable nested lists the nesting rules that describe the missing-value pattern,

```
. misstable nested
    1.  female(3) <-> age(3) -> dept(9)
```

This line says that in observations in which female is missing, so is age missing, and vice versa, and in observations in which age (or female) is missing, so is dept. The numbers in parentheses are counts of the missing values. The female variable happens to be missing in 3 observations, and the same is true for age; the dept variable is missing in 9 observations. Thus dept is missing in the 3 observations for which age and female are missing, and in 6 more observations, too.

In these data, it turns out that the missing-value pattern can be summarized in one statement. In a larger dataset, you might see something like this:

```
. misstable nested
    1.  female(50) <-> age(50) -> dept(120)
    2.  female(50) -> m1(58)
    3.  offcampus(11)
```

misstable nested accounts for every missing value. In the above, in addition to female <-> age -> dept, we have that female -> m1, and we have offcampus, the last all by itself. The last line says that the 11 missing values in offcampus are not themselves nested in the missing value of any other variable, nor do they imply the missing values in another variable. In some datasets, all the statements will be of this last form.

In our data, however, we have one statement:

```
. misstable nested
    1.  female(3) <-> age(3) -> dept(9)
```

When the missing-value pattern can be summarized in one misstable nested statement, the pattern of missing values in the data is said to be monotone.

◁

Execution time of misstable nested

The execution time of misstable nested is affected little by the number of observations but can grow quickly with the number of variables, depending on the fraction of missing values within variable. The execution time of the example above, which has 3 variables containing missing, is instant. In worst-case scenarios, with 500 variables, the time might be 25 seconds; with 1,000 variables, the execution time might be closer to an hour.

In situations where misstable nested takes a long time to complete, it will produce thousands of rules that will defy interpretation. A 523-variable dataset we have seen ran in 20 seconds and produced 8,040 rules. Although we spotted a few rules in the output that did not surprise us, such as the year of the date being missing implied that the month and the day were also missing, mostly the output was not helpful.

If you have such a dataset, we recommend you run misstable on groups of variables that you have reason to believe the pattern of missing values might be related.

Saved results

misstable summarize saves the following values of the last variable summarized in r():

Scalars
r(N_eq_dot)	number of observations containing .
r(N_gt_dot)	number of observations containing .a, .b, ..., .z
r(N_lt_dot)	number of observations containing nonmissing
r(K_uniq)	number of unique, nonmissing values
r(min)	variable's minimum value
r(max)	variable's maximum value

Macros
r(vartype)	numeric, string, or none

r(K_uniq) contains . if the number of unique, nonmissing values is greater than 500. r(vartype) contains none if no variables are summarized, and in that case, the value of the scalars are all set to missing (.). Programmers intending to access results after misstable summarize should specify the all option.

misstable patterns saves the following in r():

Scalars
r(N_complete)	number of complete observations
r(N_incomplete)	number of incomplete observations
r(K)	number of patterns

Macros
r(vars)	variables used in order presented

r(N_complete) and r(N_incomplete) are defined with respect to the variables specified if variables were specified and otherwise, defined with respect to all the numeric variables in the dataset. r(N_complete) is the number of observations that contain no missing values.

misstable tree saves the following in r():

Macros
r(vars)	variables used in order presented

misstable nested saves the following in r():

Scalars
r(K)	number of statements

Macros
r(stmt1)	first statement
r(stmt2)	second statement
.	.
.	.
r(stmt'r(K)')	last statement
r(stmt1wc)	r(stmt1) with missing-value counts
r(vars)	variables considered

A statement is encoded "*varname*", "*varname op varname*", or "*varname op varname op varname*", and so on; *op* is either "->" or "<->".

Methods and formulas

misstable is implemented as an ado-file.

Also see

[MI] **mi misstable** — Tabulate pattern of missing values

[R] **summarize** — Summary statistics

[R] **tabulate oneway** — One-way tables of frequencies

[R] **tabulate twoway** — Two-way tables of frequencies

Title

> **mkspline** — Linear and restricted cubic spline construction

Syntax

Linear spline with knots at specified points

> mkspline *newvar*$_1$ #$_1$ [*newvar*$_2$ #$_2$ [...]] *newvar*$_k$ = *oldvar* [*if*] [*in*] [, $\underline{\text{m}}$arginal
>
> $\underline{\text{di}}$splayknots]

Linear spline with knots equally spaced or at percentiles of data

> mkspline *stubname* # = *oldvar* [*if*] [*in*] [*weight*] [, $\underline{\text{m}}$arginal $\underline{\text{p}}$ctile
>
> $\underline{\text{di}}$splayknots]

Restricted cubic spline

> mkspline *stubname* = *oldvar* [*if*] [*in*] [*weight*], cubic [$\underline{\text{nk}}$nots(#) $\underline{\text{k}}$nots(*numlist*)
>
> $\underline{\text{di}}$splayknots]

fweights are allowed with the second and third syntax; see [U] **11.1.6 weight**.

Menu

Data > Create or change data > Other variable-creation commands > Linear and cubic spline construction

Description

mkspline creates variables containing a linear spline or a restricted cubic spline of *oldvar*.

In the first syntax, mkspline creates *newvar*$_1$, ..., *newvar*$_k$ containing a linear spline of *oldvar* with knots at the specified #$_1$, ..., #$_{k-1}$.

In the second syntax, mkspline creates # variables named *stubname*1, ..., *stubname*# containing a linear spline of *oldvar*. The knots are equally spaced over the range of *oldvar* or are placed at the percentiles of *oldvar*.

In the third syntax, mkspline creates variables containing a restricted cubic spline of *oldvar*. This is also known as a natural spline. The location and spacing of the knots is determined by the specification of the nknots() and knots() options.

Options

___ Options ___

marginal is allowed with the first or second syntax. It specifies that the new variables be constructed so that, when used in estimation, the coefficients represent the change in the slope from the preceding interval. The default is to construct the variables so that, when used in estimation, the coefficients measure the slopes for the interval.

1174

displayknots displays the values of the knots that were used in creating the linear or restricted cubic spline.

pctile is allowed only with the second syntax. It specifies that the knots be placed at percentiles of the data rather than being equally spaced over the range.

nknots(*#*) is allowed only with the third syntax. It specifies the number of knots that are to be used for a restricted cubic spline. This number must be between 3 and 7 unless the knot locations are specified using knots(). The default number of knots is 5.

knots(*numlist*) is allowed only with the third syntax. It specifies the exact location of the knots to be used for a restricted cubic spline. The values of these knots must be given in increasing order. When this option is omitted, the default knot values are based on Harrell's recommended percentiles with the additional restriction that the smallest knot may not be less than the fifth-smallest value of *oldvar* and the largest knot may not be greater than the fifth-largest value of *oldvar*. If both nknots() and knots() are given, they must specify the same number of knots.

Remarks

Remarks are presented under the following headings:

> *Linear splines*
> *Restricted cubic splines*

Linear splines

Linear splines allow estimating the relationship between y and x as a piecewise linear function, which is a function composed of linear segments—straight lines. One linear segment represents the function for values of x below x_0, another linear segment handles values between x_0 and x_1, and so on. The linear segments are arranged so that they join at x_0, x_1, ..., which are called the knots. An example of a piecewise linear function is shown below.

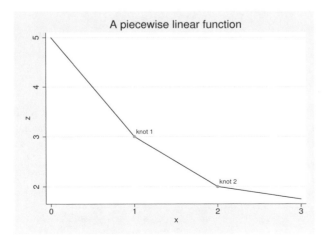

▷ Example 1

We wish to fit a model of log income on education and age by using a piecewise linear function for age:

$$\texttt{lninc} = b_0 + b_1\,\texttt{educ} + f(\texttt{age}) + u$$

The knots are to be placed at 10-year intervals: 20, 30, 40, 50, and 60.

```
. use http://www.stata-press.com/data/r12/mksp1
. mkspline age1 20 age2 30 age3 40 age4 50 age5 60 age6 = age, marginal
. regress lninc educ age1-age6
(output omitted)
```

Because we specified the `marginal` option, we could test whether the age effect is the same in the 30–40 and 40–50 intervals by asking whether the `age4` coefficient is zero. With the `marginal` option, coefficients measure the change in slope from the preceding group. Specifying `marginal` changes only the interpretation of the coefficients; the same model is fit in either case. Without the `marginal` option, the interpretation of the coefficients would have been

$$\frac{dy}{d\texttt{age}} = \begin{cases} a_1 & \text{if age} < 20 \\ a_2 & \text{if } 20 \le \text{age} < 30 \\ a_3 & \text{if } 30 \le \text{age} < 40 \\ a_4 & \text{if } 40 \le \text{age} < 50 \\ a_5 & \text{if } 50 \le \text{age} < 60 \\ a_6 & \text{otherwise} \end{cases}$$

With the `marginal` option, the interpretation is

$$\frac{dy}{d\texttt{age}} = \begin{cases} a_1 & \text{if age} < 20 \\ a_1 + a_2 & \text{if } 20 \le \text{age} < 30 \\ a_1 + a_2 + a_3 & \text{if } 30 \le \text{age} < 40 \\ a_1 + a_2 + a_3 + a_4 & \text{if } 40 \le \text{age} < 50 \\ a_1 + a_2 + a_3 + a_4 + a_5 & \text{if } 50 \le \text{age} < 60 \\ a_1 + a_2 + a_3 + a_4 + a_5 + a_6 & \text{otherwise} \end{cases}$$

◁

▷ Example 2

Say that we have a binary outcome variable called `outcome`. We are beginning an analysis and wish to parameterize the effect of dosage on outcome. We wish to divide the data into five equal-width groups of dosage for the piecewise linear function.

```
. use http://www.stata-press.com/data/r12/mksp2
. mkspline dose 5 = dosage, displayknots
```

	knot1	knot2	knot3	knot4
dosage	20	40	60	80

```
. logistic outcome dose1-dose5
(output omitted)
```

mkspline dose 5 = dosage creates five variables—dose1, dose2, ..., dose5—equally spacing the knots over the range of dosage. Because dosage varied between 0 and 100, the mkspline command above has the same effect as typing

    . mkspline dose1 20 dose2 40 dose3 60 dose4 80 dose5 = dosage

The pctile option sets the knots to divide the data into five equal sample-size groups rather than five equal-width ranges. Typing

    . mkspline pctdose 5 = dosage, pctile displayknots

	knot1	knot2	knot3	knot4
dosage	16	36.4	55.6	82

places the knots at the 20th, 40th, 60th, and 80th percentiles of the data.

◁

Restricted cubic splines

A linear spline can be used to fit many functions well. However, a restricted cubic spline may be a better choice than a linear spline when working with a very curved function. When using a restricted cubic spline, one obtains a continuous smooth function that is linear before the first knot, a piecewise cubic polynomial between adjacent knots, and linear again after the last knot.

▶ Example 3

Returning to the data from example 1, we may feel that a curved function is a better fit. First, we will use the knots() option to specify the five knots that we used previously.

    . use http://www.stata-press.com/data/r12/mksp1, clear
    . mkspline agesp = age, cubic knots(20 30 40 50 60)
    . regress lninc educ agesp*
    (output omitted)

Harrell (2001, 23) recommends placing knots at equally spaced percentiles of the original variable's marginal distribution. If we do not specify the knots() option, variables will be created containing a restricted cubic spline with five knots determined by Harrell's default percentiles.

    . use http://www.stata-press.com/data/r12/mksp1, clear
    . mkspline agesp = age, cubic displayknots
    . regress lninc educ agesp*
    (output omitted)

◁

Methods and formulas

mkspline is implemented as an ado-file.

Methods and formulas are presented under the following headings:

> *Linear splines*
> *Restricted cubic splines*

Linear splines

Let V_i, $i = 1, \ldots, n$, be the variables to be created; k_i, $i = 1, \ldots, n - 1$, be the corresponding knots; and V be the original variable (the command is `mkspline` V_1 k_1 V_2 k_2 ... V_n = V). Then

$$V_1 = \min(V, k_1)$$
$$V_i = \max\left\{\min(V, k_i), k_{i-1}\right\} - k_{i-1} \quad i = 2, \ldots, n$$

If the `marginal` option is specified, the definitions are

$$V_1 = V$$
$$V_i = \max(0, V - k_{i-1}) \quad i = 2, \ldots, n$$

In the second syntax, `mkspline` *stubname* # = V, so let m and M be the minimum and maximum of V. Without the `pctile` option, knots are set at $m + (M - m)(i/n)$ for $i = 1, \ldots, n - 1$. If `pctile` is specified, knots are set at the $100(i/n)$ percentiles, for $i = 1, \ldots, n - 1$. Percentiles are calculated by `centile`; see [R] **centile**.

Restricted cubic splines

Let k_i, $i = 1, \ldots, n$, be the knot values; V_i, $i = 1, \ldots, n - 1$, be the variables to be created; and V be the original variable. Then

$$V_1 = V$$
$$V_{i+1} = \frac{(V - k_i)_+^3 - (k_n - k_{n-1})^{-1}\{(V - k_{n-1})_+^3(k_n - k_i) - (V - k_n)_+^3(k_{n-1} - k_i)\}}{(k_n - k_1)^2}$$
$$i = 1, \ldots, n - 2$$

where

$$(u)_+ = \begin{cases} u, & \text{if } u > 0 \\ 0, & \text{if } u \leq 0 \end{cases}$$

Without the `knots()` option, the locations of the knots are determined by the percentiles recommended in Harrell (2001, 23). These percentiles are based on the chosen number of knots as follows:

No. of knots	Percentiles						
3	10	50	90				
4	5	35	65	95			
5	5	27.5	50	72.5	95		
6	5	23	41	59	77	95	
7	2.5	18.33	34.17	50	65.83	81.67	97.5

Harrell provides default percentiles when the number of knots is between 3 and 7. When using a number of knots outside this range, the location of the knots must be specified in `knots()`.

Acknowledgment

The restricted cubic spline portion of `mkspline` is based on the `rc_spline` command by William Dupont.

References

Gould, W. W. 1993. sg19: Linear splines and piecewise linear functions. *Stata Technical Bulletin* 15: 13–17. Reprinted in *Stata Technical Bulletin Reprints*, vol. 3, pp. 98–104. College Station, TX: Stata Press.

Greene, W. H. 2012. *Econometric Analysis*. 7th ed. Upper Saddle River, NJ: Prentice Hall.

Harrell, F. E., Jr. 2001. *Regression Modeling Strategies: With Applications to Linear Models, Logistic Regression, and Survival Analysis*. New York: Springer.

Newson, R. 2000. sg151: B-splines and splines parameterized by their values at reference points on the x-axis. *Stata Technical Bulletin* 57: 20–27. Reprinted in *Stata Technical Bulletin Reprints*, vol. 10, pp. 221–230. College Station, TX: Stata Press.

Orsini, N., and S. Greenland. 2011. A procedure to tabulate and plot results after flexible modeling of a quantitative covariate. *Stata Journal* 11: 1–29.

Panis, C. 1994. sg24: The piecewise linear spline transformation. *Stata Technical Bulletin* 18: 27–29. Reprinted in *Stata Technical Bulletin Reprints*, vol. 3, pp. 146–149. College Station, TX: Stata Press.

Also see

[R] **fracpoly** — Fractional polynomial regression

Title

ml — Maximum likelihood estimation

Syntax

ml model in interactive mode

> ml <u>mod</u>el   *method progname eq* $\begin{bmatrix} eq \dots \end{bmatrix}$ $\begin{bmatrix} if \end{bmatrix}$ $\begin{bmatrix} in \end{bmatrix}$ $\begin{bmatrix} weight \end{bmatrix}$
>
> $\begin{bmatrix} , & model\_options & \text{svy} & diparm\_options \end{bmatrix}$

> ml <u>mod</u>el   *method funcname*() *eq* $\begin{bmatrix} eq \dots \end{bmatrix}$ $\begin{bmatrix} if \end{bmatrix}$ $\begin{bmatrix} in \end{bmatrix}$ $\begin{bmatrix} weight \end{bmatrix}$
>
> $\begin{bmatrix} , & model\_options & \text{svy} & diparm\_options \end{bmatrix}$

ml model in noninteractive mode

> ml <u>mod</u>el   *method progname eq* $\begin{bmatrix} eq \dots \end{bmatrix}$ $\begin{bmatrix} if \end{bmatrix}$ $\begin{bmatrix} in \end{bmatrix}$ $\begin{bmatrix} weight \end{bmatrix}$, <u>max</u>imize
>
> $\begin{bmatrix} model\_options & \text{svy} & diparm\_options & noninteractive\_options \end{bmatrix}$

> ml <u>mod</u>el   *method funcname*() *eq* $\begin{bmatrix} eq \dots \end{bmatrix}$ $\begin{bmatrix} if \end{bmatrix}$ $\begin{bmatrix} in \end{bmatrix}$ $\begin{bmatrix} weight \end{bmatrix}$, <u>max</u>imize
>
> $\begin{bmatrix} model\_options & \text{svy} & diparm\_options & noninteractive\_options \end{bmatrix}$

Noninteractive mode is invoked by specifying the maximize option. Use maximize when ml will be used as a subroutine of another ado-file or program and you want to carry forth the problem, from definition to posting of results, in one command.

> ml clear

> ml <u>q</u>uery

> ml check

> ml <u>sea</u>rch $\begin{bmatrix} \begin{bmatrix} / \end{bmatrix} eqname \begin{bmatrix} : \end{bmatrix} \#_{lb} \#_{ub} \end{bmatrix}$ $\begin{bmatrix} \dots \end{bmatrix}$ $\begin{bmatrix} , & search\_options \end{bmatrix}$

> ml <u>p</u>lot $\begin{bmatrix} eqname : \end{bmatrix} name$ $\begin{bmatrix} \# & \begin{bmatrix} \# & \begin{bmatrix} \# \end{bmatrix} \end{bmatrix} \end{bmatrix}$ $\begin{bmatrix} , & \underline{\text{sav}}\text{ing}(filename \begin{bmatrix} , & \text{replace} \end{bmatrix}) \end{bmatrix}$

> ml init $\{ \begin{bmatrix} eqname : \end{bmatrix} name \text{=} \# \mid /eqname \text{=} \# \}$ $\begin{bmatrix} \dots \end{bmatrix}$
> ml init $\# \begin{bmatrix} \# \dots \end{bmatrix}$, copy
> ml init   *matname* $\begin{bmatrix} , & \text{copy skip} \end{bmatrix}$

> ml <u>re</u>port

> ml trace $\{ \text{on} \mid \text{off} \}$

> ml count $\begin{bmatrix} \text{clear} \mid \text{on} \mid \text{off} \end{bmatrix}$

> ml <u>max</u>imize $\begin{bmatrix} , & ml\_maximize\_options & display\_options & eform\_option \end{bmatrix}$

ml graph [#] [, saving(*filename*[, replace])]

ml display [, *display_options eform_option*]

ml footnote

ml score *newvar* [*if*] [*in*] [, equation(*eqname*) missing]

ml score *newvarlist* [*if*] [*in*] [, missing]

ml score [*type*] *stub*∗ [*if*] [*in*] [, missing]

where *method* is one of

lf	d0	lf0	gf0
	d1	lf1	
	d1debug	lf1debug	
	d2	lf2	
	d2debug	lf2debug	

or *method* can be specified using one of the longer, more descriptive names

method	Longer name
lf	linearform
d0	derivative0
d1	derivative1
d1debug	derivative1debug
d2	derivative2
d2debug	derivative2debug
lf0	linearform0
lf1	linearform1
lf1debug	linearform1debug
lf2	linearform2
lf2debug	linearform2debug
gf0	generalform0

eq is the equation to be estimated, enclosed in parentheses, and optionally with a name to be given to the equation, preceded by a colon,

   ([*eqname*:] [*varlist_y* =] [*varlist_x*] [, *eq_options*])

or *eq* is the name of a parameter, such as sigma, with a slash in front

   /*eqname*   which is equivalent to   (*eqname*:)

and *diparm_options* is one or more diparm(*diparm_args*) options where *diparm_args* is either __sep__ or anything accepted by the "undocumented" _diparm command; see help _diparm.

eq_options	Description
noconstant	do not include an intercept in the equation
offset(*varname_o*)	include *varname_o* in model with coefficient constrained to 1
exposure(*varname_e*)	include ln(*varname_e*) in model with coefficient constrained to 1

model_options	Description
group(*varname*)	use *varname* to identify groups
vce(*vcetype*)	*vcetype* may be <u>r</u>obust, <u>cl</u>uster *clustvar*, oim, or opg
<u>constr</u>aints(*numlist*)	constraints by number to be applied
<u>constr</u>aints(*matname*)	matrix that contains the constraints to be applied
<u>nocns</u>notes	do not display notes when constraints are dropped
<u>ti</u>tle(*string*)	place a title on the estimation output
<u>nopr</u>eserve	do not preserve the estimation subsample in memory
<u>coll</u>inear	keep collinear variables within equations
<u>miss</u>ing	keep observations containing variables with missing values
lf0($\#_k$ $\#_{ll}$)	number of parameters and log-likelihood value of the constant-only model
<u>cont</u>inue	specifies that a model has been fit and sets the initial values $\mathbf{b}_0$ for the model to be fit based on those results
<u>wald</u>test(*#*)	perform a Wald test; see *Options for use with ml model in interactive or noninteractive mode* below
obs(*#*)	number of observations
crittype(*string*)	describe the criterion optimized by ml
<u>subpop</u>(*varname*)	compute estimates for the single subpopulation
nosvyadjust	carry out Wald test as $W/k \sim F(k, d)$
<u>techn</u>ique(nr)	Stata's modified Newton–Raphson (NR) algorithm
<u>techn</u>ique(bhhh)	Berndt–Hall–Hall–Hausman (BHHH) algorithm
<u>techn</u>ique(dfp)	Davidon–Fletcher–Powell (DFP) algorithm
<u>techn</u>ique(bfgs)	Broyden–Fletcher–Goldfarb–Shanno (BFGS) algorithm

noninteractive_options	Description
<u>init</u>(*ml_init_args*)	set the initial values $\mathbf{b}_0$
<u>sea</u>rch(on)	equivalent to ml search, repeat(0); the default
<u>sea</u>rch(norescale)	equivalent to ml search, repeat(0) norescale
<u>sea</u>rch(quietly)	same as search(on), except that output is suppressed
<u>sea</u>rch(off)	prevents calling ml search
<u>r</u>epeat(*#*)	ml search's repeat() option; see below
<u>bounds</u>(*ml_search_bounds*)	specify bounds for ml search
<u>nowarn</u>ing	suppress "convergence not achieved" message of iterate(0)
novce	substitute the zero matrix for the variance matrix
negh	indicates that the evaluator returns the negative Hessian matrix
<u>sc</u>ore(*newvars*)	new variables containing the contribution to the score
maximize_options	control the maximization process; seldom used

search_options	Description
repeat(#)	number of random attempts to find better initial-value vector; default is repeat(10) in interactive mode and repeat(0) in noninteractive mode
restart	use random actions to find starting values; not recommended
norescale	do not rescale to improve parameter vector; not recommended
maximize_options	control the maximization process; seldom used

ml_maximize_options	Description
nowarning	suppress "convergence not achieved" message of iterate(0)
novce	substitute the zero matrix for the variance matrix
negh	indicates that the evaluator returns the negative Hessian matrix
score(newvars | stub*)	new variables containing the contribution to the score
nooutput	suppress display of final results
noclear	do not clear ml problem definition after model has converged
maximize_options	control the maximization process; seldom used

display_options	Description
noheader	suppress header display above the coefficient table
nofootnote	suppress footnote display below the coefficient table
level(#)	set confidence level; default is level(95)
first	display coefficient table reporting results for first equation only
neq(#)	display coefficient table reporting first # equations
showeqns	display equation names in the coefficient table
plus	display coefficient table ending in dashes–plus-sign–dashes
nocnsreport	suppress constraints display above the coefficient table
noomitted	suppress display of omitted variables
vsquish	suppress blank space separating factor-variable terms or time-series–operated variables from other variables
noemptycells	suppress empty cells for interactions of factor variables
baselevels	report base levels of factor variables and interactions
allbaselevels	display all base levels of factor variables and interactions
cformat(%fmt)	format the coefficients, standard errors, and confidence limits in the coefficient table
pformat(%fmt)	format the p-values in the coefficient table
sformat(%fmt)	format the test statistics in the coefficient table
nolstretch	do not automatically widen the coefficient table to accommodate longer variable names
coeflegend	display legend instead of statistics

eform_option	Description
<u>ef</u>orm(*string*)	display exponentiated coefficients; column title is "*string*"
<u>ef</u>orm	display exponentiated coefficients; column title is "exp(b)"
hr	report hazard ratios
shr	report subhazard ratios
<u>irr</u>	report incidence-rate ratios
or	report odds ratios
<u>rrr</u>	report relative-risk ratios

fweights, aweights, iweights, and pweights are allowed; see [U] **11.1.6 weight**. With all but method lf, you must write your likelihood-evaluation program carefully if pweights are to be specified, and pweights may not be specified with method d0, d1, d1debug, d2, or d2debug. See Gould, Pitblado, and Poi (2010, chap. 6) for details.

See [U] **20 Estimation and postestimation commands** for more capabilities of estimation commands. To redisplay results, type ml display.

Syntax of subroutines for use by evaluator programs

mleval   *newvar* = *vecname* $\big[$, eq(#) $\big]$

mleval   *scalarname* = *vecname* , scalar $\big[$ eq(#) $\big]$

mlsum   *scalarname*$_{\text{lnf}}$ = *exp* $\big[$ *if* $\big]$ $\big[$, <u>noweight</u> $\big]$

mlvecsum   *scalarname*$_{\text{lnf}}$ *rowvecname* = *exp* $\big[$ *if* $\big]$ $\big[$, eq(#) $\big]$

mlmatsum   *scalarname*$_{\text{lnf}}$ *matrixname* = *exp* $\big[$ *if* $\big]$ $\big[$, eq(#$\big[$,#$\big]$) $\big]$

mlmatbysum   *scalarname*$_{\text{lnf}}$ *matrixname* *varname*$_a$ *varname*$_b$ $\big[$ *varname*$_c$ $\big]$ $\big[$ *if* $\big]$,
            by(*varname*) $\big[$ eq(#$\big[$,#$\big]$) $\big]$

Syntax of user-written evaluator

Summary of notation

The log-likelihood function is $\ln L(\theta_{1j}, \theta_{2j}, \ldots, \theta_{Ej})$, where $\theta_{ij} = \mathbf{x}_{ij}\mathbf{b}_i$, $j = 1, \ldots, N$ indexes observations, and $i = 1, \ldots, E$ indexes the linear equations defined by ml model. If the likelihood satisfies the linear-form restrictions, it can be decomposed as $\ln L = \sum_{j=1}^{N} \ln \ell(\theta_{1j}, \theta_{2j}, \ldots, \theta_{Ej})$.

Method-lf evaluators

```
program progname
        version 12
        args lnfj theta1 theta2 ...
        // if you need to create any intermediate results:
        tempvar tmp1 tmp2 ...
        quietly gen double 'tmp1' = ...
        ...
        quietly replace 'lnfj' = ...
end
```

where
      'lnfj'   variable to be filled in with observation-by-observation values of $\ln\ell_j$
      'theta1'   variable containing evaluation of first equation $\theta_{1j}=\mathbf{x}_{1j}\mathbf{b}_1$
      'theta2'   variable containing evaluation of second equation $\theta_{2j}=\mathbf{x}_{2j}\mathbf{b}_2$
      . . .

Method-d0 evaluators

```
program progname
        version 12
        args todo b lnf

        tempvar theta1 theta2 ...
        mleval 'theta1' = 'b', eq(1)
        mleval 'theta2' = 'b', eq(2) // if there is a $\theta_2$
        ...

        // if you need to create any intermediate results:
        tempvar tmp1 tmp2 ...
        gen double 'tmp1' = ...
        ...

        mlsum 'lnf' = ...
end
```

where
      'todo'   always contains 0 (may be ignored)
      'b'   full parameter row vector $\mathbf{b}=(\mathbf{b}_1,\mathbf{b}_2,...,\mathbf{b}_E)$
      'lnf'   scalar to be filled in with overall $\ln L$

Method-d1 evaluators

```
program progname
        version 12
        args todo b lnf g

        tempvar theta1 theta2 ...
        mleval 'theta1' = 'b', eq(1)
        mleval 'theta2' = 'b', eq(2) // if there is a $\theta_2$
        ...

        // if you need to create any intermediate results:
        tempvar tmp1 tmp2 ...
        gen double 'tmp1' = ...
        ...

        mlsum 'lnf' = ...
        if ('todo'==0 | 'lnf'>=.) exit

        tempname d1 d2 ...
        mlvecsum 'lnf' 'd1' = formula for $\partial\ln\ell_j/\partial\theta_{1j}$, eq(1)
        mlvecsum 'lnf' 'd2' = formula for $\partial\ln\ell_j/\partial\theta_{2j}$, eq(2)
        ...
        matrix 'g' = ('d1','d2', ... )
end
```

where
      'todo'   contains 0 or 1
                        $0\Rightarrow$'lnf' to be filled in;
                        $1\Rightarrow$'lnf' and 'g' to be filled in
      'b'   full parameter row vector $\mathbf{b}=(\mathbf{b}_1,\mathbf{b}_2,...,\mathbf{b}_E)$
      'lnf'   scalar to be filled in with overall $\ln L$
      'g'   row vector to be filled in with overall $\mathbf{g}=\partial\ln L/\partial\mathbf{b}$

Method-d2 evaluators

```
program progname
        version 12
        args todo b lnf g H

        tempvar theta1 theta2 ...
        mleval 'theta1' = 'b', eq(1)
        mleval 'theta2' = 'b', eq(2)  // if there is a θ₂
        ...

        // if you need to create any intermediate results:
        tempvar tmp1 tmp2 ...
        gen double 'tmp1' = ...
        ...

        mlsum 'lnf' = ...
        if ('todo'==0 | 'lnf'>=.) exit

        tempname d1 d2 ...
        mlvecsum 'lnf' 'd1' = formula for ∂ lnℓⱼ/∂θ₁ⱼ, eq(1)
        mlvecsum 'lnf' 'd2' = formula for ∂ lnℓⱼ/∂θ₂ⱼ, eq(2)
        ...
        matrix 'g' = ('d1','d2', ... )
        if ('todo'==1 | 'lnf'>=.) exit

        tempname d11 d12 d22 ...
        mlmatsum 'lnf' 'd11' = formula for ∂² lnℓⱼ/∂θ₁ⱼ², eq(1)
        mlmatsum 'lnf' 'd12' = formula for ∂² lnℓⱼ/∂θ₁ⱼ∂θ₂ⱼ, eq(1,2)
        mlmatsum 'lnf' 'd22' = formula for ∂² lnℓⱼ/∂θ₂ⱼ², eq(2)
        ...
        matrix 'H' = ('d11','d12', ... \ 'd12'','d22', ... )
end
```

where

'todo'	contains 0, 1, or 2
	0⇒'lnf' to be filled in;
	1⇒'lnf' and 'g' to be filled in;
	2⇒'lnf', 'g', and 'H' to be filled in
'b'	full parameter row vector $\mathbf{b}=(\mathbf{b}_1,\mathbf{b}_2,...,\mathbf{b}_E)$
'lnf'	scalar to be filled in with overall $\ln L$
'g'	row vector to be filled in with overall $\mathbf{g}=\partial \ln L/\partial \mathbf{b}$
'H'	matrix to be filled in with overall Hessian $\mathbf{H}=\partial^2 \ln L/\partial \mathbf{b}\partial \mathbf{b}'$

Method-lf0 evaluators

```
program progname
        version 12
        args todo b lnfj

        tempvar theta1 theta2 ...
        mleval 'theta1' = 'b', eq(1)
        mleval 'theta2' = 'b', eq(2)  // if there is a θ₂
        ...

        // if you need to create any intermediate results:
        tempvar tmp1 tmp2 ...
        gen double 'tmp1' = ...
        ...

        quietly replace 'lnfj' = ...
end
```

where

'todo'	always contains 0 (may be ignored)
'b'	full parameter row vector $\mathbf{b}=(\mathbf{b}_1,\mathbf{b}_2,...,\mathbf{b}_E)$
'lnfj'	variable to be filled in with observation-by-observation values of $\ln\ell_j$

Method-lf1 evaluators

```
program progname
        version 12
        args todo b lnfj g1 g2 ...

        tempvar theta1 theta2 ...
        mleval 'theta1' = 'b', eq(1)
        mleval 'theta2' = 'b', eq(2) // if there is a θ₂
        ...

        // if you need to create any intermediate results:
        tempvar tmp1 tmp2 ...
        gen double 'tmp1' = ...
        ...

        quietly replace 'lnfj' = ...
        if ('todo'==0) exit

        quietly replace 'g1' = formula for ∂lnℓⱼ/∂θ₁ⱼ
        quietly replace 'g2' = formula for ∂lnℓⱼ/∂θ₂ⱼ
        ...

        end
```

where

'todo'	contains 0 or 1
	$0\Rightarrow$'lnfj'to be filled in;
	$1\Rightarrow$'lnfj', 'g1', 'g2', ..., to be filled in
'b'	full parameter row vector $\mathbf{b}=(\mathbf{b}_1,\mathbf{b}_2,...,\mathbf{b}_E)$
'lnfj'	variable to be filled in with observation-by-observation values of $\ln\ell_j$
'g1'	variable to be filled in with $\partial\ln\ell_j/\partial\theta_{1j}$
'g2'	variable to be filled in with $\partial\ln\ell_j/\partial\theta_{2j}$
...	

Method-lf2 evaluators

```
program progname
        version 12
        args todo b lnfj g1 g2 ... H

        tempvar theta1 theta2 ...
        mleval 'theta1' = 'b', eq(1)
        mleval 'theta2' = 'b', eq(2) // if there is a θ₂
        ...

        // if you need to create any intermediate results:
        tempvar tmp1 tmp2 ...
        gen double 'tmp1' = ...
        ...

        quietly replace 'lnfj' = ...
        if ('todo'==0) exit

        quietly replace 'g1' = formula for ∂lnℓⱼ/∂θ₁ⱼ
        quietly replace 'g2' = formula for ∂lnℓⱼ/∂θ₂ⱼ
        ...

        if ('todo'==1) exit

        tempname d11 d12 d22 lnf ...
        mlmatsum 'lnf' 'd11' = formula for ∂²lnℓⱼ/∂θ₁ⱼ², eq(1)
        mlmatsum 'lnf' 'd12' = formula for ∂²lnℓⱼ/∂θ₁ⱼ∂θ₂ⱼ, eq(1,2)
        mlmatsum 'lnf' 'd22' = formula for ∂²lnℓⱼ/∂θ₂ⱼ², eq(2)
        ...
        matrix 'H' = ('d11','d12', ... \ 'd12'','d22', ... )

        end
```

where

'todo'	contains 0 or 1
	$0 \Rightarrow$ 'lnfj' to be filled in;
	$1 \Rightarrow$ 'lnfj', 'g1', 'g2', ..., to be filled in
	$2 \Rightarrow$ 'lnfj', 'g1', 'g2', ..., and 'H' to be filled in
'b'	full parameter row vector $\mathbf{b} = (\mathbf{b}_1, \mathbf{b}_2, ..., \mathbf{b}_E)$
'lnfj'	scalar to be filled in with observation-by-observation $\ln L$
'g1'	variable to be filled in with $\partial \ln \ell_j / \partial \theta_{1j}$
'g2'	variable to be filled in with $\partial \ln \ell_j / \partial \theta_{2j}$
...	
'H'	matrix to be filled in with overall Hessian $\mathbf{H} = \partial^2 \ln L / \partial \mathbf{b} \partial \mathbf{b}'$

Method-gf0 evaluators

```
program progname
        version 12
        args todo b lnfj

        tempvar theta1 theta2 ...
        mleval 'theta1' = 'b', eq(1)
        mleval 'theta2' = 'b', eq(2)  // if there is a θ₂
        ...

        // if you need to create any intermediate results:
        tempvar tmp1 tmp2 ...
        gen double 'tmp1' = ...
        ...

        quietly replace 'lnfj' = ...
end
```

where

'todo'	always contains 0 (may be ignored)
'b'	full parameter row vector $\mathbf{b} = (\mathbf{b}_1, \mathbf{b}_2, ..., \mathbf{b}_E)$
'lnfj'	variable to be filled in with the values of the log-likelihood $\ln \ell_j$

Global macros for use by all evaluators

\$ML_y1	name of first dependent variable
\$ML_y2	name of second dependent variable, if any
...	
\$ML_samp	variable containing 1 if observation to be used; 0 otherwise
\$ML_w	variable containing weight associated with observation or 1 if no weights specified

Method-lf evaluators can ignore \$ML_samp, but restricting calculations to the \$ML_samp==1 subsample will speed execution. Method-lf evaluators must ignore \$ML_w; application of weights is handled by the method itself.

Methods d0, d1, d2, lf0, lf1, lf2, and gf0 can ignore \$ML_samp as long as ml model's nopreserve option is not specified. These methods will run more quickly if nopreserve is specified. These evaluators can ignore \$ML_w only if they use mlsum, mlvecsum, mlmatsum, and mlmatbysum to produce all final results.

Description

ml model defines the current problem.

ml clear clears the current problem definition. This command is rarely used because when you type ml model, any previous problem is automatically cleared.

ml query displays a description of the current problem.

ml check verifies that the log-likelihood evaluator you have written works. We strongly recommend using this command.

ml search searches for (better) initial values. We recommend using this command.

ml plot provides a graphical way of searching for (better) initial values.

ml init provides a way to specify initial values.

ml report reports $\ln L$'s values, gradient, and Hessian at the initial values or current parameter estimates, $\mathbf{b}_0$.

ml trace traces the execution of the user-defined log-likelihood evaluation program.

ml count counts the number of times the user-defined log-likelihood evaluation program is called; this command is seldom used. ml count clear clears the counter. ml count on turns on the counter. ml count without arguments reports the current values of the counter. ml count off stops counting calls.

ml maximize maximizes the likelihood function and reports results. Once ml maximize has successfully completed, the previously mentioned ml commands may no longer be used unless noclear is specified. ml graph and ml display may be used whether or not noclear is specified.

ml graph graphs the log-likelihood values against the iteration number.

ml display redisplays results.

ml footnote displays a warning message when the model did not converge within the specified number of iterations.

ml score creates new variables containing the equation-level scores. The variables generated by ml score are equivalent to those generated by specifying the score() option of ml maximize (and ml model ..., ... maximize).

progname is the name of a Stata program you write to evaluate the log-likelihood function.

funcname() is the name of a Mata function you write to evaluate the log-likelihood function.

In this documentation, *progname* and *funcname*() are referred to as the user-written evaluator, the likelihood evaluator, or sometimes simply as the evaluator. The program you write is written in the style required by the method you choose. The methods are lf, d0, d1, d2, lf0, lf1, lf2, and gf0. Thus, if you choose to use method lf, your program is called a method-lf evaluator.

Method-lf evaluators are required to evaluate the observation-by-observation log likelihood $\ln \ell_j$, $j = 1, \ldots, N$.

Method-d0 evaluators are required to evaluate the overall log likelihood $\ln L$. Method-d1 evaluators are required to evaluate the overall log likelihood and its gradient vector $\mathbf{g} = \partial \ln L / \partial \mathbf{b}$. Method-d2 evaluators are required to evaluate the overall log likelihood, its gradient, and its Hessian matrix $H = \partial^2 \ln L / \partial \mathbf{b} \partial \mathbf{b}'$.

Method-lf0 evaluators are required to evaluate the observation-by-observation log likelihood $\ln \ell_j$, $j = 1, \ldots, N$. Method-lf1 evaluators are required to evaluate the observation-by-observation log likelihood and its equation-level scores $g_{ji} = \partial \ln \ell / \partial \mathbf{x}_{ji} \mathbf{b}_i$. Method-lf2 evaluators are required to evaluate the observation-by-observation log likelihood, its equation-level scores, and its Hessian matrix $H = \partial^2 \ln \ell / \partial \mathbf{b} \partial \mathbf{b}'$.

Method-gf0 evaluators are required to evaluate the summable pieces of the log likelihood $\ln \ell_k$, $k = 1, \ldots, K$.

mleval is a subroutine used by evaluators of methods d0, d1, d2, lf0, lf1, lf2, and gf0 to evaluate the coefficient vector, $\mathbf{b}$, that they are passed.

mlsum is a subroutine used by evaluators of methods d0, d1, and d2 to define the value, $\ln L$, that is to be returned.

mlvecsum is a subroutine used by evaluators of methods d1 and d2 to define the gradient vector, **g**, that is to be returned. It is suitable for use only when the likelihood function meets the linear-form restrictions.

mlmatsum is a subroutine used by evaluators of methods d2 and lf2 to define the Hessian matrix, **H**, that is to be returned. It is suitable for use only when the likelihood function meets the linear-form restrictions.

mlmatbysum is a subroutine used by evaluator of method d2 to help define the Hessian matrix, **H**, that is to be returned. It is suitable for use when the likelihood function contains terms made up of grouped sums, such as in panel-data models. For such models, use mlmatsum to compute the observation-level outer products and mlmatbysum to compute the group-level outer products. mlmatbysum requires that the data be sorted by the variable identified in the by() option.

Options for use with ml model in interactive or noninteractive mode

group(*varname*) specifies the numeric variable that identifies groups. This option is typically used to identify panels for panel-data models.

vce(*vcetype*) specifies the type of standard error reported, which includes types that are robust to some kinds of misspecification, that allow for intragroup correlation, and that are derived from asymptotic theory; see [R] *vce_option*.

vce(robust), vce(cluster *clustvar*), pweight, and svy will work with evaluators of methods lf, lf0, lf1, lf2, and gf0; all you need do is specify them.

These options will not work with evaluators of methods d0, d1, or d2, and specifying these options will produce an error message.

constraints(*numlist* | *matname*) specifies the linear constraints to be applied during estimation. constraints(*numlist*) specifies the constraints by number. Constraints are defined by using the constraint command; see [R] **constraint**. constraint(*matname*) specifies a matrix that contains the constraints.

nocnsnotes prevents notes from being displayed when constraints are dropped. A constraint will be dropped if it is inconsistent, contradicts other constraints, or causes some other error when the constraint matrix is being built. Constraints are checked in the order in which they are specified.

title(*string*) specifies the title for the estimation output when results are complete.

nopreserve specifies that ml need not ensure that only the estimation subsample is in memory when the user-written likelihood evaluator is called. nopreserve is irrelevant when you use method lf.

For the other methods, if nopreserve is not specified, ml saves the data in a file (preserves the original dataset) and drops the irrelevant observations before calling the user-written evaluator. This way, even if the evaluator does not restrict its attentions to the $ML_samp==1 subsample, results will still be correct. Later, ml automatically restores the original dataset.

ml need not go through these machinations for method lf because the user-written evaluator calculates observation-by-observation values, and ml itself sums the components.

ml goes through these machinations if and only if the estimation sample is a subsample of the data in memory. If the estimation sample includes every observation in memory, ml does not preserve the original dataset. Thus programmers must not alter the original dataset unless they preserve the data themselves.

We recommend that interactive users of ml not specify nopreserve; the speed gain is not worth the possibility of getting incorrect results.

We recommend that programmers specify nopreserve, but only after verifying that their evaluator really does restrict its attentions solely to the $ML\_samp==1$ subsample.

collinear specifies that ml not remove the collinear variables within equations. There is no reason to leave collinear variables in place, but this option is of interest to programmers who, in their code, have already removed collinear variables and do not want ml to waste computer time checking again.

missing specifies that observations containing variables with missing values not be eliminated from the estimation sample. There are two reasons you might want to specify missing:

Programmers may wish to specify missing because, in other parts of their code, they have already eliminated observations with missing values and do not want ml to waste computer time looking again.

You may wish to specify missing if your model explicitly deals with missing values. Stata's heckman command is a good example of this. In such cases, there will be observations where missing values are allowed and other observations where they are not—where their presence should cause the observation to be eliminated. If you specify missing, it is your responsibility to specify an if *exp* that eliminates the irrelevant observations.

lf0($\#_k$ $\#_{ll}$) is typically used by programmers. It specifies the number of parameters and log-likelihood value of the constant-only model so that ml can report a likelihood-ratio test rather than a Wald test. These values may have been analytically determined, or they may have been determined by a previous fitting of the constant-only model on the estimation sample.

Also see the continue option directly below.

If you specify lf0(), it must be safe for you to specify the missing option, too, else how did you calculate the log likelihood for the constant-only model on the same sample? You must have identified the estimation sample, and done so correctly, so there is no reason for ml to waste time rechecking your results. All of which is to say, do not specify lf0() unless you are certain your code identifies the estimation sample correctly.

lf0(), even if specified, is ignored if vce(robust), vce(cluster *clustvar*), pweight, or svy is specified because, in that case, a likelihood-ratio test would be inappropriate.

continue is typically specified by programmers and does two things:

First, it specifies that a model has just been fit by either ml or some other estimation command, such as logit, and that the likelihood value stored in e(ll) and the number of parameters stored in e(b) as of that instant are the relevant values of the constant-only model. The current value of the log likelihood is used to present a likelihood-ratio test unless vce(robust), vce(cluster *clustvar*), pweight, svy, or constraints() is specified. A likelihood-ratio test is inappropriate when vce(robust), vce(cluster *clustvar*), pweight, or svy is specified. We suggest using lrtest when constraints() is specified; see [R] **lrtest**.

Second, continue sets the initial values, b_0, for the model about to be fit according to the e(b) currently stored.

The comments made about specifying missing with lf0() apply equally well here.

waldtest(#) is typically specified by programmers. By default, ml presents a Wald test, but that is overridden if the lf0() or continue option is specified. A Wald test is performed if vce(robust), vce(cluster *clustvar*), or pweight is specified.

waldtest(0) prevents even the Wald test from being reported.

waldtest(-1) is the default. It specifies that a Wald test be performed by constraining all coefficients except the intercept to 0 in the first equation. Remaining equations are to be unconstrained.

A Wald test is performed if neither lf0() nor continue was specified, and a Wald test is forced if vce(robust), vce(cluster *clustvar*), or pweight was specified.

waldtest(k) for $k \leq -1$ specifies that a Wald test be performed by constraining all coefficients except intercepts to 0 in the first $|k|$ equations; remaining equations are to be unconstrained. A Wald test is performed if neither lf0() nor continue was specified, and a Wald test is forced if vce(robust), vce(cluster *clustvar*), or pweight was specified.

waldtest(k) for $k \geq 1$ works like the options above, except that it forces a Wald test to be reported even if the information to perform the likelihood-ratio test is available and even if none of vce(robust), vce(cluster *clustvar*), or pweight was specified. waldtest(k), $k \geq 1$, may not be specified with lf0().

obs(#) is used mostly by programmers. It specifies that the number of observations reported and ultimately stored in e(N) be #. Ordinarily, ml works that out for itself. Programmers may want to specify this option when, for the likelihood evaluator to work for N observations, they first had to modify the dataset so that it contained a different number of observations.

crittype(*string*) is used mostly by programmers. It allows programmers to supply a string (up to 32 characters long) that describes the criterion that is being optimized by ml. The default is "log likelihood" for nonrobust and "log pseudolikelihood" for robust estimation.

svy indicates that ml is to pick up the svy settings set by svyset and use the robust variance estimator. This option requires the data to be svyset; see [SVY] **svyset**. svy may not be specified with vce() or *weights*.

subpop(*varname*) specifies that estimates be computed for the single subpopulation defined by the observations for which *varname* $\neq$ 0. Typically, *varname* = 1 defines the subpopulation, and *varname* = 0 indicates observations not belonging to the subpopulation. For observations whose subpopulation status is uncertain, *varname* should be set to missing ('.'). This option requires the svy option.

nosvyadjust specifies that the model Wald test be carried out as $W/k \sim F(k, d)$, where W is the Wald test statistic, k is the number of terms in the model excluding the constant term, d is the total number of sampled PSUs minus the total number of strata, and $F(k, d)$ is an F distribution with k numerator degrees of freedom and d denominator degrees of freedom. By default, an adjusted Wald test is conducted: $(d - k + 1)W/(kd) \sim F(k, d - k + 1)$. See Korn and Graubard (1990) for a discussion of the Wald test and the adjustments thereof. This option requires the svy option.

technique(*algorithm_spec*) specifies how the likelihood function is to be maximized. The following algorithms are currently implemented in ml. For details, see Gould, Pitblado, and Poi (2010).

technique(nr) specifies Stata's modified Newton–Raphson (NR) algorithm.

technique(bhhh) specifies the Berndt–Hall–Hall–Hausman (BHHH) algorithm.

technique(dfp) specifies the Davidon–Fletcher–Powell (DFP) algorithm.

technique(bfgs) specifies the Broyden–Fletcher–Goldfarb–Shanno (BFGS) algorithm.

The default is technique(nr).

You can switch between algorithms by specifying more than one in the technique() option. By default, ml will use an algorithm for five iterations before switching to the next algorithm. To specify a different number of iterations, include the number after the technique in the option. For example, technique(bhhh 10 nr 1000) requests that ml perform 10 iterations using the BHHH algorithm, followed by 1,000 iterations using the NR algorithm, and then switch back to BHHH for 10 iterations, and so on. The process continues until convergence or until reaching the maximum number of iterations.

Options for use with ml model in noninteractive mode

The following extra options are for use with `ml model` in noninteractive mode. Noninteractive mode is for programmers who use `ml` as a subroutine and want to issue one command that will carry forth the estimation from start to finish.

`maximize` is required. It specifies noninteractive mode.

`init(`*ml_init_args*`)` sets the initial values, b_0. *ml_init_args* are whatever you would type after the `ml init` command.

`search(on | norescale | quietly | off)` specifies whether `ml search` is to be used to improve the initial values. `search(on)` is the default and is equivalent to separately running `ml search`, `repeat(0)`. `search(norescale)` is equivalent to separately running `ml search`, `repeat(0)` `norescale`. `search(quietly)` is equivalent to `search(on)`, except that it suppresses `ml search`'s output. `search(off)` prevents calling `ml search`.

`repeat(`*#*`)` is `ml search`'s `repeat()` option. `repeat(0)` is the default.

`bounds(`*ml_search_bounds*`)` specifies the search bounds. *ml_search_bounds* is specified as

$$\left[\,eqn\_name\,\right]\ lower\_bound\ upper\_bound\ \dots\ \left[\,eqn\_name\,\right]\ lower\_bound\ upper\_bound$$

for instance, `bounds(100 100 lnsigma 0 10)`. The `ml model` command issues `ml search` *ml_search_bounds*, `repeat(`*#*`)`. Specifying search bounds is optional.

`nowarning`, `novce`, `negh`, and `score()` are `ml maximize`'s equivalent options.

maximize_options: <u>diff</u>icult, <u>tech</u>nique(*algorithm_spec*), <u>iter</u>ate(*#*), $\left[\underline{no}\right]$<u>log</u>, <u>tr</u>ace, gradient, showstep, <u>hess</u>ian, showtolerance, <u>tol</u>erance(*#*), <u>ltol</u>erance(*#*), <u>nrtol</u>erance(*#*), nonrtolerance, and from(*init_specs*); see [R] **maximize**. These options are seldom used.

Options for use when specifying equations

`noconstant` specifies that the equation not include an intercept.

`offset(`*varname_o*`)` specifies that the equation be $xb + varname_o$—that it include $varname_o$ with coefficient constrained to be 1.

`exposure(`*varname_e*`)` is an alternative to `offset(`*varname_o*`)`; it specifies that the equation be $xb + \ln(varname_e)$. The equation is to include $\ln(varname_e)$ with coefficient constrained to be 1.

Options for use with ml search

`repeat(`*#*`)` specifies the number of random attempts that are to be made to find a better initial-value vector. The default is `repeat(10)`.

`repeat(0)` specifies that no random attempts be made. More precisely, `repeat(0)` specifies that no random attempts be made if the first initial-value vector is a feasible starting point. If it is not, `ml search` will make random attempts, even if you specify `repeat(0)`, because it has no alternative. The `repeat()` option refers to the number of random attempts to be made to improve the initial values. When the initial starting value vector is not feasible, `ml search` will make up to 1,000 random attempts to find starting values. It stops when it finds one set of values that works and then moves into its improve-initial-values logic.

`repeat(`*k*`)`, $k > 0$, specifies the number of random attempts to be made to improve the initial values.

restart specifies that random actions be taken to obtain starting values and that the resulting starting values not be a deterministic function of the current values. Generally, you should not specify this option because, with restart, ml search intentionally does not produce as good a set of starting values as it could. restart is included for use by the optimizer when it gets into serious trouble. The random actions ensure that the optimizer and ml search, working together, do not cause an endless loop.

restart implies norescale, which is why we recommend that you do not specify restart. In testing, sometimes rescale worked so well that, even after randomization, the rescaler would bring the starting values right back to where they had been the first time and thus defeat the intended randomization.

norescale specifies that ml search not engage in its rescaling actions to improve the parameter vector. We do not recommend specifying this option because rescaling tends to work so well.

maximize_options: [no]log and trace; see [R] **maximize**. These options are seldom used.

Option for use with ml plot

saving(*filename*[, replace]) specifies that the graph be saved in *filename*.gph. See [G-3] *saving_option*.

Options for use with ml init

copy specifies that the list of numbers or the initialization vector be copied into the initial-value vector by position rather than by name.

skip specifies that any parameters found in the specified initialization vector that are not also found in the model be ignored. The default action is to issue an error message.

Options for use with ml maximize

nowarning is allowed only with iterate(0). nowarning suppresses the "convergence not achieved" message. Programmers might specify iterate(0) nowarning when they have a vector **b** already containing the final estimates and want ml to calculate the variance matrix and postestimation results. Then specify init(b) search(off) iterate(0) nowarning nolog.

novce is allowed only with iterate(0). novce substitutes the zero matrix for the variance matrix, which in effect posts estimation results as fixed constants.

negh indicates that the evaluator returns the negative Hessian matrix. By default, ml assumes d2 and lf2 evaluators return the Hessian matrix.

score(*newvars* | *stub**) creates new variables containing the contributions to the score for each equation and ancillary parameter in the model; see [U] **20.21 Obtaining scores**.

If score(*newvars*) is specified, the *newvars* must contain k new variables. For evaluators of methods lf, lf0, lf1, and lf2, k is the number of equations. For evaluators of method gf0, k is the number of parameters. If score(*stub**) is specified, variables named *stub*1, *stub*2, . . . , *stub*k are created.

For evaluators of methods lf, lf0, lf1, and lf2, the first variable contains $\partial \ln \ell_j / \partial(\mathbf{x}_{1j}\mathbf{b}_1)$, the second variable contains $\partial \ln \ell_j / \partial(\mathbf{x}_{2j}\mathbf{b}_2)$, and so on.

For evaluators of method gf0, the first variable contains $\partial \ln \ell_j / \partial \mathbf{b}_1$, the second variable contains $\partial \ln \ell_j / \partial \mathbf{b}_2$, and so on.

nooutput suppresses display of results. This option is different from prefixing ml maximize with quietly in that the iteration log is still displayed (assuming that nolog is not specified).

noclear specifies that the ml problem definition not be cleared after the model has converged. Perhaps you are having convergence problems and intend to run the model to convergence. If so, use ml search to see if those values can be improved, and then restart the estimation.

maximize_options: difficult, iterate(#), [no]log, trace, gradient, showstep, hessian, showtolerance, tolerance(#), ltolerance(#), nrtolerance(#), nonrtolerance; see [R] **maximize**. These options are seldom used.

display_options; see *Options for use with ml display* below.

eform_option; see *Options for use with ml display* below.

Option for use with ml graph

saving(*filename*[, replace]) specifies that the graph be saved in *filename*.gph.
See [G-3] **saving_option**.

Options for use with ml display

noheader suppresses the header display above the coefficient table that displays the final log-likelihood value, the number of observations, and the model significance test.

nofootnote suppresses the footnote display below the coefficient table, which displays a warning if the model fit did not converge within the specified number of iterations. Use ml footnote to display the warning if 1) you add to the coefficient table using the plus option or 2) you have your own footnotes and want the warning to be last.

level(#) is the standard confidence-level option. It specifies the confidence level, as a percentage, for confidence intervals of the coefficients. The default is level(95) or as set by set level; see [U] **20.7 Specifying the width of confidence intervals**.

first displays a coefficient table reporting results for the first equation only, and the report makes it appear that the first equation is the only equation. This option is used by programmers who estimate ancillary parameters in the second and subsequent equations and who wish to report the values of such parameters themselves.

neq(#) is an alternative to first. neq(#) displays a coefficient table reporting results for the first # equations. This option is used by programmers who estimate ancillary parameters in the #+1 and subsequent equations and who wish to report the values of such parameters themselves.

showeqns is a seldom-used option that displays the equation names in the coefficient table. ml display uses the numbers stored in e(k_eq) and e(k_aux) to determine how to display the coefficient table. e(k_eq) identifies the number of equations, and e(k_aux) identifies how many of these are for ancillary parameters. The first option is implied when showeqns is not specified and all but the first equation are for ancillary parameters.

plus displays the coefficient table, but rather than ending the table in a line of dashes, ends it in dashes–plus-sign–dashes. This is so that programmers can write additional display code to add more results to the table and make it appear as if the combined result is one table. Programmers typically specify plus with the first or neq() options. This option implies nofootnote.

nocnsreport suppresses the display of constraints above the coefficient table. This option is ignored if constraints were not used to fit the model.

noomitted specifies that variables that were omitted because of collinearity not be displayed. The default is to include in the table any variables omitted because of collinearity and to label them as "(omitted)".

vsquish specifies that the blank space separating factor-variable terms or time-series–operated variables from other variables in the model be suppressed.

noemptycells specifies that empty cells for interactions of factor variables not be displayed. The default is to include in the table interaction cells that do not occur in the estimation sample and to label them as "(empty)".

baselevels and **allbaselevels** control whether the base levels of factor variables and interactions are displayed. The default is to exclude from the table all base categories.

> **baselevels** specifies that base levels be reported for factor variables and for interactions whose bases cannot be inferred from their component factor variables.

> **allbaselevels** specifies that all base levels of factor variables and interactions be reported.

cformat(%*fmt***)** specifies how to format coefficients, standard errors, and confidence limits in the coefficient table.

pformat(%*fmt***)** specifies how to format p-values in the coefficient table.

sformat(%*fmt***)** specifies how to format test statistics in the coefficient table.

nolstretch specifies that the width of the coefficient table not be automatically widened to accommodate longer variable names. The default, **lstretch**, is to automatically widen the coefficient table up to the width of the Results window. To change the default, use **set lstretch off**. **nolstretch** is not shown in the dialog box.

coeflegend specifies that the legend of the coefficients and how to specify them in an expression be displayed rather than displaying the statistics for the coefficients.

eform_option: **eform(***string***)**, **eform**, **hr**, **shr**, **irr**, **or**, and **rrr** display the coefficient table in exponentiated form: for each coefficient, $\exp(b)$ rather than b is displayed, and standard errors and confidence intervals are transformed. *string* is the table header that will be displayed above the transformed coefficients and must be 11 characters or shorter in length—for example, **eform("Odds ratio")**. The options **eform**, **hr**, **shr**, **irr**, **or**, and **rrr** provide a default *string* equivalent to "exp(b)", "Haz. Ratio", "SHR", "IRR", "Odds Ratio", and "RRR", respectively. These options may not be combined.

> **ml display** looks at **e(k_eform)** to determine how many equations are affected by an *eform_option*; by default, only the first equation is affected. Type **ereturn list, all** to view **e(k_eform)**; see [P] **ereturn**.

Options for use with mleval

eq(*#***)** specifies the equation number, i, for which $\theta_{ij} = \mathbf{x}_{ij}\mathbf{b}_i$ is to be evaluated. **eq(1)** is assumed if **eq()** is not specified.

scalar asserts that the ith equation is known to evaluate to a constant, meaning that the equation was specified as **()**, **(***name***:)**, or **/***name* on the **ml model** statement. If you specify this option, the new variable created is created as a scalar. If the ith equation does not evaluate to a scalar, an error message is issued.

Option for use with mlsum

noweight specifies that weights ($ML_w) be ignored when summing the likelihood function.

Option for use with mlvecsum

eq(#) specifies the equation for which a gradient vector $\partial \ln L / \partial \mathbf{b}_i$ is to be constructed. The default is eq(1).

Option for use with mlmatsum

eq(#[,#]) specifies the equations for which the Hessian matrix is to be constructed. The default is eq(1), which is the same as eq(1,1), which means $\partial^2 \ln L / \partial \mathbf{b}_1 \partial \mathbf{b}_1'$. Specifying eq($i,j$) results in $\partial^2 \ln L / \partial \mathbf{b}_i \partial \mathbf{b}_j'$.

Options for use with mlmatbysum

by(*varname*) is required and specifies the group variable.

eq(#[,#]) specifies the equations for which the Hessian matrix is to be constructed. The default is eq(1), which is the same as eq(1,1), which means $\partial^2 \ln L / \partial \mathbf{b}_1 \partial \mathbf{b}_1'$. Specifying eq($i,j$) results in $\partial^2 \ln L / \partial \mathbf{b}_i \partial \mathbf{b}_j'$.

Options for use with ml score

equation(*eqname*) identifies from which equation the observation scores are to come. This option may be used only when generating one variable.

missing specifies that observations containing variables with missing values not be eliminated from the estimation sample.

Remarks

For a thorough discussion of ml, see the fourth edition of *Maximum Likelihood Estimation with Stata* (Gould, Pitblado, and Poi 2010). The book provides a tutorial introduction to ml, notes on advanced programming issues, and a discourse on maximum likelihood estimation from both theoretical and practical standpoints. See *Survey options and ml* at the end of *Remarks* for examples of the new svy options. For more information about survey estimation, see [SVY] **survey**, [SVY] **svy estimation**, and [SVY] **variance estimation**.

ml requires that you write a program that evaluates the log-likelihood function and, possibly, its first and second derivatives. The style of the program you write depends upon the method you choose. Methods lf, lf0, d0, and gf0 require that your program evaluate the log likelihood only. Methods d1 and lf1 require that your program evaluate the log likelihood and its first derivatives. Methods d2 and lf2 requires that your program evaluate the log likelihood and its first and second derivatives. Methods lf, lf0, d0, and gf0 differ from each other in that, with methods lf and lf0, your program is required to produce observation-by-observation log-likelihood values $\ln \ell_j$ and it is assumed that $\ln L = \sum_j \ln \ell_j$; with method d0, your program is required to produce only the overall value $\ln L$; and with method gf0, your program is required to produce the summable pieces of the log likelihood, such as those in panel-data models.

Once you have written the program—called an evaluator—you define a model to be fit using `ml model` and obtain estimates using `ml maximize`. You might type

```
. ml model ...
. ml maximize
```

but we recommend that you type

```
. ml model ...
. ml check
. ml search
. ml maximize
```

`ml check` verifies your evaluator has no obvious errors, and `ml search` finds better initial values.

You fill in the `ml model` statement with 1) the method you are using, 2) the name of your program, and 3) the "equations". You write your evaluator in terms of θ_1, θ_2, ..., each of which has a linear equation associated with it. That linear equation might be as simple as $\theta_i = b_0$, it might be $\theta_i = b_1\mathtt{mpg} + b_2\mathtt{weight} + b_3$, or it might omit the intercept b_3. The equations are specified in parentheses on the `ml model` line.

Suppose that you are using method lf and the name of your evaluator program is `myprog`. The statement

```
. ml model lf myprog (mpg weight)
```

would specify one equation with $\theta_i = b_1\mathtt{mpg} + b_2\mathtt{weight} + b_3$. If you wanted to omit b_3, you would type

```
. ml model lf myprog (mpg weight, nocons)
```

and if all you wanted was $\theta_i = b_0$, you would type

```
. ml model lf myprog ()
```

With multiple equations, you list the equations one after the other; so, if you typed

```
. ml model lf myprog (mpg weight) ()
```

you would be specifying $\theta_1 = b_1\mathtt{mpg} + b_2\mathtt{weight} + b_3$ and $\theta_2 = b_4$. You would write your likelihood in terms of θ_1 and θ_2. If the model was linear regression, θ_1 might be the **xb** part and θ_2 the variance of the residuals.

When you specify the equations, you also specify any dependent variables. If you typed

```
. ml model lf myprog (price = mpg weight) ()
```

`price` would be the one and only dependent variable, and that would be passed to your program in `$ML_y1`. If your model had two dependent variables, you could type

```
. ml model lf myprog (price displ = mpg weight) ()
```

Then `$ML_y1` would be `price` and `$ML_y2` would be `displ`. You can specify however many dependent variables are necessary and specify them on any equation. It does not matter on which equation you specify them; the first one specified is placed in `$ML_y1`, the second in `$ML_y2`, and so on.

Example 1: Method lf

Using method lf, we want to produce observation-by-observation values of the log likelihood. The probit log-likelihood function is

$$\ln \ell_j = \begin{cases} \ln \Phi(\theta_{1j}) & \text{if } y_j = 1 \\ \ln \Phi(-\theta_{1j}) & \text{if } y_j = 0 \end{cases}$$

$$\theta_{1j} = \mathbf{x}_j \mathbf{b}_1$$

The following is the method-lf evaluator for this likelihood function:

```
program myprobit
        version 12
        args lnf theta1
        quietly replace 'lnf' = ln(normal('theta1')) if $ML_y1==1
        quietly replace 'lnf' = ln(normal(-'theta1')) if $ML_y1==0
end
```

If we wanted to fit a model of `foreign` on `mpg` and `weight`, we would type

```
. use http://www.stata-press.com/data/r12/auto
(1978 Automobile Data)
. ml model lf myprobit (foreign = mpg weight)
. ml maximize
```

The 'foreign =' part specifies that y is `foreign`. The 'mpg weight' part specifies that $\theta_{1j} = b_1 \text{mpg}_j + b_2 \text{weight}_j + b_3$. The result of running this is

```
. ml model lf myprobit (foreign = mpg weight)

. ml maximize
initial:       log likelihood = -51.292891
alternative:   log likelihood = -45.055272
rescale:       log likelihood = -45.055272
Iteration 0:   log likelihood = -45.055272
Iteration 1:   log likelihood = -27.904114
Iteration 2:   log likelihood = -26.858048
Iteration 3:   log likelihood = -26.844198
Iteration 4:   log likelihood = -26.844189
Iteration 5:   log likelihood = -26.844189
```

			Number of obs	=	74
			Wald chi2(2)	=	20.75
Log likelihood = -26.844189			Prob > chi2	=	0.0000

foreign	Coef.	Std. Err.	z	P>|z|	[95% Conf. Interval]
mpg	-.1039503	.0515689	-2.02	0.044	-.2050235 -.0028772
weight	-.0023355	.0005661	-4.13	0.000	-.003445 -.0012261
_cons	8.275464	2.554142	3.24	0.001	3.269438 13.28149

◁

Example 2: Method lf for two-equation, two-dependent-variable model

A two-equation, two-dependent-variable model is a little different. Rather than receiving one θ, our program will receive two. Rather than there being one dependent variable in $ML_y1, there will be dependent variables in $ML_y1 and $ML_y2. For instance, the Weibull regression log-likelihood function is

$$\ln \ell_j = -(t_j e^{-\theta_{1j}})^{\exp(\theta_{2j})} + d_j\{\theta_{2j} - \theta_{1j} + (e^{\theta_{2j}} - 1)(\ln t_j - \theta_{1j})\}$$
$$\theta_{1j} = \mathbf{x}_j \mathbf{b}_1$$
$$\theta_{2j} = s$$

where t_j is the time of failure or censoring and $d_j = 1$ if failure and 0 if censored. We can make the log likelihood a little easier to program by introducing some extra variables:

$$p_j = \exp(\theta_{2j})$$
$$M_j = \{t_j \exp(-\theta_{1j})\}^{p_j}$$
$$R_j = \ln t_j - \theta_{1j}$$
$$\ln \ell_j = -M_j + d_j\{\theta_{2j} - \theta_{1j} + (p_j - 1)R_j\}$$

The method-lf evaluator for this is

```
program myweib
        version 12
        args lnf theta1 theta2

        tempvar p M R
        quietly gen double 'p' = exp('theta2')
        quietly gen double 'M' = ($ML_y1*exp(-'theta1'))^'p'
        quietly gen double 'R' = ln($ML_y1)-'theta1'

        quietly replace 'lnf' = -'M' + $ML_y2*('theta2'-'theta1' + ('p'-1)*'R')
end
```

We can fit a model by typing

```
. ml model lf myweib (studytime died = i.drug age) ()
. ml maximize
```

Note that we specified '()' for the second equation. The second equation corresponds to the Weibull shape parameter s, and the linear combination we want for s contains just an intercept. Alternatively, we could type

```
. ml model lf myweib (studytime died = i.drug age) /s
```

Typing /s means the same thing as typing (s:), and both really mean the same thing as (). The s, either after a slash or in parentheses before a colon, labels the equation. It makes the output look prettier, and that is all:

```
. use http://www.stata-press.com/data/r12/cancer, clear
(Patient Survival in Drug Trial)

. ml model lf myweib (studytime died = i.drug age) /s

. ml maximize
initial:         log likelihood =          -744
alternative:     log likelihood = -356.14276
rescale:         log likelihood = -200.80201
rescale eq:      log likelihood = -136.69232
Iteration 0:     log likelihood = -136.69232  (not concave)
Iteration 1:     log likelihood = -124.11726
Iteration 2:     log likelihood = -113.89828
Iteration 3:     log likelihood = -110.30451
Iteration 4:     log likelihood = -110.26747
Iteration 5:     log likelihood = -110.26736
Iteration 6:     log likelihood = -110.26736
```

				Number of obs	=	48
				Wald chi2(3)	=	35.25
Log likelihood = -110.26736				Prob > chi2	=	0.0000

	Coef.	Std. Err.	z	P>|z|	[95% Conf.	Interval]
eq1						
drug						
2	1.012966	.2903917	3.49	0.000	.4438086	1.582123
3	1.45917	.2821195	5.17	0.000	.9062261	2.012114
age	-.0671728	.0205688	-3.27	0.001	-.1074868	-.0268587
_cons	6.060723	1.152845	5.26	0.000	3.801188	8.320259
s						
_cons	.5573333	.1402154	3.97	0.000	.2825162	.8321504

◁

▷ Example 3: Method d0

Method-d0 evaluators receive $\mathbf{b} = (\mathbf{b}_1, \mathbf{b}_2, \ldots, \mathbf{b}_E)$, the coefficient vector, rather than the already evaluated $\theta_1, \theta_2, \ldots, \theta_E$, and they are required to evaluate the overall log-likelihood $\ln L$ rather than $\ln \ell_j$, $j = 1, \ldots, N$.

Use `mleval` to produce the thetas from the coefficient vector.

Use `mlsum` to sum the components that enter into $\ln L$.

In the case of Weibull, $\ln L = \sum \ln \ell_j$, and our method-d0 evaluator is

```
program weib0
        version 12
        args todo b lnf

        tempvar theta1 theta2
        mleval 'theta1' = 'b', eq(1)
        mleval 'theta2' = 'b', eq(2)

        local t "$ML_y1"             // this is just for readability
        local d "$ML_y2"

        tempvar p M R
        quietly gen double 'p' = exp('theta2')
        quietly gen double 'M' = ('t'*exp(-'theta1'))^'p'
        quietly gen double 'R' = ln('t')-'theta1'

        mlsum 'lnf' = -'M' + 'd'*('theta2'-'theta1' + ('p'-1)*'R')
end
```

To fit our model using this evaluator, we would type

```
. ml model d0 weib0 (studytime died = i.drug age) /s
. ml maximize
```                                                                ◁

❑ Technical note

Method d0 does not require $\ln L = \sum_j \ln \ell_j$, $j = 1, \ldots, N$, as method lf does. Your likelihood function might have independent components only for groups of observations. Panel-data estimators have a log-likelihood value $\ln L = \sum_i \ln L_i$, where i indexes the panels, each of which contains multiple observations. Conditional logistic regression has $\ln L = \sum_k \ln L_k$, where k indexes the risk pools. Cox regression has $\ln L = \sum_{(t)} \ln L_{(t)}$, where (t) denotes the ordered failure times.

To evaluate such likelihood functions, first calculate the within-group log-likelihood contributions. This usually involves `generate` and `replace` statements prefixed with by, as in

```
tempvar sumd
by group: gen double 'sumd' = sum($ML_y1)
```

Structure your code so that the log-likelihood contributions are recorded in the last observation of each group. Say that a variable is named 'cont'. To sum the contributions, code

```
tempvar last
quietly by group: gen byte 'last' = (_n==_N)
mlsum 'lnf' = 'cont' if 'last'
```

You must inform `mlsum` which observations contain log-likelihood values to be summed. First, you do not want to include intermediate results in the sum. Second, `mlsum` does not skip missing values. Rather, if `mlsum` sees a missing value among the contributions, it sets the overall result, 'lnf', to missing. That is how `ml maximize` is informed that the likelihood function could not be evaluated at the particular value of **b**. `ml maximize` will then take action to escape from what it thinks is an infeasible area of the likelihood function.

When the likelihood function violates the linear-form restriction $\ln L = \sum_j \ln \ell_j$, $j = 1, \ldots, N$, with $\ln \ell_j$ being a function solely of values within the jth observation, use method d0. In the following examples, we will demonstrate methods d1 and d2 with likelihood functions that meet this linear-form restriction. The d1 and d2 methods themselves do not require the linear-form restriction, but the utility routines `mlvecsum` and `mlmatsum` do. Using method d1 or d2 when the restriction is violated is difficult; however, `mlmatbysum` may be of some help for method-d2 evaluators.   ❑

Example 4: Method d1

Method-d1 evaluators are required to produce the gradient vector $\mathbf{g} = \partial \ln L/\partial \mathbf{b}$, as well as the overall log-likelihood value. Using $\texttt{mlvecsum}$, we can obtain $\partial \ln L/\partial \mathbf{b}$ from $\partial \ln L/\partial \theta_i$, $i = 1, \ldots, E$. The derivatives of the Weibull log-likelihood function are

$$\frac{\partial \ln \ell_j}{\partial \theta_{1j}} = p_j(M_j - d_j)$$

$$\frac{\partial \ln \ell_j}{\partial \theta_{2j}} = d_j - R_j p_j(M_j - d_j)$$

The method-d1 evaluator for this is

```
program weib1
        version 12
        args todo b lnf g                                    // g is new

        tempvar t1 t2
        mleval 't1' = 'b', eq(1)
        mleval 't2' = 'b', eq(2)

        local t "$ML_y1"
        local d "$ML_y2"

        tempvar p M R
        quietly gen double 'p' = exp('t2')
        quietly gen double 'M' = ('t'*exp(-'t1'))^'p'
        quietly gen double 'R' = ln('t')-'t1'

        mlsum 'lnf' = -'M' + 'd'*('t2'-'t1' + ('p'-1)*'R')
        if ('todo'==0 | 'lnf'>=.) exit                       /* <-- new */

        tempname d1 d2                                        /* <-- new */
        mlvecsum 'lnf' 'd1' = 'p'*('M'-'d'), eq(1)           /* <-- new */
        mlvecsum 'lnf' 'd2' = 'd' - 'R'*'p'*('M'-'d'), eq(2) /* <-- new */
        matrix 'g' = ('d1','d2')                             /* <-- new */
end
```

We obtained this code by starting with our method-d0 evaluator and then adding the extra lines that method d1 requires. To fit our model using this evaluator, we could type

```
. ml model d1 weib1 (studytime died = drug2 drug3 age) /s
. ml maximize
```

but we recommend substituting method d1debug for method d1 and typing

```
. ml model d1debug weib1 (studytime died = drug2 drug3 age) /s
. ml maximize
```

Method d1debug will compare the derivatives we calculate with numerical derivatives and thus verify that our program is correct. Once we are certain the program is correct, then we would switch from method d1debug to method d1.

◁

Example 5: Method d2

Method-d2 evaluators are required to produce $\mathbf{H} = \partial^2 \ln L/\partial \mathbf{b}\partial \mathbf{b}'$, the Hessian matrix, as well as the gradient and log-likelihood value. $\texttt{mlmatsum}$ will help calculate $\partial^2 \ln L/\partial \mathbf{b}\partial \mathbf{b}'$ from the second derivatives with respect to θ. For the Weibull model, these second derivatives are

$$\frac{\partial^2 \ln \ell_j}{\partial \theta_{1j}^2} = -p_j^2 M_j$$

$$\frac{\partial^2 \ln \ell_j}{\partial \theta_{1j} \partial \theta_{2j}} = p_j(M_j - d_j + R_j p_j M_j)$$

$$\frac{\partial^2 \ln \ell_j}{\partial \theta_{2j}^2} = -p_j R_j(R_j p_j M_j + M_j - d_j)$$

The method-d2 evaluator is

```
program weib2
        version 12
        args todo b lnf g H                        // H added

        tempvar t1 t2
        mleval 't1' = 'b', eq(1)
        mleval 't2' = 'b', eq(2)

        local t "$ML_y1"
        local d "$ML_y2"

        tempvar p M R
        quietly gen double 'p' = exp('t2')
        quietly gen double 'M' = ('t'*exp(-'t1'))^'p'
        quietly gen double 'R' = ln('t')-'t1'
        mlsum 'lnf' = -'M' + 'd'*('t2'-'t1' + ('p'-1)*'R')
        if ('todo'==0 | 'lnf'>=.) exit

        tempname d1 d2
        mlvecsum 'lnf' 'd1' = 'p'*('M'-'d'), eq(1)
        mlvecsum 'lnf' 'd2' = 'd' - 'R'*'p'*('M'-'d'), eq(2)
        matrix 'g' = ('d1','d2')
        if ('todo'==1 | 'lnf'>=.) exit             // new from here down

        tempname d11 d12 d22
        mlmatsum 'lnf' 'd11' = -'p'^2 * 'M', eq(1)
        mlmatsum 'lnf' 'd12' = 'p'*('M'-'d' + 'R'*'p'*'M'), eq(1,2)
        mlmatsum 'lnf' 'd22' = -'p'*'R'*('R'*'p'*'M' + 'M' - 'd') , eq(2)
        matrix 'H' = ('d11','d12' \ 'd12','d22')
end
```

We started with our previous method-d1 evaluator and added the lines that method d2 requires. We could now fit a model by typing

```
. ml model d2 weib2 (studytime died = drug2 drug3 age) /s
. ml maximize
```

but we would recommend substituting method d2debug for method d2 and typing

```
. ml model d2debug weib2 (studytime died = drug2 drug3 age) /s
. ml maximize
```

Method d2debug will compare the first and second derivatives we calculate with numerical derivatives and thus verify that our program is correct. Once we are certain the program is correct, then we would switch from method d2debug to method d2. ◁

As we stated earlier, to produce the robust variance estimator with method lf, there is nothing to do except specify vce(robust), vce(cluster *clustvar*), or pweight. For methods d0, d1, and d2, these options do not work. If your likelihood function meets the linear-form restrictions, you can use methods lf0, lf1, and lf2, then these options will work. The equation scores are defined as

$$\frac{\partial \ln \ell_j}{\partial \theta_{1j}}, \quad \frac{\partial \ln \ell_j}{\partial \theta_{2j}}, \quad \ldots$$

Your evaluator will be passed variables, one for each equation, which you fill in with the equation scores. For *both* method lf1 and lf2, these variables are passed in the fourth and subsequent positions of the argument list. That is, you must process the arguments as

```
args todo b lnf g1 g2 ... H
```

Note that for method lf1, the 'H' argument is not used and can be ignored.

Example 6: Robust variance estimates

If you have used `mlvecsum` in your evaluator of method d1 or d2, it is easy to turn it into evaluator of method lf1 or lf2 that allows the computation of the robust variance estimator. The expression that you specified on the right-hand side of `mlvecsum` is the equation score.

Here we turn the program that we gave earlier in the method-d1 example into a method-lf1 evaluator that allows `vce(robust)`, `vce(cluster clustvar)`, or pweight.

```
program weib1
        version 12
        args todo b lnfj g1 g2          // g1 and g2 are new
        tempvar t1 t2
        mleval 't1' = 'b', eq(1)
        mleval 't2' = 'b', eq(2)
        local t "$ML_y1"
        local d "$ML_y2"
        tempvar p M R
        quietly gen double 'p' = exp('t2')
        quietly gen double 'M' = ('t'*exp(-'t1'))^'p'
        quietly gen double 'R' = ln('t')-'t1'
        quietly replace 'lnfj' = -'M' + 'd'*('t2'-'t1' + ('p'-1)*'R')
        if ('todo'==0) exit
        quietly replace 'g1' = 'p'*('M'-'d')            /* <-- new    */
        quietly replace 'g2' = 'd' - 'R'*'p'*('M'-'d')  /* <-- new    */
end
```

To fit our model and get the robust variance estimates, we type

```
. ml model lf1 weib1 (studytime died = drug2 drug3 age) /s, vce(robust)
. ml maximize
```

◁

Survey options and ml

`ml` can handle stratification, poststratification, multiple stages of clustering, and finite population corrections. Specifying the `svy` option implies that the data come from a survey design and also implies that the survey linearized variance estimator is to be used; see [SVY] **variance estimation**.

Example 7

Suppose that we are interested in a probit analysis of data from a survey in which q1 is the answer to a yes/no question and x1, x2, x3 are demographic responses. The following is a lf2 evaluator for the probit model that meets the requirements for `vce(robust)` (linear form and computes the scores).

```
program mylf2probit
      version 12
      args todo b lnfj g1 H
      tempvar z Fz lnf
      mleval 'z' = 'b'
      quietly gen double 'Fz'   = normal( 'z')  if $ML_y1 == 1
      quietly replace    'Fz'   = normal(-'z')  if $ML_y1 == 0
      quietly replace    'lnfj' = log('Fz')
      if ('todo'==0) exit
      quietly replace 'g1' =  normalden('z')/'Fz'  if $ML_y1 == 1
      quietly replace 'g1' = -normalden('z')/'Fz'  if $ML_y1 == 0
      if ('todo'==1) exit
      mlmatsum 'lnf' 'H' = -'g1'*('g1'+'z'), eq(1,1)
end
```

To fit a model, we svyset the data, then use svy with ml.

```
. svyset psuid [pw=w], strata(strid)
. ml model lf2 myd2probit (q1 = x1 x2 x3), svy
. ml maximize
```

We could also use the subpop() option to make inferences about the subpopulation identified by the variable sub:

```
. svyset psuid [pw=w], strata(strid)
. ml model lf2 myd2probit (q1 = x1 x2 x3), svy subpop(sub)
. ml maximize
```

◁

Saved results

For results saved by ml without the svy option, see [R] **maximize**.

For results saved by ml with the svy option, see [SVY] **svy**.

Methods and formulas

ml is implemented using moptimize; see [M-5] **moptimize( )**.

References

Gould, W. W., J. S. Pitblado, and B. P. Poi. 2010. *Maximum Likelihood Estimation with Stata*. 4th ed. College Station, TX: Stata Press.

Korn, E. L., and B. I. Graubard. 1990. Simultaneous testing of regression coefficients with complex survey data: Use of Bonferroni t statistics. *American Statistician* 44: 270–276.

Royston, P. 2007. Profile likelihood for estimation and confidence intervals. *Stata Journal* 7: 376–387.

Also see

[R] **maximize** — Details of iterative maximization

[R] **nl** — Nonlinear least-squares estimation

[M-5] **moptimize( )** — Model optimization

[M-5] **optimize( )** — Function optimization

mlogit — Multinomial (polytomous) logistic regression

yntax

mlogit *depvar* [*indepvars*] [*if*] [*in*] [*weight*] [, *options*]

options	Description
Model	
noconstant	suppress constant term
baseoutcome(#)	value of *depvar* that will be the base outcome
constraints(*clist*)	apply specified linear constraints; *clist* has the form #[-#] [, #[-#] ...]
collinear	keep collinear variables
SE/Robust	
vce(*vcetype*)	*vcetype* may be oim, robust, cluster *clustvar*, bootstrap, or jackknife
Reporting	
level(#)	set confidence level; default is level(95)
rrr	report relative-risk ratios
nocnsreport	do not display constraints
display_options	control column formats, row spacing, line width, and display of omitted variables and base and empty cells
Maximization	
maximize_options	control the maximization process; seldom used
coeflegend	display legend instead of statistics

indepvars may contain factor variables; see [U] **11.4.3 Factor variables**.
indepvars may contain time-series operators; see [U] **11.4.4 Time-series varlists**.
bootstrap, by, fracpoly, jackknife, mfp, mi estimate, rolling, statsby, and svy are allowed; see [U] **11.1.10 Prefix commands**.
vce(bootstrap) and vce(jackknife) are not allowed with the mi estimate prefix; see [MI] **mi estimate**.
Weights are not allowed with the bootstrap prefix; see [R] **bootstrap**.
vce() and weights are not allowed with the svy prefix; see [SVY] **svy**.
fweights, iweights, and pweights are allowed; see [U] **11.1.6 weight**.
coeflegend does not appear in the dialog box.
See [U] **20 Estimation and postestimation commands** for more capabilities of estimation commands.

Menu

Statistics > Categorical outcomes > Multinomial logistic regression

Description

mlogit fits maximum-likelihood multinomial logit models, also known as polytomous logistic regression. You can define constraints to perform constrained estimation. Some people refer to conditional logistic regression as multinomial logit. If you are one of them, see [R] **clogit**.

See [R] **logistic** for a list of related estimation commands.

Options

┌─── Model ┐

noconstant; see [R] **estimation options**.

baseoutcome(#) specifies the value of *depvar* to be treated as the base outcome. The default is to choose the most frequent outcome.

constraints(*clist*), collinear; see [R] **estimation options**.

┌─── SE/Robust ┐

vce(*vcetype*) specifies the type of standard error reported, which includes types that are derived from asymptotic theory, that are robust to some kinds of misspecification, that allow for intragroup correlation, and that use bootstrap or jackknife methods; see [R] **vce_option**.

If specifying vce(bootstrap) or vce(jackknife), you must also specify baseoutcome().

┌─── Reporting ┐

level(#); see [R] **estimation options**.

rrr reports the estimated coefficients transformed to relative-risk ratios, that is, e^b rather than b; see *Description of the model* below for an explanation of this concept. Standard errors and confidence intervals are similarly transformed. This option affects how results are displayed, not how they are estimated. rrr may be specified at estimation or when replaying previously estimated results.

nocnsreport; see [R] **estimation options**.

display_options: noomitted, vsquish, noemptycells, baselevels, allbaselevels, cformat(%*fmt*), pformat(%*fmt*), sformat(%*fmt*), and nolstretch; see [R] **estimation options**.

┌─── Maximization ┐

maximize_options: difficult, technique(*algorithm_spec*), iterate(#), [no]log, trace, gradient, showstep, hessian, showtolerance, tolerance(#), ltolerance(#), nrtolerance(#), nonrtolerance, and from(*init_specs*); see [R] **maximize**. These options are seldom used.

The following option is available with mlogit but is not shown in the dialog box:

coeflegend; see [R] **estimation options**.

Remarks

Remarks are presented under the following headings:

>   *Description of the model*
>   *Fitting unconstrained models*
>   *Fitting constrained models*

mlogit fits maximum likelihood models with discrete dependent (left-hand-side) variables when the dependent variable takes on more than two outcomes and the outcomes have no natural ordering. If the dependent variable takes on only two outcomes, estimates are identical to those produced by logistic or logit; see [R] **logistic** or [R] **logit**. If the outcomes are ordered, see [R] **ologit**.

Description of the model

For an introduction to multinomial logit models, see Greene (2012, 763–766), Hosmer and Lemeshow (2000, 260–287), Long (1997, chap. 6), Long and Freese (2006, chap. 6 and 7), and Treiman (2009, 336–341). For a description emphasizing the difference in assumptions and data requirements for conditional and multinomial logit, see Davidson and MacKinnon (1993).

Consider the outcomes $1, 2, 3, \ldots, m$ recorded in y, and the explanatory variables X. Assume that there are $m = 3$ outcomes: "buy an American car", "buy a Japanese car", and "buy a European car". The values of y are then said to be "unordered". Even though the outcomes are coded 1, 2, and 3, the numerical values are arbitrary because $1 < 2 < 3$ does not imply that outcome 1 (buy American) is less than outcome 2 (buy Japanese) is less than outcome 3 (buy European). This unordered categorical property of y distinguishes the use of mlogit from regress (which is appropriate for a continuous dependent variable), from ologit (which is appropriate for ordered categorical data), and from logit (which is appropriate for two outcomes, which can be thought of as ordered).

In the multinomial logit model, you estimate a set of coefficients, $\beta^{(1)}$, $\beta^{(2)}$, and $\beta^{(3)}$, corresponding to each outcome:

$$\Pr(y = 1) = \frac{e^{X\beta^{(1)}}}{e^{X\beta^{(1)}} + e^{X\beta^{(2)}} + e^{X\beta^{(3)}}}$$

$$\Pr(y = 2) = \frac{e^{X\beta^{(2)}}}{e^{X\beta^{(1)}} + e^{X\beta^{(2)}} + e^{X\beta^{(3)}}}$$

$$\Pr(y = 3) = \frac{e^{X\beta^{(3)}}}{e^{X\beta^{(1)}} + e^{X\beta^{(2)}} + e^{X\beta^{(3)}}}$$

The model, however, is unidentified in the sense that there is more than one solution to $\beta^{(1)}$, $\beta^{(2)}$, and $\beta^{(3)}$ that leads to the same probabilities for $y = 1$, $y = 2$, and $y = 3$. To identify the model, you arbitrarily set one of $\beta^{(1)}$, $\beta^{(2)}$, or $\beta^{(3)}$ to 0—it does not matter which. That is, if you arbitrarily set $\beta^{(1)} = 0$, the remaining coefficients $\beta^{(2)}$ and $\beta^{(3)}$ will measure the change relative to the $y = 1$ group. If you instead set $\beta^{(2)} = 0$, the remaining coefficients $\beta^{(1)}$ and $\beta^{(3)}$ will measure the change relative to the $y = 2$ group. The coefficients will differ because they have different interpretations, but the predicted probabilities for $y = 1$, 2, and 3 will still be the same. Thus either parameterization will be a solution to the same underlying model.

Setting $\beta^{(1)} = 0$, the equations become

$$\Pr(y = 1) = \frac{1}{1 + e^{X\beta^{(2)}} + e^{X\beta^{(3)}}}$$

$$\Pr(y = 2) = \frac{e^{X\beta^{(2)}}}{1 + e^{X\beta^{(2)}} + e^{X\beta^{(3)}}}$$

$$\Pr(y = 3) = \frac{e^{X\beta^{(3)}}}{1 + e^{X\beta^{(2)}} + e^{X\beta^{(3)}}}$$

The relative probability of $y = 2$ to the base outcome is

$$\frac{\Pr(y = 2)}{\Pr(y = 1)} = e^{X\beta^{(2)}}$$

Let's call this ratio the relative risk, and let's further assume that X and $\beta_k^{(2)}$ are vectors equal to $(x_1, x_2, \ldots, x_k)$ and $(\beta_1^{(2)}, \beta_2^{(2)}, \ldots, \beta_k^{(2)})'$, respectively. The ratio of the relative risk for a one-unit change in x_i is then

$$\frac{e^{\beta_1^{(2)}x_1 + \cdots + \beta_i^{(2)}(x_i+1) + \cdots + \beta_k^{(2)}x_k}}{e^{\beta_1^{(2)}x_1 + \cdots + \beta_i^{(2)}x_i + \cdots + \beta_k^{(2)}x_k}} = e^{\beta_i^{(2)}}$$

Thus the exponentiated value of a coefficient is the relative-risk ratio for a one-unit change in the corresponding variable (risk is measured as the risk of the outcome relative to the base outcome).

Fitting unconstrained models

▷ Example 1

We have data on the type of health insurance available to 616 psychologically depressed subjects in the United States (Tarlov et al. 1989; Wells et al. 1989). The insurance is categorized as either an indemnity plan (that is, regular fee-for-service insurance, which may have a deductible or coinsurance rate) or a prepaid plan (a fixed up-front payment allowing subsequent unlimited use as provided, for instance, by an HMO). The third possibility is that the subject has no insurance whatsoever. We wish to explore the demographic factors associated with each subject's insurance choice. One of the demographic factors in our data is the race of the participant, coded as white or nonwhite:

```
. use http://www.stata-press.com/data/r12/sysdsn1
(Health insurance data)

. tabulate insure nonwhite, chi2 col
```

Key
frequency
column percentage

insure	nonwhite 0	1	Total
Indemnity	251	43	294
	50.71	35.54	47.73
Prepaid	208	69	277
	42.02	57.02	44.97
Uninsure	36	9	45
	7.27	7.44	7.31
Total	495	121	616
	100.00	100.00	100.00

Pearson chi2(2) =   9.5599   Pr = 0.008

Although `insure` appears to take on the values Indemnity, Prepaid, and Uninsure, it actually takes on the values 1, 2, and 3. The words appear because we have associated a value label with the numeric variable `insure`; see [U] **12.6.3 Value labels**.

When we fit a multinomial logit model, we can tell `mlogit` which outcome to use as the base outcome, or we can let `mlogit` choose. To fit a model of `insure` on `nonwhite`, letting `mlogit` choose the base outcome, we type

```
. mlogit insure nonwhite
Iteration 0:   log likelihood = -556.59502
Iteration 1:   log likelihood = -551.78935
Iteration 2:   log likelihood = -551.78348
Iteration 3:   log likelihood = -551.78348
Multinomial logistic regression                  Number of obs   =         616
                                                 LR chi2(2)      =        9.62
                                                 Prob > chi2     =      0.0081
Log likelihood = -551.78348                      Pseudo R2       =      0.0086
```

insure	Coef.	Std. Err.	z	P>|z|	[95% Conf. Interval]
Indemnity	(base outcome)				
Prepaid					
nonwhite	.6608212	.2157321	3.06	0.002	.2379942 1.083648
_cons	-.1879149	.0937644	-2.00	0.045	-.3716896 -.0041401
Uninsure					
nonwhite	.3779586	.407589	0.93	0.354	-.4209011 1.176818
_cons	-1.941934	.1782185	-10.90	0.000	-2.291236 -1.592632

`mlogit` chose the indemnity outcome as the base outcome and presented coefficients for the outcomes prepaid and uninsured. According to the model, the probability of prepaid for whites (`nonwhite` = 0) is

$$\Pr(\texttt{insure} = \texttt{Prepaid}) = \frac{e^{-.188}}{1 + e^{-.188} + e^{-1.942}} = 0.420$$

Similarly, for nonwhites, the probability of prepaid is

$$\Pr(\texttt{insure} = \texttt{Prepaid}) = \frac{e^{-.188+.661}}{1 + e^{-.188+.661} + e^{-1.942+.378}} = 0.570$$

These results agree with the column percentages presented by `tabulate` because the `mlogit` model is fully saturated. That is, there are enough terms in the model to fully explain the column percentage in each cell. The model chi-squared and the `tabulate` chi-squared are in almost perfect agreement; both test that the column percentages of `insure` are the same for both values of `nonwhite`.

◁

Example 2

By specifying the `baseoutcome()` option, we can control which outcome of the dependent variable is treated as the base. Left to its own, `mlogit` chose to make outcome 1, indemnity, the base outcome. To make outcome 2, prepaid, the base, we would type

```
. mlogit insure nonwhite, base(2)
Iteration 0:   log likelihood = -556.59502
Iteration 1:   log likelihood = -551.78935
Iteration 2:   log likelihood = -551.78348
Iteration 3:   log likelihood = -551.78348
```

Multinomial logistic regression				Number of obs	=	616
				LR chi2(2)	=	9.62
				Prob > chi2	=	0.0081
Log likelihood = -551.78348				Pseudo R2	=	0.0086

insure	Coef.	Std. Err.	z	P>|z|	[95% Conf. Interval]	
Indemnity						
nonwhite	-.6608212	.2157321	-3.06	0.002	-1.083648	-.2379942
_cons	.1879149	.0937644	2.00	0.045	.0041401	.3716896
Prepaid	(base outcome)					
Uninsure						
nonwhite	-.2828627	.3977302	-0.71	0.477	-1.0624	.4966742
_cons	-1.754019	.1805145	-9.72	0.000	-2.107821	-1.400217

The baseoutcome() option requires that we specify the numeric value of the outcome, so we could not type base(Prepaid).

Although the coefficients now appear to be different, the summary statistics reported at the top are identical. With this parameterization, the probability of prepaid insurance for whites is

$$\Pr(\texttt{insure} = \texttt{Prepaid}) = \frac{1}{1 + e^{.188} + e^{-1.754}} = 0.420$$

This is the same answer we obtained previously.

◁

▷ Example 3

By specifying rrr, which we can do at estimation time or when we redisplay results, we see the model in terms of relative-risk ratios:

```
. mlogit, rrr
```

Multinomial logistic regression				Number of obs	=	616
				LR chi2(2)	=	9.62
				Prob > chi2	=	0.0081
Log likelihood = -551.78348				Pseudo R2	=	0.0086

insure	RRR	Std. Err.	z	P>|z|	[95% Conf. Interval]	
Indemnity						
nonwhite	.516427	.1114099	-3.06	0.002	.3383588	.7882073
_cons	1.206731	.1131483	2.00	0.045	1.004149	1.450183
Prepaid	(base outcome)					
Uninsure						
nonwhite	.7536233	.2997387	-0.71	0.477	.3456255	1.643247
_cons	.1730769	.0312429	-9.72	0.000	.1215024	.2465434

Looked at this way, the relative risk of choosing an indemnity over a prepaid plan is 0.516 for nonwhites relative to whites.

To illustrate, from the output and discussions of examples 1 and 2 we find that

$$\Pr\left(\text{insure} = \text{Indemnity} \mid \text{white}\right) = \frac{1}{1 + e^{-.188} + e^{-1.942}} = 0.507$$

and thus the relative risk of choosing indemnity over prepaid (for whites) is

$$\frac{\Pr\left(\text{insure} = \text{Indemnity} \mid \text{white}\right)}{\Pr\left(\text{insure} = \text{Prepaid} \mid \text{white}\right)} = \frac{0.507}{0.420} = 1.207$$

For nonwhites,

$$\Pr\left(\text{insure} = \text{Indemnity} \mid \text{not white}\right) = \frac{1}{1 + e^{-.188+.661} + e^{-1.942+.378}} = 0.355$$

and thus the relative risk of choosing indemnity over prepaid (for nonwhites) is

$$\frac{\Pr\left(\text{insure} = \text{Indemnity} \mid \text{not white}\right)}{\Pr\left(\text{insure} = \text{Prepaid} \mid \text{not white}\right)} = \frac{0.355}{0.570} = 0.623$$

The ratio of these two relative risks, hence the name "relative-risk ratio", is $0.623/1.207 = 0.516$, as given in the output under the heading "RRR".

◁

❑ Technical note

In models where only two categories are considered, the mlogit model reduces to standard logit. Consequently the exponentiated regression coefficients, labeled as RRR within mlogit, are equal to the odds ratios as given when the or option is specified under logit; see [R] **logit**.

As such, always referring to mlogit's exponentiated coefficients as odds ratios may be tempting. However, the discussion in example 3 demonstrates that doing so would be incorrect. In general mlogit models, the exponentiated coefficients are ratios of relative risks, not ratios of odds.

❑

▷ Example 4

One of the advantages of mlogit over tabulate is that we can include continuous variables and multiple categorical variables in the model. In examining the data on insurance choice, we decide that we want to control for age, gender, and site of study (the study was conducted in three sites):

```
. mlogit insure age male nonwhite i.site
Iteration 0:   log likelihood = -555.85446
Iteration 1:   log likelihood = -534.67443
Iteration 2:   log likelihood = -534.36284
Iteration 3:   log likelihood = -534.36165
Iteration 4:   log likelihood = -534.36165
```

Multinomial logistic regression				Number of obs	=	615
				LR chi2(10)	=	42.99
				Prob > chi2	=	0.0000
Log likelihood = -534.36165				Pseudo R2	=	0.0387

insure	Coef.	Std. Err.	z	P>|z|	[95% Conf. Interval]	
Indemnity	(base outcome)					
Prepaid						
age	-.011745	.0061946	-1.90	0.058	-.0238862	.0003962
male	.5616934	.2027465	2.77	0.006	.1643175	.9590693
nonwhite	.9747768	.2363213	4.12	0.000	.5115955	1.437958
site						
2	.1130359	.2101903	0.54	0.591	-.2989296	.5250013
3	-.5879879	.2279351	-2.58	0.010	-1.034733	-.1412433
_cons	.2697127	.3284422	0.82	0.412	-.3740222	.9134476
Uninsure						
age	-.0077961	.0114418	-0.68	0.496	-.0302217	.0146294
male	.4518496	.3674867	1.23	0.219	-.268411	1.17211
nonwhite	.2170589	.4256361	0.51	0.610	-.6171725	1.05129
site						
2	-1.211563	.4705127	-2.57	0.010	-2.133751	-.2893747
3	-.2078123	.3662926	-0.57	0.570	-.9257327	.510108
_cons	-1.286943	.5923219	-2.17	0.030	-2.447872	-.1260134

These results suggest that the inclination of nonwhites to choose prepaid care is even stronger than it was without controlling. We also see that subjects in site 2 are less likely to be uninsured.

◁

Fitting constrained models

mlogit can fit models with subsets of coefficients constrained to be zero, with subsets of coefficients constrained to be equal both within and across equations, and with subsets of coefficients arbitrarily constrained to equal linear combinations of other estimated coefficients.

Before fitting a constrained model, you define the constraints with the constraint command; see [R] **constraint**. Once the constraints are defined, you estimate using mlogit, specifying the constraint() option. Typing constraint(4) would use the constraint you previously saved as 4. Typing constraint(1,4,6) would use the previously stored constraints 1, 4, and 6. Typing constraint(1-4,6) would use the previously stored constraints 1, 2, 3, 4, and 6.

Sometimes you will not be able to specify the constraints without knowing the omitted outcome. In such cases, assume that the omitted outcome is whatever outcome is convenient for you, and include the baseoutcome() option when you specify the mlogit command.

▷ Example 5

We can use constraints to test hypotheses, among other things. In our insurance-choice model, let's test the hypothesis that there is no distinction between having indemnity insurance and being uninsured. Indemnity-style insurance was the omitted outcome, so we type

```
. test [Uninsure]
 ( 1)   [Uninsure]age = 0
 ( 2)   [Uninsure]male = 0
 ( 3)   [Uninsure]nonwhite = 0
 ( 4)   [Uninsure]1b.site = 0
 ( 5)   [Uninsure]2.site = 0
 ( 6)   [Uninsure]3.site = 0
        Constraint 4 dropped
              chi2(  5) =       9.31
            Prob > chi2 =     0.0973
```

If indemnity had not been the omitted outcome, we would have typed test [Uninsure=Indemnity].

The results produced by test are an approximation based on the estimated covariance matrix of the coefficients. Because the probability of being uninsured is low, the log likelihood may be nonlinear for the uninsured. Conventional statistical wisdom is not to trust the asymptotic answer under these circumstances but to perform a likelihood-ratio test instead.

To use Stata's lrtest (likelihood-ratio test) command, we must fit both the unconstrained and constrained models. The unconstrained model is the one we have previously fit. Following the instruction in [R] **lrtest**, we first save the unconstrained model results:

```
. estimates store unconstrained
```

To fit the constrained model, we must refit our model with all the coefficients except the constant set to 0 in the Uninsure equation. We define the constraint and then refit:

```
. constraint 1 [Uninsure]

. mlogit insure age male nonwhite i.site, constraints(1)

Iteration 0:    log likelihood = -555.85446
Iteration 1:    log likelihood = -539.80523
Iteration 2:    log likelihood = -539.75644
Iteration 3:    log likelihood = -539.75643
```

Multinomial logistic regression

Number of obs =   615
Wald chi2(5)   =   29.70
Prob > chi2   = 0.0000

Log likelihood = -539.75643

```
( 1)    [Uninsure]o.age = 0
( 2)    [Uninsure]o.male = 0
( 3)    [Uninsure]o.nonwhite = 0
( 4)    [Uninsure]2o.site = 0
( 5)    [Uninsure]3o.site = 0
```

insure	Coef.	Std. Err.	z	P>|z|	[95% Conf. Interval]	
Indemnity	(base outcome)					
Prepaid						
age	-.0107025	.0060039	-1.78	0.075	-.0224699	.0010649
male	.4963616	.1939683	2.56	0.010	.1161907	.8765324
nonwhite	.9421369	.2252094	4.18	0.000	.5007346	1.383539
site						
2	.2530912	.2029465	1.25	0.212	-.1446767	.6508591
3	-.5521773	.2187237	-2.52	0.012	-.9808678	-.1234869
_cons	.1792752	.3171372	0.57	0.572	-.4423023	.8008527
Uninsure						
age	0	(omitted)				
male	0	(omitted)				
nonwhite	0	(omitted)				
site						
2	0	(omitted)				
3	0	(omitted)				
_cons	-1.87351	.1601099	-11.70	0.000	-2.18732	-1.5597

We can now perform the likelihood-ratio test:

```
. lrtest unconstrained .
```

Likelihood-ratio test   LR chi2(5)   =   10.79
(Assumption: . nested in unconstrained)   Prob > chi2 =   0.0557

The likelihood-ratio chi-squared is 10.79 with 5 degrees of freedom—just slightly greater than the magic $p = 0.05$ level—so we should not call this difference significant.   ◁

❑ Technical note

In certain circumstances, you should fit a multinomial logit model with conditional logit; see [R] **clogit**. With substantial data manipulation, clogit can handle the same class of models with some interesting additions. For example, if we had available the price and deductible of the most competitive insurance plan of each type, mlogit could not use this information, but clogit could.

❑

Saved results

mlogit saves the following in e():

Scalars
e(N)	number of observations
e(k_out)	number of outcomes
e(k)	number of parameters
e(k_eq)	number of equations in e(b)
e(k_eq_model)	number of equations in overall model test
e(k_dv)	number of dependent variables
e(df_m)	model degrees of freedom
e(r2_p)	pseudo-R-squared
e(ll)	log likelihood
e(ll_0)	log likelihood, constant-only model
e(N_clust)	number of clusters
e(chi2)	χ^2
e(p)	significance
e(k_eq_base)	equation number of the base outcome
e(baseout)	the value of *depvar* to be treated as the base outcome
e(ibaseout)	index of the base outcome
e(rank)	rank of e(V)
e(ic)	number of iterations
e(rc)	return code
e(converged)	1 if converged, 0 otherwise

Macros
e(cmd)	mlogit
e(cmdline)	command as typed
e(depvar)	name of dependent variable
e(wtype)	weight type
e(wexp)	weight expression
e(title)	title in estimation output
e(clustvar)	name of cluster variable
e(chi2type)	Wald or LR; type of model χ^2 test
e(vce)	*vcetype* specified in vce()
e(vcetype)	title used to label Std. Err.
e(eqnames)	names of equations
e(baselab)	value label corresponding to base outcome
e(opt)	type of optimization
e(which)	max or min; whether optimizer is to perform maximization or minimization
e(ml_method)	type of ml method
e(user)	name of likelihood-evaluator program
e(technique)	maximization technique
e(properties)	b V
e(predict)	program used to implement predict
e(marginsnotok)	predictions disallowed by margins
e(asbalanced)	factor variables fvset as asbalanced
e(asobserved)	factor variables fvset as asobserved

Matrices
    e(b)   coefficient vector
    e(out)   outcome values
    e(Cns)   constraints matrix
    e(ilog)   iteration log (up to 20 iterations)
    e(gradient)   gradient vector
    e(V)   variance–covariance matrix of the estimators
    e(V_modelbased)   model-based variance
Functions
    e(sample)   marks estimation sample

Methods and formulas

mlogit is implemented as an ado-file.

The multinomial logit model is described in Greene (2012, 763–766).

Suppose that there are k categorical outcomes and—without loss of generality—let the base outcome be 1. The probability that the response for the jth observation is equal to the ith outcome is

$$
p_{ij} = \Pr(y_j = i) =
\begin{cases}
\dfrac{1}{1 + \sum\limits_{m=2}^{k} \exp(\mathbf{x}_j \boldsymbol{\beta}_m)}, & \text{if } i = 1 \\[3ex]
\dfrac{\exp(\mathbf{x}_j \boldsymbol{\beta}_i)}{1 + \sum\limits_{m=2}^{k} \exp(\mathbf{x}_j \boldsymbol{\beta}_m)}, & \text{if } i > 1
\end{cases}
$$

where $\mathbf{x}_j$ is the row vector of observed values of the independent variables for the jth observation and $\boldsymbol{\beta}_m$ is the coefficient vector for outcome m. The log pseudolikelihood is

$$
\ln L = \sum_j w_j \sum_{i=1}^{k} I_i(y_j) \ln p_{ik}
$$

where w_j is an optional weight and

$$
I_i(y_j) =
\begin{cases}
1, & \text{if } y_j = i \\
0, & \text{otherwise}
\end{cases}
$$

Newton–Raphson maximum likelihood is used; see [R] **maximize**.

For constrained equations, the set of constraints is orthogonalized, and a subset of maximizable parameters is selected. For example, a parameter that is constrained to zero is not a maximizable parameter. If two parameters are constrained to be equal to each other, only one is a maximizable parameter.

Let $\mathbf{r}$ be the vector of maximizable parameters. $\mathbf{r}$ is physically a subset of the solution parameters, $\mathbf{b}$. A matrix, $\mathbf{T}$, and a vector, $\mathbf{m}$, are defined as

$$
\mathbf{b} = \mathbf{Tr} + \mathbf{m}
$$

so that

$$\frac{\partial f}{\partial \mathbf{b}} = \frac{\partial f}{\partial \mathbf{r}} \mathbf{T}'$$

$$\frac{\partial^2 f}{\partial \mathbf{b}^2} = \mathbf{T} \frac{\partial^2 f}{\partial \mathbf{r}^2} \mathbf{T}'$$

$\mathbf{T}$ consists of a block form in which one part is a permutation of the identity matrix and the other part describes how to calculate the constrained parameters from the maximizable parameters.

This command supports the Huber/White/sandwich estimator of the variance and its clustered version using `vce(robust)` and `vce(cluster clustvar)`, respectively. See [P] **_robust**, particularly *Maximum likelihood estimators* and *Methods and formulas*.

`mlogit` also supports estimation with survey data. For details on VCEs with survey data, see [SVY] **variance estimation**.

References

Davidson, R., and J. G. MacKinnon. 1993. *Estimation and Inference in Econometrics*. New York: Oxford University Press.

Freese, J., and J. S. Long. 2000. sg155: Tests for the multinomial logit model. *Stata Technical Bulletin* 58: 19–25. Reprinted in *Stata Technical Bulletin Reprints*, vol. 10, pp. 247–255. College Station, TX: Stata Press.

Greene, W. H. 2012. *Econometric Analysis*. 7th ed. Upper Saddle River, NJ: Prentice Hall.

Haan, P., and A. Uhlendorff. 2006. Estimation of multinomial logit models with unobserved heterogeneity using maximum simulated likelihood. *Stata Journal* 6: 229–245.

Hamilton, L. C. 1993. sqv8: Interpreting multinomial logistic regression. *Stata Technical Bulletin* 13: 24–28. Reprinted in *Stata Technical Bulletin Reprints*, vol. 3, pp. 176–181. College Station, TX: Stata Press.

———. 2009. *Statistics with Stata (Updated for Version 10)*. Belmont, CA: Brooks/Cole.

Hendrickx, J. 2000. sbe37: Special restrictions in multinomial logistic regression. *Stata Technical Bulletin* 56: 18–26. Reprinted in *Stata Technical Bulletin Reprints*, vol. 10, pp. 93–103. College Station, TX: Stata Press.

Hole, A. R. 2007. Fitting mixed logit models by using maximum simulated likelihood. *Stata Journal* 7: 388–401.

Hosmer, D. W., Jr., and S. Lemeshow. 2000. *Applied Logistic Regression*. 2nd ed. New York: Wiley.

Kleinbaum, D. G., and M. Klein. 2010. *Logistic Regression: A Self-Learning Text*. 3rd ed. New York: Springer.

Long, J. S. 1997. *Regression Models for Categorical and Limited Dependent Variables*. Thousand Oaks, CA: Sage.

Long, J. S., and J. Freese. 2006. *Regression Models for Categorical Dependent Variables Using Stata*. 2nd ed. College Station, TX: Stata Press.

Tarlov, A. R., J. E. Ware, Jr., S. Greenfield, E. C. Nelson, E. Perrin, and M. Zubkoff. 1989. The medical outcomes study. An application of methods for monitoring the results of medical care. *Journal of the American Medical Association* 262: 925–930.

Treiman, D. J. 2009. *Quantitative Data Analysis: Doing Social Research to Test Ideas*. San Francisco, CA: Jossey-Bass.

Wells, K. B., R. D. Hays, M. A. Burnam, W. H. Rogers, S. Greenfield, and J. E. Ware, Jr. 1989. Detection of depressive disorder for patients receiving prepaid or fee-for-service care. Results from the Medical Outcomes Survey. *Journal of the American Medical Association* 262: 3298–3302.

Xu, J., and J. S. Long. 2005. Confidence intervals for predicted outcomes in regression models for categorical outcomes. *Stata Journal* 5: 537–559.

Also see

[R] **mlogit postestimation** — Postestimation tools for mlogit

[R] **clogit** — Conditional (fixed-effects) logistic regression

[R] **logistic** — Logistic regression, reporting odds ratios

[R] **logit** — Logistic regression, reporting coefficients

[R] **mprobit** — Multinomial probit regression

[R] **nlogit** — Nested logit regression

[R] **ologit** — Ordered logistic regression

[R] **rologit** — Rank-ordered logistic regression

[R] **slogit** — Stereotype logistic regression

[MI] **estimation** — Estimation commands for use with mi estimate

[SVY] **svy estimation** — Estimation commands for survey data

[U] **20 Estimation and postestimation commands**

mlogit postestimation — Postestimation tools for mlogit

Description

The following postestimation commands are available after mlogit:

Command	Description
contrast	contrasts and ANOVA-style joint tests of estimates
estat	AIC, BIC, VCE, and estimation sample summary
estat (svy)	postestimation statistics for survey data
estimates	cataloging estimation results
hausman	Hausman's specification test
lincom	point estimates, standard errors, testing, and inference for linear combinations of coefficients
lrtest[1]	likelihood-ratio test
margins	marginal means, predictive margins, marginal effects, and average marginal effects
marginsplot	graph the results from margins (profile plots, interaction plots, etc.)
nlcom	point estimates, standard errors, testing, and inference for nonlinear combinations of coefficients
predict	predictions, residuals, influence statistics, and other diagnostic measures
predictnl	point estimates, standard errors, testing, and inference for generalized predictions
pwcompare	pairwise comparisons of estimates
suest	seemingly unrelated estimation
test	Wald tests of simple and composite linear hypotheses
testnl	Wald tests of nonlinear hypotheses

[1] lrtest is not appropriate with svy estimation results.

See the corresponding entries in the *Base Reference Manual* for details, but see [SVY] **estat** for details about estat (svy).

Syntax for predict

predict $[type]$ { *stub* | *newvar* | *newvarlist* } $[if]$ $[in]$ [, *statistic* <u>ou</u>tcome(*outcome*)]

predict $[type]$ { *stub* | *newvarlist* } $[if]$ $[in]$, <u>sc</u>ores

statistic	Description
Main	
<u>pr</u>	probability of a positive outcome; the default
xb	linear prediction
stdp	standard error of the linear prediction
stddp	standard error of the difference in two linear predictions

If you do not specify outcome(), pr (with one new variable specified), xb, and stdp assume outcome(#1). You must specify outcome() with the stddp option.

You specify one or k new variables with pr, where k is the number of outcomes.

You specify one new variable with xb, stdp, and stddp.

These statistics are available both in and out of sample; type predict ... if e(sample) ... if wanted only for the estimation sample.

Menu

Statistics > Postestimation > Predictions, residuals, etc.

Options for predict

⌐ Main ⌐

pr, the default, calculates the probability of each of the categories of the dependent variable or the probability of the level specified in outcome(*outcome*). If you specify the outcome(*outcome*) option, you need to specify only one new variable; otherwise, you must specify a new variable for each category of the dependent variable.

xb calculates the linear prediction. You must also specify the outcome(*outcome*) option.

stdp calculates the standard error of the linear prediction. You must also specify the outcome(*outcome*) option.

stddp calculates the standard error of the difference in two linear predictions. You must specify the outcome(*outcome*) option, and here you specify the two particular outcomes of interest inside the parentheses, for example, predict sed, stddp outcome(1,3).

outcome(*outcome*) specifies the outcome for which the statistic is to be calculated. equation() is a synonym for outcome(): it does not matter which you use. outcome() or equation() can be specified using

   #1, #2, ..., where #1 means the first category of the dependent variable, #2 means the second category, etc.;

   the values of the dependent variable; or

   the value labels of the dependent variable if they exist.

scores calculates equation-level score variables. The number of score variables created will be one less than the number of outcomes in the model. If the number of outcomes in the model were k, then

   the first new variable will contain $\partial \ln L/\partial(\mathbf{x}_j\boldsymbol{\beta}_1)$;

   the second new variable will contain $\partial \ln L/\partial(\mathbf{x}_j\boldsymbol{\beta}_2)$;

   ...

   the $(k-1)$th new variable will contain $\partial \ln L/\partial(\mathbf{x}_j\boldsymbol{\beta}_{k-1})$.

Remarks

Remarks are presented under the following headings:

   Obtaining predicted values
   Calculating marginal effects
   Testing hypotheses about coefficients

Obtaining predicted values

▷ Example 1

After estimation, we can use `predict` to obtain predicted probabilities, index values, and standard errors of the index, or differences in the index. For instance, in example 4 of [R] **mlogit**, we fit a model of insurance choice on various characteristics. We can obtain the predicted probabilities for outcome 1 by typing

```
. use http://www.stata-press.com/data/r12/sysdsn1
(Health insurance data)
. mlogit insure age i.male i.nonwhite i.site
  (output omitted )
. predict p1 if e(sample), outcome(1)
(option pr assumed; predicted probability)
(29 missing values generated)
. summarize p1
```

Variable	Obs	Mean	Std. Dev.	Min	Max
p1	615	.4764228	.1032279	.1698142	.71939

We added the `i.` prefix to the `male`, `nonwhite`, and `site` variables to explicitly identify them as factor variables. That makes no difference in the estimated results, but we will take advantage of it in later examples. We also included `if e(sample)` to restrict the calculation to the estimation sample. In example 4 of [R] **mlogit**, the multinomial logit model was fit on 615 observations, so there must be missing values in our dataset.

Although we typed `outcome(1)`, specifying 1 for the indemnity outcome, we could have typed `outcome(Indemnity)`. For instance, to obtain the probabilities for prepaid, we could type

```
. predict p2 if e(sample), outcome(prepaid)
(option pr assumed; predicted probability)
outcome prepaid not found
r(303);
. predict p2 if e(sample), outcome(Prepaid)
(option pr assumed; predicted probability)
(29 missing values generated)
. summarize p2
```

Variable	Obs	Mean	Std. Dev.	Min	Max
p2	615	.4504065	.1125962	.1964103	.7885724

We must specify the label exactly as it appears in the underlying value label (or how it appears in the `mlogit` output), including capitalization.

Here we have used `predict` to obtain probabilities for the same sample on which we estimated. That is not necessary. We could use another dataset that had the independent variables defined (in our example, `age`, `male`, `nonwhite`, and `site`) and use `predict` to obtain predicted probabilities; here, we would not specify `if e(sample)`.

◁

▷ Example 2

predict can also be used to obtain the index values—the $\sum x_i \widehat{\beta}_i^{(k)}$—as well as the probabilities:

```
. predict idx1, outcome(Indemnity) xb
(1 missing value generated)

. summarize idx1
```

Variable	Obs	Mean	Std. Dev.	Min	Max
idx1	643	0	0	0	0

The indemnity outcome was our base outcome—the outcome for which all the coefficients were set to 0—so the index is always 0. For the prepaid and uninsured outcomes, we type

```
. predict idx2, outcome(Prepaid) xb
(1 missing value generated)

. predict idx3, outcome(Uninsure) xb
(1 missing value generated)

. summarize idx2 idx3
```

Variable	Obs	Mean	Std. Dev.	Min	Max
idx2	643	-.0566113	.4962973	-1.298198	1.700719
idx3	643	-1.980747	.6018139	-3.112741	-.8258458

We can obtain the standard error of the index by specifying the stdp option:

```
. predict se2, outcome(Prepaid) stdp
(1 missing value generated)

. list p2 idx2 se2 in 1/5
```

	p2	idx2	se2
1.	.3709022	-.4831167	.2437772
2.	.4977667	.055111	.1694686
3.	.4113073	-.1712106	.1793498
4.	.5424927	.3788345	.2513701
5.	.	-.0925817	.1452616

We obtained the probability, p2, in the previous example.

Finally, predict can calculate the standard error of the difference in the index values between two outcomes with the stddp option:

```
. predict se_2_3, outcome(Prepaid,Uninsure) stddp
(1 missing value generated)

. list idx2 idx3 se_2_3 in 1/5
```

	idx2	idx3	se_2_3
1.	-.4831167	-3.073253	.5469354
2.	.055111	-2.715986	.4331918
3.	-.1712106	-1.579621	.3053815
4.	.3788345	-1.462007	.4492552
5.	-.0925817	-2.814022	.4024784

In the first observation, the difference in the indexes is $-0.483 - (-3.073) = 2.59$. The standard error of that difference is 0.547.

◁

▷ Example 3

It is more difficult to interpret the results from `mlogit` than those from `clogit` or `logit` because there are multiple equations. For example, suppose that one of the independent variables in our model takes on the values 0 and 1, and we are attempting to understand the effect of this variable. Assume that the coefficient on this variable for the second outcome, $\beta^{(2)}$, is positive. We might then be tempted to reason that the probability of the second outcome is higher if the variable is 1 rather than 0. Most of the time, that will be true, but occasionally we will be surprised. The probability of some other outcome could increase even more (say, $\beta^{(3)} > \beta^{(2)}$), and thus the probability of outcome 2 would actually fall relative to that outcome. We can use `predict` to help interpret such results.

Continuing with our previously fit insurance-choice model, we wish to describe the model's predictions by race. For this purpose, we can use the method of predictive margins (also known as recycled predictions), in which we vary characteristics of interest across the whole dataset and average the predictions. That is, we have data on both whites and nonwhites, and our individuals have other characteristics as well. We will first pretend that all the people in our data are white but hold their other characteristics constant. We then calculate the probabilities of each outcome. Next we will pretend that all the people in our data are nonwhite, still holding their other characteristics constant. Again we calculate the probabilities of each outcome. The difference in those two sets of calculated probabilities, then, is the difference due to race, holding other characteristics constant.

```
. gen byte nonwhold = nonwhite                // save real race
. replace nonwhite = 0                        // make everyone white
(126 real changes made)
. predict wpind, outcome(Indemnity)           // predict probabilities
(option pr assumed; predicted probability)
(1 missing value generated)
. predict wpp, outcome(Prepaid)
(option pr assumed; predicted probability)
(1 missing value generated)
. predict wpnoi, outcome(Uninsure)
(option pr assumed; predicted probability)
(1 missing value generated)
. replace nonwhite=1                           // make everyone nonwhite
(644 real changes made)
. predict nwpind, outcome(Indemnity)
(option pr assumed; predicted probability)
(1 missing value generated)
. predict nwpp, outcome(Prepaid)
(option pr assumed; predicted probability)
(1 missing value generated)
. predict nwpnoi, outcome(Uninsure)
(option pr assumed; predicted probability)
(1 missing value generated)
. replace nonwhite=nonwhold                    // restore real race
(518 real changes made)
```

```
. summarize wp* nwp*, sep(3)
```

Variable	Obs	Mean	Std. Dev.	Min	Max
wpind	643	.5141673	.0872679	.3092903	.71939
wpp	643	.4082052	.0993286	.1964103	.6502247
wpnoi	643	.0776275	.0360283	.0273596	.1302816
nwpind	643	.3112809	.0817693	.1511329	.535021
nwpp	643	.630078	.0979976	.3871782	.8278881
nwpnoi	643	.0586411	.0287185	.0209648	.0933874

In example 1 of [R] **mlogit**, we presented a cross-tabulation of insurance type and race. Those values were unadjusted. The means reported above are the values adjusted for age, sex, and site. Combining the results gives

	Unadjusted white	Unadjusted nonwhite	Adjusted white	Adjusted nonwhite
Indemnity	0.51	0.36	0.51	0.31
Prepaid	0.42	0.57	0.41	0.63
Uninsured	0.07	0.07	0.08	0.06

We find, for instance, after adjusting for age, sex, and site, that although 57% of nonwhites in our data had prepaid plans, 63% of nonwhites chose prepaid plans.

Computing predictive margins by hand was instructive, but we can compute these values more easily using the margins command (see [R] **margins**). The two margins for the indemnity outcome can be estimated by typing

```
. margins nonwhite, predict(outcome(Indemnity)) noesample
Predictive margins                              Number of obs   =        643
Model VCE    : OIM

Expression   : Pr(insure==Indemnity), predict(outcome(Indemnity))
```

	Margin	Delta-method Std. Err.	z	P>|z|	[95% Conf. Interval]
nonwhite					
0	.5141673	.0223485	23.01	0.000	.470365	.5579695
1	.3112809	.0418049	7.45	0.000	.2293448	.393217

margins also estimates the standard errors and confidence intervals of the margins. By default, margins uses only the estimation sample. We added the noesample option so that margins would use the entire sample and produce results comparable to our earlier analysis.

We can use `marginsplot` to graph the results from `margins`:

```
. marginsplot
Variables that uniquely identify margins: nonwhite
```

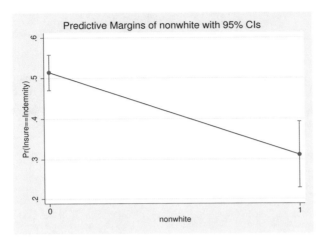

The margins for the other two outcomes can be computed by typing

```
. margins nonwhite, predict(outcome(Prepaid)) noesample
(output omitted )
. margins nonwhite, predict(outcome(Uninsure)) noesample
(output omitted )
```

◁

◻ Technical note

You can use `predict` to classify predicted values and compare them with the observed outcomes to interpret a multinomial logit model. This is a variation on the notions of sensitivity and specificity for logistic regression. Here we will classify indemnity and prepaid as definitely predicting indemnity, definitely predicting prepaid, and ambiguous.

```
. predict indem, outcome(Indemnity) index          // obtain indexes
(1 missing value generated)
. predict prepaid, outcome(Prepaid) index
(1 missing value generated)
. gen diff = prepaid-indem                          // obtain difference
(1 missing value generated)
. predict sediff, outcome(Indemnity,Prepaid) stddp  // & its standard error
(1 missing value generated)
. gen type = 1 if diff/sediff < -1.96               // definitely indemnity
(504 missing values generated)
. replace type = 3 if diff/sediff > 1.96            // definitely prepaid
(100 real changes made)
. replace type = 2 if type>=. & diff/sediff < .     // ambiguous
(404 real changes made)
. label def type 1 "Def Ind" 2 "Ambiguous" 3 "Def Prep"
. label values type type                            // label results
```

```
. tabulate insure type
```

insure	Def Ind	type Ambiguous	Def Prep	Total
Indemnity	78	183	33	294
Prepaid	44	177	56	277
Uninsure	12	28	5	45
Total	134	388	94	616

We can see that the predictive power of this model is modest. There are many misclassifications in both directions, though there are more correctly classified observations than misclassified observations.

Also the uninsured look overwhelmingly as though they might have come from the indemnity system rather than from the prepaid system.

❑

Calculating marginal effects

▷ Example 4

We have already noted that the coefficients from multinomial logit can be difficult to interpret because they are relative to the base outcome. Another way to evaluate the effect of covariates is to examine the marginal effect of changing their values on the probability of observing an outcome.

The `margins` command can be used for this too. We can estimate the marginal effect of each covariate on the probability of observing the first outcome—indemnity insurance—by typing

```
. margins, dydx(*) predict(outcome(Indemnity))
```

```
Average marginal effects                          Number of obs    =        615
Model VCE    : OIM

Expression   : Pr(insure==Indemnity), predict(outcome(Indemnity))
dy/dx w.r.t. : age 1.male 1.nonwhite 2.site 3.site
```

	dy/dx	Delta-method Std. Err.	z	P>|z|	[95% Conf. Interval]	
age	.0026655	.001399	1.91	0.057	-.0000765	.0054074
1.male	-.1295734	.0450945	-2.87	0.004	-.2179571	-.0411898
1.nonwhite	-.2032404	.0482554	-4.21	0.000	-.2978192	-.1086616
site						
2	.0070995	.0479993	0.15	0.882	-.0869775	.1011765
3	.1216165	.0505833	2.40	0.016	.022475	.220758

Note: dy/dx for factor levels is the discrete change from the base level.

By default, `margins` estimates the average marginal effect over the estimation sample, and that is what we see above. Being male decreases the average probability of having indemnity insurance by 0.130. We also see, from the note at the bottom of the table, that the marginal effect was computed as a discrete change in the probability of being male rather than female. That is why we made `male` a factor variable when fitting the model.

The `dydx(*)` option requested that `margins` estimate the marginal effect for each regressor, `dydx(age)` would have produced estimates only for the effect of `age`. `margins` has many options for controlling how the marginal effect is computed, including the ability to average over subgroups or to compute estimates for specified values of the regressors; see [R] **margins**.

We could evaluate the marginal effects on the other two outcomes by typing

. margins, dydx(*) predict(outcome(Prepaid))
(output omitted)

. margins, dydx(*) predict(outcome(Uninsure))
(output omitted)

◁

esting hypotheses about coefficients

Example 5

test tests hypotheses about the coefficients just as after any estimation command; see [R] **test**. Note, however, test's syntax for dealing with multiple-equation models. Because test bases its results on the estimated covariance matrix, we might prefer a likelihood-ratio test; see example 5 in [R] **mlogit** for an example of lrtest.

If we simply list variables after the test command, we are testing that the corresponding coefficients are zero across all equations:

```
. test 2.site 3.site

 ( 1)   [Indemnity]2.site = 0
 ( 2)   [Prepaid]2.site = 0
 ( 3)   [Uninsure]2.site = 0
 ( 4)   [Indemnity]3.site = 0
 ( 5)   [Prepaid]3.site = 0
 ( 6)   [Uninsure]3.site = 0
        Constraint 1 dropped
        Constraint 4 dropped

            chi2(  4) =     19.74
          Prob > chi2 =     0.0006
```

We can test that all the coefficients (except the constant) in an equation are zero by simply typing the outcome in square brackets:

```
. test [Uninsure]

 ( 1)   [Uninsure]age = 0
 ( 2)   [Uninsure]0b.male = 0
 ( 3)   [Uninsure]1.male = 0
 ( 4)   [Uninsure]0b.nonwhite = 0
 ( 5)   [Uninsure]1.nonwhite = 0
 ( 6)   [Uninsure]1b.site = 0
 ( 7)   [Uninsure]2.site = 0
 ( 8)   [Uninsure]3.site = 0
        Constraint 2 dropped
        Constraint 4 dropped
        Constraint 6 dropped

            chi2(  5) =      9.31
          Prob > chi2 =     0.0973
```

We specify the outcome just as we do with predict; we can specify the label if the outcome variable is labeled, or we can specify the numeric value of the outcome. We would have obtained the same test as above if we had typed test [3] because 3 is the value of insure for the outcome uninsured.

We can combine the two syntaxes. To test that the coefficients on the site variables are 0 in the equation corresponding to the outcome prepaid, we can type

```
. test [Prepaid]: 2.site 3.site
( 1)  [Prepaid]2.site = 0
( 2)  [Prepaid]3.site = 0
           chi2(  2) =    10.78
         Prob > chi2 =     0.0046
```

We specified the outcome and then followed that with a colon and the variables we wanted to test.

We can also test that coefficients are equal across equations. To test that all coefficients except the constant are equal for the prepaid and uninsured outcomes, we can type

```
. test [Prepaid=Uninsure]
( 1)  [Prepaid]age - [Uninsure]age = 0
( 2)  [Prepaid]0b.male - [Uninsure]0b.male = 0
( 3)  [Prepaid]1.male - [Uninsure]1.male = 0
( 4)  [Prepaid]0b.nonwhite - [Uninsure]0b.nonwhite = 0
( 5)  [Prepaid]1.nonwhite - [Uninsure]1.nonwhite = 0
( 6)  [Prepaid]1b.site - [Uninsure]1b.site = 0
( 7)  [Prepaid]2.site - [Uninsure]2.site = 0
( 8)  [Prepaid]3.site - [Uninsure]3.site = 0
       Constraint 2 dropped
       Constraint 4 dropped
       Constraint 6 dropped
           chi2(  5) =    13.80
         Prob > chi2 =     0.0169
```

To test that only the site variables are equal, we can type

```
. test [Prepaid=Uninsure]: 2.site 3.site
( 1)  [Prepaid]2.site - [Uninsure]2.site = 0
( 2)  [Prepaid]3.site - [Uninsure]3.site = 0
           chi2(  2) =    12.68
         Prob > chi2 =     0.0018
```

Finally, we can test any arbitrary constraint by simply entering the equation and specifying the coefficients as described in [U] **13.5 Accessing coefficients and standard errors**. The following hypothesis is senseless but illustrates the point:

```
. test ([Prepaid]age+[Uninsure]2.site)/2 = 2-[Uninsure]1.nonwhite
( 1)  .5*[Prepaid]age + [Uninsure]1.nonwhite + .5*[Uninsure]2.site = 2
           chi2(  1) =    22.45
         Prob > chi2 =     0.0000
```

See [R] **test** for more information about test. The information there about combining hypotheses across test commands (the accumulate option) also applies after mlogit.

◁

Methods and formulas

All postestimation commands listed above are implemented as ado-files.

Also see

[R] **mlogit** — Multinomial (polytomous) logistic regression

[U] **20 Estimation and postestimation commands**

Title

> **more** — The —more— message

Syntax

Tell Stata to pause or not pause for —more— messages

> set <u>mor</u>e { on | off } [, <u>perm</u>anently]

Set number of lines between —more— messages

> set <u>pagesize</u> #

Description

set more on, which is the default, tells Stata to wait until you press a key before continuing when a —more— message is displayed.

set more off tells Stata not to pause or display the —more— message.

set pagesize # sets the number of lines between —more— messages. The permanently option is not allowed with set pagesize.

Option

permanently specifies that, in addition to making the change right now, the more setting be remembered and become the default setting when you invoke Stata.

Remarks

When you see —more— at the bottom of the screen,

Press ...	and Stata ...
letter *l* or *Enter*	displays the next line
letter *q*	acts as if you pressed *Break*
Spacebar or any other key	displays the next screen

You can also click on the **More** button or click on —more— to display the next screen.

—more— is Stata's way of telling you that it has something more to show you but that showing it to you will cause the information on the screen to scroll off.

If you type set more off, —more— conditions will never arise, and Stata's output will scroll by at full speed.

If you type set more on, —more— conditions will be restored at the appropriate places.

Programmers should see [P] **more** for information on the more programming command.

Also see

[R] **query** — Display system parameters

[P] **creturn** — Return c-class values

[P] **more** — Pause until key is pressed

[U] **7 —more— conditions**

Title

> **mprobit** — Multinomial probit regression

Syntax

mprobit *depvar* [*indepvars*] [*if*] [*in*] [*weight*] [, *options*]

options	Description
Model	
<u>nocon</u>stant	suppress constant terms
<u>baseo</u>utcome(*# | lbl*)	outcome used to normalize location
<u>probit</u>param	use the probit variance parameterization
<u>constr</u>aints(*constraints*)	apply specified linear constraints
<u>coll</u>inear	keep collinear variables
SE/Robust	
vce(*vcetype*)	*vcetype* may be oim, <u>r</u>obust, <u>cl</u>uster *clustvar*, opg, <u>boot</u>strap, or <u>jack</u>knife
Reporting	
<u>l</u>evel(*#*)	set confidence level; default is level(95)
<u>nocns</u>report	do not display constraints
display_options	control column formats, row spacing, line width, and display of omitted variables and base and empty cells
Integration	
<u>intp</u>oints(*#*)	number of quadrature points
Maximization	
maximize_options	control the maximization process; seldom used
<u>coefl</u>egend	display legend instead of statistics

indepvars may contain factor variables; see [U] **11.4.3 Factor variables**.
bootstrap, by, jackknife, mi estimate, rolling, statsby, and svy are allowed; see
  [U] **11.1.10 Prefix commands**.
vce(bootstrap) and vce(jackknife) are not allowed with the mi estimate prefix; see [MI] **mi estimate**.
Weights are not allowed with the bootstrap prefix; see [R] **bootstrap**.
vce() and weights are not allowed with the svy prefix; see [SVY] **svy**.
fweights, iweights, and pweights are allowed; see [U] **11.1.6 weight**.
coeflegend does not appear in the dialog box.
See [U] **20 Estimation and postestimation commands** for more capabilities of estimation commands.

Menu

Statistics > Categorical outcomes > Independent multinomial probit

Description

mprobit fits multinomial probit (MNP) models via maximum likelihood. *depvar* contains the outcome for each observation, and *indepvars* are the associated covariates. The error terms are assumed to be independent, standard normal, random variables. See [R] **asmprobit** for the case where the latent-variable errors are correlated or heteroskedastic and you have alternative-specific variables.

Options

▭ Model ▭

noconstant suppresses the $J - 1$ constant terms.

baseoutcome(*# | lbl*) specifies the outcome used to normalize the location of the latent variable. The base outcome may be specified as a number or a label. The default is to use the most frequent outcome. The coefficients associated with the base outcome are zero.

probitparam specifies to use the probit variance parameterization by fixing the variance of the differenced latent errors between the scale and the base alternatives to be one. The default is to make the variance of the base and scale latent errors one, thereby making the variance of the difference to be two.

constraints(*constraints*), collinear; see [R] **estimation options**.

▭ SE/Robust ▭

vce(*vcetype*) specifies the type of standard error reported, which includes types that are derived from asymptotic theory, that are robust to some kinds of misspecification, that allow for intragroup correlation, and that use bootstrap or jackknife methods; see [R] **vce_option**.

If specifying vce(bootstrap) or vce(jackknife), you must also specify baseoutcome().

▭ Reporting ▭

level(*#*); see [R] **estimation options**.

nocnsreport; see [R] **estimation options**.

display_options: noomitted, vsquish, noemptycells, baselevels, allbaselevels, cformat(*%fmt*), pformat(*%fmt*), sformat(*%fmt*), and nolstretch; see [R] **estimation options**.

▭ Integration ▭

intpoints(*#*) specifies the number of Gaussian quadrature points to use in approximating the likelihood. The default is 15.

▭ Maximization ▭

maximize_options: difficult, technique(*algorithm_spec*), iterate(*#*), [no]log, trace, gradient, showstep, hessian, showtolerance, tolerance(*#*), ltolerance(*#*), nrtolerance(*#*), nonrtolerance, and from(*init_specs*); see [R] **maximize**. These options are seldom used.

Setting the optimization type to technique(bhhh) resets the default *vcetype* to vce(opg).

The following option is available with mprobit but is not shown in the dialog box:

coeflegend; see [R] **estimation options**.

Remarks

The MNP model is used with discrete dependent variables that take on more than two outcomes that do not have a natural ordering. The stochastic error terms for this implementation of the model are assumed to have independent, standard normal distributions. To use `mprobit`, you must have one observation for each decision maker in the sample. See [R] **asmprobit** for another implementation of the MNP model that permits correlated and heteroskedastic errors and is suitable when you have data for each alternative that a decision maker faced.

The MNP model is frequently motivated using a latent-variable framework. The latent variable for the jth alternative, $j = 1, \ldots, J$, is

$$\eta_{ij} = \mathbf{z}_i \boldsymbol{\alpha}_j + \xi_{ij}$$

where the $1 \times q$ row vector $\mathbf{z}_i$ contains the observed independent variables for the ith decision maker. Associated with $\mathbf{z}_i$ are the J vectors of regression coefficients $\boldsymbol{\alpha}_j$. The $\xi_{i,1}, \ldots, \xi_{i,J}$ are distributed independently and identically standard normal. The decision maker chooses the alternative k such that $\eta_{ik} \geq \eta_{im}$ for $m \neq k$.

Suppose that case i chooses alternative k, and take the difference between latent variable η_{ik} and the $J - 1$ others:

$$\begin{aligned} v_{ijk} &= \eta_{ij} - \eta_{ik} \\ &= \mathbf{z}_i(\boldsymbol{\alpha}_j - \boldsymbol{\alpha}_k) + \xi_{ij} - \xi_{ik} \\ &= \mathbf{z}_i \boldsymbol{\gamma}_{j'} + \epsilon_{ij'} \end{aligned} \tag{1}$$

where $j' = j$ if $j < k$ and $j' = j-1$ if $j > k$ so that $j' = 1, \ldots, J-1$. $\mathrm{Var}(\epsilon_{ij'}) = \mathrm{Var}(\xi_{ij} - \xi_{ik}) = 2$ and $\mathrm{Cov}(\epsilon_{ij'}, \epsilon_{il'}) = 1$ for $j' \neq l'$. The probability that alternative k is chosen is

$$\begin{aligned} \Pr(i \text{ chooses } k) &= \Pr(v_{i1k} \leq 0, \ldots, v_{i,J-1,k} \leq 0) \\ &= \Pr(\epsilon_{i1} \leq -\mathbf{z}_i \boldsymbol{\gamma}_1, \ldots, \epsilon_{i,J-1} \leq -\mathbf{z}_i \boldsymbol{\gamma}_{J-1}) \end{aligned}$$

Hence, evaluating the likelihood function involves computing probabilities from the multivariate normal distribution. That all the covariances are equal simplifies the problem somewhat; see *Methods and formulas* for details.

In (1), not all J of the $\boldsymbol{\alpha}_j$ are identifiable. To remove the indeterminacy, $\boldsymbol{\alpha}_l$ is set to the zero vector, where l is the base outcome as specified in the `baseoutcome()` option. That fixes the lth latent variable to zero so that the remaining variables measure the attractiveness of the other alternatives relative to the base.

▷ Example 1

As discussed in example 1 of [R] **mlogit**, we have data on the type of health insurance available to 616 psychologically depressed subjects in the United States (Tarlov et al. 1989; Wells et al. 1989). Patients may have either an indemnity (fee-for-service) plan or a prepaid plan such as an HMO, or the patient may be uninsured. Demographic variables include age, gender, race, and site. Indemnity insurance is the most popular alternative, so `mprobit` will choose it as the base outcome by default.

```
. use http://www.stata-press.com/data/r12/sysdsn1
(Health insurance data)

. mprobit insure age male nonwhite i.site

Iteration 0:   log likelihood = -535.89424
Iteration 1:   log likelihood = -534.56173
Iteration 2:   log likelihood = -534.52835
Iteration 3:   log likelihood = -534.52833

Multinomial probit regression                   Number of obs   =       615
                                                 Wald chi2(10)   =     40.18
Log likelihood = -534.52833                      Prob > chi2     =    0.0000
```

insure	Coef.	Std. Err.	z	P>|z|	[95% Conf. Interval]	
Indemnity	(base outcome)					
Prepaid						
age	-.0098536	.0052688	-1.87	0.061	-.0201802	.000473
male	.4774678	.1718316	2.78	0.005	.1406841	.8142515
nonwhite	.8245003	.1977582	4.17	0.000	.4369013	1.212099
site						
2	.0973956	.1794546	0.54	0.587	-.2543289	.4491201
3	-.495892	.1904984	-2.60	0.009	-.869262	-.1225221
_cons	.22315	.2792424	0.80	0.424	-.324155	.7704549
Uninsure						
age	-.0050814	.0075327	-0.67	0.500	-.0198452	.0096823
male	.3332637	.2432986	1.37	0.171	-.1435929	.8101203
nonwhite	.2485859	.2767734	0.90	0.369	-.29388	.7910518
site						
2	-.6899485	.2804497	-2.46	0.014	-1.23962	-.1402771
3	-.1788447	.2479898	-0.72	0.471	-.6648957	.3072063
_cons	-.9855917	.3891873	-2.53	0.011	-1.748385	-.2227986

◁

The likelihood function for mprobit is derived under the assumption that all decision-making units face the same choice set, which is the union of all outcomes observed in the dataset. If that is not true for your model, then an alternative is to use the asmprobit command, which does not require this assumption. To do that, you will need to expand the dataset so that each decision maker has k_i observations, where k_i is the number of alternatives in the choice set faced by decision maker i. You will also need to create a binary variable to indicate the choice made by each decision maker. Moreover, you will need to use the correlation(independent) and stddev(homoskedastic) options with asmprobit unless you have alternative-specific variables.

Saved results

mprobit saves the following in e():

Scalars
e(N)	number of observations
e(k_out)	number of outcomes
e(k_points)	number of quadrature points
e(k)	number of parameters
e(k_eq)	number of equations in e(b)
e(k_eq_model)	number of equations in overall model test
e(k_indvars)	number of independent variables
e(k_dv)	number of dependent variables
e(df_m)	model degrees of freedom
e(ll)	log simulated-likelihood
e(N_clust)	number of clusters
e(chi2)	χ^2
e(p)	significance
e(i_base)	base outcome index
e(const)	0 if noconstant is specified, 1 otherwise
e(probitparam)	1 if probitparam is specified, 0 otherwise
e(rank)	rank of e(V)
e(ic)	number of iterations
e(rc)	return code
e(converged)	1 if converged, 0 otherwise

Macros
e(cmd)	mprobit
e(cmdline)	command as typed
e(depvar)	name of dependent variable
e(indvars)	independent variables
e(wtype)	weight type
e(wexp)	weight expression
e(title)	title in estimation output
e(clustvar)	name of cluster variable
e(chi2type)	Wald, type of model χ^2 test
e(vce)	*vcetype* specified in vce()
e(vcetype)	title used to label Std. Err.
e(outeqs)	outcome equations
e(out#)	outcome labels, #=1,...,e(k_out)
e(opt)	type of optimization
e(which)	max or min; whether optimizer is to perform maximization or minimization
e(ml_method)	type of ml method
e(user)	name of likelihood-evaluator program
e(technique)	maximization technique
e(properties)	b V
e(predict)	program used to implement predict
e(marginsnotok)	predictions disallowed by margins
e(asbalanced)	factor variables fvset as asbalanced
e(asobserved)	factor variables fvset as asobserved

Matrices
    e(b)   coefficient vector
    e(outcomes)   outcome values
    e(Cns)   constraints matrix
    e(ilog)   iteration log (up to 20 iterations)
    e(gradient)   gradient vector
    e(V)   variance–covariance matrix of the estimators
    e(V_modelbased)   model-based variance
Functions
    e(sample)   marks estimation sample

Methods and formulas

mprobit is implemented as an ado-file.

See Cameron and Trivedi (2005, chap. 15) for a discussion of multinomial models, including multinomial probit. Long and Freese (2006, chap. 6) discuss the multinomial logistic, multinomial probit, and stereotype logistic regression models, with examples using Stata.

As discussed in *Remarks*, the latent variables for a J-alternative model are $\eta_{ij} = \mathbf{z}_i \boldsymbol{\alpha}_j + \xi_{ij}$, for $j = 1, \ldots, J$, $i = 1, \ldots, n$, and $\{\xi_{i,1}, \ldots, \xi_{i,J}\} \sim$ i.i.d. $N(0,1)$. The experimenter observes alternative k for the ith observation if $\eta_{ik} > \eta_{il}$ for $l \neq k$. For $j \neq k$, let

$$
\begin{aligned}
v_{ij'} &= \eta_{ij} - \eta_{ik} \\
&= \mathbf{z}_i(\boldsymbol{\alpha}_j - \boldsymbol{\alpha}_k) + \xi_{ij} - \xi_{ik} \\
&= \mathbf{z}_i \boldsymbol{\gamma}_{j'} + \epsilon_{ij'}
\end{aligned}
$$

where $j' = j$ if $j < k$ and $j' = j - 1$ if $j > k$ so that $j' = 1, \ldots, J-1$. $\epsilon_i = (\epsilon_{i1}, \ldots, \epsilon_{i,J-1}) \sim MVN(\mathbf{0}, \boldsymbol{\Sigma})$, where

$$
\boldsymbol{\Sigma} = \begin{pmatrix} 2 & 1 & 1 & \ldots & 1 \\ 1 & 2 & 1 & \ldots & 1 \\ 1 & 1 & 2 & \ldots & 1 \\ \vdots & \vdots & \vdots & \ddots & \vdots \\ 1 & 1 & 1 & \ldots & 2 \end{pmatrix}
$$

Denote the deterministic part of the model as $\lambda_{ij'} = \mathbf{z}_i \boldsymbol{\gamma}_{j'}$; the probability that subject i chooses outcome k is

$$
\begin{aligned}
\Pr(y_i = k) &= \Pr(v_{i1} \leq 0, \ldots, v_{i,J-1} \leq 0) \\
&= \Pr(\epsilon_{i1} \leq -\lambda_{i1}, \ldots, \epsilon_{i,J-1} \leq -\lambda_{i,J-1}) \\
&= \frac{1}{(2\pi)^{(J-1)/2} |\boldsymbol{\Sigma}|^{1/2}} \int_{-\infty}^{-\lambda_{i1}} \cdots \int_{-\infty}^{-\lambda_{i,J-1}} \exp\left(-\tfrac{1}{2} \mathbf{z}' \boldsymbol{\Sigma}^{-1} \mathbf{z}\right) d\mathbf{z}
\end{aligned}
$$

Because of the exchangeable correlation structure of $\boldsymbol{\Sigma}$ ($\rho_{ij} = 1/2$ for all $i \neq j$), we can use Dunnett's (1989) result to reduce the multidimensional integral to one dimension:

$$
\Pr(y_i = k) = \frac{1}{\sqrt{\pi}} \int_0^\infty \left\{ \prod_{j=1}^{J-1} \Phi\left(-z\sqrt{2} - \lambda_{ij}\right) + \prod_{j=1}^{J-1} \Phi\left(z\sqrt{2} - \lambda_{ij}\right) \right\} e^{-z^2} dz
$$

Gaussian quadrature is used to approximate this integral, resulting in the K-point quadrature formula

$$\Pr(y_i = k) \approx \frac{1}{2} \sum_{k=1}^{K} w_k \left\{ \prod_{j=1}^{J-1} \Phi\left(-\sqrt{2x_k} - \lambda_{ij}\right) + \prod_{j=1}^{J-1} \Phi\left(\sqrt{2x_k} - \lambda_{ij}\right) \right\}$$

where w_k and x_k are the weights and roots of the Laguerre polynomial of order K. In mprobit, K is specified by the intpoints() option.

This command supports the Huber/White/sandwich estimator of the variance and its clustered version using vce(robust) and vce(cluster *clustvar*), respectively. See [P] **_robust**, particularly *Maximum likelihood estimators* and *Methods and formulas*.

mprobit also supports estimation with survey data. For details on VCEs with survey data, see [SVY] **variance estimation**.

References

Cameron, A. C., and P. K. Trivedi. 2005. *Microeconometrics: Methods and Applications*. New York: Cambridge University Press.

Dunnett, C. W. 1989. Algorithm AS 251: Multivariate normal probability integrals with product correlation structure. *Journal of the Royal Statistical Society, Series C* 38: 564–579.

Haan, P., and A. Uhlendorff. 2006. Estimation of multinomial logit models with unobserved heterogeneity using maximum simulated likelihood. *Stata Journal* 6: 229–245.

Hole, A. R. 2007. Fitting mixed logit models by using maximum simulated likelihood. *Stata Journal* 7: 388–401.

Long, J. S., and J. Freese. 2006. *Regression Models for Categorical Dependent Variables Using Stata*. 2nd ed. College Station, TX: Stata Press.

Tarlov, A. R., J. E. Ware, Jr., S. Greenfield, E. C. Nelson, E. Perrin, and M. Zubkoff. 1989. The medical outcomes study. An application of methods for monitoring the results of medical care. *Journal of the American Medical Association* 262: 925–930.

Wells, K. B., R. D. Hays, M. A. Burnam, W. H. Rogers, S. Greenfield, and J. E. Ware, Jr. 1989. Detection of depressive disorder for patients receiving prepaid or fee-for-service care. Results from the Medical Outcomes Survey. *Journal of the American Medical Association* 262: 3298–3302.

Also see

[R] **mprobit postestimation** — Postestimation tools for mprobit

[R] **asmprobit** — Alternative-specific multinomial probit regression

[R] **mlogit** — Multinomial (polytomous) logistic regression

[R] **clogit** — Conditional (fixed-effects) logistic regression

[R] **nlogit** — Nested logit regression

[R] **ologit** — Ordered logistic regression

[R] **oprobit** — Ordered probit regression

[MI] **estimation** — Estimation commands for use with mi estimate

[SVY] **svy estimation** — Estimation commands for survey data

[U] **20 Estimation and postestimation commands**

Title

> **mprobit postestimation** — Postestimation tools for mprobit

Description

The following postestimation commands are available after `mprobit`:

Command	Description
contrast	contrasts and ANOVA-style joint tests of estimates
estat	AIC, BIC, VCE, and estimation sample summary
estat (svy)	postestimation statistics for survey data
estimates	cataloging estimation results
hausman	Hausman's specification test
lincom	point estimates, standard errors, testing, and inference for linear combinations of coefficients
lrtest[1]	likelihood-ratio test
margins	marginal means, predictive margins, marginal effects, and average marginal effects
marginsplot	graph the results from margins (profile plots, interaction plots, etc.)
nlcom	point estimates, standard errors, testing, and inference for nonlinear combinations of coefficients
predict	predicted probabilities, linear predictions, and standard errors
predictnl	point estimates, standard errors, testing, and inference for generalized predictions
pwcompare	pairwise comparisons of estimates
suest	seemingly unrelated estimation
test	Wald tests of simple and composite linear hypotheses
testnl	Wald tests of nonlinear hypotheses

[1] `lrtest` is not appropriate with svy estimation results.

See the corresponding entries in the *Base Reference Manual* for details, but see [SVY] **estat** for details about `estat` (svy).

Syntax for predict

predict [*type*] { *stub** | *newvar* | *newvarlist* } [*if*] [*in*] [, *statistic* <u>ou</u>tcome(*outcome*)]

predict [*type*] { *stub** | *newvarlist* } [*if*] [*in*], <u>sc</u>ores

statistic	Description
Main	
<u>pr</u>	probability of a positive outcome; the default
xb	linear prediction
stdp	standard error of the linear prediction

If you do not specify `outcome()`, `pr` (with one new variable specified), `xb`, and `stdp` assume `outcome(#1)`.

You specify one or k new variables with pr, where k is the number of outcomes.

You specify one new variable with xb and stdp.

These statistics are available both in and out of sample; type predict ... if e(sample) ... if wanted only for the estimation sample.

Menu

Statistics > Postestimation > Predictions, residuals, etc.

Options for predict

⌐ Main ⌐

pr, the default, calculates the probability of each of the categories of the dependent variable or the probability of the level specified in outcome(*outcome*). If you specify the outcome(*outcome*) option, you need to specify only one new variable; otherwise, you must specify a new variable for each category of the dependent variable.

xb calculates the linear prediction, $\mathbf{x}_i\boldsymbol{\alpha}_j$, for alternative j and individual i. The index, j, corresponds to the outcome specified in outcome().

stdp calculates the standard error of the linear prediction.

outcome(*outcome*) specifies the outcome for which the statistic is to be calculated. equation() is a synonym for outcome(): it does not matter which you use. outcome() or equation() can be specified using

> #1, #2, ..., where #1 means the first category of the dependent variable, #2 means the second category, etc.;

> the values of the dependent variable; or

> the value labels of the dependent variable if they exist.

scores calculates the equation-level score variables. The jth new variable will contain the scores for the jth fitted equation.

Remarks

Once you have fit a multinomial probit model, you can use predict to obtain probabilities that an individual will choose each of the alternatives for the estimation sample, as well as other samples; see [U] **20 Estimation and postestimation commands** and [R] **predict**.

▷ Example 1

In example 1 of [R] **mprobit**, we fit the multinomial probit model to a dataset containing the type of health insurance available to 616 psychologically depressed subjects in the United States (Tarlov et al. 1989; Wells et al. 1989). We can obtain the predicted probabilities by typing

```
. use http://www.stata-press.com/data/r12/sysdsn1
(Health insurance data)
. mprobit insure age male nonwhite i.site
  (output omitted )
. predict p1-p3
(option pr assumed; predicted probabilities)
```

. list p1-p3 insure in 1/10

	p1	p2	p3	insure
1.	.5961306	.3741824	.029687	Indemnity
2.	.4719296	.4972289	.0308415	Prepaid
3.	.4896086	.4121961	.0981953	Indemnity
4.	.3730529	.5416623	.0852848	Prepaid
5.	.5063069	.4629773	.0307158	.
6.	.4768125	.4923548	.0308327	Prepaid
7.	.5035672	.4657016	.0307312	Prepaid
8.	.3326361	.5580404	.1093235	.
9.	.4758165	.4384811	.0857024	Uninsure
10.	.5734057	.3316601	.0949342	Prepaid

insure contains a missing value for observations 5 and 8. Because of that, those two observations were not used in the estimation. However, because none of the independent variables is missing, predict can still calculate the probabilities. Had we typed

. predict p1-p3 if e(sample)

predict would have filled in missing values for p1, p2, and p3 for those observations because they were not used in the estimation.

◁

Methods and formulas

All postestimation commands listed above are implemented as ado-files.

References

Tarlov, A. R., J. E. Ware, Jr., S. Greenfield, E. C. Nelson, E. Perrin, and M. Zubkoff. 1989. The medical outcomes study. An application of methods for monitoring the results of medical care. *Journal of the American Medical Association* 262: 925–930.

Wells, K. B., R. D. Hays, M. A. Burnam, W. H. Rogers, S. Greenfield, and J. E. Ware, Jr. 1989. Detection of depressive disorder for patients receiving prepaid or fee-for-service care. Results from the Medical Outcomes Survey. *Journal of the American Medical Association* 262: 3298–3302.

Also see

[R] **mprobit** — Multinomial probit regression

[U] **20 Estimation and postestimation commands**

Title

> **mvreg** — Multivariate regression

Syntax

mvreg *depvars* = *indepvars* [*if*] [*in*] [*weight*] [, *options*]

options	Description
Model	
<u>nocon</u>stant	suppress constant term
Reporting	
<u>level</u>(#)	set confidence level; default is level(95)
<u>corr</u>	report correlation matrix
display_options	control column formats, row spacing, line width, and display of omitted variables and base and empty cells
<u>nohe</u>ader	suppress header table from above coefficient table
<u>not</u>able	suppress coefficient table
<u>coefl</u>egend	display legend instead of statistics

indepvars may contain factor variables; see [U] **11.4.3 Factor variables**.

depvars and *indepvars* may contain time-series operators; see [U] **11.4.4 Time-series varlists**.

bootstrap, by, jackknife, mi estimate, rolling, and statsby are allowed; see [U] **11.1.10 Prefix commands**.

Weights are not allowed with the bootstrap prefix; see [R] **bootstrap**.

aweights are not allowed with the jackknife prefix; see [R] **jackknife**.

aweights and fweights are allowed; see [U] **11.1.6 weight**.

noheader, notable, and coeflegend do not appear in the dialog box.

See [U] **20 Estimation and postestimation commands** for more capabilities of estimation commands.

Menu

Statistics > Linear models and related > Multiple-equation models > Multivariate regression

Description

mvreg fits multivariate regression models.

Options

⌐ Model ⌐

noconstant suppresses the constant term (intercept) in the model.

⌐ Reporting ⌐

level(#) specifies the confidence level, as a percentage, for confidence intervals. The default is level(95) or as set by set level; see [U] **20.7 Specifying the width of confidence intervals**.

corr displays the correlation matrix of the residuals between the equations.

display_options: <u>noomit</u>ted, vsquish, <u>noempty</u>cells, <u>base</u>levels, <u>allbase</u>levels,
  cformat(% *fmt*), pformat(% *fmt*), sformat(% *fmt*), and nolstretch; see [R] **estimation options**.

The following options are available with mvreg but are not shown in the dialog box:

noheader suppresses display of the table reporting F statistics, R-squared, and root mean squared
  error above the coefficient table.

notable suppresses display of the coefficient table.

coeflegend; see [R] **estimation options**.

Remarks

  Multivariate regression differs from multiple regression in that *several* dependent variables are
jointly regressed on the same independent variables. Multivariate regression is related to Zellner's
seemingly unrelated regression (see [R] **sureg**), but because the same set of independent variables is
used for each dependent variable, the syntax is simpler, and the calculations are faster.

  The individual coefficients and standard errors produced by mvreg are identical to those that would
be produced by regress estimating each equation separately. The difference is that mvreg, being a
joint estimator, also estimates the between-equation covariances, so you can test coefficients across
equations and, in fact, the test syntax makes such tests more convenient.

▷ Example 1

  Using the automobile data, we fit a multivariate regression for space variables (headroom, trunk,
and turn) in terms of a set of other variables, including three performance variables (displacement,
gear_ratio, and mpg):

```
. use http://www.stata-press.com/data/r12/auto
(1978 Automobile Data)
. mvreg headroom trunk turn = price mpg displ gear_ratio length weight
```

Equation	Obs	Parms	RMSE	"R-sq"	F	P
headroom	74	7	.7390205	0.2996	4.777213	0.0004
trunk	74	7	3.052314	0.5326	12.7265	0.0000
turn	74	7	2.132377	0.7844	40.62042	0.0000

	Coef.	Std. Err.	t	P>|t|	[95% Conf. Interval]	
headroom						
price	-.0000528	.000038	-1.39	0.168	-.0001286	.0000229
mpg	-.0093774	.0260463	-0.36	0.720	-.061366	.0426112
displacement	.0031025	.0024999	1.24	0.219	-.0018873	.0080922
gear_ratio	.2108071	.3539588	0.60	0.553	-.4956976	.9173119
length	.015886	.012944	1.23	0.224	-.0099504	.0417223
weight	-.0000868	.0004724	-0.18	0.855	-.0010296	.0008561
_cons	-.4525117	2.170073	-0.21	0.835	-4.783995	3.878972
trunk						
price	.0000445	.0001567	0.28	0.778	-.0002684	.0003573
mpg	-.0220919	.1075767	-0.21	0.838	-.2368159	.1926322
displacement	.0032118	.0103251	0.31	0.757	-.0173971	.0238207
gear_ratio	-.2271321	1.461926	-0.16	0.877	-3.145149	2.690885
length	.170811	.0534615	3.20	0.002	.0641014	.2775206
weight	-.0015944	.001951	-0.82	0.417	-.0054885	.0022997
_cons	-13.28253	8.962868	-1.48	0.143	-31.17249	4.607429
turn						
price	-.0002647	.0001095	-2.42	0.018	-.0004833	-.0000462
mpg	-.0492948	.0751542	-0.66	0.514	-.1993031	.1007136
displacement	.0036977	.0072132	0.51	0.610	-.0106999	.0180953
gear_ratio	-.1048432	1.021316	-0.10	0.919	-2.143399	1.933712
length	.072128	.0373487	1.93	0.058	-.0024204	.1466764
weight	.0027059	.001363	1.99	0.051	-.0000145	.0054264
_cons	20.19157	6.261549	3.22	0.002	7.693467	32.68968

We should have specified the `corr` option so that we would also see the correlations between the residuals of the equations. We can correct our omission because `mvreg`—like all estimation commands—typed without arguments redisplays results. The `noheader` and `notable` (read "no-table") options suppress redisplaying the output we have already seen:

```
. mvreg, notable noheader corr
Correlation matrix of residuals:

          headroom     trunk      turn
headroom    1.0000
   trunk    0.4986    1.0000
    turn   -0.1090   -0.0628    1.0000
Breusch-Pagan test of independence: chi2(3) =    19.566, Pr = 0.0002
```

The Breusch–Pagan test is significant, so the residuals of these three space variables are not independent of each other.

The three performance variables among our independent variables are `mpg`, `displacement`, and `gear_ratio`. We can jointly test the significance of these three variables in all the equations by typing

```
. test mpg displacement gear_ratio

( 1)  [headroom]mpg = 0
( 2)  [trunk]mpg = 0
( 3)  [turn]mpg = 0
( 4)  [headroom]displacement = 0
( 5)  [trunk]displacement = 0
( 6)  [turn]displacement = 0
( 7)  [headroom]gear_ratio = 0
( 8)  [trunk]gear_ratio = 0
( 9)  [turn]gear_ratio = 0

        F( 9,    67) =     0.33
           Prob > F =     0.9622
```

These three variables are not, as a group, significant. We might have suspected this from their individual significance in the individual regressions, but this multivariate test provides an overall assessment with one p-value.

We can also perform a test for the joint significance of all three equations:

```
. test [headroom]
(output omitted )

. test [trunk], accum
(output omitted )

. test [turn], accum

( 1)  [headroom]price = 0
( 2)  [headroom]mpg = 0
( 3)  [headroom]displacement = 0
( 4)  [headroom]gear_ratio = 0
( 5)  [headroom]length = 0
( 6)  [headroom]weight = 0
( 7)  [trunk]price = 0
( 8)  [trunk]mpg = 0
( 9)  [trunk]displacement = 0
(10)  [trunk]gear_ratio = 0
(11)  [trunk]length = 0
(12)  [trunk]weight = 0
(13)  [turn]price = 0
(14)  [turn]mpg = 0
(15)  [turn]displacement = 0
(16)  [turn]gear_ratio = 0
(17)  [turn]length = 0
(18)  [turn]weight = 0

        F( 18,    67) =    19.34
           Prob > F =     0.0000
```

The set of variables as a whole is strongly significant. We might have suspected this, too, from the individual equations.

◁

Technical note

The mvreg command provides a good way to deal with multiple comparisons. If we wanted to assess the effect of length, we might be dissuaded from interpreting any of its coefficients except that in the trunk equation. [trunk]length—the coefficient on length in the trunk equation—has a p-value of 0.002, but in the other two equations, it has p-values of only 0.224 and 0.058.

A conservative statistician might argue that there are 18 tests of significance in mvreg's output (not counting those for the intercept), so p-values more than $0.05/18 = 0.0028$ should be declared insignificant at the 5% level. A more aggressive but, in our opinion, reasonable approach would be to first note that the three equations are jointly significant, so we are justified in making some interpretation. Then we would work through the individual variables using `test`, possibly using $0.05/6 = 0.0083$ (6 because there are six independent variables) for the 5% significance level. For instance, examining `length`:

```
. test length

 ( 1)  [headroom]length = 0
 ( 2)  [trunk]length = 0
 ( 3)  [turn]length = 0

       F(  3,    67) =    4.94
            Prob > F =    0.0037
```

The reported significance level of 0.0037 is less than 0.0083, so we will declare this variable significant. `[trunk]length` is certainly significant with its p-value of 0.002, but what about in the remaining two equations with p-values 0.224 and 0.058? We perform a joint test:

```
. test [headroom]length [turn]length

 ( 1)  [headroom]length = 0
 ( 2)  [turn]length = 0

       F(  2,    67) =    2.91
            Prob > F =    0.0613
```

At this point, reasonable statisticians could disagree. The 0.06 significance value suggests no interpretation, but these were the two least-significant values out of three, so we would expect the p-value to be a little high. Perhaps an equivocal statement is warranted: there seems to be an effect, but chance cannot be excluded.

❑

Saved results

mvreg saves the following in e():

Scalars
e(N)	number of observations
e(k)	number of parameters in each equation
e(k_eq)	number of equations in e(b)
e(df_r)	residual degrees of freedom
e(chi2)	Breusch–Pagan χ^2 (corr only)
e(df_chi2)	degrees of freedom for Breusch–Pagan χ^2 (corr only)
e(rank)	rank of e(V)

Macros
e(cmd)	mvreg
e(cmdline)	command as typed
e(depvar)	names of dependent variables
e(eqnames)	names of equations
e(wtype)	weight type
e(wexp)	weight expression
e(r2)	R-squared for each equation
e(rmse)	RMSE for each equation
e(F)	F statistic for each equation
e(p_F)	significance of F for each equation
e(properties)	b V
e(estat_cmd)	program used to implement estat
e(predict)	program used to implement predict
e(marginsok)	predictions allowed by margins
e(marginsnotok)	predictions disallowed by margins
e(asbalanced)	factor variables fvset as asbalanced
e(asobserved)	factor variables fvset as asobserved

Matrices
e(b)	coefficient vector
e(Sigma)	$\widehat{\Sigma}$ matrix
e(V)	variance–covariance matrix of the estimators

Functions
e(sample)	marks estimation sample

Methods and formulas

mvreg is implemented as an ado-file.

Given q equations and p independent variables (including the constant), the parameter estimates are given by the $p \times q$ matrix

$$\mathbf{B} = (\mathbf{X}'\mathbf{W}\mathbf{X})^{-1}\mathbf{X}'\mathbf{W}\mathbf{Y}$$

where $\mathbf{Y}$ is an $n \times q$ matrix of dependent variables and $\mathbf{X}$ is a $n \times p$ matrix of independent variables. $\mathbf{W}$ is a weighting matrix equal to $\mathbf{I}$ if no weights are specified. If weights are specified, let $\mathbf{v}$: $1 \times n$ be the specified weights. If fweight frequency weights are specified, $\mathbf{W} = \mathrm{diag}(\mathbf{v})$. If aweight analytic weights are specified, $\mathbf{W} = \mathrm{diag}\{\mathbf{v}/(\mathbf{1}'\mathbf{v})(\mathbf{1}'\mathbf{1})\}$, meaning that the weights are normalized to sum to the number of observations.

The residual covariance matrix is

$$\mathbf{R} = \{\mathbf{Y}'\mathbf{W}\mathbf{Y} - \mathbf{B}'(\mathbf{X}'\mathbf{W}\mathbf{X})\mathbf{B}\}/(n - p)$$

The estimated covariance matrix of the estimates is $\mathbf{R} \otimes (\mathbf{X}'\mathbf{W}\mathbf{X})^{-1}$. These results are identical to those produced by sureg when the same list of independent variables is specified repeatedly; see [R] sureg.

The Breusch and Pagan (1980) χ^2 statistic—a Lagrange multiplier statistic—is given by

$$\lambda = n \sum_{i=1}^{q} \sum_{j=1}^{i-1} r_{ij}^2$$

where r_{ij} is the estimated correlation between the residuals of the equations and n is the number of observations. It is distributed as χ^2 with $q(q-1)/2$ degrees of freedom.

Reference

Breusch, T. S., and A. R. Pagan. 1980. The Lagrange multiplier test and its applications to model specification in econometrics. *Review of Economic Studies* 47: 239–253.

Also see

[R] **mvreg postestimation** — Postestimation tools for mvreg

[MV] **manova** — Multivariate analysis of variance and covariance

[R] **nlsur** — Estimation of nonlinear systems of equations

[R] **reg3** — Three-stage estimation for systems of simultaneous equations

[R] **regress** — Linear regression

[R] **regress postestimation** — Postestimation tools for regress

[R] **sureg** — Zellner's seemingly unrelated regression

[MI] **estimation** — Estimation commands for use with mi estimate

Stata Structural Equation Modeling Reference Manual

[U] **20 Estimation and postestimation commands**

escription

The following postestimation commands are available after `mvreg`:

Command	Description
contrast	contrasts and ANOVA-style joint tests of estimates
estat	VCE and estimation sample summary
estimates	cataloging estimation results
lincom	point estimates, standard errors, testing, and inference for linear combinations of coefficients
margins	marginal means, predictive margins, marginal effects, and average marginal effects
marginsplot	graph the results from margins (profile plots, interaction plots, etc.)
nlcom	point estimates, standard errors, testing, and inference for nonlinear combinations of coefficients
predict	predictions, residuals, influence statistics, and other diagnostic measures
predictnl	point estimates, standard errors, testing, and inference for generalized predictions
pwcompare	pairwise comparisons of estimates
test	Wald tests of simple and composite linear hypotheses
testnl	Wald tests of nonlinear hypotheses

See the corresponding entries in the *Base Reference Manual* for details.

yntax for predict

predict [*type*] *newvar* [*if*] [*in*] [, equation(*eqno* [, *eqno*]) *statistic*]

statistic	Description
Main	
xb	linear prediction; the default
stdp	standard error of the linear prediction
residuals	residuals
difference	difference between the linear predictions of two equations
stddp	standard error of the difference in linear predictions

These statistics are available both in and out of sample; type `predict ... if e(sample) ...` if wanted only for the estimation sample.

Menu

Statistics > Postestimation > Predictions, residuals, etc.

Options for predict

⌐ Main ⌐

equation(*eqno* [, *eqno*]) specifies the equation to which you are referring.

equation() is filled in with one *eqno* for the xb, stdp, and residuals options. equation(#1) would mean the calculation is to be made for the first equation, equation(#2) would mean the second, and so on. You could also refer to the equations by their names. equation(income) would refer to the equation named income and equation(hours), to the equation named hours.

If you do not specify equation(), results are the same as if you specified equation(#1).

difference and stddp refer to between-equation concepts. To use these options, you must specify two equations, for example, equation(#1,#2) or equation(income,hours). When two equations must be specified, equation() is required. With equation(#1,#2), difference computes the prediction of equation(#1) minus the prediction of equation(#2).

xb, the default, calculates the fitted values—the prediction of $\mathbf{x}_j \mathbf{b}$ for the specified equation.

stdp calculates the standard error of the prediction for the specified equation (the standard error of the predicted expected value or mean for the observation's covariate pattern). The standard error of the prediction is also referred to as the standard error of the fitted value.

residuals calculates the residuals.

difference calculates the difference between the linear predictions of two equations in the system.

stddp is allowed only after you have previously fit a multiple-equation model. The standard error of the difference in linear predictions $(\mathbf{x}_{1j}\mathbf{b} - \mathbf{x}_{2j}\mathbf{b})$ between equations 1 and 2 is calculated.

For more information on using predict after multiple-equation estimation commands, see [R] **predict**.

Methods and formulas

All postestimation commands listed above are implemented as ado-files.

Also see

[R] **mvreg** — Multivariate regression

[U] **20 Estimation and postestimation commands**